This text provides a thorough treatment of the fundamental principles of fluid mechanics and convection heat transfer and shows how to apply the principles to a wide variety of fluid flow problems.

This book is intended for use primarily by senior and first-year graduate engineering students. The focus is on incompressible viscous flows with special applications to non-Newtonian fluid flows, turbulent flows, and free or forced convection flows. A special feature of the text is its coverage of generalized mass, momentum, and heat transfer equations; cartesian tensor manipulations; scale analyses; mathematical modeling techniques; and practical solution methods. The final chapter is unique in its case-study approach, applying general modeling principles to analyze nonisothermal flow systems found in a wide range of engineering disciplines.

Numerous end-of-chapter sample problem solutions, homework assignments, and mathematical aids are provided to enhance the reader's understanding and problem-solving skills.

Engineering Fluid Dynamics

Engineering
Fluid Dynamics

An Interdisciplinary Systems Approach

CLEMENT KLEINSTREUER
North Carolina State University

CAMBRIDGE UNIVERSITY PRESS
Cambridge, New York, Melbourne, Madrid, Cape Town, Singapore, São Paulo

Cambridge University Press
The Edinburgh Building, Cambridge CB2 2RU, UK

Published in the United States of America by Cambridge University Press, New York

www.cambridge.org
Information on this title: www.cambridge.org/9780521496704

© Cambridge University Press 1997

First published 1997
This digitally printed first paperback version 2005

A catalogue record for this publication is available from the British Library

ISBN-13 978-0-521-49670-4 hardback
ISBN-10 0-521-49670-5 hardback

ISBN-13 978-0-521-01917-0 paperback
ISBN-10 0-521-01917-6 paperback

*Dedicated to the memory of Mother Barbara
and to my family, Christin, Nicole, and Joshua*

Contents

Preface

Engineering fluid dynamics is considered (here) to be synonymous with fluid mechanics and convection heat transfer with engineering applications. The textbook is written for intermediate to advanced readers: for professionals as well as selected seniors and first-year graduate students in mechanical, biomedical, nuclear, and chemical engineering.

The main objective of the textbook is to provide the reader with sufficient background to enable him/her

(i) to bridge fluid mechanics and convection heat transfer material on an introductory graduate level with specialized advancements in hydrodynamic instability, turbulence, multiphase flows, or computational fluid dynamics; and

(ii) to tackle basic research projects in fluid mechanics and convection-heat-transfer related fields.

Although the text contains the basic engineering concepts, physical explanations, exercises, and mathematical aids necessary to succeed, it is the experience students will gain with the homework assignments, in-class discussions, journal article reviews, and course project reports that will move them to a deeper understanding and a higher level of proficiency. Specifically, the in-depth understanding of (the) *basics* in fluid mechanics/convection heat transfer and the skills to apply fundamental knowledge to the solution of interdisciplinary fluid dynamics problems are more valuable than presentation of large amounts of material within, typically, a very restricted time frame. Thus, in addition to the potentially unique learning experience provided with the text, the advanced student is given powerful tools, including computational fluid dynamics (CFD) software (cf. App. F), which may lead him/her to a level of maturity to solve challenging (industrial) fluid flow and heat transfer problems. In order to complete this task, a few interesting topics that usually overload a first-year graduate course had to be omitted; for example, three-dimensional boundary-layer flows, stability theory and statistical applications in turbulence, inviscid fluid flow theory, and advanced concepts of vorticity dynamics. Instead, topics such as scale analysis, non-Newtonian fluid flows; mathematical aids, including computer programs for CFD assignments; and selected (real-world) case studies were included. In summary, the text stresses pedagogical aspects of achieving a good understanding of basic viscous flows *balanced* by mathematical skills necessary to set up and solve or simulate interdisciplinary flow problems of increasing complexity.

The book consists of five major chapters plus several appendixes. Every chapter closes with illustrative sample problem solutions, practical problem assignments, and useful references. Chapter 1 reviews fluid kinematics and fluid dynamics aspects as well as fluid flow properties. The preliminary concepts of Chapter 1 in conjunction with the tensor applications of Appendix A provide the background for the main material presented.

Chapter 2 introduces the conservation laws in integral and differential forms. In discussing a number of illustrative examples, the mathematical/physical understanding of the basic transport equations is advanced. Instead of following the general trend toward completeness in listing solutions to variations of (the) basic thermal flow problems, some space is devoted to the repeated analyses and discussions of *fundamental* problems employing *different* solution approaches.

After very brief reviews of compressible flow and ideal flow, Chapter 3 treats incompressible viscous flows following the common thread of low-, moderate-, and high-Reynolds number flows, including turbulent flows. Again, sample problem solutions and specific problem assignments may enlighten the reader and strengthen his/her problem-solving skills. *A written review of an appropriate journal article and subsequently oral (i.e., viewgraph) presentation should be a course requirement. End-of-the-semester reports on challenging computer projects using the listed ordinary differential equation (ODE) and partial differential equation (PDE) solvers are highly recommended.* The information and mathematical aids given in Appendixes A–F make the book rather self-contained.

Many engineering fluid mechanics problems are nonisothermal and hence Chapter 4 provides the basics in forced and free convection heat transfer. The emphasis is on free or forced external (i.e., thin shear-layer) flows and forced internal flows, considering both the laminar and the fully turbulent regimes. The material presented in Chapter 4 and several sections of Chapter 5 should be sufficient to cover a first-year graduate course in convection heat transfer if, again, reviewing and presenting selected journal articles as well as computer course projects are part of the course requirements.

Chapter 5 is quite unique when compared to existing texts, in applying mathematical modeling approaches to set up and simulate real-world thermal flow systems. Selected applications and case studies, drawn primarily from the author's publications, illustrate the use of general principles to solve complex, multidisciplinary viscous flow as well as convection heat transfer problems. Hence, the engineering equations and solution methods introduced, the problem solutions presented, and the modeling steps discussed in Chapters 2–4 and Section 5.1 as well as the material provided in Appendixes A–F, form the background for Chapter 5. Furthermore, the project solutions in Chapter 5 stress the similarities between seemingly very different fluid flow systems ranging from biomedical to nuclear engineering applications. Specifically, the topics include laminar, moving boundary flows, turbulent boundary-layer flows, biofluid mechanics applications, and thermal flow of power-law fluids. Background information is provided in introductory sections for every case study. Whatever is selected from Chapter 5, it should help in the review and presentation of refereed journal articles as well as in the development of computer projects that are important requirements for each course.

The text can be used for a one-semester course in fluid mechanics (i.e., Ch. 1–3 plus Appendixes and Ch. 5 selectively), or, more economically, for a two-course sequence in fluid mechanics *and* convection heat transfer, where the latter course relies on Chapters 2, 4, and 5. Prerequisites include an introductory knowledge of fluid mechanics (e.g., texts by Sabersky et al. (1989), White (1986), or Gerhart et al. (1992)), thermodynamics (cf. Cengel and Boles (1993)), heat/mass transfer (e.g., texts by White (1988) or Bird et al. (1960)), and applied mathematics (e.g., texts by Boyce & DiPrima (1977), Habib (1975), Rieder and Busby (1986), or Wylie and Barrett (1982)). Corequisite courses, typically in numerical analysis and partial differential equations, are highly recommended.

The present book is based on lecture notes that were periodically extended and improved while I taught first-year graduate courses in fluid mechanics and convection heat transfer at Rensselaer Polytechnic Institute (RPI), Troy, New York, and at North Carolina State

University (NCSU), Raleigh. Naturally, I have borrowed (and referenced) some material from introductory texts, advanced reference works, and complementary books. In addition, homework set solutions, prepared by my former students in chemical, mechanical, environmental, nuclear, and biomedical engineering, were modified and used in Sections 1.4, 2.4, 3.3, and 4.3.

The author has received many helpful comments and suggestions during the preparation of the manuscript from former students and from colleagues in academic circles, research labs, and industry. I would like to mention Mr. Al Powell, who helped to put together Appendixes F.1 through F.4, and Mr. Kenneth Comer who greatly assisted in proofreading the manuscript and who finalized the nomenclature and index. I am especially grateful to Ms. Mary Molly Taylor, who did an excellent job in typing the manuscript, and to Mr. David Farmer, who expertly generated the figures, graphs, and sketches on his lap-top computer. Of the many people who were instrumental in publishing this book, I would like to mention Ms. Florence Padgett, Physical Sciences Editor, Ms. Katharita Lamoza, Production Editor, and Ms. Ellen Tirpak, Project Manager.

Please address technical correspondence to: C. Kleinstreuer, N.C. State University, Raleigh, NC 27695-7910.

Symbols and Abbreviations

Abbreviations

F	Force (N)
J	Joules (J)
L	Length (m)
M	Mass (kg)
T	Time (sec, s)
θ	Temperature (°C, K)
W	Watts (J/s)
1	dimensionless
—	not applicable

Roman Symbols

a	acceleration	LT^{-2}
A	area	L^2
A	Van Driest dampening length	L
A	aspect ratio	1
$\vec{a}_p$	particle acceleration vector	LT^{-2}
A_D^+	Van Driest constant (inner variable)	1
B	bulk modulus	1
B	empirical constant	various
BC	boundary condition	various
Bi	Biot number	1
BL	boundary layer	—
BVP	boundary value problem	—
C, c	heat capacity	$J\theta^{-1}$
C, c	constant	various
C_d, C_D	drag coefficient	1
C_f	local skin friction coefficient	1
C_p	specific heat (constant pressure)	$JM^{-1}\theta^{-1}$
C_v	specific heat (constant volume)	$JM^{-1}\theta^{-1}$
CFD	computational fluid dynamics	—
CVM	control volume method	—
d	diameter	L
D_{AB}	binary mass diffusion coefficient	L^2T^{-1}
D_{ij}	diffusive transport term	L^2T^{-1}
D_h	hydraulic diameter	L
D_m	mass diffusivity	L^2T^{-1}
DNS	direct numerical simulation	—

$\hat{e}_i$	unit vector in i-direction	1
erf	error function	1
$erfc$	complementary error function	1
exp	exponential (math operator)	—
E or $\dot{E}$	energy or thermal energy rate	J or W
Ec	Eckert number	1
EVM	eddy viscosity modeling	—
F	force	F
f	friction factor	1
fct, f	function	1
$\vec{f}_b$	body force	FL^{-3}
$\vec{f}_{ext}$	external body force	FL^{-3}
$f(\eta)$	similarity function	1
F	Helmholtz free energy	J
FDM	finite difference method	—
FEM	finite element method	—
g	gravitational constant	LT^{-2}
$\vec{g}$	gravitational vector	LT^{-2}
G	extensive system property	various
G	Gibbs free energy	J
Gr	Grashof number	1
GTE	General Transport Equation	—
h_f	friction head loss	L
h	specific enthalpy	JM^{-1}
h	small gap (position)	L
h	heat transfer coefficient	$JL^{-2}\theta^{-1}$
$\hat{h}$	specific enthalpy	JM^{-1}
H	total enthalpy	J
H	shape factor	1
H	height	L
HFD	hydrodynamically fully developed	—
$\hat{i}$	unit vector for the x-direction	1
I.C.	initial condition	various
IVP	initial value problem	—
J	momentum flux	MLT^{-1}
$\hat{j}$	unit vector for y-direction	1
$\hat{k}$	unit vector for z-direction	1
k	thermal conductivity	$JL^{-1}\theta^{-1}$
k	turbulence kinetic energy	J
k	permeability	LT^{-1}
l	Prandtl's mixing length	L
l_e	entry length	L
l_k	Kolmogorov length scale	L
l_m	mixing length	L
$\ln$	natural log	—
l_τ	viscous wall length	L

l^+	dimensionless mixing length	L
L	length	L
L	biharmonic differential operator	—
Le	Lewis number	1
LES	large eddy simulation	1
LHS	left-hand side	—
m	mass	M
$\dot{m}$	mass flow rate	MT^{-1}
$\dot{m}_0$	initial mass flow rate	MT^{-1}
M	total momentum	MLT^{-1}
Ma	Mach number	1
MLH	Prandtl's mixing length hypothesis	—
n	power-law exponent	1
N	mole number	1
Nu	Nusselt number	1
ODE	ordinary differential equation	—
p	pressure	FL^{-2}
P	perimeter	L
p	pressure parameter	1
P_{ij}	stress production term	$L^2 T^{-3}$
PDE	partial differential equation	—
Pe	Peclet number	1
Pr	Prandtl number	1
Pr_t	turbulent Prandtl number	1
$p\infty$	pressure in a free approach stream	FL^{-2}
q	heat flux	JL^{-2}
Q	(heat) flow rate	J, W/s
$\dot{Q}$	fluid flow rate	$L^3 T^{-1}$
Q_z	flow rate in z-direction	$L^3 T^{-1}$
r	radial position	L
r_0^+	dimensionless radius	1
$\vec{r}_p$	position vector	L
$\hat{r}$	dimensionless radial position	1
R	specific gas constant	$JM^{-1}\theta^{-1}$
Ra	Rayleigh number	1
Re	Reynolds number	1
REV, REV	representative elementary volume	—
RHS	right-hand side	—
Ri	Richardson number	1
ROMA	relative order of magnitude analysis	—
RSM	Reynolds stress modeling	—
RSTE	Reynolds stress transport equation	—
S	surface	L^2
S, s	entropy	$JM^{-1}\theta^{-1}$
S	source term	various
Sc	Schmidt number	1

SGSM	subgrid scale modeling	—
t	time	T
t_c	convection time scale	T
T_{ij}	turbulent diffusion transport term	$L^2 T^{-3}$
t_h	heat diffusion time scale	T
t_k	Kolmogorov time scale	T
t_e	representative fluid element travel time	T
t_v	viscous diffusion time scale	T
T	temperature	θ
TFD	thermally fully developed	—
TSL	thin shear layer	—
T_b	bulk temperature	θ
T_f	reference temperature of fluid	θ
T_∞	free stream or ambient temperature	θ
T_m	mean or mixing cup temperature	θ
T_w	wall temperature	θ
T^+	inner variable (turbulent temperature)	1
tr	transpose (math operator)	—
u	velocity component in x-direction	LT^{-1}
u_{av}	average axial velocity	LT^{-1}
u^+	turbulent boundary layer velocity	LT^{-1}
u_e	boundary-layer edge velocity	LT^{-1}
u_m	mean velocity	LT^{-1}
u_0	average inlet velocity	LT^{-1}
u_τ	friction velocity	LT^{-1}
u^+	inner variable (turbulent velocity)	1
u_∞	free stream or approach velocity	LT^{-1}
U	outer (potential flow) velocity	LT^{-1}
U	internal energy	$J, J/M$
$\hat{u}$	specific internal energy	—
v	velocity component in y-direction	LT^{-1}
$\vec{v}$	velocity vector	LT^{-1}
v_r	radial velocity	LT^{-1}
v_z	axial velocity	LT^{-1}
v_θ	circumferential velocity	LT^{-1}
$V, \forall$	volume	L^3
$\hat{v}$	specific volume	$L^3 M^{-1}$
w	velocity component in z-direction	LT^{-1}
W, w	width	L
w.r.t.	with respect to	—
W	work	W, FL
W	wake function	various
x	axial position	L
x_m	unheated starting length	L
$\hat{x}$	dimensionless position	L
y	position	L
y^+	inner variable (wall Reynolds number)	1
z	axial or vertical position	L

Greek Symbols

α	coefficient	various
α	angle	radians or degrees
α	thermal diffusivity	L^2T^{-1}
β	angle	radians or degrees
β	pressure gradient parameter	1
β	volumetric expansion coefficient	θ^{-1}
Γ	circulation	L^2T^{-1}
γ	ratio of specific heats	1
γ	intermittency factor	1
$\overset{=}{\dot{\gamma}}$	rate of deformation tensor	T^{-1}
δ	small displacement	L
δ	small increment	—
δ	boundary layer or thin shear layer thickness	L
δ_1, δ_2	displacement, momentum thicknesses	L
δ_{ij}	Kronecker delta	1
δ_{th}	thermal boundary layer	L
$\vec{\delta}$	unit vector	1
$\overset{=}{\delta}$	unit tensor	1
$\Delta(\ )$	change of ( )	—
ε	turbulent eddy diffusivity	1
ε_{ij}	viscous dissipation function	L^2T^{-3}
ε_{ijk}	permutation symbol	1
$\vec{\varepsilon}$	internal source function	—
$\overset{=}{\varepsilon}$	strain rate tensor	T^{-1}
$\zeta, \vec{\zeta}$	vorticity, vorticity vector	T^{-1}
η_0	zero shear rate viscosity	$ML^{-1}T^{-1}$
η	non-Newtonian fluid viscosity	$ML^{-1}T^{-1}$
η	similarity or combined variable	1
η	dimensionless coordinate	1
θ	angle	rad
θ	dimensionless temperature field	1
θ_m	mean nondimensionless temperature	1
θ_{th}	enthalpy thickness	L
κ	Von Karman universal constant	1
λ	dimensionless system parameter	1
$-\lambda^2$	arbitrary separation constant	—
λ_n	eigenvalues	various
μ	chemical potential	$J\,\text{mol}^{-2}$
μ	dynamic viscosity	$ML^{-1}T^{-1}$
ν	kinematic viscosity	L^2T^{-1}
ν_t	turbulent eddy viscosity	L^2T^{-1}
$\overset{=}{\pi}$	total stress tensor	FL^{-2}
π	pi (3.14159......)	1
Π	wake parameter	—
Π_{ij}	pressure–strain correlation term	L^2T^{-3}

ρ	density	ML^{-3}
$\vec{\sigma}$	normal stress vector	FL^{-2}
$\sum$	math operator (to sum)	—
$\vec{\sum}$	source term (vector)	various
τ	dissipation time scale	T
$\vec{\tau}$	shear stress vector	FL^{-2}
$\vec{\vec{\tau}}$	shear stress tensor	FL^{-2}
ϕ	velocity potential	$L^2 T^{-1}$
ϕ, Φ	viscous dissipation function	various
ξ	dimensionless streamwise distance	1
ξ	integration dummy variable	various
ψ	generalized transport V function	various
ψ	stream function, arbitrary quantity	$L^2 T^{-1}$, various
$\vec{\Omega}$	diffusional flux vector	1
$\omega, \vec{\omega}$	angular velocity (vector)	rad/time

Special Symbols

:=	equivalent to, here equal to
$\nabla( )$	gradient of ( )
$\nabla \cdot ( )$	divergence of ( )
$\nabla \times ( )$	curl, rotation of ( )
$\perp$	perpendicular to
1-*D*	one-dimensional flow
2-*D*	two-dimensional flow
3-*D*	three-dimensional flow
∞	infinite
$O( )$	order of ( )
$\propto, \sim$	proportional to
#	arbitrary variable (various units)
$\approx$	approximately
$( )^{tr}$	transpose of
$\triangleleft$	angle between
$\hat{=}$	equivalent to

Subscripts

o	initial (time or location); wall
abs	absolute
b	body
c	concentration
cL, $\pounds$	centerline
d	with respect to diameter
ext	external
f	fluid
g, grav	gravity

Hf	Helmholtz free energy
kin	kinetic
max	maximum value
n	normal
pr	pressure
r	radial
rel	relative
s	solid or surface
t	turbulent
T	temperature
visc	viscous
w, o	wall
x	with respect to the x-coordinate
y	with respect to the y-coordinate
z	with respect to the z-coordinate
,t	derivative with respect to time
,x	derivative with respect to x
,y	derivative with respect to y
,z	derivative with respect to z
θ	with respect to the θ-coordinate
ϕ	with respect to the ϕ-coordinate
∞	free stream, far field or infty

Superscripts

$^-$	mean value
$\char`\^$	distinguishes variables; nondimensional
$+$	turbulent, law of the wall variable
$\prime$	differentiation (with respect to η)

CHAPTER 1

Preliminary Concepts

1.1   Introduction and Overview

As indicated in the Preface, the main objectives of this book are to provide the reader, in a unifying way, with classical material in fluid mechanics and convection heat transfer and to introduce him or her to basic techniques for modeling engineering fluid dynamics systems. Thus, studying the book in a formal graduate course setting may enhance the student's physical understanding, increase problem-solving skills, and build up confidence to solve other thermal flow problems not discussed in this text. The approach and objectives of problem-solving steps, or in more complex cases "model development" in the engineering sciences, are summarized in Fig. 1.1. This sequence will be highlighted throughout.

The material in Chapters 1 and 2 together with Appendices A and B may equalize readers' different entry levels in fluid dynamics, systems analysis, and engineering mathematics. Specifically, in Section 1.2.1, the two fundamental flow field descriptions (i.e., Lagrange vs. Euler) are reviewed; Sections 1.2.2–1.2.4 discuss the *kinematics* of shear flow (i.e., fluid element translation, rotation, and deformation), *thermodynamic* properties (e.g., pressure, temperature, density, and entropy), and *transport* properties (e.g., viscosity, conductivity, and diffusivity). Some basics of particle dynamics are extended to *fluid particle* dynamics in Section 1.3. Differential operators and cartesian tensor applications, useful for Sections 1.2.2 and 1.3 as well as for Chapter 2 are summarized in Appendix A. Fluid flow systems under consideration in Chapters 1–5 are *restricted* to single-phase flow, continuum mechanics, deterministic processes, and Eulerian flow descriptions (cf. Sect. 1.4). Selected problem solutions illustrating the material presented in Chapter 1, are given in Section 1.4.

The material in Appendix A should be frequently consulted and the different notations (vector, tensor, etc.) should be swiftly absorbed.

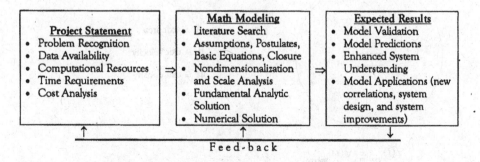

Project Statement	**Math Modeling**	**Expected Results**
• Problem Recognition	• Literature Search	• Model Validation
• Data Availability	• Assumptions, Postulates,	• Model Predictions
• Computational Resources	Basic Equations, Closure	• Enhanced System
• Time Requirements	• Nondimensionalization	Understanding
• Cost Analysis	and Scale Analysis	• Model Applications (new
	• Fundamental Analytic	correlations, system
	Solution	design, and system
	• Numerical Solution	improvements)

F e e d - b a c k

Fig. 1.1. Sequence of model development.

1

1.2   Fluid Particle Kinematics and Fluid Flow Properties

The *kinematics* of fluid motion is described in terms of velocity fields and their derivatives such as acceleration and vorticity. In contrast, the *dynamics* of fluid motion describes the forces acting on a fluid element. *Fluid properties*, such as density and viscosity, "correlate" forces with fluid motion. Typically for flow field descriptions, the *Eulerian* framework is preferred by engineers, whereas the *Lagrangian* point of view is often taken by (atmospheric) scientists.

1.2.1   Flow Field Descriptions

The method for describing fluid flow systems associated with Lagrange considers a *closed* system (cf. Fig. 1.2), which consists of an identifiable quantity of mass. The system boundaries may be fixed or movable but no mass crosses them. An example is the nonleaking piston–cylinder device, where the system mass is the enclosed gas, that is, $m_{\text{gas}} = \text{constant}$. The Lagrangian viewpoint of fluid mechanics is an extension of particle mechanics. The independent variables are the position vector and the time. Position vectors, $\vec{r}_p(t)$, give

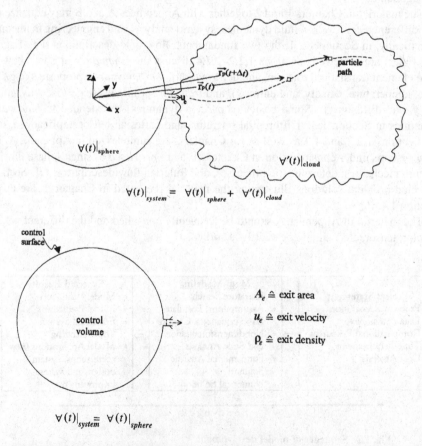

Fig. 1.2. Closed versus open system: (a) Closed system (Lagrange) (system mass is constant); (b) Open system (Euler) (control volume approach).

the paths of all particles with time as a parameter. Hence even "steady" processes are time-dependent in this framework. Once $\vec{r}_p(t)$ measured from a fixed origin is known, the particle velocity or particle ensemble velocity, $\vec{v}_p = (d\vec{r}_p/dt)$, and acceleration, $\vec{a}_p = (d\vec{v}_p/dt) = (d^2\vec{r}_p/dt^2)$, can be readily computed.

In contrast, the *open* system or *control volume* approach, associated with Euler, considers an arbitrary, typically fixed region in space. Its boundary is called a control surface through which flow quantities (mass, momentum, energy) enter and leave. Examples include all mass, force, or energy balances taken for a finite (control) volume, which usually is a slice (or shell) of the given system. From the Eulerian viewpoint, all flow properties are functions of a fixed point in space and time. Hence, the particle position vector in Eulerian variables is simply the fixed space coordinates. In order to express the time rate of change of a particle property in Eulerian variables, the substantial, Stokes, or material derivative $(D\#/Dt) \equiv (\partial\#/\partial t) + (\vec{v} \cdot \nabla)\#$ is employed (cf. App. A). This is a shorthand expression that states that the (total) time derivative of variable "#" (i.e., a vector or scalar flow property), evaluated as we follow a material particle, fluid element, or closed system, is equal to the *local* time rate of change of "#" and a convective or spatial change in "#" (cf. Problem Solution 1.3.3 in Sect. 1.4). Clearly, the operator (D/Dt) in fluid mechanics should be distinguished from (d/dt) employed in solid particle dynamics (cf. Chapter 2).

In summary, the Lagrangian viewpoint is most beneficial when tracking the history of a particle or a particle cloud as required, for example, in air pollution analyses. The Eulerian viewpoint is, in general, more useful because physical laws written in this framework do not contain the position vector, and the velocity appears as the major (dependent) variable.

1.2.2 Fluid Kinematics

Forces (i.e., body and surface forces) set fluids into motion as in turn fluid motion may exert forces. For example, stresses that are surface forces per unit area continuously deform a fluid, that is, generate fluid motion. On the other hand, wind and waves produce periodic forces on structures. The types of fluid motion and the relationships between imposed shear stress and the rate of fluid element deformation are of interest in kinematics. Certain materials such as Bingham plastics are actually not fluids since they can sustain, like solids, a *limited* shear stress while remaining in static equilibrium. Fluid motion created by changes (i.e., gradients) in pressure and stresses can be decomposed into

- translation, which is fluid element displacement or convection expressed by the velocity vector $\vec{v} = (u, v, w)$ or (v_r, v_θ, v_z), and so on;
- rotation, where a fluid element spins about its own axis causing a changing *orientation*, measured by the vorticity vector $\vec{\zeta} \sim \vec{\omega} \sim \nabla \times \vec{v}$; and
- deformation, which is the dilatation plus distortion of a fluid element, measured with the strain rate tensor $\vec{\vec{\varepsilon}} \sim \nabla\vec{v}$ or $\vec{\vec{\gamma}}$, which is the rate-of-deformation tensor.

Clearly, once the velocity field, $\vec{v}$, is known, all forms of fluid motion can be readily computed. Observing a plane fluid element in rectangular coordinates at time t and then after forces acted upon it, at time $t + \Delta t$, we can describe the resulting motion as follows (cf. Fig. 1.3a).

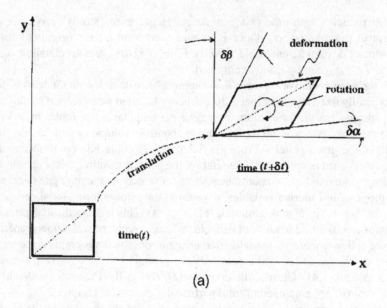

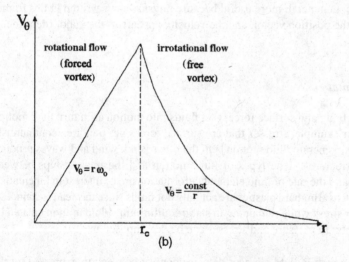

Fig. 1.3. Elements of fluid motion: (a) translating, rotating, and deforming fluid element during time δt; (b) rotational versus irrotational flow.

A   Translation

Rate of fluid element displacement in rectangular coordinates.

$$u = \lim_{\Delta t \to 0} \frac{\Delta x}{\Delta t}; \qquad v = \lim_{\Delta t \to 0} \frac{\Delta y}{\Delta t} \quad \text{and} \quad w = \lim_{\Delta t \to 0} \frac{\Delta z}{\Delta t}$$

♦   where $u\hat{i} + v\hat{j} + w\hat{k} = \vec{v}$ (1.1)

B   Rotation

An angular velocity about the z-axis, ω_z, can be defined as the *net* average rate of counterclockwise spin of the fluid element.

$$\omega_z = \frac{1}{2}\left(\frac{d\alpha}{dt} - \frac{d\beta}{dt}\right)$$

where $d\alpha$ and $d\beta$ are related to the planar velocity derivatives as (cf. Fig. 1.3a and Sect. 1.4)

$$d\alpha = \frac{\partial v}{\partial x}\,dt \quad \text{and} \quad d\beta = \frac{\partial u}{\partial y}\,dt$$

so that

$$\omega_z = \frac{1}{2}\left(\frac{\partial v}{\partial x} - \frac{\partial u}{\partial y}\right)$$

or, in general,

$$\blacklozenge \quad \vec{\omega} = \omega_x\hat{i} + \omega_y\hat{j} + \omega_z\hat{k} = \frac{1}{2}\nabla \times \vec{v} = \frac{1}{2}\vec{\zeta} = \frac{1}{2}\begin{vmatrix} \hat{i} & \hat{j} & \hat{k} \\ \frac{\partial}{\partial x} & \frac{\partial}{\partial y} & \frac{\partial}{\partial z} \\ u & v & w \end{vmatrix} \qquad (1.2)$$

where $\vec{\zeta}$ is the vorticity vector. An example of rotational versus irrotational flow is given in Fig. 1.3b.

Example: "Idealized Hurricane"

(i)   Eye of the hurricane is rotational, i.e., with solid body-like rotation, $v_\theta \sim r$
(ii) Outside a critical radius r_c, the flow field is basically irrotational, i.e., $v_\theta \sim r^{-1}$

C   Deformation

Fluid elements may experience a volume change (i.e., elongation or compression) due to pressure and normal stresses as well as shape distortion due to shear stresses, that is, tangential forces. The possible fluid element *elongation* or *compression* is expressed in terms of the tensional/compressional strain rates

$$\varepsilon_{xx} = \frac{\partial u}{\partial x}, \qquad \varepsilon_{yy} = \frac{\partial v}{\partial y} \quad \text{and} \quad \varepsilon_{zz} = \frac{\partial w}{\partial z}, \qquad \text{i.e.,} \quad \varepsilon_{ii} = \frac{\partial u_i}{\partial x_i}$$

For example, in unidirectional flow a fluid volume rate change in x-direction can be written as

$$\frac{1}{A_x}\frac{\Delta \dot{V}}{\Delta x} \approx \frac{\Delta u}{\Delta x} \approx \varepsilon_{xx}$$

The average *distortion* rate is the arithmetic mean of the rate change in angles α and β, that is, $1/2(d\alpha/dt + d\beta/dt)$, looking onto the x–y plane as given in Fig. 1.3a. Hence, with

$(\partial\alpha/\partial t) \approx (\partial v/\partial x)$ and $(\partial\beta/\partial t) \approx (\partial u/\partial y)$,

$$\varepsilon_{yx} = \varepsilon_{xy} = \frac{1}{2}\left(\frac{\partial v}{\partial x} + \frac{\partial u}{\partial y}\right)$$

Combining fluid element dilatation and distortion and generalizing the analysis for three-dimensional elements yield the rate-of-strain tensor

$$\blacklozenge \qquad \varepsilon_{ij} = \frac{1}{2}\left(\frac{\partial u_i}{\partial x_j} + \frac{\partial u_j}{\partial x_i}\right) := \frac{1}{2}\left[\nabla\vec{v} + (\nabla\vec{v})^{\text{tr}}\right] \qquad (1.3)$$

where $\nabla\vec{v} \equiv \text{grad }\vec{v}$ is a dyadic product, usually called the velocity field gradient, and $(\nabla\vec{v})^{\text{tr}}$ is its transpose. Note that $\nabla \times \vec{v} \sim \vec{\zeta}$ is a vector, whereas $\nabla\vec{v} \sim \vec{\vec{\varepsilon}}$ is a second-rank tensor with nine components (cf. App. A). It should be already transparent that $\nabla\vec{v} \sim \vec{\vec{\varepsilon}}$ can be directly related to the stress tensor $\vec{\vec{\tau}}$ (cf. Sect. 1.3.1).

1.2.3 Thermodynamic Properties

Thermodynamic properties of a system such as internal energy, entropy, pressure, and temperature are related to each other and define the *state* of the system. A *process*, being considered either quasi-reversible or irreversible, is any mechanism by which the system state is changed. *Extensive properties* such as volume, energy, and mole number are system-size-dependent, whereas *intensive* properties (e.g., pressure and temperature) are size-independent. In order to determine the state of simple compressible substances, three independent properties have to be fixed:

- internal energy U,
- volume V,

and, of special importance in chemical engineering, the

- mole number of a substance, N.

For example, knowing the functional form for the entropy S (i.e., an equation of state for a given system)

$$S = S(U, V, N) \qquad (1.4)$$

we can form the total differential of S as

$$dS = \frac{\partial S}{\partial U}\bigg|_{V,N} dU + \frac{\partial S}{\partial V}\bigg|_{U,N} dV + \frac{\partial S}{\partial N}\bigg|_{U,V} dN$$

where $(\partial S/\partial U)|_{V,N} \equiv (1/T)$ which implies that the temperature $T = T(U, V, N)$ is also an equation of state of the substance. With two other equations of state for the pressure $p = p(U, V, N)$ and chemical potential $\mu = \mu(U, V, N)$ we can define

$$p \equiv T\frac{\partial S}{\partial V}\bigg|_{U,N} \quad \text{and} \quad \mu \equiv -T\frac{\partial S}{\partial N}\bigg|_{U,V}$$

so that the total change in entropy reads

$$dS = \frac{1}{T} dU + \frac{p}{T} dV - \frac{\mu}{T} dN$$

or

♦ $TdS = dU + pdV - \mu dN$ (1.5)

which is the *second law of thermodynamics in* a different form.

Alternative forms of the basic differential equation of thermodynamics can be obtained by introducing the

enthalpy $H \equiv U + pV$

Gibbs free energy $G \equiv H - TS$

Helmholtz free energy $F \equiv U - TS$

Differentiating, for example, the enthalpy definition, we have

$$dH = dU + p\,dV + V\,dp$$

or, substituting dU from Eq. (1.5) results in

$$dH = T\,dS + V\,dp + \mu\,dN$$ (1.6)

that is, $H = H(S, p, N)$, which contains all thermodynamic information.

In classical thermodynamics, the theory starts directly from experimental results, using heat and work as fundamental concepts. The *first law of thermodynamics* for a *closed stationary system* undergoing a process of infinitesimally small changes received is

♦ $dU = \delta Q + \delta W$ (1.7)

where the use of δ as a differential sign indicates that an infinitesimal amount of, say, work done, $\delta W = -p\,dV$, is an inexact differential. An alternative form of the first law would be

$$dH = \delta Q + \delta W + d(pV)$$

Defining the heat capacity C as

$$C = \lim_{T_2 \to T_1} \frac{Q}{T_2 - T_1}$$

we can write

$$C = \frac{\delta Q}{dT}$$ (1.8a)

or for the case of constant V and N

$$TdS = \delta Q$$

so that

$$\frac{\delta Q}{dT}\bigg|_{V,N} \equiv C_V = T\frac{\partial S}{\partial T}\bigg|_{V,N} \quad \text{and} \quad C_p \equiv \frac{\delta Q}{dT}\bigg|_{P,N} \tag{1.8b,c}$$

is the specific heat at constant volume and constant pressure.

It is apparent that different choices of independent and dependent variables can be made in thermodynamics depending upon which is most advantageous for a given problem. The most common way to specify thermodynamics information is via the equations of state of a so-called pVT system. Any *isotropic*, that is, direction-invariant, system of constant mass and constant composition that exerts a uniform hydrostatic pressure on the surroundings, in the absence of surface, gravitational, electrical, and magnetic effects, is called a pVT system. Examples include a pure substance, that is, a single chemical species in any solid, liquid, or gaseous form, or a homogeneous mixture of different chemical species such as a mixture of gases (e.g., air) or liquids (e.g., water). Using simple theorems in partial differential calculus, one can express any thermodynamic property in terms of these three coordinates (pressure, volume, and temperature) of such a system, provided that it is in a state of equilibrium. The most common example is the *ideal gas law*,

$$\blacklozenge \qquad \frac{p}{\rho} = RT \quad \text{or} \quad pV = mRT \tag{1.9a,b}$$

where R is (the) specific gas constant.

As all this holds for a *fluid at rest*, we extend these ideas to a *moving continuum* by assuming that the bulk motion of the fluid does not affect the thermodynamics state. Thus, the thermodynamic properties are determined by an observer moving with the local velocity (cf. Sects. 1.2.1 and 2.1.1).

1.2.4   Transport Properties

The transport properties of interest include the fluid viscosity, μ, the thermal conductivity, k, and the binary mass diffusivity, D_{AB}. For laminar flow of Newtonian isotropic fluid, these properties are typically a function of temperature only. However, in practical applications, transport properties are also functions of pressure and concentration and they may exhibit (nonlinear) directional dependencies. Additional complexities arise when the *fluid* is non-Newtonian, such as particle suspensions and polymeric liquids, and when the *flow* is turbulent. In such cases, transport properties are not only dependent on the type of fluid but predominantly a function of local flow characteristics, such as the gradient of the velocity field, grad $\vec{v} \equiv \nabla \vec{v}$, and the presence of solid surfaces or other interfaces.

A solution of the transient or steady one-dimensional form of the transport equations is often employed to measure the transport properties for various fluids at different temperatures (cf. App. E). Theoretical expressions for μ, k, and D_{AB} have been established for fluids in shear flow; however, the transport properties are taken to be independent of pressure and density variations (cf. Kay and Nedderman 1985; Bird, Stewart, and Lightfoot 1960; Bird, Armstrong, and Hassager 1987; White 1991).

1.3　Fluid Particle Dynamics

As indicated in Section 1.2.2, surface and body forces acting on fluid elements generate fluid motion, and vice versa. (cf. illustration):

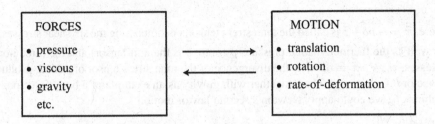

One could define *fluid flow* as a substance in a process of continuous deformation due to the interaction of driving forces (e.g., pressure, gravity, stress) and resisting forces (e.g., friction or drag). Some substances appear to be borderline solids–fluids, such as toothpaste, tar, sand, foodstuff, depending upon the magnitude of the forces applied. Clearly, *stresses* on fluid elements, either a normal force per surface area, like normal stress and pressure, or a tangential force per surface area, are flow properties of utmost importance. In mathematical terms, the interactions of *net* surface forces and change in total stress can be described as follows. Starting with a one-dimensional force balance for a rectangular fluid element depicted in Figure 1.4 yields

$$\Delta F_x = \left[\frac{\partial}{\partial x}\pi_{xx} + \frac{\partial}{\partial y}\pi_{yx} + \frac{\partial}{\partial z}\pi_{zx} \right] \Delta x\, \Delta y\, \Delta z$$

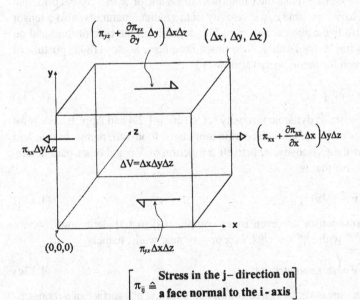

Fig. 1.4. Selected surface forces acting on a fluid element.

or, extended to three dimensions, we can write in the limit:

$$\left.\frac{d\vec{F}}{dV}\right|_{\text{surface}} = \left.\frac{d\vec{F}}{dV}\right|_{\text{pressure}} + \left.\frac{d\vec{F}}{dV}\right|_{\text{viscous}} = \nabla \cdot \vec{\vec{\pi}} = \underbrace{-\nabla p}_{\vec{f}_p} + \underbrace{\nabla \cdot \vec{\vec{\tau}}}_{\vec{f}_v} \tag{1.10}$$

Here, $\vec{\vec{\pi}} = -p\vec{\vec{\delta}} + \vec{\vec{\tau}}$ is called the *total* stress tensor, encompassing the six shear stresses, τ_{ij}, as well as the thermodynamic pressure p (note, $\vec{\vec{\delta}}$ is the unit tensor) and the three normal stresses, τ_{ii}; $\nabla \cdot \vec{\vec{\pi}} \equiv \text{div } \vec{\vec{\pi}}$ is the divergence of the total stress tensor of rank 2, producing a vector (cf. App. A). Thus, together with gravity as an example of a body force per unit volume $\vec{f}_b$, we could apply Newton's second law of motion

♦ $$\qquad m\vec{a} = \sum \vec{F}_{\text{ext}} \quad \text{or} \quad \rho\frac{d\vec{v}}{dt} = \sum \vec{f}_{\text{ext}} \tag{1.11a,b}$$

to an incompressible fluid flow field as

$$\rho \underbrace{\frac{D\vec{v}}{Dt}}_{\vec{a}_{\text{total}}} = \vec{f}_p + \vec{f}_v + \vec{f}_b \tag{1.12a}$$

or

♦ $$\rho\left[\underbrace{\frac{\partial \vec{v}}{\partial t}}_{\vec{a}_{\text{local}}} + \underbrace{(\vec{v} \cdot \nabla)\vec{v}}_{\vec{a}_{\text{convective}}}\right] = -\nabla p + \nabla \cdot \vec{\vec{\tau}} + \vec{g} \tag{1.12b}$$

The question now is, how can we relate the (unknown) stress tensor, $\vec{\vec{\pi}}$ or $\vec{\vec{\tau}}$, to the principal unknown, $\vec{v}$, or at least to $\nabla\vec{v} \equiv \text{grad }\vec{v}$, the velocity field gradient, naturally, also a tensor of rank 2. Such a constitutive equation, say, $\vec{\vec{\tau}} \sim \nabla\vec{v}$, depends on the type of fluid and on the flow regime. For example, for laminar Newtonian fluid flow, Stokes (1845) postulated a *linear* correlation, which for *incompressible* flow is

♦ $$\vec{\vec{\tau}} = \mu[\nabla\vec{v} + (\nabla\vec{v})^{\text{tr}}] \tag{1.13a}$$

where μ is the fluid's (constant) dynamic viscosity (cf. Section 1.2.4 and App. E). For non-Newtonian fluids such as polymeric liquids, exotic lubricants, foodstuff, paints, pastes, and slurries, the generalized fluid viscosity, η, is itself a function of $\nabla\vec{v}$ and other parameters, rendering $\vec{\vec{\tau}}(\nabla\vec{v})$ *nonlinear*, that is,

♦ $$\vec{\vec{\tau}} \approx \eta(\nabla\vec{v})[\nabla\vec{v} + (\nabla\vec{v})^{\text{tr}}] \tag{1.13b}$$

Expressions for the stress tensor are even more complicated in turbulent flows, where $\tau_{ij}^{\text{total}} = \tau_{ij}^{\text{laminar}} + \tau_{ij}^{\text{turbulent}}$ with $\tau_{ij}^{\text{turb}} \gg \tau_{ij}^{\text{lam}}$, except very near walls, namely,

$$\vec{\vec{\tau}}_{\text{turb}} = \text{fct}(\vec{v}, \nabla\vec{v}, \mu, \text{geometry, etc.}) \tag{1.13c}$$

In the next section, the stress vector (three components) acting at a surface of a representative fluid element and the stress tensor (nine components) acting on a three-dimensional (3-D) fluid element are discussed. One should recall that fluid element *volume changes* and

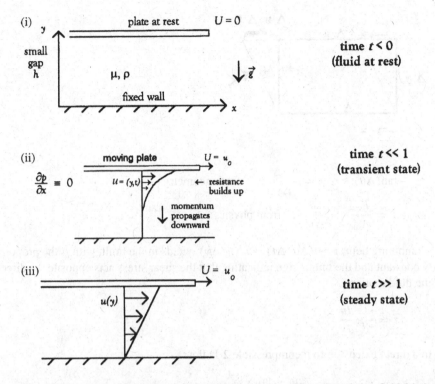

Fig. 1.5. Onset of momentum transfer for simple shear flow: (i) resisting fluid friction; (ii) diffusive momentum flux perpendicular to the plate motion; (iii) steady-state flow (after Bird et al. 1960).

shape distortions are described by the total stress tensor $\vec{\vec{\pi}}$, whereas fluid element *rotation* is described by the vorticity vector, introduced in Section 1.2.2 and further discussed in Sect. 1.3.2.

1.3.1 The Stress Tensor

Consider the development of *simple shear flow* (cf. Fig. 1.5), where the upper plate is suddenly at $t = 0$ brought to a speed of $u_0 =$ constant. The unsteady force required to pull the plate is $\tau_{\text{wall}} \cdot A_{\text{surface}}$ where τ_{wall} is proportional to the (changing) velocity gradient, $(\partial u / \partial y)$, at the plate surface, that is, at $y = h$. During the time-dependent state and thereafter, frictions between the fluid layers affect each other: that is, momentum flux is "diffusing" *normal* to the plate/fluid motion. The fluid elements at the plate surface always move with the plate velocity, obeying the so-called no-slip condition. When steady-state is reached, the pulling force, equal in magnitude to the drag force, is constant and a linear velocity profile has been established, called Couette flow (cf. Sects. 2.2.2 and 3.2.1A).

If a tiny planar fluid element, $\Delta x \cdot \Delta y$, of that simple shear flow field is considered at times t and $t + \Delta t$, we can make the following observations (cf. sketch):

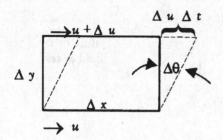

$$\tan(\Delta\theta) \approx \Delta\theta = \frac{\Delta u \Delta t}{\Delta y} \qquad \text{from geometry}$$

$$\tau = \frac{F_{\text{visc}}}{A_{\text{surf}}} \sim \frac{\Delta\theta}{\Delta t} \qquad \text{from physics}$$

Combining both, $\tau \sim (\Delta\theta/\Delta t) \sim (\Delta u/\Delta y)$, yields in the limit, with μ the proportionality constant and the minus sign indicating that the shear stress acts opposite the direction of the fluid motion

$$\tau = -\mu\frac{du}{dy}$$

or in a direct extension to incompressible 2-D flow

$$\tau_{yx} = \tau_{xy} = -\mu\left(\frac{\partial u}{\partial y} + \frac{\partial v}{\partial x}\right)$$

For every fluid element *surface* an arbitrary force can be decomposed into three components generating a stress vector: one normal stress and two shear stresses (cf. Fig. 1.6). For a representative fluid element, there are 3 stresses × 3 orthogonal surfaces = 9 stress components. The resulting *stress tensor* is symmetric, that is, $\tau_{ij} = \tau_{ji}$, and there are (three) mutually perpendicular directions or planes for which the shearing stress vanishes and only normal, that is, maximum and minimum, *principal* stresses remain (cf. Problem Solution 1.2(d) in Sect. 1.4). For example, in rectangular coordinates the (viscous) stress tensor, occasionally called the deviatoric stress, can be written in matrix notation as

$$\blacklozenge \qquad \vec{\vec{\tau}} := \tau_{ij} = \begin{bmatrix} \tau_{xx} & \tau_{xy} & \tau_{xz} \\ \tau_{yx} & \tau_{yy} & \tau_{yz} \\ \tau_{zx} & \tau_{zy} & \tau_{zz} \end{bmatrix}$$

Alternatively, the stress tensor can be viewed as a second-rank tensor that associates a (stress) vector, $\vec{\tau}_k$, with each direction $\vec{\delta}_k$, $k = 1, 2, 3$, as can be deduced from Fig. 1.6:

$$\vec{\vec{\tau}} = \vec{\delta}_1\vec{\tau}_1 + \vec{\delta}_2\vec{\tau}_2 + \vec{\delta}_3\vec{\tau}_3 \tag{1.14}$$

Returning to the question posed previously – How can we relate the stress tensor $\vec{\vec{\tau}}$ to the principal unknown $\vec{v}$ or at least to $\nabla\vec{v}$? – we first consider Newtonian fluids such as air, water, or oil and then power-law fluids, such as high-molecular liquids, pastes, synthetic lubricants, and blood in capillaries.

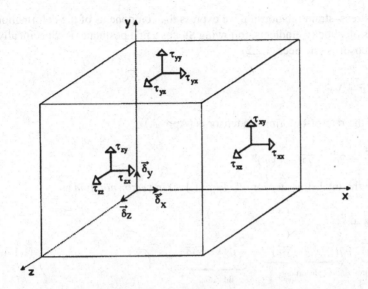

Fig. 1.6. Normal and tangential stress components of a representative fluid element. Note: The first index indicates the orientation of the test surface and the second subscript indicates the direction or component of the stress.

A Newtonian Fluids

Stokes (1845) stated three postulates for Newtonian fluids:

(i)   The fluid is continuous and τ_{ij} is a *linear* function of the strain rates ε_{ij} (or rate-of-deformation tensor $\dot{\gamma}_{ij} \equiv \text{def } \vec{v} = 2\varepsilon_{ij}$). Hence, $\tau_{ij} \propto \dot{\gamma}_{ij}$ where def $\vec{v} \equiv \dot{\gamma}_{ij} \propto (\partial u_i / \partial x_j)$ as shown later.

(ii) The fluid is *isotropic*. That is, the principal stress axes are identical with the principal strain-rate axes so that the deformation law is invariant (cf. Sect. 1.4 and App. A).

(iii) When the strain rates are zero $\pi_{ij} = -p\delta_{ij}$, that is, the thermodynamic or hydrostatic pressure is obtained.

Mathematically, we postulate

$$\pi_{ij} = \underbrace{-p\delta_{ij}}_{\text{pressure}} + \overbrace{\underbrace{A\varepsilon_{ij}}_{\text{distortion}} + \underbrace{B(\nabla \cdot \vec{v})\delta_{ij}}_{\text{dilatation}}}^{\tau_{ij}} \quad \text{where} \quad \delta_{ij} = \begin{cases} 1 & \text{if } i = j \\ 0 & \text{if } i \neq j \end{cases} \quad (1.15)$$

$A \equiv 2\mu$ is twice the dynamic (shear) viscosity and $B \equiv \lambda = (2/3)\mu$ is the bulk (dilatational) viscosity. Thus, the components of the total stress tensor are

$$\pi_{ij} = \begin{vmatrix} -p + \tau_{xx} & \tau_{xy} & \tau_{xz} \\ \tau_{yx} & -p + \tau_{yy} & \tau_{yz} \\ \tau_{zx} & \tau_{zy} & -p + \tau_{zz} \end{vmatrix} \quad (1.16)$$

As mentioned earlier, the unknown stresses, τ_{ij}, have to be eliminated by using appropriate constitutive equations that relate stresses to strains, to velocity gradients. Hence, in order to obtain a linear stress–strain relationship we express the components of the deformation rate tensor in terms of velocity gradients, following Stokes's first postulate (i). Specifically, the rate-of-strain tensor is (cf. Sect. 1.2.2)

$$\vec{\vec{\varepsilon}} = \frac{1}{2}(\nabla\vec{v} + \nabla\vec{v}^{\text{tr}})$$

so that in terms of the rate-of-deformation tensor is (App. A)

$$\text{def}\,\vec{v} \equiv \vec{\vec{\gamma}} = 2\vec{\vec{\varepsilon}} = \nabla\vec{v} + (\nabla\vec{v})^{\text{tr}}$$

Now, we can write the total stress tensor (cf. Eq. (1.15)) in *vector notation* as

$$\vec{\vec{\pi}} = -p\vec{\vec{\delta}} + \vec{\vec{\tau}}$$

$$\quad = \underbrace{-p\vec{\vec{\delta}}}_{\text{pressure}} + \underbrace{\mu[\nabla\vec{v} + (\nabla\vec{v})^{\text{tr}}]}_{\text{distortion}} - \underbrace{\frac{2}{3}\mu(\nabla\cdot\vec{v})\vec{\vec{\delta}}}_{\text{dilatation}} \tag{1.17a}$$

or in *tensor notation* as

$$\pi_{ij} = -p\delta_{ij} + \mu\left(\frac{\partial u_i}{\partial x_j} + \frac{\partial u_j}{\partial x_i} - \frac{2}{3}\frac{\partial u_k}{\partial x_k}\delta_{ij}\right) \tag{1.17b}$$

where $\vec{\vec{\delta}}$ is the unit tensor; $i, j, k = 1, 2, 3$; and the Kronecker delta is

$$\delta_{ij} = \begin{cases} 1 & \text{if } i = j \\ 0 & \text{if } i \neq j \end{cases}$$

The tensor has two interpretations (cf. Figs. 1.5 and 1.6):

  (i) Momentum flux tensor: momentum is transferred in a flow field by molecular friction, resulting in momentum diffusion.
 (ii) total stress tensor: momentum is transferred in a flow field as a result of surface forces (i.e., normal and tangential) exerted by the surrounding fluid.

Note: For *incompressible* fluid flow the density is constant; that implies that the divergence of the velocity field is zero (cf. Sect. 2.2.1), that is, div $\vec{v} \equiv \nabla\cdot\vec{v} = 0$, so that the total stress tensor reduces to

$$\vec{\vec{\pi}} = -p\vec{\vec{\delta}} + \mu\vec{\vec{\gamma}} = -p\vec{\vec{\delta}} + \mu[\nabla\vec{v} + (\nabla\vec{v})^{\text{tr}}] \tag{1.18}$$

Hence, for a Newtonian fluid with constant density, $\tau_{ij} = 2\mu\varepsilon_{ij}$ so that the nine components of $\vec{\vec{\tau}}$ are in rectangular coordinates:

 • normal stresses

$$\tau_{xx} = 2\mu\frac{\partial u}{\partial x}, \qquad \tau_{yy} = 2\mu\frac{\partial v}{\partial y}, \quad \text{and} \quad \tau_{zz} = 2\mu\frac{\partial w}{\partial z} \tag{1.19}$$

- (symmetric) shear or tangential stresses

$$\tau_{xy} = \tau_{yx} = \mu\left(\frac{\partial u}{\partial y} + \frac{\partial v}{\partial x}\right); \qquad \tau_{xz} = \tau_{zx} = \mu\left(\frac{\partial w}{\partial x} + \frac{\partial u}{\partial z}\right);$$

and

$$\tau_{yz} = \tau_{zy} = \mu\left(\frac{\partial v}{\partial z} + \frac{\partial w}{\partial y}\right)$$

where the negative sign (cf. the beginning of this section) is recovered with an appropriate force or momentum balance (cf. Chapter 2).

B Power-Law Fluids

Fluids of interest in engineering that exhibit unusual flow behavior include exotic lubricants, blood in capillaries, syrup and other foodstuff, pastes, and slurries, as well as a wide range of polymeric liquids. In general, these liquids are composed of macromolecules in which the molecular weight (MW) is high, that is, $MW > 10^4$. Because of their structural complexity and diversity, new expressions for the stress tensor τ_{ij} have to be found. It is necessary to define such (nonlinear) expressions for various classes of non-Newtonian fluids in order to simulate/interpret their specific molecular structures, flow phenomena, and material function measurements. Thus, even for rather simple "power-law" or generalized Newtonian fluids, the equation of motion in terms of τ_{ij} is the starting point rather than the Navier–Stokes equations, which assume constant fluid properties.

Two approaches for the development of such "rheological equations of state" are commonly used: the continuum theory and the molecular theory (cf. Bird et al. 1987). We only deal with the continuum approach and focus on simple polymeric liquids and plastics with a viscosity that is shear-rate-dependent. Flow phenomena such as rod-climbing and jet-swelling or viscoelastic effects like fluid–recoil, stress–relaxation, and stress–overshoot are discussed elsewhere (Bird et al. 1987). The steady-state shear flows of interest here can be described with a *generalized* Newtonian model where a specific analytical representation of the shear-rate-dependent viscosity, $\eta = \eta(\dot{\gamma})$, fits experimental data sets. Recall that the constitutive equation for an incompressible Newtonian fluid is

♦ $$\vec{\vec{\tau}} = \mu\vec{\vec{\gamma}}$$ (1.20a)

where, typically, $\mu = \mu(T, p)$ or $\mu = $ constant, and the rate-of-deformation tensor is $\vec{\vec{\gamma}} = \nabla\vec{v} + (\nabla\vec{v})^{\mathrm{tr}}$. Now, analogous to Equation (1.20a), a non-Newtonian viscosity, which is a function of the scalar invariant of $\vec{\vec{\gamma}}$, is introduced:

♦ $$\vec{\vec{\tau}} = \eta\vec{\vec{\gamma}}$$ (1.20b)

where (cf. Apps. A.2 and B.1)

$$\eta = \eta(\dot{\gamma}) \qquad \text{with } \dot{\gamma} = |\vec{\vec{\gamma}}| = \sqrt{\frac{1}{2}\sum_i\sum_j \dot{\gamma}_{ij}\dot{\gamma}_{ji}}$$

For many industrial (i.e., macromolecular) fluids, if log η is plotted as a function of log $\dot{\gamma}$, the relationship appears as a tilted straight line. Thus a "power law" for $\eta(\dot{\gamma})$ can be obtained as (Fig. 1.7)

$$\eta = m\dot{\gamma}^{n-1} \tag{1.20c}$$

where m and n are material constants that have to be measured for each polymeric solution. It has to be noted that the region where $\dot{\gamma} \to 0$, in particular $\eta(\dot{\gamma} = 0) = \eta_0$ with η_0 the zero-shear-rate viscosity, is not described by the power law. For $n = 1$, Equation (1.20c) collapses to $\eta = m$ and with $m \equiv \mu$, Equations (1.20a) and (1.20b) are identical. If $n < 1$, the fluid is called *pseudoplastic* or shear-thinning, whereas for $n > 1$, the fluid is called *dilatant*, that is, expanding or shear-thickening (cf. Fig. 1.7a).

For fluids that do not obey the power law (1.20c), other empirical models have been developed. One of the more accessible relationships is the *Bingham model*, which describes pastes, paints, plastics, ketchup, and thick slurries. Mathematically (cf. Fig. 1.7a)

$$\eta = \infty \quad \text{for } \tau \leq \tau_0 \tag{1.21a}$$

and

$$\eta = \mu_p + \frac{\tau_0}{\dot{\gamma}} \quad \text{for } \tau > \tau_0 \tag{1.21b}$$

Here, τ_0 is a yield stress that has to be exceeded in order to start moving the Bingham fluid, such as toothpaste. Additional models can be found in the rheological literature.

A few analytic solutions to basic problems concerning non-Newtonian fluid flow are given in Sects. 1.3.1B and 3.3.4. In addition, two case studies outlined in Section 5.3 deal with power-law fluids, where rheological and thermal effects change the fluid flow behavior dramatically.

1.3.2 Vorticity Dynamics

When fluid elements are subjected to moments, they rotate. In a real flow field such torques and the resulting vorticity may be produced by *unbalanced* shear stresses and/or body forces. The major source of vorticity appears in wall-bounded flows because of the associated "no-slip" boundary condition. For example, when a rotation-free approach stream "encounters" a solid stationary boundary, say a horizontal plate, fluid particles at and near the plate surface start spinning, a frictional (or viscous) effect that propagates normal to the plate (cf. Fig. 1.5).

In Sect. 1.2.2, vorticity was introduced as the net spin, $\zeta_z = 2\omega_z = (d\alpha/dt) - (d\beta/dt) := (\partial v/\partial x) - (\partial u/\partial y)$, of a fluid element in the x–y plane. An example of vortical flows are the idealized flow fields of a hurricane that has a rotational core, a so-called forced vortex (cf. Fig. 1.3b):

$$v_\theta = \omega r \quad 0 \leq r \leq r_c$$

and an irrotational outer region, a *free* vortex

$$v_\theta = \frac{\Gamma}{(2\pi r)} \quad r_c \leq r < \infty$$

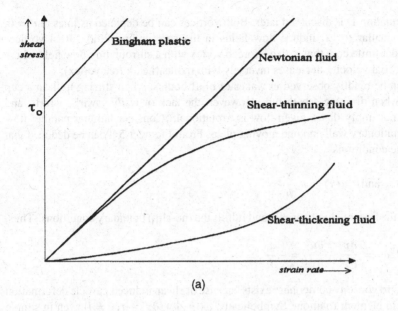

(a)

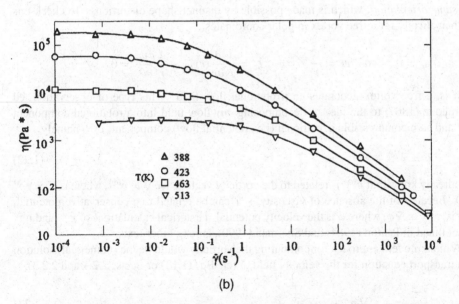

(b)

Fig. 1.7. Typical non-Newtonian viscosity dependencies for polymeric liquids: (a) shear stress as a function of strain rate for Newtonian and non-Newtonian fluids; (b) example of a non-Newtonian viscosity η of a low-density polyethylene melt at several different temperatures (after Bird et al. 1987) Note: Almost all macromolecular fluids show this shear thinning or pseudoplastic behavior.

where the circulation Γ is discussed later. Both vortices can be depicted as lines at $r = 0$, which, perpendicular to v_θ, induce flow fields in the surrounding fluid. Real vortices possess a slender finite core that is like a forced vortex with a surrounding flow field where the circumferential velocity decreases inversely with radius (i.e., a free vortex).

Vortices can be readily observed in wakes of bluff bodies, when stirring milk in a cup of coffee, or when draining a bath tub. However, the lack of *visible* swirls, whirls, and vortices does not imply that a given flow is irrotational. Consider laminar parallel flow between one stationary wall and one moving plate. From Figure 1.5 it can be deduced that for steady-state conditions

$$v = 0 \quad \text{and} \quad u(y) = u_0 \frac{y}{h}$$

which reflects the linear velocity profile and fulfills the (no-slip) boundary conditions. Thus,

$$\zeta_z = 2\omega_z = \frac{\partial v}{\partial x} - \frac{\partial u}{\partial y} := -\frac{u_0}{h}$$

That is, a nonzero vorticity component exists because of shear-induced particle deformation that causes fluid element rotation. Symbolically, $\omega_z \sim \partial u/\partial y \sim \tau_{yx} \neq 0$ even in simple shear flow. In contrast, in a free vortex, fluid particles circle around but keep all the time the *same orientation*, which is made possible by distinct shape distortions. To check this mathematically for a free vortex in polar coordinates,

$$v_r = 0 \quad \text{and} \quad v_\theta = \frac{C}{r} \quad \Rightarrow \quad \zeta_z = \frac{1}{r}\frac{\partial}{\partial r}(rv_\theta) - \frac{1}{r}\frac{\partial v_r}{\partial \theta} := 0$$

That is, a free vortex generates an irrotational flow field. This type of observation led Helmholtz (1867) to the idea of decomposing *any* flow field into a rotational component $\vec{v}^{(\omega)}$ and an incompressible irrotational (i.e., potential flow) component, $\vec{v}^{(\phi)}$, namely,

$$\vec{v} = \vec{v}^{(\omega)} + \vec{v}^{(\phi)} \tag{1.22}$$

We already know that $\vec{v}^{(\omega)}$ is related to the vorticity vector $\vec{\zeta} = \nabla \times \vec{v}^{(\omega)}$, whereas $\nabla \times v^{(\phi)} = 0$. Because of the absence of viscosity, $\vec{v}^{(\phi)}$ can be related to a conservative potential, that is, $\vec{v}^{(\phi)} \equiv \nabla\phi$, where ϕ is the velocity potential. Historical calculations of $\vec{v}^{(\omega)}$ and $\vec{v}^{(\phi)}$ are outlined in Panton (1984) and Sherman (1990), among other texts.

A more modern approach for computing the vorticity would be the (numerical) solution of a transport equation for the velocity field, $\vec{v}$ (cf. Eq. (1.12) or Sects. 2.2.1 and 2.2.3):

$$\underbrace{\frac{D\vec{v}}{Dt}}_{\substack{\text{total}\\\text{accel.}}} = \underbrace{-\frac{1}{\rho}\nabla p}_{\sim \text{pressure f.}} + \underbrace{\nu\nabla^2\vec{v}}_{\sim \text{viscous f.}} + \underbrace{\vec{g}}_{\sim \text{gravit. f.}} \tag{1.23}$$

and then forming $\nabla \times \vec{v} \equiv \vec{\zeta}$ to obtain the vorticity field. Alternatively, taking the "curl" or "cross–product" of Eq. (1.23) yields

$$\blacklozenge \qquad \frac{D\vec{\zeta}}{Dt} = (\vec{\zeta} \cdot \nabla)\vec{v} + \nu\nabla^2\vec{\zeta} \tag{1.24}$$

Although the mathematical details of Eq. (1.24) will be discussed in Chapter 3, it is of interest to examine now the physical meaning of some terms in the *vorticity transport equation* (1.24). Specifically, it has to be noted that (cf. App. A)

(i) $\quad \nabla \times \dfrac{D\vec{v}}{Dt} = \nabla \times \left[\dfrac{\partial \vec{v}}{\partial t} + (\vec{v} \cdot \nabla)\vec{v} \right] = \dfrac{\partial \vec{\zeta}}{\partial t} + (\vec{v} \cdot \nabla)\vec{\zeta} - (\vec{\zeta} \cdot \nabla)\vec{v}$

or

$\quad \text{curl}\left(\dfrac{D\vec{v}}{Dt} \right) = \dfrac{D\vec{\zeta}}{Dt} - (\vec{\zeta} \cdot \nabla)\vec{v}$

where $\partial \vec{\zeta}/\partial t$ is the local change in vorticity;

$(\vec{v} \cdot \nabla)\vec{\zeta}$ depicts vorticity convection; and

$(\vec{\zeta} \cdot \nabla)\vec{v}$ describes vorticity affected by velocity variations resulting in vortex stretching or vorticity production. For example,

$$|(\vec{\zeta} \cdot \nabla)\vec{v}|_z := \zeta_z \frac{\partial w}{\partial z} \triangleq \left[(\text{spin}) \times \left(\begin{array}{c} \text{volume} \\ \text{change} \end{array} \right) \right]_z$$

Thus, an incompressible fluid element elongates in the z-direction, $\partial w/\partial z$, and has to contract at the same time, in order to conserve mass (i.e., $\nabla \cdot \vec{v} = 0$), causing an increase in rotation or vorticity production (e.g., spin increase of a "contracting" ice dancer).

(ii) $\nabla \times [-1/\rho \nabla p + \vec{g}] \equiv 0$, because pressure and gravity forces do not generate a torque on the fluid elements and hence generate no rotation as long as mass center and geometric center are identical (i.e., exceptions include stratified flows and Coriolis flows).

(iii) $\nu \nabla^2 \vec{\zeta}$ represents vorticity diffusion due to viscous action. An example is the transient diffusion equation describing the balance of

$$\left(\begin{array}{c} \text{rate of change of} \\ \text{vorticity at a "point"} \end{array} \right) = \left(\begin{array}{c} \text{net diffusion flux} \\ \text{of vorticity into the point} \end{array} \right)$$

that is,

$$\frac{\partial \vec{\zeta}}{\partial t} = \nu \nabla^2 \vec{\zeta} \tag{1.25}$$

From the definition of vorticity, $\vec{\zeta} = \nabla \times \vec{v}$, it is transparent that the direction of the vectors $\vec{\zeta}$ is perpendicular to the "plane" of ∇ and $\vec{v}$. When ∇ and $\vec{v}$ are parallel, then $\zeta \equiv 0$. The vectors ∇, $\vec{v}$, and $\vec{\zeta}$ form a right-handed system, which implies $\nabla \times \vec{v} = -\vec{v} \times \nabla$. As shown in Section 1.2.2, in rectangular coordinates

♦ $\quad \vec{\zeta} = \nabla \times \vec{v} = \begin{vmatrix} \hat{i} & \hat{j} & \hat{k} \\ \frac{\partial}{\partial x} & \frac{\partial}{\partial y} & \frac{\partial}{\partial z} \\ u & v & w \end{vmatrix}$ $\hspace{2cm}$ (1.26a)

Similar to the definition of a streamline, a *vortex line* is a line that is everywhere tangent to the vorticity vector. Near walls and for most two-dimensional or axisymmetric flows, $\vec{v} \perp \vec{\zeta}$: That is, vortex lines are perpendicular to streamlines. A bundle of vortex lines may form a *vortex* tube. Such vortex tubes move with the flow, and their instantaneous strengths can be expressed as the *circulation* of velocity along a closed contour. Thus, the flux of vorticity through any section of a vortex tube is

$$\iint_A \hat{n} \cdot \vec{\zeta}\, dA = \oint_C \vec{v} \cdot d\vec{r} = \text{const} \tag{1.26b}$$

A nice example is Hill's spherical vortex, which is a good approximation for the liquid-phase velocity field of droplets induced by low gas-phase Reynolds number flow. The kidney-shaped streamlines are forming a ring to which the vortex lines, forming complete circles, are perpendicular (cf. Sect. 3.4). In external flows (e.g., airfoil boundary layers), vortex lines are concentrated near the surface and in the wake. In internal, say, pipe flow, vortex lines are rings that propagate throughout the changing flow field. A vortex line for an *inviscid* fluid consists always of the same fluid element.

Summary
- *Vorticity dynamics* is

  (i)   a method to separate a flow into viscous ($\vec{\zeta} \neq 0$) and inviscid ($\vec{\zeta} = 0$) effects because only shear, if unbalanced, sets an isotropic fluid particle into *rotation*;

  (ii)   a tool to investigate and depict flow patterns or flow phenomena, not easily described via velocity and pressure fields alone, especially in boundary-layer flows (cf. Sect. 3.4.3) and turbulent flow (cf. Sect. 3.4.4). Graphs of vortex lines

$$y(x) = \int \frac{\zeta_y}{\zeta_x}\, dx + C \tag{1.27a}$$

and vorticity contours

$$|\vec{\zeta}| = \text{const} \tag{1.27b}$$

are very instructive in 2-D flow visualization. For 3-D flow, the helicity density

$$h = \vec{\zeta} \cdot \vec{v} \tag{1.27c}$$

or its integral form is quite popular.

- *Vorticity* can be interpreted as (cf. illustrations):

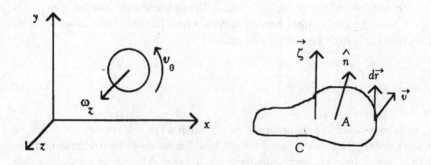

(i)   a point function that measures the solid-body-like rotation

$$\nabla \times \vec{v} \equiv \text{curl } \vec{v} = \vec{\zeta} = 2\vec{\omega}$$

that corresponds to changing orientation of a fluid particle in space;

(ii)   the *circulation* Γ per unit area for a surface A is perpendicular to the vorticity vector, $\vec{\zeta}$, that is,

$$\Gamma \equiv \oint_C \vec{v} \cdot d\vec{r} := \int\int_A \hat{n} \cdot \vec{\zeta}\, dA$$

with Stokes's theorem or

$$\hat{n} \cdot \vec{\zeta} = \frac{d\Gamma}{dA}$$

1.4   Sample-Problem Solutions

The problem solutions in this section illuminate the material presented in Chapter 1 and in Appendix A. Sometimes merely mathematical exercises may help to enhance the reader's background with respect to basic concepts in fluid flow properties, tensor manipulations, and flow systems analysis. Occasionally, a few aspects of UG fluid mechanics are also being reviewed.

1.0   Consider buoyant gas flow from a chimney. Sketch how gas velocity and temperature measurements would be carried out following (a) the Eulerian approach and (b) the Lagrangian approach. Comment.

Concepts
(a)   Compare Section 1.2.1. With the Eulerian method, we obtain information about a flow field in terms of velocity, pressure, temperature, and so on, at fixed points in space as the fluid flows past those points.
(b)   In the Lagrangian method, individual particles are "tagged" via the position vector, and their properties are determined in space and time as they move about.

Solution

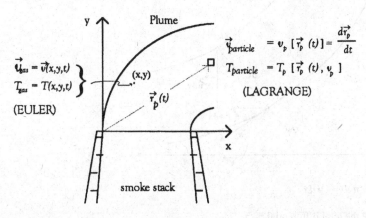

1.1 Consider two-dimensional incompressible flows

(a) Given $\vec{v} = (u, v, 0)$ where $u = cx + 2\omega_0 y + u_o$ and $v = cy + v_o$, calculate the kinematic properties $\vec{\omega}$ and $\vec{\vec{\varepsilon}}$.

Solution

   (a) Angular velocity $\omega_z = \frac{1}{2}(\partial v/\partial x - \partial u/\partial y) := -\omega_0$, which implies *rotational* flow (cf. Sect. 1.2.2).

   (b) Normal strain rates $\varepsilon_{xx} = \partial u/\partial x := c$ and $\varepsilon_{yy} = \partial v/\partial y := c$
   Tangential strain rates $\varepsilon_{xy} = \varepsilon_{yx} = \frac{1}{2}(\partial v/\partial x + \partial u/\partial y) := \omega_0$ $\square$

(b) Given $u = ax^2 + by$ and $v = -2axy + ct$, check for compressibility, calculate the stream function, and develop an equation for the streamlines.

Solution

   (a) Compare Section 1.2.2. Checking whether for this 2-D flow field the *compressional strain rates* are zero, that is,

$\varepsilon_{ii} = (\partial u_i/\partial x_i) = 0$, we compute

$$\frac{\partial u}{\partial x} + \frac{\partial v}{\partial y} = 0 \quad \Rightarrow \quad 2ax + (-2ax) \equiv 0$$

That is, the flow is incompressible. In general, $\operatorname{div} \vec{v} = \nabla \bullet \vec{v} = 0$ for incompressible flow.

   (b) For incompressible planar flow the stream function ψ can be defined via

$$u = \frac{\partial \psi}{\partial y} \quad \text{and} \quad v = -\frac{\partial \psi}{\partial x}$$

which fulfills the continuity equation automatically. Alternatively, equating the differential amount of volumetric flow, $d\psi$, across curve segment dC to the flows across control surfaces dx and dy yields (cf. sketch)

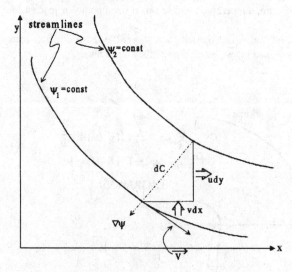

$$d\psi = u\, dy - v\, dx \tag{S1.1a}$$

where the total differential is

$$d\psi = \frac{\partial \psi}{\partial x} dx + \frac{\partial \psi}{\partial y} dy \qquad\qquad\qquad (S1.1b)$$

so that

$$\frac{\partial \psi}{\partial y} = u \quad \Rightarrow \quad \psi = \int u \, dy + f(x, t) \qquad\qquad (S1.2a)$$

and then

$$\frac{\partial \psi}{\partial x} = -v \quad \Rightarrow \quad \frac{\partial}{\partial x}\left[\int u \, dy\right] + \frac{\partial f}{\partial x} = -v \qquad (S1.2b)$$

For our case, Equation (S1.2a) yields

$$\psi = ax^2 y + \frac{b}{2} y^2 + f(x, t)$$

and Equation (S1.2b) yields

$$2axy + \frac{\partial f}{\partial x} = 2axy - ct$$

or

$$f(x, t) = -ctx + g(t)$$

If we assign $\psi(x = 0, y = 0) = 0$ and $g(t) = 0$, we obtain

$$\blacklozenge \qquad \psi(x, y, t) = ax^2 y + \frac{b}{2} y^2 - ctx$$

(c) *Streamlines:* Since $\nabla\psi \perp \vec{v}$ (cf. sketch), their dot product is by definition zero, $(\nabla\psi) \cdot \vec{v} \equiv 0$, which implies that $\psi = $ constant on a streamline, that is, with Eq. (S1.1a)

$$d\psi = -v dx + u dy = 0$$

or

$$\frac{u}{v} = \frac{dx}{dy} \qquad\qquad\qquad\qquad (S1.3a)$$

Here, with $dy/dx \equiv y'$, Equation (S1.3a) yields

$$y' = \frac{ax^2 + by}{-2axy + ct} \qquad\qquad\qquad (S1.3b)$$

which is a first-order nonlinear inhomogeneous ordinary differential equation (ODE) where the parameter t may represent the time. In general, is it possible that streamlines exist for 2-D flows without a stream function? Yes! □

1.2 Tensor identification and manipulations (cf. Appendix A)

(a) If $\vec{v}$ and ∇ are vectors, that is, tensors of rank 1 and $\overset{\Rightarrow}{\sigma}$, $\overset{\Rightarrow}{\tau}$, and $\overset{\Rightarrow}{\delta}$ are tensors of rank 2, what will be the results of the following operations? That is, are scalars, vectors, or tensors produced?

$$\overset{\Rightarrow}{\tau} \cdot \vec{v}, \ \overset{\Rightarrow}{\sigma} \cdot \overset{\Rightarrow}{\tau}, \ \nabla \cdot \overset{\Rightarrow}{\tau}, \ \overset{\Rightarrow}{\sigma} : \overset{\Rightarrow}{\tau}, \ \overset{\Rightarrow}{\delta} \cdot \vec{v}, \ \nabla \cdot \left[\overset{\Rightarrow}{\tau} \cdot \vec{v}\right] \quad \text{and} \quad \vec{v} \cdot \left[\nabla \cdot \overset{\Rightarrow}{\tau}\right]$$

It has to be noted that these types of terms appear in the vector notation description of the conservation laws (cf. Chapter 2).

Solution

With the help of App. A we identify

$$\left[\vec{\vec{\tau}} \cdot \vec{v}\right] \Rightarrow \text{vector}; \quad \left\{\vec{\vec{\sigma}} \cdot \vec{\vec{\tau}}\right\} \Rightarrow \text{tensor}; \quad \left[\nabla \cdot \vec{\vec{\tau}}\right] \Rightarrow \text{vector}; \quad \left(\vec{\vec{\sigma}} : \vec{\vec{\tau}}\right) \Rightarrow \text{scalar}$$

$$\left[\vec{\vec{\delta}} \cdot \vec{v}\right] \Rightarrow \text{vector}; \quad \left(\nabla \cdot \left[\vec{\vec{\tau}} \cdot \vec{v}\right]\right) \Rightarrow \text{scalar}; \quad \text{and} \quad \left(\vec{v} \cdot \left[\nabla \cdot \vec{\vec{\tau}}\right]\right) \Rightarrow \text{scalar} \qquad \square$$

(b) Prove the given sample identities adopting the index notation (cf. Bird et al. 1987) that follows. This is a useful exercise for transformations of the transport equations derived in Chapter 2.

Notation

$$\text{grad } s \equiv \nabla s = \frac{\partial s}{\partial x_i} \hat{e}_i \tag{S1.4}$$

$$\text{div } \vec{v} \equiv \nabla \cdot \vec{v} = \frac{\partial v_i}{\partial x_i} \tag{S1.5}$$

$$\text{curl } \vec{v} \equiv \nabla \times \vec{v} = \varepsilon_{ijk} \hat{e}_i \frac{\partial v_k}{\partial x_j} \tag{S1.6}$$

where the Einstein convention (sum over double indices) is implied; $\hat{e}_i$, $i = 1, 2, 3$, are unit vectors; and ε_{ijk} is the permutation symbol given in the following.

Note the following:

- $\hat{e}_i \cdot \hat{e}_j = \delta_{ij}$ (Kronecker delta)
- $\hat{e}_i \times \hat{e}_j = \varepsilon_{ijk} \hat{e}_k$
- The Kronecker delta δ_{ij} is defined as

$$\delta_{ij} = \begin{cases} 1 & \text{if } i = j \\ 0 & \text{if } i \neq j \end{cases}$$

- The permutation symbol ε_{ijk} is defined as

$$\varepsilon_{ijk} = \begin{cases} +1 & \text{if } ijk = 123, 231, \text{ or } 312 \\ -1 & \text{if } ijk = 321, 132, \text{ or } 213 \\ 0 & \text{if any two indices are alike} \end{cases}$$

so that

$$\varepsilon_{ijk}\varepsilon_{mnk} \equiv \sum_{k=1}^{3} \varepsilon_{ijk}\varepsilon_{mnk} = \delta_{im}\delta_{jn} - \delta_{in}\delta_{jm}$$

Examples and Proofs

(a) $\text{div}(\phi\vec{v}) = \phi \, \text{div } \vec{v} + \vec{v} \cdot \text{grad } \phi$ (product rule)

or

$$\nabla \cdot (\phi\vec{v}) = \phi \nabla \cdot \vec{v} + \vec{v} \cdot \nabla \phi$$

Proof

$$\nabla \cdot (\phi \vec{v}) \equiv \left\{ \frac{\partial}{\partial x_i} \hat{e}_i \right\} \cdot (\phi \hat{e}_j v_j) = \phi (\hat{e}_i \cdot \hat{e}_j) \frac{\partial}{\partial x_i} v_j$$

$$+ (v_j \hat{e}_j) \cdot \frac{\partial \phi}{\partial x_i} \hat{e}_i = \phi \delta_{ij} \frac{\partial}{\partial x_i} v_j + (v_j \hat{e}_j) \cdot \frac{\partial \phi}{\partial x_i} \hat{e}_i$$

$$= \phi \frac{\partial v_i}{\partial x_i} + (v_j \hat{e}_j) \cdot \frac{\partial \phi}{\partial x_i} \hat{e}_i$$

◆ $\nabla \cdot (\phi v) = \phi \nabla \cdot \vec{v} + \vec{v} \cdot \nabla \phi$

(b) $\text{div}(\vec{u} \times \vec{v}) = \vec{v} \cdot \text{curl}\, \vec{u} - \vec{u} \cdot \text{curl}\, \vec{v}$
or

$$\nabla \cdot (\vec{u} \times \vec{v}) = \vec{v} \cdot (\nabla \times \vec{u}) - \vec{u} \cdot (\nabla \times \vec{v})$$

Proof

$$\nabla \cdot (\vec{u} \times \vec{v}) \equiv \left\{ \frac{\partial}{\partial x_i} \hat{e}_i \right\} \cdot \{[\hat{e}_j u_j] \times [\hat{e}_k v_k]\}$$

$$= \left\{ \frac{\partial}{\partial x_i} \hat{e}_i \right\} \cdot \{[\hat{e}_j \times \hat{e}_k] u_j v_k\}$$

$$= \left\{ \frac{\partial}{\partial x_i} \hat{e}_i \right\} \cdot \{\varepsilon_{ijk} \hat{e}_i u_j v_k\}$$

$$= \varepsilon_{ijk} \frac{\partial}{\partial x_i} (u_j v_k) = \varepsilon_{ijk} \left[v_i \frac{\partial}{\partial x_i} u_k - u_i \frac{\partial}{\partial x_j} v_k \right]$$

$$= v_i \varepsilon_{ijk} \frac{\partial}{\partial x_j} u_k - u_i \varepsilon_{ijk} \frac{\partial}{\partial x_j} v_k$$

◆ $\nabla \cdot (\vec{u} \times \vec{v}) = \vec{v} \cdot (\nabla \cdot \vec{u}) - \vec{u} \cdot (\nabla \cdot \vec{v})$

(c) $\text{curl}(\vec{u} \times \vec{v}) = \vec{v} \cdot \text{grad}\, \vec{u} - \vec{u} \cdot \text{grad}\, \vec{v} + \vec{u}\, \text{div}\, \vec{v} - \vec{v}\, \text{div}\, \vec{u}$
or

$$\nabla \times (\vec{u} \times \vec{v}) = (\vec{v} \cdot \nabla)\vec{u} - (\vec{u} \cdot \nabla)\vec{v} + \vec{u}(\nabla \cdot \vec{v}) - \vec{v}(\nabla \cdot \vec{u})$$

Proof

$$\nabla \times (\vec{u} \times \vec{v}) \equiv \varepsilon_{lmk} \frac{\partial}{\partial x_l} [\{e_j u_j\} \times \{\hat{e}_k v_k\}]$$

$$= \varepsilon_{lmk} \frac{\partial}{\partial x_l} [\varepsilon_{ijk} u_j v_k]$$

$$= (\varepsilon_{lmk} \varepsilon_{ijk}) \left[\frac{\partial}{\partial x_l} (u_j v_k) \right]$$

$$= (\delta_{li} \delta_{mj} - \delta_{lj} \delta_{mi}) \left(u_j \frac{\partial v_k}{\partial x_l} + v_k \frac{\partial u_j}{\partial x_l} \right)$$

$$= \delta_{li} \delta_{mj} u_j \frac{\partial v_k}{\partial x_l} - \delta_{lj} \delta_{mi} u_j + \delta_{li} \delta_{mj} v_k \frac{\partial u_j}{\delta x_l} - \delta_{lj} \delta_{mk} v_k \frac{\partial u_j}{\delta x_l}$$

◆ $\nabla \times (\vec{u} \times \vec{v}) = \vec{u}(\nabla \cdot \vec{v}) - (\vec{v} \cdot \nabla)\vec{u} - \vec{v}(\nabla \cdot \vec{u}) + (\vec{v} \cdot \nabla)\vec{u}$ □

(c) Show that the viscous dissipation term $(\bar{\bar{\tau}} : \nabla\vec{v}) = (\nabla \cdot [\bar{\bar{\tau}} \cdot \vec{v}]) - (\vec{v} \cdot [\nabla \cdot \bar{\bar{\tau}}])$.

Solution
In terms of components (index notation), the left-hand side (LHS) is

$$(\bar{\bar{\tau}} : \nabla\vec{v}) = \sum_i \sum_j \tau_{ij} \frac{\partial}{\partial x_j} v_i \qquad \text{where } i, j = 1, 2, 3$$

In terms of components, the right-hand side (RHS) reads

$$(\nabla \cdot [\bar{\bar{\tau}} \cdot \vec{v}]) = \sum_i \frac{\partial}{\partial x_i} [\bar{\bar{\tau}} \cdot \vec{v}]_i = \sum_i \sum_j \frac{\partial}{\partial x_i} (\tau_{ij} v_j)$$

and

$$(\vec{v} \cdot [\nabla \cdot \bar{\bar{\tau}}]) = \sum_j v_j [\nabla \cdot \bar{\bar{\tau}}]_j = \sum_j \sum_i v_j \frac{\partial}{\partial x_i} \tau_{ij}$$

Subtracting the last two expressions yields the LHS.   □

(d) Given the components of a symmetric stress tensor

$$\bar{\bar{\tau}} := \tau_{ij} = \begin{bmatrix} -1 & 4 & 5 \\ 4 & -2 & 6 \\ 5 & 6 & -3 \end{bmatrix}$$

find the magnitudes and directions of the (three) principal stresses (cf. Sect. 1.3.1).

Recall
In general, the stress vector on a representative surface of a fluid element can be decomposed into a normal stress and tangential stresses. However, because of the symmetry of the stress tensor, there must be three mutually orthogonal orientations $\hat{n}_1$, $\hat{n}_2$, and $\hat{n}_3$, where in the limit, the tangential stresses vanish (cf. Timoshenko and Goodier, 1970 or 1990). The remaining normal stresses are called *principal stresses*.

Solution
Let S be the magnitude of the (normal) principal stress acting on the plane of a tetrahedron whose sides are parallel to the 123 or xyz-coordinates. Now the determinant has to be equal to zero, that is,

$$\begin{vmatrix} \tau_{11} - S & \tau_{21} & \tau_{31} \\ \tau_{12} & \tau_{22} - S & \tau_{32} \\ \tau_{13} & \tau_{23} & \tau_{33} - S \end{vmatrix} = 0$$

in order to obtain nontrivial solutions for the direction cosines. Thus, a cubic equation in S results in

$$(\tau_{11} - S)(\tau_{22} - S)(\tau_{33} - S) + 2\tau_{12}\tau_{23}\tau_{31} - (\tau_{11} - S)\tau_{23}^2$$
$$- (\tau_{22} - S)\tau_{13}^2 - (\tau_{33} - S)\tau_{12}^2 = 0$$

The solutions S_i, $i = 1, 2, 3$, are the magnitudes of the principal stresses. In this case,

$$S^3 + 6S^2 - 66S - 368 = 0$$

or

$$(S - 8)(S^2 + 145 + 46) = 0$$

$$\therefore S_i = 8 \quad \text{(extension)} \quad \text{and} \quad S_{23} = -7 \pm \sqrt{3} \quad \text{(compression)}$$

Each eigenvalue $S^{(i)} = (8, -7+\sqrt{3}, -7-\sqrt{3})$ corresponds to one of the three principal directions that can be computed from

$$\frac{n_2}{n_1} = \frac{(\tau_{11} - S)\tau_{32} - \tau_{12}\tau_{31}}{(\tau_{22} - S)\tau_{31} - \tau_{21}\tau_{32}}$$

$$\frac{n_3}{n_1} = \frac{(\tau_{11} - S)\tau_{23} - \tau_{13}\tau_{21}}{(\tau_{33} - S)\tau_{21} - \tau_{31}\tau_{23}}$$

and

$$n_1^2 + n_2^2 + n_3^2 = 1$$

Now, for $S^{(1)} = 8 \rightarrow$

$$\frac{n_2^{(1)}}{n_1^{(1)}} = 1, \quad \frac{n_3^{(1)}}{n_1^{(1)}} = 1, \quad \text{and} \quad 3\left[n_1^{(1)}\right]^2 = 1$$

so that $S^{(1)} = 8$ is associated with

$$n_1^{(1)} = n_2^{(1)} = n_3^{(1)} = \frac{1}{\sqrt{3}} \approx 0.5774$$

Similarly,

$$S^{(2)} \approx -5.268 \rightarrow n_1^{(2)} = 0.7887; \qquad n_2^{(2)} = -0.5773; \qquad n_3^{(2)} = -0.2113$$

and

$$S^{(3)} \approx -8.732 \rightarrow n_1^{(3)} = 0.2113; \qquad n_2^{(3)} = 0.5773; \qquad n_3^{(3)} = -0.7887$$

Note that

$$\hat{n}^{(1)} \cdot \hat{n}^{(2)} = \hat{n}^{(1)} \cdot \hat{n}^{(3)} \approx 0$$

indicating orthogonality.   □

1.3 Kinematics of laminar flows

(a) Consider laminar flow in a tube where $v_z = v_0[1 - (r/R)^2]$ with v_o the centerline velocity. Compute the vorticity of the fluid flow field (cf. Sect. 1.3.2).

Solution
Recall that the vorticity vector in cylindrical coordinates reads

$$\vec{\zeta} = \nabla \times \vec{v} = \frac{1}{r} \begin{vmatrix} \hat{e}_r & r\hat{e}_\theta & \hat{e}_z \\ \frac{\partial}{\partial r} & \frac{\partial}{\partial \theta} & \frac{\partial}{\partial z} \\ v_r & rv_\theta & v_z \end{vmatrix} \tag{S1.7}$$

Here, $\vec{\zeta}$ reduces to $\zeta_\theta = 1/r(-r\hat{e}_\theta[\partial v_z/\partial r]) := 2v_0 r/R^2 \hat{e}_\theta$, which implies that the fluid element spin is zero at the centerline and has a maximum at the pipe wall.   □

(b) Consider steady, *inviscid*, incompressible 1-D flow in a converging section where $A(x) = (A_0/(1 + [x/l]))$ and at $x = 0$, $u = u_o$. Find the velocity and acceleration in (i) Eulerian form and (ii) a Lagrangian framework (cf. Sect. 1.2.1).

Solution
(i) Euler's field description (control volume approach): A simple mass balance reveals that the volumetric flow rate is constant, (cf. sketch):

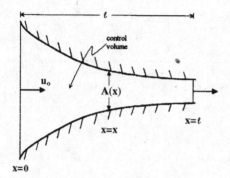

$$A_o u_o = Au \tag{S1.8a}$$

or

$$u(x) = u_o \frac{A_0}{A} := u_o\left(1 + \frac{x}{\ell}\right) \tag{S1.8b}$$

Now, the total acceleration

$$\vec{a} = \frac{D\vec{v}}{Dt} \tag{S1.9}$$

reduces here to

$$a_x = \frac{Du}{Dt} := \frac{\partial u}{\partial t} + u\frac{\partial u}{\partial x}$$

or with the given $u(x)$,

$$a_x = \frac{u_0^2}{\ell}\left(1 + \frac{x}{\ell}\right)$$

(ii) Lagrangian field description: Of interest is $u(t)$, the velocity of a representative fluid particle tracked in time starting at $x = 0$. In general,

$$\vec{v} = \vec{v}[t, \vec{r}(t)] := \frac{dx}{dx}\hat{i} + \frac{dy}{dt}\hat{j} + \frac{dz}{dt}\hat{k} \tag{S1.10}$$

Here, $x\hat{i} \equiv \vec{r}(t)$ has to be eliminated from Euler's $u(x)$ expression (S1.8b).

$$u(x) = \frac{dx}{dt} = u_o\left(1 + \frac{x}{\ell}\right)$$

or

$$\int_0^x \frac{dx}{1+(x/\ell)} = \int_0^t u_o \, dt$$

$$\therefore \ell \ln\left(1+\frac{x}{\ell}\right) = u_o t \Rightarrow 1+\frac{x}{\ell} = \exp\left(\frac{u_o t}{\ell}\right)$$

Substitution yields

$$u = u_0 \exp\left(\frac{u_o t}{\ell}\right)$$

Now,

$$a_x = \frac{du}{dt} = \frac{u_o^2}{\ell} \exp\left(\frac{u_o t}{\ell}\right) \qquad \square$$

(c) Transient 1-D air-velocity measurements have been recorded as follows (cf. table):

| Time | | Axial Position | |
	$x = 0$	$x = 10$ m	$x = 20$ m
$t = 0$ s	$v = 0.0$ m/s	$v = 0.0$ m/s	$v = 0.0$ m/s
$t = 1$ s	$= 1.0$	$= 1.20$	$\doteq 1.4$
$t = 2$ s	$= 1.7$	$= 1.80$	$= 1.9$
$t = 3$ s	$= 2.1$	$= 2.15$	$= 2.2$

Find the local and convective accelerations as well as Dv/Dt at $t = 1.0$ s and $x = 10$ m.

Solution

$$\vec{a}_{total} = \frac{D\vec{v}}{Dt} = \vec{a}_{local} + \vec{a}_{convective} = \frac{\partial \vec{v}}{\partial t} + (\vec{v} \cdot \nabla)\vec{v} \qquad (S1.11)$$

Here, for 1-D incompressible flow,

$$a_{loc} = \frac{\partial v}{\partial t} \quad \text{and} \quad a_{conv} = v\frac{\partial v}{\partial x}$$

A simple central finite difference approximation yields

$$\left.\frac{\partial v}{\partial t}\right|_{\substack{t=1 \\ x=10}} \approx \frac{1.8 \text{ m/s} - 0 \text{ m/s}}{2 \text{ s} - 0 \text{ s}} = \underline{\underline{0.90 \text{ m/s}^2}}$$

and

$$\left.v\frac{\partial v}{\partial x}\right|_{\substack{t=1 \\ x=10}} \approx (1.20)\frac{1.40 - 1.00}{20 - 0.0} = \underline{\underline{0.24 \text{ m/s}^2}} \qquad \square$$

(d) Consider the following velocity field in the xy-plane:

$$\vec{v} = v_\theta(r)\hat{e}_\theta \qquad \text{where} \quad v_\theta = \omega_o r$$

Calculate the vorticity vector $\vec{\zeta}$ (cf. Sect. 1.3.2).

Solution
The vorticity vector

$$\vec{\zeta} = \text{curl}\ \vec{v} = \nabla \times \vec{v} \quad \text{or} \quad \zeta_k = \frac{\partial u_j}{\partial x_i} - \frac{\partial u_i}{\partial x_j}$$

In cylindrical coordinates (cf. App. B)

$$\nabla \times \vec{v} = \frac{1}{r}\left[\frac{\partial v_z}{\partial \theta} - \frac{\partial(rv_\theta)}{\partial z}\right]\hat{e}_r$$

$$+ \left[\frac{\partial v_r}{\partial z} - \frac{\partial v_z}{\partial r}\right]\hat{e}_\theta + \left[\frac{1}{r}\left(\frac{\partial(rv_\theta)}{\partial r} - \frac{\partial v_r}{\partial \theta}\right)\right]\hat{e}_z$$

Now, with $\vec{v} = [0, v_\theta(r), 0]$,

$$\vec{\zeta} = \left[\frac{1}{r}\frac{\partial(rv_\theta)}{\partial r}\right]\hat{e}_z$$

or

$$\vec{\zeta} = 2\omega_0 \hat{e}_z \qquad\qquad\qquad\qquad\qquad \square$$

1.4 **Flow characteristics, flow patterns, and streamlines**

(a) Consider the velocity field given by $u = x^2 - y^2$ and $v = -2xy$. Characterize the flow.

Observations

- Absence of time: This implies *steady* flow.
- Number of directions the velocity can vary: two-dimensional flow.
- Divergence-free velocity field:

$$\frac{\partial u}{\partial x} + \frac{\partial v}{\partial y} = 0 \quad \Rightarrow \quad 2x - 2x = 0$$

that is, incompressible flow.
- Vorticity vector:

$$\zeta_z = \frac{\partial v}{\partial x} - \frac{\partial u}{\partial y} \quad \Rightarrow \quad -2y + 2y = 0$$

that is, irrotational flow.
- Streamlines: Since every arc length, $d\vec{r}$, along a streamline, $\psi = \text{constant}$, must be tangent to the velocity vector, $\vec{v}$, we have

$$\frac{d\vec{r}}{v} = \frac{dx}{u} = \frac{dy}{v} = \frac{dz}{w} \qquad\qquad\qquad\qquad \text{(S1.12)}$$

Streamline Solution

Here,

$$\frac{dx}{x^2 - y^2} = -\frac{dy}{2xy} \quad \text{or} \quad 2xy\,dx + (x^2 - y^2)\,dy = df$$

where $d\psi$ is the exact differential so that

$$\psi(x, y) = x^2 y - \frac{y^3}{3} = \text{const}$$

or

$$3x^2 y - y^3 = C \qquad\qquad \square$$

(b) Establish the connection between circulation and vorticity and evaluate the stream function of a free vortex (cf. Fig. 1.3b and Summary of Sect. 1.3.2).

Definitions

Circulation, Γ, is the line integral around a closed path of the velocity component along the contour in the flow field. The vorticity, ζ, is the circulation per unit area at a point, say, a circular element of radius r. In general (cf. sketch),

$$\Gamma = \oint v \cos\theta\,ds$$

and for a circular area

$$\Gamma_0 = \int_0^{2\pi} \omega r\,ds$$

where $ds = rd\theta$ and hence $\zeta = (\Gamma_0/A) = (2\pi r^2 \omega)/(\pi r^2) = 2\omega$.

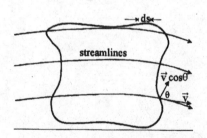

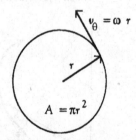

In cylindrical coordinates, $u = \partial\psi/r\partial\theta$ and $v = -\partial\psi/\partial r$ from continuity. As previously established, the circulation for a free vortex is $\Gamma = 2\pi rv$, where $v \equiv v_\theta \equiv \omega r$. Now,

$$v = \frac{\Gamma}{2\pi r} = -\frac{\partial\psi}{\partial r}$$

Integrating from $r = a$ when $\psi = 0$ to $r = r$ when $\psi = \psi$, we obtain

$$\psi = -\frac{\Gamma}{2\pi}\ln\left(\frac{r}{a}\right) \qquad\qquad \square$$

(c) The velocity distributions of an idealized flow field are given by

$$v_r = -\frac{C_1}{r} \quad \text{and} \quad v_\theta = -\frac{C_2}{r}$$

Determine and sketch the streamlines $r(\theta)$.

Solution

Because of the fact that the velocity vector $\vec{v}$ is everywhere tangent to a streamline and recalling that

$$v_r = \frac{dr}{dt} \quad \text{and} \quad v_\theta = r\frac{d\theta}{dt}$$

we obtain the equation of a representative streamline in polar coordinates as

$$\frac{v_r}{v_\theta} = \frac{dr}{rd\theta}$$

Substituting the given velocity components yields

$$\frac{C_1}{C_2} \equiv C = \frac{dr}{rd\theta}$$

or

$$\theta = \frac{1}{C}\ln r + \ln C_3$$

so that

$$r(\theta) = C_4 e^{C\theta}$$

which represents a log-spiral as shown (see graph):

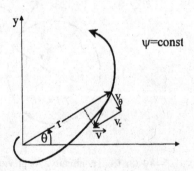

$\psi=\text{const}$

□

1.5 Physical interpretation of terms in the equation of motion (cf. Sect. 1.3 and App. B)

(a) Does the total stress tensor, $\vec{\vec{\pi}}$, account for "rotation" of fluid elements due to surface forces?

Solution

Recall that $\vec{\vec{\pi}} = -p\vec{\vec{\delta}} + \vec{\vec{\tau}}$ where for incompressible Newtonian fluids $\vec{\vec{\tau}} = \mu\vec{\vec{\gamma}}$ with $\vec{\vec{\gamma}} = (\nabla\vec{v} + \nabla\vec{v}^{\text{tr}})$. Now compare the magnitude of the deformation tensor $|\vec{\vec{\gamma}}|$ with the divergence of the vorticity field $(\nabla \cdot \vec{\zeta})$ (cf. App. A and Sect. 1.3.2). □

(b) What is physically implied when the inertia or convective term, $\nabla \cdot (\psi \vec{v})$, is written as $\vec{v} \nabla \cdot \psi$, where ψ is an arbitrary system property?

Solution

ψ is any dependent variable such as the velocity, temperature, or species concentration. Applying the product rule, we have

$$\text{div}(\psi \vec{v}) = \nabla \cdot (\psi \vec{v}) = \psi \nabla \cdot \vec{v} + \vec{v} \nabla \cdot \psi$$

with $\rho = \text{const}$, which implies $\nabla \cdot \vec{v} = 0$. We obtain

$$\nabla \cdot (\psi \vec{v}) \quad \Rightarrow \quad \vec{v} \nabla \cdot \psi \qquad \text{for incompressible flow only.} \qquad \square$$

(c) Interpret $p = -1/3 \text{ trace } \vec{\vec{\tau}}$.

Solution

$$p = -\frac{1}{3} \text{ trace } \vec{\vec{\tau}} \equiv -\frac{1}{3}(\tau_{xx} + \tau_{yy} + \tau_{zz})$$

which is the mechanical pressure induced by normal stresses. $\qquad \square$

(d) Interpret $(\nabla \cdot [\nabla \times \vec{v}])$.

Solution

$\nabla \times \vec{v} = \vec{\zeta}$ is the vorticity vector, so that $\nabla \cdot \vec{\zeta}$ is the divergence of the vorticity field. $\qquad \square$

(e) State the physical significance of $\nabla \cdot \vec{v}$ versus $\nabla \vec{v}$.

Solution

$\nabla \cdot \vec{v} = \text{div} \vec{v}$ is the divergence of a velocity field; $\nabla \cdot \vec{v} = 0$ indicates incompressible flow. In contrast, $\nabla \vec{v}$ is often called "velocity gradient"; is actually a dyadic product, representing components of the stress tensor (cf. Sect 1.3.1). $\qquad \square$

1.6 Thermodynamic properties (cf. Sect. 1.2.3)

Consider the *isentropic bulk modulus* of a fluid, which is defined as the change in pressure per change in density, that is,

$$B = \rho \left(\frac{\partial p}{\partial \rho} \right)_s$$

Estimate B of (i) a perfect gas and (ii) water at standard conditions as a multiple of the atmospheric pressure.

Solution

(i) For a perfect gas, the pressure–density relation is

$$p = A \rho \gamma \qquad \text{where } \gamma = \frac{c_p}{c_v} = \gamma(T) \geq 1 \tag{S1.13}$$

$$\therefore B_{\text{gas}} = \rho \gamma A = p := 142 \times 10^3 \text{ Pa} \qquad \text{for air}$$

(ii) For water, a good $p(\rho)$-correlation is (cf. White 1991)

$$p = p_a \left[(C + 1) \left(\frac{\rho}{\rho_a} \right)^n - C \right] \tag{S1.14}$$

where $C = 3,000$ and $n = 7$.

$$\therefore B_{\text{liq}} = n(C+1)p_a \left(\frac{\rho}{\rho_a}\right)^n$$

or $B = \rho c^2$, where $c^2 = (\partial p/\partial \rho)_s$ is the speed of sound squared.
At atmospheric pressure, $\rho = \rho_a$, and

$$B_{\text{water}} \approx 21007 p_a := \underline{\underline{2.2 \times 10^9 \text{ Pa}}}$$

Problem Assignments

1.0 Elaborate on derivations given in Chapter 1 and answer questions related to problem solutions in Section 1.4.

1.1 Show that $\nabla \vec{v} \neq \vec{v}\nabla$.

1.2 Given the velocity fields ($v_x = by$, $v_y = bx$, $v_z = 0$) and ($v_x = -by$, $v_y = bx$, $v_z = 0$), evaluate $\nabla \cdot \vec{v}$ and $\nabla \vec{v}$ and check whether the fields are irrotational or not.

1.3 Given the momentum equation in vector form

$$\frac{\partial}{\partial t}(\rho \vec{v}) + \nabla \cdot (\rho \vec{v}\vec{v}) = -\nabla p - \nabla \cdot \vec{\vec{\tau}} + \vec{f}$$

state briefly the physical meaning of each term and write its components in rectangular coordinates.

1.4 Derive $\nabla \cdot \vec{v} = 0$ in cylindrical coordinates starting with a small fluid element $\Delta V = \Delta r \cdot r\Delta\theta \cdot \Delta z$.

1.5 Find Stokes derivative Du/Dt for $u = x^2 y^2/(zt)$ and for $u = yz + t$ and interpret the results.

1.6 Given the velocity fields $\vec{V}_1 = [u = cx, v = cy, w = -2cz]$ and $\vec{v}_2 = [u = u(x, y), v = v(x, y), w = 0]$, find their rates of strain and stress components and check for rotation and dilatation.

1.7 The velocity vector for an incompressible flow is

$$\vec{v} = 3x^2\hat{i} - 6xy\hat{j} + 16xy^2\hat{k}$$

Check for mass conservation and calculate stresses at $x = 6m$, $y = 3m$, and $z = 0$ when

$$\tau_{zz} = 100 \text{ N/m}^2 \text{ and } \mu = 1 \text{ kg/(ms)}.$$

1.8 Given steady planar flow in the Eulerian description, that is, $\vec{v} = (kx, -ky, 0)$, find the Lagrangian description

$$\frac{d\vec{r}(\vec{r}_0, t)}{dt} = \vec{v}[\vec{r}(\vec{r}_0, t), t]$$

where $\vec{r}(\vec{r}_0, t) = \vec{r}(a, b, c, t)$ are the path lines.

Answer

$$x = a \exp[k(t - t_0)], \qquad y = b \exp[-k(t - t_0)], \quad \text{and} \quad z = c$$

where $\quad x = a, \qquad y = b, \qquad z = c \qquad$ at $t = t_0 \qquad$ is the initial position

1.9 Consider at a fixed instant the motion of fluid elements on a circle with radius $dr = [(dx)^2 + (dy)^2]^{1/2} \triangleq \overline{PQ}$ where $P(x, y)$ is the center point P and $Q(x + dx, y + dy)$ lies on the circle. What is the angular speed of the line $\overline{PQ}$, and what is the average rotational speed about P? that is,

$$\frac{1}{2\pi} \int_0^{2\pi} \omega_{PQ} \, d\theta := \frac{1}{2}\left(\frac{\partial v}{\partial x} - \frac{\partial u}{\partial y}\right)$$

where $\quad \omega_{PQ} = \dfrac{\partial v}{\partial x}\cos^2\theta - \dfrac{\partial u}{\partial y}\sin^2\theta + \left(\dfrac{\partial v}{\partial y} - \dfrac{\partial u}{\partial x}\right)\sin\theta\cos\theta$

1.10 Consider two neighboring fluid particles located at time t at $P(x, y)$ and $Q(x+dx, y+dy)$. Show that the relative velocity components are

$$du = \left.\frac{\partial u}{\partial x}\right|_P dx + \left.\frac{\partial u}{\partial y}\right|_P dy \quad \text{and} \quad dv = \left.\frac{\partial v}{\partial x}\right|_P dx + \left.\frac{\partial v}{\partial y}\right|_P dy$$

Decompose du and dv into a fluid rotation part and a fluid deformation part and indicate shear and normal strain components.

1.11 An axisymmetric steady flow field generated by a rotating vertical cylinder can be described as

$$v_r = 0 \quad \text{and} \quad v_\theta = \frac{c_1 r}{2} + \frac{c_2}{r}$$

(a) Calculate the *circulation* for a fluid particle moving in a closed circular path of radius a with the velocity v_θ.

(b) What is the *potential circulation*, $\Gamma_\infty = \Gamma(a \to 0) = $ const, for an *irrotational flow*?

1.12 Consider a Rankin vortex

$$v_\theta = \omega r \quad \text{for } 0 \le r \le r_c \quad \text{and} \quad v_\theta = \frac{\omega r_c^2}{r} \quad \text{for } r > r_c$$

(a) Compute the vorticity about the z-axis and the circulation.

(b) Plot $v_\theta(r)$, $\zeta_z(r)$, and $\Gamma(r^2)$ for $0 \le r \le \infty$.

A Typical First Homework Set

1.1 Using Appendix A, write out in detail the tensor field components in rectangular coordinates for the following expressions: (a)$\nabla \cdot \vec{v}$, (b) $\nabla\vec{v}$, (c)$\nabla \times \vec{v}$, (d)$(\vec{v} \cdot \nabla)\vec{v}$, (e)$\nabla^2\vec{v}$. Provide an explanatory definition for (a)–(e) and briefly sketch applications wherever appropriate.

1.2 Show that $\nabla \cdot \vec{v} \ne \vec{v} \cdot \nabla$ and that $\nabla \cdot s\vec{v} \ne \nabla s \cdot \vec{v}$.

1.3 Depict the (nine) components of the stress tensor in cylindrical coordinates and write them in matrix form.

1.4 Do Sample-Problems 1.2, 1.5, 1.6, 1.8, 1.10, and 1.12.

References and Further Reading Material

Citations

Aris, R. 1967. *Vector Tensors, and the Basic Equations of Fluid Mechanics*. Prentice-Hall, Englewood Cliffs, NJ.

Bird, R. B., R. C. Armstrong, and O. Hassager. 1987. *Dynamics of Polymeric Liquids*. Vol. I. *Fluid Mechanics*, 2nd ed. Wiley-Interscience, New York.

Bird, R. B., W. E. Stewart, and E. N. Lightfoot. 1960. *Transport Phenomena*. John Wiley, New York.

Boyce, W. E., and R. C. DiPrima. 1977. *Elementary Differential Equations and Boundary Value Problems*. John Wiley, New York.

Bradshaw, P., T. Cebeci, and J. H. Whitelaw. 1981. *Engineering Calculation Methods for Turbulent Flow*. Academic Press, New York.

Cengel, Y. A., and M. A. Boles. 1993. *Thermodynamics: An Engineering Approach*, 2nd ed. McGraw-Hill, New York.

Colebrook, C. F. 1938–1939. "Turbulent Flow in Pipes with Particular Reference to the Transition Between the Smooth and Rough Pipe Laws," *J. Inst. Civ. Eng. Lond.*, **11**, 133–156.

Gerhart, P. H., Gross, R. J., and Hochstein, J. I. 1992. *Fundamentals of Fluid Mechanics*. Addison-Wesley, New York.

Habib, I. S. 1975. *Engineering Analysis Methods*. Lexington Books, D. C. Heath, Lexington, MA.

Helmholtz, H. 1867. *Phil. Mag.*, **33**(4), 485.

Kay, J. M., and R. M. Nedderman. 1985. *Fluid Mechanics and Transfer Processes*. Cambridge University Press, Cambridge.

Laufer, J. 1954. "The Structure of Turbulence in Fully Developed Pipe Flow," *NACA* Rep. No. 1174.

Lin, C. C., and L. A. Segal. 1975. *Mathematics Applied to Deterministic Problems in the Natural Sciences*. Macmillan, New York.

Moody, L. F. 1944. "Friction Factors for Pipe Flow," *ASME Trans.*, **66**, 671–684.

Panton, R. L. 1984. *Incompressible Flow*. Wiley-Interscience, New York.

Rieder, W. G., and H. R. Busby. 1986. *Introductory Engineering Modeling*. John Wiley, New York.

Sabersky, R. H., A. J. Acosta, and E. G. Hamptman. 1989. *Fluid Flow*. Macmillan, New York.

Sherman, F. S. 1990. *Viscous Flow*. McGraw-Hill, New York.

Spiegel, M. R. 1972. *Advanced Mathematics for Engineers and Scientists*. Schaum's Outline, McGraw-Hill, New York.

Stokes, G. G. 1845. "On the Theories of Internal Friction of Fluids in Motion," *Trans. Cambridge Phil. Soc.*, **8**, 287–305.

Stokes, G. G. 1885. *Trans. Camb. Phil. Soc.*, **8**, 287.

Timoshenko, S., and J. N. Goodier. 1990. *Theory of Elasticity*. McGraw-Hill, New York.

White, F. M. 1986. *Fluid Mechanics*, 2nd ed. McGraw-Hill, New York.

White, F. M. 1991. *Viscous Fluid Flow*. McGraw-Hill, New York.

Wylie, C. R., and L. C. Barrrett. 1982. *Advanced Engineering Mathematices*. McGraw-Hill, New York.

Further Reading

Batchelor, G. K. 1967. *An Introduction to Fluid Dynamics*. Cambridge University Press, Cambridge.

Cebeci, T., and P. Bradshaw. 1977. *Momentum Transfer in Boundary Layers*. McGraw-Hill, New York.

Churchill, S. W. 1988. *Viscous Flows: The Practical Use of Theory*. Butterworth-Heinemann, Boston.

Drazin, P. G., and W. H. Reid. 1981. *Hydrodynamic Stability*. Cambridge University Press, Cambridge.

Eskinazi, S. 1975. *Fluid Mechanics and Thermodynamics of Our Environment*. Academic Press, New York.

Karamcheti, K. 1968. *Principles of Ideal Fluid Aerodynamics*. Krieger, Malabar, FL.

Lai, W. M., et al. 1993. *Introduction to Continuum Mechanics*. Pergamon Press, Oxford.

Lam, H. 1932. *Hydrodynamics*. Dover, New York.

Landau, L. D., and E. M. Lifschitz. 1987. *Fluid Mechanics*, 2nd ed. Pergamon Press, Oxford.

Leal, L. G. 1992. *Laminar Flow and Convective Transport Processes: Scaling Processes and Asymptotic Analysis*. Butterworth-Heinemann, Boston.

LéMehauté, B. 1976. *An Introduction to Hydrodynamics and Water Waves*. Springer-Verlag, New York.

Lighthill, J. 1986. *An Informal Introduction to Theoretical Fluid Mechanics*. Virgile Clarendon Press, Oxford.

Lu, P.-C. 1977. *Introduction to the Mechanics of Viscous Fluids*. McGraw-Hill, New York.

Nunn, R. H. 1989. *Intermediate Fluid Mechanics*. Hemisphere, New York.

Probstein, R. F. 1991. *Physico-Chemical Hydrodynamics*. Butterworth-Heinemann, Boston.

Schey, H. M. 1973. *Div, Grad, Curl, and All That*. W. W. Norton, New York.

Schlichting, H. 1979. *Boundary-Layer Theory*. McGraw-Hill, New York.

Tritton, D. J. 1977. *Physical Fluid Dynamics*. Van Nostrand Reinhold, New York.

White, F. M. 1988. *Heat and Mass Transfer*. Addison-Wesley, Reading, MA.

Yih, C. S. 1969. *Fluid Dynamics*. West River Press, Ann Arbor, MI.

CHAPTER 2

Derivations and Transformations of the Conservation Equations

In order to obtain the basic transport equations, a suitable computational domain is selected and the conservation laws for mass, momentum, and energy are applied. The form of the resulting equations, including constitutive equations to gain closure, depends largely on the type of approach taken in terms of "suitable model selection." As discussed in Section 1.2.1 for the *closed system* approach, we consider a deformable volume moving with the fluid such that the individual fluid particles are always accounted for (Lagrange). In contrast, from the Eulerian point of view (open system), we consider a *fixed control volume* with the fluid moving through it. When these fixed or moving volumes are of finite extent, rate equations or integral equations can be directly obtained for the *global flow quantities* such as flow rates and forces (cf. Reynolds Transport Theorem). When such a system volume or control volume shrinks to an infinitesimally small fluid element of the flow field, differential equations can be directly obtained for the *local flow quantities* such as velocities and pressure. Alternatively, one could take a *molecular* approach (as done at supercomputing research centers) where the fundamental laws of nature are directly applied to the individual moving, and colliding molecules of a particular fluid in motion.

In summary, engineers prefer the Eulerian framework, that is, mass, momentum, and energy balances for an open system. Specifically, the *integral* or *control volume approach* is employed whenever it is sufficient to characterize the system's behavior with averaged and/or global values of the principal variables. The resulting solutions to these equations are magnitudes of engineering variables averaged over the particular control volume chosen. Modeling batch processes, impinging jets, well-mixed flow compartments, or fluid layers of finite thickness may serve as examples. In contrast, the *differential approach* is employed for the derivation of partial differential equations for instantaneous point descriptions of velocity, pressure, concentration, and temperature fields. Needless to say, with proper theorems, integral equations can usually be transformed into differential equations, and vice versa.

2.1 Derivation and Applications of the Transport Equations in Integral Form

We define an initial amount of an extensive system property G, contained at time t, within a system volume $\forall$, bounded by the surface S. Thus, we write for the arbitrary property (e.g., mass m, momentum $m\vec{v}$, energy E, entropy, S)

$$G_{\text{system}} = \int_{\forall_{\text{system}}} \rho\eta \, d\forall$$

where the intensive function $\eta \equiv G/(\text{unit mass})$ is accordingly 1, v, e, s, and so forth. With a glance at Figure 2.1, we observe that

39

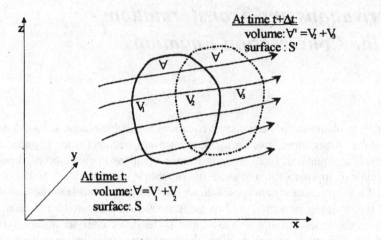

Fig. 2.1. Three-dimensional system moving and deforming in space (mass$_{system}$ = const).

- during a time interval Δt, the material volume moves and deforms;
- its position and shape at time $t + \Delta t$ are shown in dashed lines;
- the system at time t, that is, volume $\forall$ and surface S, as well as the system at time $t + \Delta t$, that is, volume $\forall'$ and surface S', are composed of two regions: $\forall = \forall_1 + \forall_2$ and $\forall' = \forall_2 + \forall_3$; that is, subregion $\forall_2$ is common to $\forall$ and $\forall'$.

The temporal rate of change of G for the moving system (Lagrangian point of view) is given by

$$\left. \frac{DG}{Dt} \right|_{system} = \lim_{\Delta t \to 0} \left\{ \frac{[G_2 + G_3]_{t+\Delta t} - [G_1 + G_2]_t}{\Delta t} \right\} \tag{2.1a}$$

Rearrangement of this first-derivative definition allows a physical interpretation of the material derivative

$$\left. \frac{DG}{Dt} \right|_{system} = \underbrace{\lim_{\Delta t \to 0} \left\{ \frac{[G_2|_{t+\Delta t} - G_2|_t]}{\Delta t} \right\}}_{\text{accumulation}} + \underbrace{\lim_{\Delta t \to 0} \frac{G_3|_{t+\Delta t}}{\Delta t}}_{\text{outflow}} - \underbrace{\lim_{\Delta t \to 0} \frac{G_1|_t}{\Delta t}}_{\text{inflow}} \tag{2.1b}$$

As $\Delta t \to 0$, the volume $\forall'$ again approaches the finite volume $\forall$, that is, the volume of the system at time t, so that in the limit the individual terms of the RHS become

$$\frac{\partial G_2}{\partial t} \equiv \frac{\partial G}{\partial t} = \frac{\partial}{\partial t} \iiint_{c.\forall.} \rho \eta \, d\forall \qquad \{\text{accumulation term}\}$$

$$\lim_{\Delta t \to 0} \left\{ \frac{G_3|_{t+\Delta t} - G_1|_t}{\Delta t} \right\} = \iint_{c.s.} \rho \eta \vec{v} \cdot d\vec{S} \qquad \left\{ \begin{array}{l} \text{net flow of } G \text{ across the} \\ \text{entire control surface} \end{array} \right\}$$

Hence, we obtain the Reynolds Transport Equation, where $c.\forall.$ = control volume and $c.s.$ = control surface, in the form

$$\frac{D}{Dt} G_{\text{system}} = \frac{\partial}{\partial t} \iiint\limits_{c.\forall.} \eta\rho \, d\forall + \iint\limits_{c.s.} \rho\eta\vec{v} \cdot d\vec{S}$$

(2.2)

$$\begin{Bmatrix} \text{time rate of} \\ \text{change of } G \text{ as} \\ \text{system moves} \end{Bmatrix} = \begin{Bmatrix} \text{accumulation of} \\ \text{specific property} \\ \text{within volume} \end{Bmatrix} + \begin{Bmatrix} \text{net efflux of} \\ \text{specific property} \\ \text{through control surface} \end{Bmatrix}$$

Note that the scalar product $\vec{v} \cdot d\vec{S} = (\vec{v} \cdot \hat{n}) \, dS = |\vec{v}||d\vec{S}| \cos\alpha$, with $\hat{n}$ the unit vector normal to $d\vec{S}$ in the outward direction, and $\alpha = \sphericalangle(\vec{v}, \hat{n})$.

In case the control volume moves (e.g., a ship in a current, a nozzle and moving-vane system, etc.), $\vec{v}$ is replaced by

$$\vec{v}_{\text{relative}} = \vec{v}_{\text{absolute}} \pm \vec{v}_{\text{control volume}}$$

The following two cases may illustrate the sign convention:

(i) $v_{\text{rel}} = \vec{v}_{\text{abs}} + \vec{v}_{c.\forall.}$, for example, when a jet from a stationary nozzle impacts a disk moving *toward* the nozzle:

$$v_{\text{rel}} = v_{\text{jet}} - (-v_{\text{disk}}) = v_j + v_d$$

and

(ii) $v_{\text{rel}} = \vec{v}_{\text{abs}} - \vec{v}_{c.\forall.}$, for example, when the disk moves *away* from the nozzle. That is, $v_{\text{rel}} = v_j - v_d$, which implies that the jet impact is zero when $v_{\text{disk}} \geq v_{\text{jet}}$.

When the control volume *accelerates*, for example, in the case of a launching rocket, a relative acceleration term $\vec{a}_{\text{rel}} = d\vec{v}_{c.\forall.}/dt$ has to be added to Eq. (2.2).

2.1.1   The Reynolds Transport Theorem

Equation (2.2) is a generalized transport equation in integral form, known as the Reynolds Transport Theorem (RTT). Steps for setting up the RTT and applying it to fluid mechanics problems include the following:

(i) Selection of

$$G_{\text{system}} = \begin{Bmatrix} \text{mass} \\ \text{linear momentum} \\ \text{energy} \end{Bmatrix} \quad \therefore \eta = \begin{Bmatrix} 1 \\ \vec{v} \\ e \end{Bmatrix}$$

(ii) Determination of

$$\frac{DG_{\text{syst}}}{Dt} = \begin{Bmatrix} 0 \\ \sum \vec{F}_{\text{ext}} \\ \dot{Q}_{\text{heat}} - \dot{W}_{\text{perf}} \end{Bmatrix}$$

which reflect the conservation laws

(iii)   Insertion of specific DG/Dt and related η into Eq. (2.2)
(iv)   Selection of "smart" control volume (cf. Sect. 2.1.2)
(v)   Solution of surface integrals and, in the case of transient system, the volume integral

2.1.2   Applications of the Control Volume Approach

A   Considering Conservation of Mass

$$G \equiv m_{system} \qquad \eta = 1 \quad \text{and} \quad \frac{DG}{Dt} \equiv 0$$

Thus, Eq. (2.2) takes on the form

$$0 = \frac{\partial}{\partial t}\left\{ \iiint\limits_{c.v.} \rho \, d\forall \right\} + \iint\limits_{c.s.} \rho(\vec{v}_{rel} \cdot \hat{n}) \, dA \tag{2.3}$$

For a *fixed* control volume and arbitrary sequence of mathematical operation, Eq. (2.3) can be rewritten as

$$\blacklozenge \qquad \iiint\limits_{c.v.} \frac{\partial \rho}{\partial t} \, d\forall + \iint\limits_{c.s.} \rho(\vec{v} \cdot \hat{n}) \, dA = 0 \tag{2.4}$$

Special cases of Eq. (2.4) include the following:

(i)   *Steady-state* within the control volume implies $\partial/\partial t = 0$, that is, no fluid mass accumulation with time, no internal fluid sources, and so forth.

$$\iint\limits_{A} \rho(\vec{v} \cdot \hat{n}) \, dA = 0 \tag{2.5a}$$

or simplified

$$\sum (\rho A \bar{v})_{in} = \sum (\rho A \bar{v})_{out} \tag{2.5b}$$

where $\bar{v}$ denotes average velocities.

(ii)   For *incompressible* fluid flow, $\rho = $ constant so that for steady or transient cases

$$\iint (\vec{v} \cdot \hat{n}) \, dA = 0 \quad \Rightarrow \quad \sum Q_{in} = \sum Q_{out} \tag{2.6a,b}$$

It has to be noted that for (cf. sketches):

Inflow: $(\vec{v} \cdot \hat{n}) \, dA$ is "negative"

and

Outflow: $(\vec{v} \cdot \hat{n}) \, dA$ is "positive"

A brief example may illustrate the application of the mass conservation law in integral form. Additional sample problems are solved in Sect. 2.4.

Example (1)
Consider steady laminar incompressible flow in the entrance region of a horizontal circular pipe of radius R. The inlet velocity profile at $x = 0$ (i.e., station ①) is $u_1 = U = $ const. The flow *develops* downstream at station ② into a parabolic profile

$$u_2 = u_{max}[1 - (r/R)^2]$$

Find u_{max} as a function of U.

This is clearly case (ii), where $\rho = $ const and the (cylindrical) control volume of radius R and entrance length $0 \le x \le l_e$ is fixed and nondeforming. Thus, Eq. (2.6a)

$$\iint_{c.s.} (\vec{v} \cdot \hat{n}) \, dA = 0$$

produces two surface integrals at sections ① and ②:

$$A_1(-U) + \iint_{A_2} (\vec{v} \cdot \hat{n}) \, dA_2 = 0$$

where $A_1 = \pi R^2$, $dA_2 = 2\pi r \, dr$, and $\vec{v} \cdot \hat{n} = u_2(r)$. Hence,

$$-\pi R^2 U + 2\pi u_{max} \int_0^R \left[1 - \left(\frac{r}{R}\right)^2\right] r \, dr = 0$$

The average velocity has to be $\bar{u} = U$, or with $Q = \bar{u} A$

$$\bar{u} = \frac{1}{A} \int_A (\vec{v} \cdot \hat{n}) \, dA$$

so that

$$\bar{u} \equiv \bar{u}_2 = \frac{u_{max}}{2}$$

B   Considering Conservation of Linear Momentum

$$G \equiv (m\vec{v})_{\text{system}}, \qquad \eta = \vec{v} \quad \text{and} \quad \frac{DG}{Dt} \equiv \sum \vec{F}_{\text{ext}} := \sum \vec{F}_{\text{surface}} + \sum \vec{F}_{\text{body}}$$

Thus, Eq. (2.2) takes on the form

$$\blacklozenge \qquad \frac{D(m\vec{v})}{Dt} = \frac{\partial}{\partial t} \iiint_{c.\forall.} \vec{v}\rho \, d\forall + \iint_{c.s.} \vec{v}\rho\vec{v} \cdot d\vec{S} = \sum \vec{F}_S + \sum \vec{F}_B \qquad (2.7)$$

The forces exerted on the control *surface* are mainly the pressure and viscous forces (cf. Sect. 1.3.1)

$$\sum \vec{F}_S = \iint_{c.s.} \left(-p\vec{\delta} + \vec{\vec{\tau}} \right) \cdot d\vec{S} \qquad (2.8a)$$

and the forces acting at the control volume center include the gravity force and the Coriolis force.

$$\sum \vec{F}_B = \iiint_{c.\forall.} \vec{f}_B \, d\forall \qquad (2.8b)$$

As expected, Eq. (2.7) is Newton's second law of motion for a particle of mass m applied to an open fluid flow system, namely,

$$\sum \vec{F}_{\text{external}} = m\vec{a}$$

Here,

$$\underbrace{\sum \vec{F}}_{①} = \underbrace{\frac{D}{Dt} \iiint_{\forall_{\text{syst.}}} \vec{v}\rho \, d\forall}_{②} := \underbrace{\frac{\partial}{\partial t} \iiint_{c.\forall.} \vec{v}\rho \, d\forall}_{③} + \underbrace{\iint_{c.s.} \rho(\vec{v} \cdot \hat{n}) \, dA}_{④}$$

where term ① represents all forces acting on the system (or control volume), and term ② is the total time rate of change of linear momentum of the (closed) system that is due to the external forces. Now, term ② is split in the Eulerian flow field description into two terms: Term ③ is the *local* time rate of change of momentum within the control volume, and term ④ is the *net* rate of flow of momentum through the control surface. It should be noted that $\rho(\vec{v} \cdot \hat{n}) \, dA \equiv d\dot{m}$.

The linear momentum equation in integral form (2.7) is a vector equation with three components. For example, in rectangular coordinates for a 2-D flow field

$$\sum F_x = \frac{\partial}{\partial t} \int u\rho \, d\forall + \int u\rho\vec{v} \cdot d\vec{A} \qquad (2.9a)$$

$$\sum F_y = \frac{\partial}{\partial t} \int v\rho \, d\forall + \int v\rho\vec{v} \cdot d\vec{A} \qquad (2.9b)$$

For steady one-dimensional flows, Eq. (2.9) collapses to

$$\sum F = \sum (\dot{m}\bar{v})_{\text{out}} - \sum (\dot{m}\bar{v})_{\text{in}} \qquad (2.10)$$

A brief example may illustrate the use of Eq. (2.10) in conjunction with Eq. (2.5).

Example (2)

Consider a steady jet impinging on a conical disk of diameter D moving toward the jet (cf. sketches):

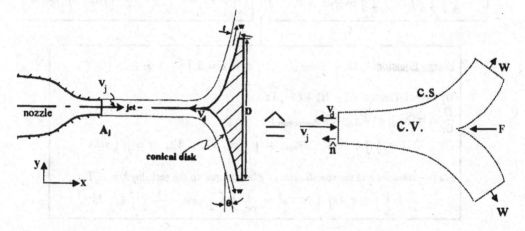

Neglecting frictional effects and pressure differentials, the reactive force on the disk, F, is the only (reactive) force of interest here; v_{jet} is the absolute velocity and v_{disk} is the control volume velocity so that

$$v_{\text{rel}} = v_j - (-v_d) = v_j + v_d$$

Thus, Eq. (2.5) reduces to $\dot{m}_{\text{in}} = \dot{m}_{\text{out}}$, or

$$\rho A_j (v_j + v_d) = (\rho A w)_{\text{out}} \approx \rho (\pi D t) w$$

so that

$$w = \frac{A_j (v_j + v_d)}{\pi D t}$$

Now, Eq. (2.10) yields

$$\sum F_x := -F = (\dot{m}w_x)_{\text{out}} - (\dot{m}w_x)_{\text{in}}$$

or with $w_{x,\text{out}} = w \sin\theta$ and $w_{x,\text{in}} = v_j + v_d$

$$F = \rho A_j (v_j + v_d)^2 \left[1 - \frac{A_j \sin\theta}{\pi D t} \right]$$

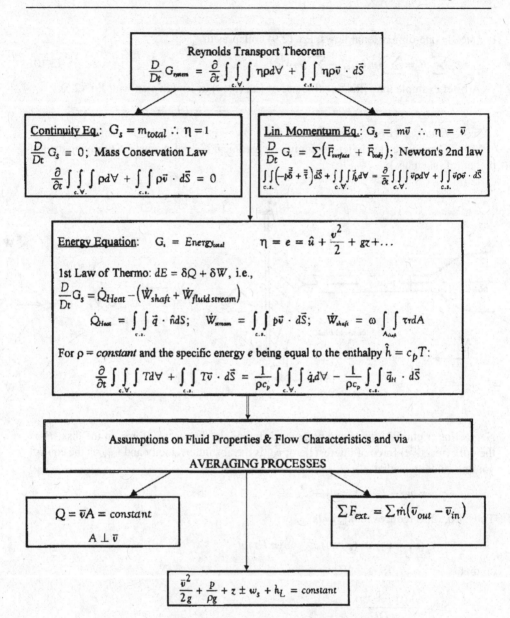

Fig. 2.2. The conservation laws in integral form and their spinoffs.

C Considering Conservation of Energy

$$G = E_{\text{total}}|_{\text{system}}, \qquad \eta = e = e_{\text{internal}} + e_{\text{kinetic}} + e_{\text{potential}} := \hat{u} + \frac{v^2}{2} + gz$$

and with $E_{\text{system}} = Q_{\text{heat}} - W_{\text{system}}$

$$\frac{DG}{Dt} = \dot{Q}_{\text{heat}} - (\dot{W}_{\text{shaft}} + \dot{W}_{\text{fluid stream}}) \qquad \text{(first law of thermodynamics)}$$

Thus, Eq. (2.2) takes on the form

$$\blacklozenge \qquad \frac{DE_{\text{total}}}{Dt} = \frac{\partial}{\partial t} \iiint_{c.\forall.} e\rho \, d\forall + \iint_{c.s.} e\rho \vec{v} \cdot d\vec{S} = \dot{Q} - \sum \dot{W} \qquad (2.11)$$

Typically, $\dot{Q}$ is the heat supplied to the system by the surroundings, whereas $\sum \dot{W}$ represents the shaft work done by the system. Although Eq. (2.11) has important applications in thermodynamics, it plays only a minor role in incompressible fluid dynamics (cf. Chapter 4). In its most useful form for steady flow along a streamline between points ① and ②, it is known as the mechanical energy equation or extended Bernoulli equation.

$$\pm w_s - \int_1^2 \frac{dp}{\rho} - gh_{\text{loss}} = \left(\frac{v^2}{2} + gz \right)_② - \left(\frac{v^2}{2} + gz \right)_①$$

where $\pm w_s$ is the work input/output of the moving fluid due to pumps or turbines, and (gh_{loss}) is an energy dissipation term due to heat loss, frictional losses, form losses, and so on.

Figure 2.2 summarizes applications of the Reynolds Transport Theorem and highlights the derivation of common engineering equations. It also illustrates the chain of necessary simplifications to be made to arrive at some frequently used "modeling" equations.

2.2   Derivations and Applications of the Transport Equations in Differential Form

There are different ways of deriving the mass, momentum, and energy transfer equations in differential form. Here, three approaches that may provide a balanced mathematical–physical understanding of the terms in the fluid dynamics equations are presented:

(i)   Application of the Reynolds Transport Theorem (cf. Sect. 2.1.1) to a control volume that shrinks to a representative elementary volume (RE∀), employing Green's Theorem.

(ii) Development of mass, force, and energy balances for an RE∀ employing the truncated Taylor series expansion.

(iii) Generalization of approach (ii) focusing on a dynamic balance of a *single* functional that represents all three conservation quantities, leading to a *generalized transport equation* (GTE) (cf. Sect. 2.2.2).

Approach (iii) relies on the fact that the basic engineering equations used in various branches of continuum physics look rather similar: Newton's second law of motion, for example, describing the dynamics of a single point–particle is used with modifications to solve mechanical engineering problems in rigid body dynamics, particle systems, vibration of structures, and elasticity theory, as well as transport phenomena in fluid mechanics, rheology, and magnetohydrodynamics (cf. Dorf 1995). In addition, the similarity between the heat transfer and the mass transfer equations is obvious. All engineering equations are based on the fundamental principle of conservation of momentum, energy, and mass. In other words, a dynamic balance for an arbitrary infinitesimal domain, considering all significant internal plus external forces, mass fluxes, and energy flows, will yield the desired field equations. Similar to the derivation of the Reynolds Transport Theorem, the uniqueness of the generalized approach given in Sect. 2.2.2 is that a dynamic balance is performed for a *single* functional rather than three different balances reflecting the conservation principles. It has to be noted that the derivation procedures rely on the material discussed in Chapter 1 and tensor manipulations summarized in Appendix A.

2.2.1   Derivations of the Individual Conservation Laws

In this section we focus on partial differential equations that represent a balance of fluid mass, linear momentum, and fluid enthalpy. The procedural steps, of course, differ between Approaches (i) and (ii), but they are the same for the three conservation laws:

A   Conservation of Fluid Mass (The Continuity Equation)

The Reynolds Transport Theorem for a fixed control volume reads (cf. Eq. (2.4))

$$0 = \iiint\limits_{c.\forall.} \frac{\partial}{\partial t} \rho \, d\forall + \iint\limits_{c.s.} \rho \vec{v} \cdot d\vec{S}$$

which can be rewritten with Green's Theorem

$$\iint\limits_{A} (\vec{K} \cdot \hat{n}) \, dA \equiv \iiint\limits_{\forall} (\nabla \cdot \vec{K}) \, d\forall$$

to

$$\iiint\limits_{c.\forall.} \frac{\partial}{\partial t} \rho \, d\forall + \iiint\limits_{c.\forall.} \nabla \cdot (\rho \vec{v}) \, d\forall = 0$$

or

$$\iiint\limits_{c.\forall.} \left\{ \frac{\partial \rho}{\partial t} + \nabla \cdot (\rho \vec{v}) \right\} d\forall = 0$$

which requires that $d\forall = 0$, or

$$\blacklozenge \qquad \frac{\partial \rho}{\partial t} + \nabla \cdot (\rho \vec{v}) = 0 \tag{2.12}$$

Alternatively, starting with the principle that

$$m_{\text{system}} = (\rho \forall)_{\text{system}} = \text{const}$$

we can formally write

$$\frac{dm}{dt} = \rho \frac{d\forall}{dt} + \forall \frac{d\rho}{dt} = 0$$

or

$$\frac{d\rho}{dt} = -\rho \left(\forall^{-1} \frac{d\forall}{dt} \right)$$

Now, for a fluid element (cf. Chapter 1)

$$\frac{D\rho}{Dt} \equiv \frac{\partial \rho}{\partial t} + (\vec{v} \cdot \nabla)\rho = -\rho(\nabla \cdot \vec{v})$$

or

$$\blacklozenge \qquad \frac{\partial \rho}{\partial t} + \nabla \cdot (\rho\vec{v}) = 0 \tag{2.12}$$

Approach (ii) starts with an REV (representative elementary volume) of size $\Delta\forall = \Delta x \, \Delta y \, \Delta z$ for which a one-dimensional mass balance yields (cf. sketch):

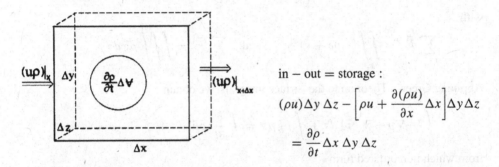

in − out = storage :

$$(\rho u)\Delta y \, \Delta z - \left[\rho u + \frac{\partial(\rho u)}{\partial x}\Delta x \right]\Delta y \Delta z$$

$$= \frac{\partial \rho}{\partial t}\Delta x \, \Delta y \, \Delta z$$

Simplification of the mass balance and division by $\Delta\forall$ yields

$$-\frac{\partial(\rho u)}{\partial x} = \frac{\partial \rho}{\partial t}$$

Extension to three-dimensional mass transfer yields in rectangular coordinates

$$-\frac{\partial(\rho u)}{\partial x} - \frac{\partial(\rho v)}{\partial y} - \frac{\partial(\rho w)}{\partial z} = \frac{\partial \rho}{\partial t}$$

The extension to any orthogonal coordinate system yields

$$\blacklozenge \qquad \frac{\partial \rho}{\partial t} + \nabla \cdot (\rho\vec{v}) = 0 \quad \text{or} \quad \frac{D\rho}{Dt} + \rho(\nabla \cdot \vec{v}) = 0 \tag{2.12a,b}$$

Notes

- The term $\nabla \cdot (\rho\vec{v})$ can be interpreted as the "net outflow $\dot{m}/\forall$" so that when $\nabla \cdot (\rho\vec{v}) > 0$, the fluid density has to decrease with time: that is, $\dot{\rho} < 0$. This scenario may occur for pressure reduction and subsequent gas density decrease inside a leaky container.

- For steady compressible fluid flow $\partial\rho/\partial t = 0$, which implies that $\nabla \cdot (\rho\vec{v}) = 0$. However, for *incompressible* flow, $\rho = $ const, that is, $(D\rho/Dt) = 0$, so that

$$\nabla \cdot \vec{v} = \operatorname{div} \vec{v} = 0 \tag{2.13}$$

- Equation (2.13) is true for steady and transient (incompressible) flows. In two dimensions there is a (stream) function, say, $\psi(x, y, t)$, that satisfies the continuity equation identically. Hence in rectangular coordinates, $u \equiv \partial\psi/\partial y$ and $v \equiv -\partial\psi/\partial x$ (cf. Sect. 2.4).

B Conservation of Linear Momentum (The Equation of Motion)

The Reynolds Transport Theorem for linear momentum transfer (cf. Eqs. (2.7 and 2.8)) reads

$$\sum \vec{F}_s + \sum \vec{F}_b = \frac{\partial}{\partial t} \iiint_{c.\forall.} \vec{v}\rho \, d\forall + \iint_{c.s.} \vec{v}\rho\vec{v} \cdot d\vec{S}$$

with

$$\sum \vec{F}_s = \iint_{c.s.} (-p\vec{\delta} + \vec{\vec{\tau}}) \cdot d\vec{S} \quad \text{and} \quad \sum \vec{F}_b = \iiint_{c.\forall.} \rho\vec{g} \, d\forall$$

Applying Green's Theorem to the surface integrals, we obtain

$$\int (-\nabla p + \nabla \cdot \vec{\vec{\tau}}) \, d\forall + \int \rho\vec{g} \, d\forall = \int \frac{\partial}{\partial t}(\rho\vec{v}) \, d\forall + \int \nabla \cdot (\rho\vec{v}\vec{v}) \, d\forall$$

from which in reordered form

$$\frac{\partial}{\partial t}(\rho\vec{v}) + \nabla \cdot (\rho\vec{v}\vec{v}) = -\nabla p + \nabla \cdot \vec{\vec{\tau}} + \rho\vec{g} \tag{2.14}$$

$$\left\{\begin{array}{l}\text{local time}\\\text{rate of change}\\\text{of momentum}\end{array}\right\} + \left\{\begin{array}{l}\text{convective}\\\text{change in}\\\text{momentum}\\\text{(inertia)}\end{array}\right\} = \left\{\begin{array}{l}\text{net}\\\text{pressure}\\\text{force}\end{array}\right\} + \left\{\begin{array}{l}\text{net}\\\text{viscous}\\\text{force}\end{array}\right\} + \left\{\begin{array}{l}\text{gravit.}\\\text{force}\end{array}\right\}$$

For Newtonian fluids with *constant properties* and Stokes's hypothesis invoked (cf. Sect. 1.3.1), the equation of motion reduces to the celebrated Navier–Stokes equations

$$\nabla \cdot \vec{v} = 0 \tag{2.15a}$$

$$\frac{D\vec{v}}{Dt} \equiv \frac{\partial\vec{v}}{\partial t} + (\vec{v} \cdot \nabla)\vec{v} = -\frac{1}{\rho}\nabla p + \nu\nabla^2\vec{v} + \vec{g} \tag{2.15b}$$

Thus, a divergence-free fluid flow field may accelerate, $\vec{a}_{\text{total}} = D\vec{v}/Dt$ as a result of various forces, that is, f_{pressure}, f_{viscous}, and/or f_{gravity}.

Approach (ii) considers again a fluid element of constant mass m, subjected to various forces. We now apply Newton's second law of motion to the accelerating fluid element (cf. sketch and Sects. 1.2.2 and 1.3.1).

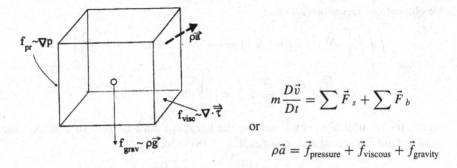

$$m\frac{D\vec{v}}{Dt} = \sum \vec{F}_s + \sum \vec{F}_b$$

or

$$\rho\vec{a} = \vec{f}_{\text{pressure}} + \vec{f}_{\text{viscous}} + \vec{f}_{\text{gravity}}$$

Hence, we can write a dynamic force balance per unit volume as

$$\rho\frac{D\vec{v}}{Dt} \equiv \rho\left[\frac{\partial \vec{v}}{\partial t} + (\vec{v} \cdot \nabla)\vec{v}\right] = -\nabla p + \nabla \cdot \vec{\vec{\tau}} + \rho\vec{g} \qquad (2.16)$$

For inviscid fluid flow, $\mu \equiv 0$, which implies $\vec{\vec{\tau}} = 0$, so that the equation of motion reduces to Euler's equation

$$\rho\vec{a}_{\text{total}} = \rho\frac{D\vec{v}}{Dt} = -\nabla p + \rho\vec{g} \qquad (2.17)$$

Notes

When a flow field is inviscid and irrotational, it is called *potential flow*. When the flow is very slow, the inertia term $(\vec{v} \cdot \nabla)\vec{v} \equiv \vec{a}_{\text{convective}}$ is negligible; it is called very-low-Reynolds-number flow or *creeping flow*. When $\vec{a}_{\text{local}} \equiv \partial\vec{v}/\partial t$ is zero, the flow is *steady*. For *boundary-layer type flows*, that is, $u \gg v$, the streamwise momentum equation is dominant and $\partial^2 u/\partial x^2 \approx 0$.

C Conservation of Energy (The Heat Transfer Equation)

The first law of thermodynamics for a closed system, reflecting the principle of energy conservation, can be stated as

$$\begin{bmatrix} \text{Net rate of energy transfer} \\ \text{to the system} \end{bmatrix} = \begin{bmatrix} \text{total rate of change of} \\ \text{the system's energy} \end{bmatrix}$$

that is,

$$\dot{Q} + \dot{W} = \frac{DE}{Dt}$$

Employing again Green's Theorem and recalling that

$$\dot{Q} \equiv \dot{Q}_{\text{conduction}} = \iint\limits_{c.s.} \vec{q} \cdot \hat{n}\, dS \qquad \text{where } \vec{q} = -k\nabla T \qquad \text{(Fourier's law)}$$

and

$$\dot{W} \equiv -\dot{W}_{\text{flow}} = -\iint\limits_{c.s.} p(\vec{v} \cdot \hat{n})\, dS$$

we obtain from Equation (2.11):

$$\iiint\limits_{c.\forall.} \left\{ -\nabla \cdot (k\nabla T) - \nabla \cdot (p\vec{v}) - \frac{\partial}{\partial t}\left[\rho\left(\hat{u} + \frac{v^2}{2} + \vec{g}\cdot\vec{r} \right) \right] \right.$$

$$\left. - \nabla \cdot \left[\rho\vec{v}\left(\hat{u} + \frac{v^2}{2} + \vec{g}\cdot\vec{r} \right) \right] \right\} d\forall = 0 \tag{2.18}$$

Again, for an arbitrary control volume the integrand must be zero so that for a moving *inviscid* fluid with constant properties, Eq. (2.18) reduces to

$$-k\nabla^2 T = \underbrace{\vec{v}\cdot\nabla p + \rho\vec{v}\cdot\frac{D\vec{v}}{Dt} - \rho\vec{v}\cdot\vec{g}}_{\vec{v}\cdot\left(\nabla p + \rho\frac{D\vec{v}}{Dt} - \rho\vec{g}\right) \equiv 0 \quad \text{(cf. Eq. (2.17))}} - \rho\frac{D\hat{u}}{Dt}$$

or

$$\rho\frac{D\hat{u}}{Dt} = k\nabla^2 T \tag{2.19}$$

where $\hat{u}$ is the specific internal energy. In real fluid flows, energy dissipation, which may cause temperature increases due to fluid friction as encountered, for example, in wall-bounded viscous flows, has to be considered (cf. Approach (ii) and Sect. 4.3).

Approach (ii) is best carried out in terms of an enthalpy balance for a representative elementary volume: that is, an REV is fixed in a flow field, where the rate of work done on the fluid by gravitational and pressure forces is neglected.

$$\hat{h} = \hat{u} + \frac{p}{\rho} \approx c_p T \quad \text{(enthalpy)}$$

$$\vec{q} = -k\nabla T \quad \text{(conductive heat flux)}$$

The verbal enthalpy balance may read

$$\left\{ \begin{array}{l} \text{Rate of enthalpy} \\ \text{accumulation} \end{array} \right\} = \left\{ \begin{array}{l} \text{net change of enthalpy} \\ \text{by convection} \end{array} \right\} + \left\{ \begin{array}{l} \text{net heat addition} \\ \text{by conduction} \end{array} \right\}$$

$$+ \left\{ \begin{array}{l} \text{heat generation by} \\ \text{viscous dissipation} \end{array} \right\}$$

which in mathematical terms can be expressed as

$$\frac{\partial}{\partial t}(\rho\hat{h}) = -\frac{\partial}{\partial x}(u\rho\hat{h}) - \frac{\partial}{\partial y}(v\rho\hat{h}) - \frac{\partial}{\partial z}(w\rho\hat{h}) - \left(\frac{\partial q_x}{\partial x} + \frac{\partial q_y}{\partial y} + \frac{\partial q_z}{\partial z} \right)$$

$$- \left[\frac{\partial}{\partial x}(\tau_{xx}u + \tau_{yx}v + \tau_{zx}w) + \frac{\partial}{\partial y}(\tau_{xy}u + \tau_{yy}v + \tau_{zy}w) \right.$$

$$\left. + \frac{\partial}{\partial z}(\tau_{xz}u + \tau_{yz}v + \tau_{zz}w) \right]$$

or

$$\frac{\partial}{\partial t}(\rho \hat{h}) = -\nabla \cdot \rho \vec{v} \hat{h} - \nabla \cdot \vec{q} - \nabla \cdot (\vec{\vec{\tau}} \cdot \vec{v}) \qquad (2.20)$$

In terms of the fluid temperature for constant-fluid-property flows using $\hat{h} = c_p T$,

$$\blacklozenge \qquad \frac{\partial T}{\partial t} + (\vec{v} \cdot \nabla)T = \alpha \nabla^2 T + \Phi \qquad (2.21)$$

where $\alpha = k/\rho c_p$ is the thermal diffusivity and the viscous dissipation function $\Phi(\nabla \vec{v})$ is discussed later.

Notes
The convective heat transfer term $(\vec{v} \cdot \nabla)T$ requires the solution of the equation of motion a priori or simultaneously. In the latter case, *both* transport equations are coupled when the gravity term in the equation of motion becomes temperature dependent as a result of buoyancy effects as encountered in free convection, that is,

$$\rho \vec{g} \Rightarrow \rho_0 \beta (T - T_o) \vec{g} \qquad \text{(Boussinesq assumption)}$$

$$\text{where} \qquad \beta = -\frac{1}{\rho}\left(\frac{\partial \rho}{\partial T}\right)_p$$

is the coefficient of thermal expansion. Another temperature dependence in the equation of motion occurs when the fluid properties ρ, ν, k, c_p and so on, are strongly affected by the local temperature, for example, $\nu(T) = \nu_0 + \nu_1 T + \nu_2 T^2 + \cdots$. In many engineering applications, the elaborate "heat source" term Φ can be reduced to

$$\Phi = \frac{\mu}{\rho c_p}\left(\frac{\partial u}{\partial y}\right)^2$$

or equivalent (cf. Ch. 4).

2.2.2 Derivation of a Generalized Transport Equation

The physical insight gained by employing different approaches for deriving the transport equations may be further enhanced in this section.

The fundamental laws of classical mechanics and thermodynamics are written in terms of mass, force, energy balances, and the entropy imbalance for a closed system interacting with its surroundings via external forces $\sum \vec{F}$, heat Q, and work W. They can be reformulated for a fluid flow field in the Eulerian framework as follows:

Mass conservation $\qquad m_{\text{Syst}} = \text{const} \rightarrow \left.\dfrac{D_m}{Dt}\right|_{\text{Syst}} = 0$

Conservation of linear momentum

$$\frac{d}{dt}(m\vec{v}) = \sum \vec{F}_{\text{external}} \rightarrow \frac{D}{Dt}(m\vec{v}) = \sum \vec{F}_{\text{Surface}} + \sum \vec{F}_{\text{Body}}$$

Energy conservation $\qquad dE = \delta Q - \delta W \rightarrow \dfrac{DE}{Dt} = \delta \dot{Q} - \delta \dot{W}$

Entropy imbalance $\qquad dS \geq \dfrac{\delta Q}{T}$

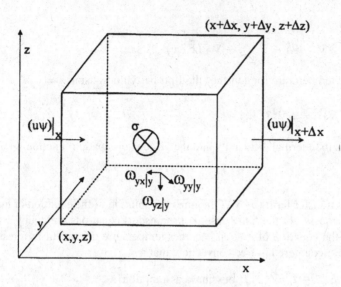

Fig. 2.3. Schematic control element with one-dimensional convective transport, internal sink/source, and external force/flux components on one control surface.

The present task is to derive the resulting engineering equations that will enable us to model arbitrary transport and conversion phenomena provided that the number of principal unknowns matches the number of equations (closure problem) and that all necessary boundary conditions are specified for a unique solution (well-posed problem). Thus, from the Eulerian viewpoint, a balance is taken over a very small open system and REV, which is fixed with respect to the local coordinates. The system's changes are described with deterministic equations; random phenomena are represented as part of the governing equation with an allowance for time-averaged, irregular fluctuation of some engineering variables. Hence, similar to the Reynolds Transport Theorem (cf. Sect. 2.1.1), it is possible to derive all necessary modeling equations from just *one fundamental tensor equation*, the so-called generalized transport equation (GTE). It should be stressed again that derivations in the engineering sciences are a means to explain and demonstrate the interrelationship between mathematics and physics rather than a painful intellectual exercise that should be executed as quickly as possible.

To derive the GTE in differential form, we define an arbitrary continuous transport function ψ, a flux tensor ω, as well as an internal force and source term σ. We then perform a comprehensive balance over a representative material element (Fig. 2.3).

The arbitrary transport function ψ might be

- a *scalar* like the specific entropy s, energy e, mass ρ, or concentration c;
- a *vector* like the specific, linear momentum $\rho\vec{v}$ or angular momentum $\vec{r} \times \rho\vec{v}$.

The specific surface flux ω might include

- tangential and normal forces in terms of the stress tensor or momentum flux $\vec{\vec{\tau}}$ or pressure $p\vec{\vec{\delta}}$;
- diffusion in terms of Fourier's heat flux $\vec{q} \propto \nabla T$ or Fick's mass flux $\vec{j} \propto \nabla c$.

The internal system's changes σ might be due to

- body forces like electromagnetic or gravitational forces $\vec{f}$;

- material/energy sinks, sources, or net production from chemical reactions, radioactive decay, sorption, evaporation, or condensation, as well as net material/energy input not accounted for elsewhere, for example, via Neumann-type boundary conditions.

The generalized material, force, and energy balance for a representative elementary volume (REV) can be conceptualized as

$$\left\{ \begin{array}{l} \text{Time rate of} \\ \psi\text{-accumulation} \end{array} \right\} = \left\{ \begin{array}{l} \text{rate of net convective} \\ \text{transport: } \psi_{\text{IN}} - \psi_{\text{OUT}} \end{array} \right\} + \left\{ \begin{array}{l} \text{surface forces and} \\ \text{diffusional fluxes} \end{array} \right\}$$

$$+ \left\{ \begin{array}{l} \text{production, sources,} \\ \text{and body forces} \end{array} \right\}$$

This generalized balance in mathematical form reads (cf. Fig. 2.3)

$$\Delta x\, \Delta y\, \Delta z \frac{\partial \psi}{\partial t} = -\{[(u\psi)_{x+\Delta x} - (u\psi)_x]\Delta y\, \Delta z$$

$$- [(v\psi)_{y+\Delta y} - (v\psi)_y]\Delta x\, \Delta z - [(w\psi)_{z+\Delta z} - (w\psi)_z]\Delta x\, \Delta y\}$$

$$+ \left\{ \begin{array}{l} [(\omega_{xx}|_x - \omega_{xx}|_{x+\Delta x})\Delta y\Delta z + (\omega_{yx}|_y - \omega_{yx}|_{y+\Delta y})\Delta x\Delta z \\ + (\omega_{zx}|_z - \omega_{zx}|_{z+\Delta z})\Delta x\, \Delta y] \\ + [\omega\text{-components in } y\text{-direction}] \\ + [\omega\text{-components in } z\text{-direction}] \end{array} \right\}$$

$$+ \{\rho\sigma_x + \rho\sigma_y + \rho\sigma_z\}\Delta x\, \Delta y\, \Delta z$$

Dividing the equation by $\Delta V = \Delta x \Delta y \Delta z$ and then letting Δx, Δy, and Δz approach zero, yields

$$\frac{\partial \psi}{\partial t} = -\left\{ \frac{\partial}{\partial x}(u\psi) + \frac{\partial}{\partial y}(v\psi) + \frac{\partial}{\partial z}(w\psi) \right\} + \left\{ \left(\frac{\partial \omega_{xx}}{\partial x} + \frac{\partial \omega_{yx}}{\partial y} + \frac{\partial \omega_{zx}}{\partial z} \right) \right.$$

$$+ \left(\frac{\partial \omega_{xy}}{\partial x} + \frac{\partial \omega_{yy}}{\partial y} + \frac{\partial \omega_{zy}}{\partial z} \right) + \left. \left(\frac{\partial \omega_{xz}}{\partial x} + \frac{\partial \omega_{yz}}{\partial y} + \frac{\partial \omega_{zz}}{\partial z} \right) \right\}$$

$$+ \{\rho(\sigma_x + \sigma_y + \sigma_z)\}$$

Or, written in a more compact form using vector notation

$$\blacklozenge \qquad \frac{\partial \psi}{\partial t} + \nabla \cdot (\vec{v}\psi) = \nabla \cdot \vec{\vec{\Omega}} + \vec{\Sigma} \qquad\qquad (2.22)$$

In order to solve a particular problem, suitable modeling equations have to be obtained first by

(i) defining the continuous transport function ψ plus the associated surface/force/diffusional flux term $\vec{\vec{\Omega}}$ as well as the internal source function $\vec{\Sigma}$;

(ii) selecting phenomenological relationships (i.e., semiempirical formulas, constitutive equations, etc.) expressing fluxes, fluid properties, sinks, and sources in terms of principal variables or their gradients; and

(iii) tailoring or reducing the resulting set of governing equations on the basis of justifiable assumptions and postulates.

2.2.3   Applications of the Generalized Transport Equation to Thermal Flow Analyses

The generalized transport equation in differential form is now applied to represent the conservation principles. In the first application, the transport function ψ is a vector, that is, $\psi \equiv \rho\vec{v}$, to generate the momentum transfer equation. In the second application ψ is a scalar property (i.e., fluid mass, species concentration, or internal energy) to generate constituent transfer equations: the continuity, mass transfer, and heat transfer equations, respectively.

A   Conservation of (Linear) Momentum

Setting

$$\psi \equiv \rho\vec{v} \triangleq \text{linear momentum/(unit volume)}$$

$$\vec{\vec{\Omega}} \equiv \vec{\vec{\pi}} = -p\vec{\vec{\delta}} + \vec{\vec{\tau}} \triangleq \text{total stress tensor = pressure + stresses}$$

and

$$\vec{\Sigma} \equiv \rho\vec{f} \triangleq \text{body forces/(unit volume)}$$

The GTE in differential form, Equation (2.22), can be written as

$$\frac{\partial}{\partial t}(\rho\vec{v}) + \nabla \cdot (\rho\vec{v}\vec{v}) = \nabla \cdot \left(-p\vec{\vec{\delta}} + \vec{\vec{\tau}}\right) + \rho\vec{f} \tag{2.23}$$

The terms in Eq. (2.23) can be interpreted as

$$\left\{ \begin{array}{l} \text{Local time rate} \\ \text{of change} \\ \text{of momentum} \end{array} \right\} + \left\{ \begin{array}{l} \text{net change} \\ \text{of momentum} \\ \text{by fluid flow} \end{array} \right\} = \left\{ \begin{array}{l} \text{net pressure and} \\ \text{stress changes acting} \\ \text{on fluid element} \end{array} \right\}$$

$$+ \left\{ \begin{array}{l} \text{gravity and buoyancy} \\ \text{or electromagnetic} \\ \text{forces on fluid element} \end{array} \right\}$$

Newton's law of motion surfaces when Equation (2.23) is rewritten employing the material time derivative (Sect. 1.2):

$$\rho\frac{D\vec{v}}{Dt} = \sum \rho\vec{f}_{\text{surface}} + \sum \rho\vec{f}_{\text{body}} = -\nabla p + \nabla \cdot \vec{\vec{\tau}} + \rho\vec{f}_b \tag{2.24}$$

As discussed in Section 1.3.1, the tensor of rank 2, $\vec{\vec{\pi}} = -p\vec{\vec{\delta}} + \vec{\vec{\tau}}$, has nine components. That is, π_{ij} represents the flux of i-momentum in the j-direction. In general, the total stress tensor is a nonlinear function of the fluid as well as the flow properties. For Newtonian fluids and homogeneous, isotropic flow fields, π_{ij} is linearly related to the deformation-rate tensor $\dot{\gamma}_{ij}$. When the Newtonian fluid is at rest, $\vec{\vec{\pi}}$ reduces to the hydrostatic pressure, which is related to the density ρ, temperature T, and concentration c via a thermodynamic equation of state $p = p(\rho, T, c)$.

From Equation (2.24), assuming constant properties of a Newtonian fluid, we obtain the Navier–Stokes equation

$$\rho\frac{D\vec{v}}{Dt} = -\nabla p + \mu\nabla^2\vec{v} + \rho\vec{f} \tag{2.25a}$$

or

$$\frac{\partial \vec{v}}{\partial t} + (\vec{v} \cdot \nabla)\vec{v} = -\frac{1}{\rho}\nabla p + \nu\nabla^2\vec{v} + \vec{f} \qquad (2.25b)$$

Equation (2.25) implies Stokes's postulates (cf. Sect. 1.3.1)

$$\vec{\vec{\tau}} = \mu[\nabla\vec{v} + (\nabla\vec{v})^{tr}]$$

The three components of the vector equation (2.25) in rectangular, cylindrical, and spherical coordinates are given in Appendix B.

B Conservation of Constituents

The constituents of interest here include

(i)   the fluid mass/(unit volume), that is, density ρ;
(ii) the species mass/(unit volume), that is, concentration c_i of a tracer or solvent following "exactly" the carrier fluid motion; or
(iii) the enthalpy/(unit volume), that is, basically the temperature T.

With respect to (i): $\psi \equiv \rho \triangleq$ density; $\vec{\vec{\Omega}} = \vec{\Sigma} \equiv 0$, that is, no change of the fluid mass due to surface forces and/or internal net sources. Hence,

$$\frac{\partial \rho}{\partial t} + \nabla \cdot (\rho\vec{v}) = 0 \qquad (2.26a)$$

or

$$\frac{D\rho}{Dt} + \rho(\nabla \cdot \vec{v}) = 0 \qquad (2.26b)$$

With respect to (ii):

$$\psi \equiv c_i \triangleq \text{concentration of species } i$$

$$\Omega \equiv \vec{j}_c = \text{mass flux of constituent } c$$

$$\Sigma \equiv S_c \triangleq \text{net production, sink and sources (chem. reactions, sorption, etc.)}$$

$$\frac{\partial c_i}{\partial t} + \nabla \cdot (c_i\vec{v}) = -\nabla \cdot \vec{j}_c \pm S_c \qquad (2.27)$$

For incompressible carrier fluids $\nabla \cdot \vec{v} = 0$, from Equation (2.26), and with Fick's law of molecular diffusion, $\vec{j}_c = -D_m\nabla c$, Equation (2.27) yields

$$\frac{\partial c_i}{\partial t} + (\vec{v} \cdot \nabla)c_i = \nabla \cdot (D_m\nabla c_i) + S_c \qquad (2.28)$$

In comparing the continuity equation (Eq. 2.26) with the mass transfer or convection–diffusion equation (Eq. 2.27), it can be seen that the material density or species concentration, c, varies additionally because of two new terms on the RHS: dispersion and net production of c.

With respect to (iii):

$$\psi \equiv \hat{h} = \rho c_p T \triangleq \text{enthalpy per unit volume}$$

$$\Omega \equiv \vec{q}_H \triangleq \text{heat flux due to conduction, i.e., thermal diffusion}$$

$\Sigma \equiv S_H \triangleq$ net heat source due to possible radiation, fluid friction, etc.

$$\frac{\partial}{\partial t}(\rho c_p T) + \nabla \cdot (\rho c_p T \vec{v}) = \nabla \cdot \vec{q}_H + S_T \tag{2.29}$$

For incompressible fluids $\nabla \cdot \vec{v} = 0$, constant heat capacity c_p, $S_T \sim \Phi(\mu, \nabla \vec{v})$, and with Fourier's law $\vec{q}_H = -k\nabla T$, Equation (2.29) yields:

$$\blacklozenge \qquad \frac{\partial T}{\partial t} + (\vec{v} \cdot \nabla)T = \nabla \cdot (\alpha \nabla T) + \Phi \tag{2.30}$$

Here, $\alpha = k/\rho c_p$ is the thermal diffusivity and Φ is the viscous dissipation function, which is a typical heat source due to internal fluid friction in very viscous, high-shear-rate flow. The physical–mathematical similarity between the heat equation, Equation (2.30), and the convection–diffusion equation, Equation (2.28), is apparent. Another classical similarity exists between the flux vectors $\vec{q}_H$ and $\vec{j}_c$ and the stress tensor $\overset{\leftrightarrow}{\tau}$ for Newtonian fluids:

$$\overset{\leftrightarrow}{\tau} \propto \nabla \vec{v}$$

$$\vec{j}_c \propto \nabla c$$

and

$$\vec{q}_H \propto \nabla T$$

Further simplifications are possible for flow problem with *constant fluid properties*: constant kinematic viscosity or momentum diffusivity ν, mass diffusivity D_M, and thermal diffusivity α. The set of four modeling equations for $\vec{v}$, p as well as T and c to be considered, reads as follows:

Continuity equation	$\nabla \cdot \vec{v} = 0$	(2.31)
Equation of motion	$\dfrac{\partial \vec{v}}{\partial t} + (\vec{v} \cdot \nabla)\vec{v} = -\dfrac{1}{\rho}\nabla p + \nu \nabla^2 \vec{v} + \vec{g}$	(2.32)
Heat transfer equation	$\dfrac{\partial T}{\partial t} + (\vec{v} \cdot \nabla)T = \alpha \nabla^2 T + S_T$	(2.33)
Mass transfer equation	$\dfrac{\partial c}{\partial t} + (\vec{v} \cdot \nabla)c = D_m \nabla^2 c + S_c$	(2.34)

Note

The preceding equations are the most frequently used transport equations for a distributed parameter (or differential modeling approach). The first three of these equations are occasionally referred to as the Navier–Stokes equations. Note that Equation (2.32) is a vector equation and hence can be decomposed into three scalar equations for the velocity components, for example, u, v, w in the rectangular or v_r, v_ϕ, v_z in the cylindrical coordinate system. *Nonlinearities* may appear via the inertia term, $(\vec{v} \cdot \nabla)\vec{v}$, with expressions for transport properties, and in source terms. The equations may be *coupled* and hence have to be solved simultaneously when the convective terms are important and fluid properties are dependent upon temperature and/or concentration (cf. Chapters 4 and 5).

The basic momentum and constituent transport equations in different notations and their reduced forms for applications in thermofluid transfer process modeling are summarized in Appendices B–D.

Table 2.1. *Appropriate boundary conditions for partial differential equations*

Type of boundary condition	Elliptic PDE Laplace, Poisson in (x, y)	Hyperbolic PDE wave equation in (x, t)	Parabolic PDE diffusion equation in (x, t)
Dirichlet: $\alpha v_{\text{wall}} = \gamma$			
Open surface	Insufficient	Insufficient	Unique, stable solution in one direction
Closed surface	Unique, stable solution	Solution not unique	Too restrictive
Neumann: $\beta \dfrac{\partial v}{\partial n}\bigg|_{\text{wall}} = \gamma$			
Open surface	Insufficient	Insufficient	Unique, stable solution in one direction
Closed surface	Unique, stable solution	Solution not unique	Too restrictive
Cauchy: $\alpha v_{\text{wall}} + \beta \dfrac{\partial v}{\partial n}\bigg|_{\text{wall}} = \gamma$			
Open surface	Nonphysical (unstable) results	Unique, stable solution	Too restrictive
Closed surface	Too restrictive	Too restrictive	Too restrictive

2.2.4   Initial and Boundary Conditions

The determination of the correct initial and boundary conditions (BCs) associated with the problem-oriented equations is one of the very important steps in systems analysis. Indeed, the type of boundary conditions of *partial* differential equations (PDEs) may add significantly to the level of problem complexity. Specifically, a change in boundary conditions for a particular PDE already solved may require a new solution approach, whereas new BCs for an *ordinary* differential equation can be readily incorporated into its general solution.

In general, *initial conditions*, required for *transient models*, prescribe the values of all principal variables and/or their derivatives throughout the flow domain at some initial time, normally $t = 0$. In mathematical terms: $\vec{v}(t = 0, \vec{x}) = f(\vec{x})$ where $\vec{x}$ is the spatial coordinate vector.

Boundary conditions determine the magnitudes of variables and/or derivatives on the system's boundaries. In mathematical terms

$$v(x_{\text{wall}}, t) = g \qquad \text{(Dirichlet-type BC)} \tag{2.35a}$$

$$\frac{\partial v}{\partial n}(x_{\text{wall}}, t) = h \qquad \text{(Neumann-type BC)} \tag{2.35b}$$

A third type, the Cauchy boundary condition, is a combination of both. A summary of the relation of these three types of boundary conditions to the three types of two-dimensional partial differential equations is given in Table 2.1.

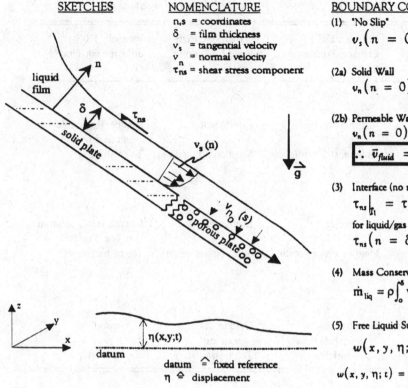

SKETCHES   NOMENCLATURE   BOUNDARY CONDITIONS

NOMENCLATURE

n,s = coordinates
δ = film thickness
v_s = tangential velocity
v_n = normal velocity
τ_{ns} = shear stress component

BOUNDARY CONDITIONS

(1) "No Slip"
$$v_s(n = 0) = 0$$

(2a) Solid Wall
$$v_n(n = 0) = 0$$

(2b) Permeable Wall
$$v_n(n = 0) = v_{n_o}(s)$$
$$\therefore \ \bar{v}_{fluid} = \bar{v}_{phase \ boundary}$$

(3) Interface (no ripples, waves, etc.)
$$\tau_{ns}\big|_{f_1} = \tau_{sm}\big|_{f_2} \ ;$$
for liquid/gas interface
$$\tau_{ns}(n = \delta) \approx 0$$

(4) Mass Conservation
$$\dot{m}_{liq} = \rho \int_o^\delta v_s(n)dn$$

(5) Free Liquid Surface (waves)
$$w(x, y, \eta; t) = \frac{D\eta}{Dt} \ ; \text{or}$$
$$w(x, y, \eta; t) = \frac{\partial\eta}{\partial t} + u\frac{\partial\eta}{\partial x} + v\frac{\partial\eta}{\partial y}$$

datum $\hat{=}$ fixed reference
$\eta \ \hat{=}$ displacement

Fig. 2.4. Illustration of boundary conditions.

Note

Parts of Table 2.1 are simply a matter of maintaining internal consistency. For instance, for Poisson's equation with a closed surface, Dirichlet conditions lead to a unique, stable solution; Neumann conditions likewise lead to a unique, stable solution independent of the Dirichlet solution. However, Cauchy boundary conditions (meaning Dirichlet plus Neumann) could lead to instabilities.

Parabolic PDEs, describing boundary-layer-type or axial-convection-dominated flows, that is, $v(\partial^2 u/\partial x^2) \ll u(\partial u/\partial x)$, commonly dominate fluid mechanics and heat transfer analyses. However, most fluid mechanics research today requires the solution of the *elliptic* Navier–Stokes equations (cf. Sect. 2.3).

Figure 2.4 illustrates appropriate physical–mathematical boundary conditions for the hypothetical case of a viscous liquid moving downward on an inclined plate. It has to be noted that the number of boundary conditions required is equal to the highest order of spatial derivatives in a given differential equation.

2.3   Scale Analysis, Transformations, and Solution Methods

2.3.1   *Scale Analysis*

Scaling can be most helpful in mathematical modeling. Examples of interest include

- the determination of similarity variables (cf. Sect. 2.3.3) with which certain PDEs can be reduced to ODEs,
- the selection of suitable profiles for the integral method (cf. Sect. 2.3.4), and
- the development of dimensionless groups characterizing the process dynamics at hand.

Scale analysis utilizes relative-order-of-magnitude analysis (ROMA), discussed in undergraduate texts on "dimensional analysis." A scaling exercise can produce for certain problems the accurate functional form of the desired system parameters, and sometimes their actual values within an order of magnitude or less (Bejan 1984). Last but not least, scaling results can be directly utilized to plot the final problem solutions efficiently. However, as with most approximation and transformation techniques, scaling requires *physical insight* in order to determine appropriate reference variables and suitable force/flux equations for a particular problem.

Specifically, given a computational domain and the governing partial differential equations describing the transfer processes within this region of interest (cf. Sect. 2.2), the anticipated magnitudes of all terms in these equations are compared for each equation. Some basic knowledge of the spatial extent of the computational domain, the dominant flux or force terms in that region, and the maximum variation of the key variables is required a priori. Thus, system-specific *scales* for length, time, velocity, temperature change, and so on, are assumed and the proportionalities of appropriate terms are established.

Example (1) demonstrates that key dimensionless groups can be directly derived from the governing equations using scale analysis. Example (2) outlines the generic multistep procedure to derive a functional dependence of a key system variable. Further scaling applications are given in Sections 2.3.3 and 2.4 as well as in Chapters 4 and 5.

Example (1)

Employing scale analysis, derive the Reynolds number Re_l from the Navier–Stokes equation in vector form (cf. Eq. (2.32)), and the Peclet number $Pe_d \equiv Re_d\, Pr$ from the heat transfer equation for, say, thermal pipe flow (cf. Eq. (2.33)).

Reynolds Number

Governing equation
$$\frac{\partial \vec{v}}{\partial t} + (\vec{v} \cdot \nabla)\vec{v} = -\frac{1}{\rho}\nabla p + \nu\nabla^2\vec{v} + \vec{g}$$

Definition $Re = \dfrac{\textbf{inertial force}}{\textbf{viscous force}} \triangleq \dfrac{(\vec{v} \cdot \nabla)\vec{v}}{\nu\nabla^2\vec{v}}$

Appropriate scale parameters   average velocity $\bar{u}$ and system length l

Scaling of force ratio $\dfrac{(\vec{v} \cdot \nabla)\vec{v}}{\nu\nabla^2\vec{v}} := \dfrac{\bar{u}(1/l)\bar{u}}{\nu(1/l)^2\bar{u}} = \dfrac{\bar{u}l}{\nu}$

Result   ◆ $Re_l = \dfrac{\bar{u}l}{\nu}$

Peclet Number

Governing equation $\dfrac{\partial T}{\partial t} + (\vec{v} \cdot \nabla)T = \alpha\nabla^2 T + S_T$

Definition $\quad$ Pe $= \dfrac{\text{convection heat transfer}}{\text{conduction heat transfer}} \triangleq \dfrac{(\vec{v} \cdot \nabla) T}{\alpha \nabla^2 T}$

Temperature scale $\overline{T}$, $\qquad$ velocity scale $\bar{u}$, $\qquad$ length scale d

Scaling of ratio $\qquad \dfrac{(\vec{v} \cdot \nabla) T}{\alpha \nabla^2 T} := \dfrac{\bar{u}(1/d)\overline{T}}{\alpha(1/d^2)\overline{T}} = \dfrac{\bar{u}d}{\alpha}$

Result $\quad \blacklozenge \quad$ Pe$_d = \dfrac{\bar{u}d}{\alpha}$

Example (2)

Steady laminar free convection on a vertical isothermal wall is an example of buoyancy-induced (or temperature-driven) flows, that is, upward motion on a vertical heated wall, as discussed in more detail in Section 4.1.3.

Objective

Find δ_{th}/x, that is, an expression for the growth of the dimensionless thermal boundary-layer thickness.

Step 1. System Geometry and Scale Considerations: Define spatial extent of the computational domain and the order of magnitude of the independent variables and dependent variables.

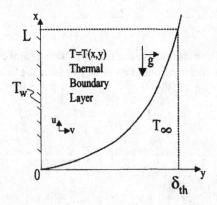

Domain $\qquad 0 \le x \le L \quad$ and $\quad 0 \le y \le \delta_{th}$

Order of Magnitude $\qquad x \sim L \quad$ and $\quad y \sim \delta_{th}$;

$$u(x, y) \sim u \quad \text{and} \quad v(x, y) \sim v$$

Notes

- In compliance with boundary-layer theory we can state $\frac{v}{u} \propto \frac{\delta_{th}}{\ell} \ll 1$.
- The temperature range is $T_\infty \le T \le T_w$, that is, $\Delta T_{\max} = T_w - T_\infty$ and $\Delta T = T - T_\infty$.

Step 2. Reduced Equations and Physical Insight: Governing equations in terms of flux and/or force balances are (cf. Eqs. (2.32) and (2.33))

$$\underbrace{u\frac{\partial T}{\partial x} + v\frac{\partial T}{\partial y}}_{\text{convection}} = \underbrace{\alpha\frac{\partial^2 T}{\partial y^2}}_{\text{conduction}} \tag{E2.1}$$

and

$$\underbrace{u\frac{\partial u}{\partial x} + v\frac{\partial u}{\partial y}}_{\text{inertia}} = \underbrace{\nu\frac{\partial^2 u}{\partial y^2}}_{\text{friction}} + \underbrace{g\beta(T - T_\infty)}_{\text{buoyancy}} \tag{E2.2}$$

Physical Insight

Near the wall, buoyancy and opposing frictional forces are dominant.

Step 3. Scale Analysis: Substitute every variable in the governing equations (E2.1 and E2.2) by its scale stated in Step 1.

$$u\frac{\Delta T}{L}, \ v\frac{\Delta T}{\delta_{\text{th}}} \sim \alpha\frac{\Delta T}{\delta_{\text{th}}^2}$$

$$\left\{\begin{array}{l}\text{heat convection terms}\\ \text{of comparable}\\ \text{magnitude}\end{array}\right\} \propto \left\{\begin{array}{l}\text{heat conduction}\\ \text{term or thermal}\\ \text{diffusivity}\end{array}\right\} \tag{E2.3}$$

and

$$u\frac{u}{L}, \ v\frac{u}{\delta_{\text{th}}} \sim \frac{\nu u}{\delta_{\text{th}}^2}, \ g\beta\Delta T$$

$$\left\{\begin{array}{l}\text{inertia}\\ \text{terms}\end{array}\right\} \propto \left\{\begin{array}{l}\text{friction or}\\ \text{buoyancy}\end{array}\right\} \tag{E2.4}$$

From boundary-layer theory as noted earlier or from scaling of the continuity equation

$$\frac{u}{\ell} \sim \frac{v}{\delta_{\text{th}}} \tag{E2.5}$$

Hence, one term in the energy balance (E2.3) can be disregarded since they are both equivalent.

Step 4. Scaled Force and/or Flux Balance: State force and flux balances by setting appropriate terms of the vector (or scalar) equations proportional to each other. From the scaled heat equation (E2.3)

$$u\frac{\Delta T}{L} \sim \alpha\frac{\Delta T}{\delta_{\text{th}}^2}$$

$$\left\{\begin{array}{l}\text{heat}\\ \text{convection}\end{array}\right\} \propto \left\{\begin{array}{l}\text{thermal}\\ \text{diffusion}\end{array}\right\} \tag{E2.6}$$

or

$$u \sim \frac{\alpha L}{\delta_{\text{th}}^2} \tag{E2.7}$$

Now, $u(x, y)$ has to be eliminated in order to obtain an expression of the form δ_{th}/x. The momentum flux balance (E2.4) is used and either the inertia terms or the friction term is compared with the buoyancy term. For example,

$$\nu \frac{u}{\delta_{th}^2} \sim g\beta\Delta T$$

(E2.8)

(friction) $\propto$ (buoyancy)

is the appropriate force balance near the wall for any fluid. For example, for heavy oils, which implies that $\text{Pr} \gg 1$ and $\delta > \delta_{th}$, this wall layer is of thickness δ_{th}, and Equation (E2.7) together with Equation (E2.8) yields

$$\nu \frac{\alpha L}{\delta_{th}^4} \sim g\beta\Delta T$$

(E2.9)

or

$$\frac{\delta_{th}}{L} \sim \text{Ra}_L^{-1/4}$$

(E2.10)

or

$$\frac{\delta_{th}}{x} \sim \text{Ra}_x^{-1/4}$$

(E2.11)

where

$$\text{Ra}_x = \text{Gr}_x\text{Pr} = \frac{g\beta\Delta T x^3}{\alpha\nu}$$

(E2.12)

is the local Rayleigh number. In other words, the thermal boundary layer δ_{th} grows with $x^{1/4}$ in natural convection of high-*Prandtl-number* fluids on vertical isothermal plates. This outcome coincides with the result obtained from similarity theory (cf. Sect. 2.3.3A).

2.3.2   Transformations

Occasionally, *transformed* equations are easier to solve than their original counterparts. For example, a given partial differential equation may be converted into an equivalent equation for which an analytic solution is known (e.g., Özisik, 1993). In some cases, the numerical effort is significantly reduced when dealing with transformed PDEs and associated boundary conditions (cf. Sects. 2.3.3 and 2.4). The functional relationships of transformation variables can be frequently deduced from dimensional analysis and/or scaling (discussed in the previous section).

Of interest here are the necessary expressions for partial derivatives to be used in transformation procedures. Considering a second-order PDE for $u = u(x, y)$, a coordinate transformation to, say, $u = u(\xi, \eta)$ is desirable, where $\xi = \xi(x, y)$ and $\eta = \eta(x, y)$. Using the notation $\partial u/\partial x = u_x$, $\partial^2 u/\partial x^2 = u_{xx}$, $\partial^2 u/\partial x \partial y = u_{xy}$, and so on, we obtain

♦ $\quad u_x = u_\xi \xi_x + u_\eta \eta_x$ (2.36a)

♦ $\quad u_y = u_\xi \xi_y + u_\eta \eta_y$ (2.36b)

♦ $\quad u_{xx} = u_{\xi\xi}\xi_x^2 + 2u_{\xi\eta}\xi_x\eta_x + u_{\eta\eta}\eta_x^2 + u_\xi\xi_{xx} + u_\eta\eta_{xx}$ (2.36c)

♦ $\quad u_{xy} = u_{\xi\xi}\xi_x\xi_y + u_{\xi\eta}(\xi_x\eta_y + \xi_y\eta_x) + u_{\eta\eta}\eta_x\eta_y + u_\xi\xi_{xy} + u_\eta\eta_{xy}$ (2.36d)

♦ $\quad u_{yy} = u_{\xi\xi}\xi_y^2 + 2u_{\xi\eta}\xi_y\eta_y + u_{\eta\eta}\eta_y^2 + u_\xi\xi_{yy} + u_\eta\eta_{xy}$ (2.36e)

These formulas are applied in the next section on similarity theory and in Chapter 5.

2.3.3 Solution Methods

A number of sophisticated but time-consuming solution methods employing complex transformations have lost their appeal with the advent of cheap and powerful calculators, desktop computers, engineering workstations, and associated numerical software. Still, the *separation-of-variables method* with special function representation, *similarity theory*, and the *methods of weighted residuals* (MWR) are very important in fluid dynamics. In reviewing the similarity theory and the integral method, which is a special case of the MWR, the balanced approach of "physical insight" and "mathematical skills" can be pursued.

First of all, the governing equations and related auxiliary conditions have to be understood and classified fully before any solution approach can be considered. These are summarized in Table 2.2, where Steps 1 to 4 imply sequential approaches of increasing complexity.

A Similarity Theory

Some engineering transport phenomena free of nonuniformities exhibit solution profiles that are similar; that is, when nondimensionalized and plotted versus a suitable similarity or combined variable, they all map onto a *single* profile. There are different approaches for finding such similarity variables that combine two independent variables, say, y and x or y and t. For example, the Buckingham Pi Theorem is a well known method of reducing a number of dimensional variables into a smaller number of dimensionless groups – an important exercise to derive a similarity variable as shown later.

Solutions of certain classes of *boundary-layer type problems* in momentum, heat, and mass transfer are similar; that is, if plotted at different points along a longitudinal axis, their profiles would only differ by a scale factor (Fig. 2.5). Hence, the requirement is that for two arbitrary sections x_1 and x_2, the principal variable (e.g., the longitudinal velocity component $u(x, y)$) must satisfy the equation (cf. Schlichting 1979):

$$\frac{u\{x_1, [y/g(x_1)]\}}{U(x_1)} = \frac{u\{x_2, [y/g(x_2)]\}}{U(x_2)}$$

where $y/g(x) \equiv \eta$ is a dimensionless variable that combines two independent variables: that is, $u/U(\eta)$ collapses to one profile for self-similar flows.

This is generally fulfilled in laminar boundary layer flow without singularities, but with specific $U(x)$ distributions. The two scale factors are (i) $U(x)$, the outer flow obtainable from potential flow theory and (ii) the function $g(x)$, which is proportional to the local boundary-layer thickness $\delta(x)$.

Table 2.2. *Outline of solution methods*[a]

- Types of equations of interest $\begin{cases} \text{Linear or nonlinear algebraic equations as well as integral} \\ \text{equations; IVPs or BVPs; nonlinear systems of ODEs; PDEs:} \\ \text{parabolic, parabolic–hyperbolic, parabolized, or elliptic} \end{cases}$

- Auxiliary conditions $\begin{cases} \text{IC in time at } t = 0 \text{ or in space at } x = 0 \\ \\ \text{Homogeneous or inhomogeneous BC} \begin{cases} \text{Neumann: } \dfrac{\partial u}{\partial n} \\ \text{Dirichlet: } u \\ \text{Cauchy: mixed} \end{cases} \end{cases}$

- Solution approaches

  Step 1:   Classify governing equations

  Type of equation $\begin{cases} \text{Algebraic} \\ \text{ODE} \\ \text{PDE} \\ \text{Integral} \\ \text{Mixed (integro-differential)} \end{cases}$

  Characteristics of, e.g., differential equations: nonlinear, variable coefficients, inhomogeneous, second-order, time dependent; multidimensional (for PDEs)

  Step 2:   Employ transformations and/or approximations. E.g., parabolic PDEs can be often transformed to ODEs via separation of variables, integral method, and/or similarity theory. E.g., asymptotic analysis using perturbation may be sufficient

  Step 3:   Look up textbook solution or solution procedure for a given type and given characteristics of your equation.

  Step 4:   Use appropriate algorithms or computer software for solving
  Linear algebraic systems: Gauss–Seidel, SOR
  Nonlinear algebraic eqs.: Newton–Raphson
  ODEs: IVP, Runge–Kutta
  ODEs: BVP, shooting methods
  Stiff ODEs: Gear
  Non-linear ODEs: special transformations
  PDEs: FDM, FEM, FVM, CVM
  PDEs solvers in CFD: FLUENT, PHOENICS, CFD 2000, FLOW 3-D, CFX, FIDAP, etc.

Note: [a]IVP, initial value problem; BVP, boundary value problem; ODE, ordinary differential equation; PDE, partial differential equation; IC, initial condition; BC, boundary condition; SOR, successive over-relaxation; FDM, finite difference method; FEM, finite element method; FVM, finite volume method; CVM, control volume method; CFD, computational fluid dynamics.

Similarity solutions are of interest since they indicate that it might be possible to transform two-dimensional *partial* differential equations into *one ordinary* differential equation (ODE). Similarity theory deals with several transformation aspects:

- The conditions under which similar solutions exist; these include that the given system be *free of any nonuniformities* (shock waves, curvature effects, flow separation, sinks or sources, and other singularities or nonuniformities) and that the problem can be represented by *parabolic* PDEs.

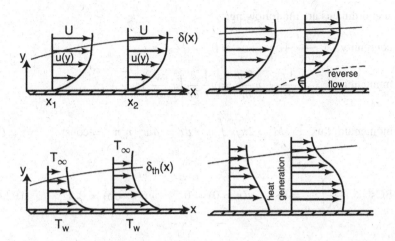

Fig. 2.5. Similar and nonsimilar velocity and temperature profiles.

- The selection of appropriate transformations within a particular coordinate system; specifically, a similarity variable, η, has to be found such that the explicit dependence on the former (two) independent variables disappears. Thus, η, the new independent coordinate, combines, for example, x and y as $\eta = y/g(x)$ into one variable where $g(x)$ could be $\delta(x)$, the boundary-layer thickness. Associated with finding this transformation is the determination of an appropriate similarity function $f(n)$, which, inserted into the two-dimensional PDE, converts the system equation into an ODE for $f(\eta)$.
- The methodologies for implementing the tasks just outlined differ in the aspects they emphasize. For example, with the intuitive–physical approach, the functional forms of the similarity variable $\eta = \eta(x, y)$ or $\eta(x, t)$ or $\eta(r, z)$ and the similarity function $f(\eta)$ are postulated on physical grounds and then nondimensionalized. In more mathematically formal methods, such as the free parameter technique, Buckingham's Pi Theorem, or the Group Theoretic Method, η and f are postulated with unknown exponents or the governing PDE is linearly transformed, again with unknown exponents. Now, requirements of fulfilling the boundary conditions, nondimensionality, and/or invariance of the original and the transformed PDEs generate equations with which the exponents and hence $\eta(x, y)$ and $f(\eta)$ can be determined (cf. Hansen 1964; Na 1979).

Three problem solutions discussed later illustrate the concepts: the intuitive approach relying on functional postulates for physical quantities (cf. Problem 1), the dimensional-analysis approach employing the Pi Theorem (cf. Problem 2), and a formal approach using linear transformations for each independent and dependent variable (cf. Problem 3).

Problem (1)

Consider a free jet far downstream from a small circular orifice of radius r_0 (cf. sketch). The problem-oriented equations for the steady axisymmetric laminar jet in cylindrical coordinates are the following:

continuity $\dfrac{\partial w}{\partial z} + \dfrac{\partial v}{\partial r} + \dfrac{v}{r} = 0$ (P2.1)

momentum $w\dfrac{\partial w}{\partial z} + v\dfrac{\partial w}{\partial r} = v\dfrac{1}{r}\dfrac{\partial}{\partial r}\left(r\dfrac{\partial w}{\partial r}\right)$ (P2.2)

momentum flux $M = 2\pi\rho \displaystyle\int_{o}^{\infty} w^2 r\, dr = \rho w_o^2\left(r_o^2\pi\right) = \text{const}$ (P2.3)

BCs $\left.\dfrac{\partial w}{\partial r}\right|_{r=0} = 0 \qquad v(r=0) = 0 \qquad w(r=\infty) = 0$ (P2.4a–c)

Assumptions

- The thin-shear-layer concept holds: $b/z \ll 1$.
- The velocity profiles $w(z, r)$ are similar.
- The width of the jet grows proportionally to z^n.
- The momentum flux is independent of z: the rate of flow of momentum M is constant across any cross section of the jet.
- The convective and diffusive terms in the z-momentum equation are of the same order of magnitude.

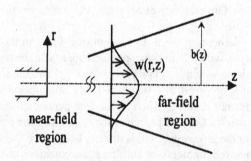

With the key assumption that the jet half-width $b(z) = z^n$, we postulate, similar to $\eta = y/\delta(x)$ in boundary-layer theory, $\eta = \frac{r}{z^n}$ and $\psi \propto z^p f(\eta)$, where ψ is the stream function, that is, $w = (1/r)\frac{\partial\psi}{\partial r}$ and $v = -\partial\psi/\partial z$. It is of interest to note that in the stream function approach, the similarity function $f(\eta)$ is a *dimensionless* stream function so that ψ has to be on dimensional grounds $\psi = (\text{velocity} \times \text{length}) f(\eta)$. The postulates lead in mathematical terms to

$$\frac{1}{r}\frac{\partial\psi}{\partial r} = w \propto z^{p-n} \qquad \frac{\partial w}{\partial z} \propto z^{p-2-1} \quad \text{and} \quad \frac{1}{r}\frac{\partial}{\partial r}\left(r\frac{\partial w}{\partial r}\right) \propto z^{p-2n}$$

Inserting now these proportionalities into the momentum flux expression $M = $ constant and into the z-momentum equation balancing inertial and frictional forces, we equate the powers of z in order to eliminate the z-dependence. Thus,

$$M \sim \int w^2 r\, dr \rightarrow (wr)^2 = \text{const} \quad \text{and} \quad w\frac{\partial w}{\partial z} \sim \frac{1}{r}\frac{\partial}{\partial r}\left(r\frac{\partial w}{\partial r}\right) \tag{P2.5}$$

so that

$$2p - 2n = 0 \quad \text{and} \quad 2p - 2n - 1 = p - 2n$$

or

$$p = n = 1 \tag{P2.6}$$

Hence, appropriately, $\psi = vzf(\eta)$ and with $\eta = r/z$, the velocity components are

$$w = \frac{v}{z}\frac{f'}{\eta} \tag{P2.7}$$

and

$$v = \frac{v}{z}\left(f' - \frac{f}{\eta}\right) \tag{P2.8}$$

Insertion of (P2.7) and (P2.8) into Equation (P2.2), where Equations (P2.1) and (P2.3) are automatically satisfied, should produce a (nonlinear) ordinary differential equation for $f(\eta)$:

$$\frac{ff'}{\eta^2} - \frac{f'^2}{\eta} - \frac{ff''}{\eta} = \frac{d}{d\eta}\left(f'' - \frac{f'}{\eta}\right) \tag{P2.9}$$

Boundary Conditions

$$w(r \rightarrow \infty) \rightarrow 0 \quad : \quad \eta = \infty \quad : \quad \frac{df}{d\eta} = 0 \tag{P2.10}$$

$$v(r = 0) = 0 \quad : \quad \eta = 0 \quad : \quad \frac{f}{\eta} = 0 \tag{P2.11}$$

Notes

- $\lim_{\eta \rightarrow 0}\left(\frac{df/d\eta}{\eta}\right) = w(0, z)z$ should stay finite $\therefore df/d\eta = 0$
- The condition $\partial w/\partial r$ at $r = 0$, i.e., $\lim_{\eta \rightarrow 0}\left(d^2f/d\eta^2\right)$ is not needed.
- If z or r had appeared in the last equation, we would have been forced to try a new postulate for η or abandon the similarity method as a transformation technique altogether.

The LHS of Equation (P2.9) is equivalent to

$$-d\frac{[ff'/\eta]}{d\eta}$$

which integrated yields

$$-\frac{f}{\eta}f'$$

so that Equation (P2.9) now reads

$$-\frac{f}{\eta} f' = f'' - \frac{f'}{\eta}$$

(P2.12)

or

◆ $$f f' = f' - \eta f''$$

(P2.13)

Equation (P2.13) is subject to $f'(0) = 0$, and $f(0) = 0$, which implies finite $w(z)$ on jet centerline and symmetry, $v(r = 0) = 0$. Equation (P2.13) can be solved with a shooting routine (cf. Sect. 2.4 or Na (1982)). Alternatively, multiplying Equation (P2.13) through by $(-\eta)$ yields Euler's equation, which can be solved by using the transformations $\eta = e^{\xi}$ and $F = df/d\xi$.

Schlichting (1979) obtained an analytical solution by using a new independent variable ξ, $\xi = \kappa \eta$; and with the integration constant $\kappa = 3\rho M/16\pi \mu^2$

$$M = \frac{16}{3} \pi \rho \kappa^2 v^2$$

(P2.14)

$$v(r, z) = \frac{1}{4} \frac{3}{\pi} \frac{M}{\rho} \frac{1}{z} \frac{\xi - 1/4\xi^3}{\left(1 + 1/4\xi^2\right)^2}$$

(P2.15)

$$u(r, z) = \frac{3}{4} \frac{M}{\pi \mu} \frac{1}{z} \frac{1}{\left(1 + 1/4\xi^2\right)^2}$$

(P2.16)

The volumetric flow rate is obtained as $Q = 8\pi vz$; that is, it increases linearly with z as a result of the additional entrainment of ambient fluid.

Self-similar *thermal* jets in the laminar and turbulent regime are discussed in Section 4.1.5.

Problem (2)

Consider an ultrathin, semi-infinite flat plate or a plane wall immersed in a viscous fluid. The wall is stationary until $t = 0$, when it is suddenly accelerated to a constant velocity u_0. Find the resulting velocity and pressure field in the fluid induced by the impulsive motion of the plane wall (Stokes's first problem; cf. Sect. 3.2.1). The solid surface moves in the x-direction and y is normal to the surface (cf. sketch).

Assumptions

- $u = u(y, t)$ only; $v = w = 0$
- $\nabla p = (0, -\rho y, 0)$
- Laminar flow with constant fluid properties
- Fully developed flow: $\partial/\partial x = 0$

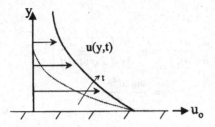

With these assumptions, the Navier–Stokes equations (cf. App. B or C) reduce to the problem-oriented equations

$$u_{,t} = \nu u_{,yy} \tag{P2.17}$$

and

$$p_{,y} = -\rho y \tag{P2.18}$$

subject to $u(y = 0) = u_o$ for $t \geq 0$ and $u(y \to \infty) \to 0$ for all t. We postulate

$$u(y, t) \cong u_o f(\eta) \tag{P2.19a}$$

where f is a dimensionless function and $\eta(y, t)$ is a combined variable. Since the velocity $u = u(y, \nu, t)$, the dimensionless variable, η, can be expressed as

$$\eta \cong y^a \nu^b t^c \tag{P2.19b}$$

We now find the exponents a, b, c so that η is *dimensionless* and thereby combines the two independent variables y and t into one (similarity) variable:

$$\Pi = y^a \nu^b t^c \qquad \text{such that its dimensions } [\Pi] = L^a (L^2 T^{-1})^b (T)^c \equiv 1$$

where L and T are characteristic length and time scales, respectively.

Following *Buckingham's π-Theorem*, the exponents are evaluated from

$$L^a L^{2b} \equiv 1 : a + 2b = 0; \quad \text{and} \quad T^{-b} T^c \equiv 1 : -b + c = 0 \qquad \text{so that } [\Pi] \equiv 1$$

Thus

$$b = c = -\frac{1}{2a}$$

Setting $a = 1$ without loss of generality, we obtain

$$\eta \cong y \nu^{-1/2} t^{-1/2}$$

or with a factor 1/2 for convenience,

$$\eta = \frac{1}{2} \frac{y}{\sqrt{\nu t}} \tag{P2.20}$$

Hence, with $u = u_o f(\eta)$ and following Eq. (2.36a)

$$\frac{\partial u}{\partial t} = u_o \frac{df}{d\eta} \frac{\partial \eta}{\partial t} = -\frac{y u_o}{4\sqrt{\nu}} t^{-3/2} \frac{df}{d\eta}$$

and with Eq. (2.36c)

$$\frac{\partial^2 u}{\partial y^2} = u_o \frac{\partial}{\partial y}\left\{\frac{\partial}{\partial y}[f(\eta)]\right\} = u_o \frac{\partial}{\partial y}\left\{\frac{df}{d\eta}\frac{\partial \eta}{\partial y}\right\} = u_o \frac{\partial}{\partial y}\left\{\frac{1}{2\sqrt{vt}}\frac{df}{d\eta}\right\}$$

$$= \frac{u_0}{4vt}\frac{d^2 f}{d\eta^2}$$

Thus, Eq. (P2.17) is transformed to

$$-\eta \frac{df}{d\eta} = \frac{1}{2}\frac{d^2 f}{d\eta^2}$$

or

♦ $\quad\quad f'' + 2\eta f' = 0$ $\hfill$ (P2.21)

subject to $f(\eta = 0) = 1$ and $f(\eta \to \infty) \to 0$.

Setting $f' \equiv g(\eta)$ and integrating the resulting ODE for $g(\eta)$ yield

$$g(\eta) = C_1 e^{-\eta^2} = f'$$

so that

$$f(\eta) = C_1 \int_0^\eta e^{-\eta^2} d\eta + C_2 \hfill (P2.22)$$

The boundary conditions produce $C_2 = 1$ and $C_1 = [\int_o^\infty e^{-\eta^2} d\eta]^{-1}$ which leads to

$$f(\eta) = 1 - \frac{2}{\pi}\int_0^\eta e^{-\eta^2} d\eta = 1 - erf(\eta) \hfill (P2.23)$$

so that

♦ $\quad\quad u(y,t) = u_0 erfc\dfrac{y}{\sqrt{4vt}}$ $\hfill$ (P2.24)

Note

A finite distance normal to the plate where u approaches 0 indicates the existence of a boundary layer beyond which the outer fluid field is undisturbed by the moving plane wall or flat plate. This thickness or penetration depth is at $\eta \approx 2$. Thus,

$$\eta(y = \delta; t) = 2 = \frac{\delta}{\sqrt{4vt}}$$

or

♦ $\quad\quad \delta = 4\sqrt{vt}$ $\hfill$ (P2.25)

Introducing $t_L = L/U_0$, which is the time a fluid particle takes to travel a distance L (in longitudinal direction), we obtain a new expression for the boundary-layer thickness

$$\delta = 4L\sqrt{\frac{v}{U_0 L}}$$

or

♦
$$\frac{\delta}{L} = \frac{4}{\sqrt{Re_L}}$$
(P2.26)

The general result, $\delta/x \sim Re_x^{-1/2}$, is a basic starting point in laminar boundary-layer theory (cf. Sect. 3.2.3).

Problem (3)

Consider a cylinder of radius r_0 rotating at ω_0 and hence inducing a flow field in a viscous fluid reservoir. Suddenly the cylinder stops and the fluid motion becomes unsteady. Transform the problem-oriented PDE to an ODE for $f(\eta) \sim v_\theta(r, t)$.

Postulating that $v_r = v_z = 0$ and $v_\theta = v_\theta(r, t)$, the θ-momentum equation reads (cf. App. B)

♦
$$\frac{\partial v_\theta}{\partial t} = \nu \frac{\partial}{\partial r} \left[\frac{1}{r} \frac{\partial}{\partial r} (r v_\theta) \right]$$
(P2.27)

Seeking similarity, a *linear transformation* is introduced:

$$t = A^{\alpha_1} \hat{t}, \qquad r = A^{\alpha_2} \hat{r} \quad \text{and} \quad v_\theta = A^{\alpha_3} \hat{v}_\theta$$
(P2.28a–c)

where A is a parameter of transformation and the α_i's are constants. Substituting (P2.28a–c) in (P2.27) yields

$$A^{\alpha_3 - \alpha_1} \frac{\partial \hat{v}_\theta}{\partial \hat{t}} = A^{\alpha_3 - 2\alpha_2} \nu \left[\frac{1}{r} \frac{\partial}{\partial r} \left(r \frac{\partial v_\theta}{\partial r} \right) - \frac{v_\theta}{r^2} \right]$$
(P2.29)

To achieve similarity, we require that (P2.29) and (P2.27) be invariant; that implies that

$$\alpha_3 - \alpha_1 = \alpha_3 - 2\alpha_2 \quad \text{or} \quad \alpha_1 = 2\alpha_2$$
(P2.30a,b)

where α_3 is kept free. Thus, from (P2.28a–c) we form

$$A = \left(\frac{t}{\hat{t}} \right)^{1/\alpha_1} = \left(\frac{r}{\hat{r}} \right)^{1/\alpha_2} = \left(\frac{v_\theta}{\hat{v}_\theta} \right)^{1/\alpha_3}$$

or with (P2.30a, b)

$$\frac{t^{1/\alpha_1}}{r^{1/\alpha_1}} = \frac{t}{r^{\alpha_1/\alpha_2}} = \frac{t}{r^2} = \frac{\hat{t}}{\hat{r}^2} \quad \text{and} \quad \frac{v_\theta}{r^{\alpha_3/\alpha_2}} = \frac{\hat{v}_\theta}{\hat{r}^{\alpha_3/\alpha_2}}$$
(P2.31a,b)

Now, the combined variable $\eta \sim t \cdot r^{-2}$ or $r^2 t^{-1}$, which nondimensionalized may have the form

♦
$$\eta = \frac{r^2}{\nu t}$$
(P2.32a)

Equation (P2.31b) indicates that

$$\frac{v_\theta}{r^{\alpha_3/\alpha_2}} = \frac{v_\theta}{r^{\alpha_3}} := C f(\eta)$$

where $\alpha_2 = 1$ when $\alpha_1 = 2$

The free exponent α_3 has to be such that continuity and the boundary conditions are fulfilled, and the constant C has to make $f(\eta)$ dimensionless. From the boundary conditions

$$v_\theta(r = r_0) = r_0\omega(t) \quad \text{and} \quad v_\theta(r \to \infty) \to 0$$

we deduce that $v_\theta \sim r_0^2\omega/r$, or

♦ $\qquad v_\theta = \dfrac{C}{r} f(\eta)$ (P2.32b)

where $C = r_0^2\omega_0$ and hence $\alpha_3 = -1$. With $v_\theta(r, t) = r_0^2\omega_0/rf(\eta)$ and $\eta = r^2/\nu t$, Eq. (P2.27) is transformed to

♦ $\qquad 4f'' + f' = 0$ (P2.33a)

subject to

$$f(\eta = 0) = 0 \quad \text{and} \quad f(\eta \to \infty) = 1$$ (P2.33b,c)

B   Integral Method

In contrast to separation of variables and similarity theory, the integral method is an *approximation* method. The Von Karman integral method is the most famous member of the family of integral relations, which in turn is a special case of the method of weighted residuals (MWR). Specifically, a transport equation in normal form can be written as (cf. Eq. (2.22))

$$L(\phi) \equiv \frac{\partial \phi}{\partial t} + \nabla \cdot (\vec{v}\phi) - \nu\nabla^2\phi - S = 0$$ (2.37)

where $L(\bullet)$ is a (nonlinear) operator, ϕ is a dependent variable, and S represents sink–source terms. Now, the unknown ϕ-function is replaced by an *approximate* expression, that is, a "profile" or functional $\tilde{\phi}$ that satisfies the boundary conditions, but contains a number of unknown coefficients or parameters. As can be expected,

$$L(\tilde{\phi}) \neq 0, \qquad \text{i.e., } L(\tilde{\phi}) \equiv R$$

where R is the residual. In requiring that

$$\int_\Omega W R \, d\Omega = 0$$

we force the weighted residual over the computational domain Ω to be zero and thereby determine the unknown coefficients or parameters in the assumed $\tilde{\phi}$-function. The type of weighing function W determines the special case of the MWR: for example, integral method, collocation method, Galerkin finite element method, or control volume method (cf. Finlayson 1978; Hirsch 1988).

The Von Karman method is most applicable to laminar–turbulent similar or *nonsimilar* boundary-layer-type flows for which appropriate velocity and temperature profile are known, that is, thin and thick wall shear layers as well as plumes, jets, and wakes. Solutions of such problems yield global or integral system parameters, such as flow rates, fluxes, forces, boundary-layer thicknesses, shape factors, or drag coefficients.

In general, a two-dimensional partial differential equation is integrated in one direction, typically normal to the main flow, and thereby transformed into an ordinary differential equation, which is then solved analytically or numerically. Implementation of the integral method rests on two general characteristics of boundary-layer-type problems: (i) the boundary conditions for a particular system simplify the integration process significantly so that a simpler differential equation is obtained, and (ii) all extra unknown functions, or parameters, remaining in the governing differential equation are approximated on physical grounds or by empirical relationships. Thus, closure is gained by using, for example, the entrainment concept for plumes, jets, and wakes or by expressing velocity and temperature profiles with power expansions for high Reynolds number flows past submerged bodies.

Features of the methods are best illustrated in treating the two-dimensional boundary-layer equations for laminar or turbulent flow past a porous plate (cf. Sect. 3.2.3c). The point of departure is the differential form of the continuity and momentum equations for transient two-dimensional boundary-layer flows, which are first-order approximations of Eq. (2.23) for momentum transfer and of Eq. (2.29) for heat transfer.

$$\frac{\partial u}{\partial x} + \frac{\partial v}{\partial y} = 0 \tag{2.38}$$

$$\frac{\partial u}{\partial t} + u\frac{\partial u}{\partial x} + v\frac{\partial u}{\partial y} = -\frac{1}{\rho}\frac{\partial p}{\partial x} + \frac{1}{\rho}\frac{\partial \tau_{yx}}{\partial y} \tag{2.39}$$

where the axial pressure gradient is related to the outer flow $U(x, t)$ along the boundary-layer edge via Euler's equation, namely,

$$-\frac{1}{\rho}\frac{\partial p}{\partial x} = \frac{\partial U}{\partial t} + U\frac{\partial U}{\partial x} \tag{2.40}$$

The continuity equation can be manipulated in two ways:

(a) Integration of the continuity equation (2.38) yields

$$v(x, y; t) = -\int \frac{\partial u}{\partial x}\,dy + f(x, t) \tag{2.41a}$$

or simply

$$v = -\int_0^y \frac{\partial u}{\partial x}\,dy \tag{2.41b}$$

(b) Multiplication of Equation (2.23) by $(U - u)$, which is the common integrand of the fictitious boundary layer thicknesses, δ_i ($i = 1 \rightarrow$ displacement, $i = 2 \rightarrow$ momentum, and $i = 3 \rightarrow$ energy dissipation thickness).

Recall,

$$\delta_1 = \int_{y=0}^{\infty} \left(1 - \frac{u}{U}\right)dy \tag{2.42a}$$

$$\delta_2 U^2 = \int\limits_{y=0}^{\infty} u(U - u) \, dy \tag{2.42b}$$

$$\delta_3 U^3 = \int\limits_{y=0}^{\infty} u(U^2 - u^2) \, dy \tag{2.42c}$$

Thus the expanded version of the continuity equation then reads

$$U \frac{\partial u}{\partial x} - u \frac{\partial u}{\partial x} + U \frac{\partial v}{\partial y} - u \frac{\partial v}{\partial y} = 0 \tag{2.43}$$

These two possibilities trigger two approaches to derive the desired results.

Approach (i)
Steady, two-dimensional laminar or turbulent boundary layer flows: Integrating the momentum equation (2.39) directly for *steady* flow yields

$$\int\limits_{y=0}^{y=\delta(x)} \left(u \frac{\partial u}{\partial x} + v \frac{\partial u}{\partial y} - U \frac{dU}{dx} \right) dy = \frac{1}{\rho} \int\limits_{0}^{\delta} \frac{\partial \tau_{yx}}{\partial y} \, dy = -\frac{\tau_{\text{wall}}}{\rho} \tag{2.44}$$

Inserting Equation (2.41b) into (2.44) and integrating by parts, $\int u \, dv = uv - \int v \, du$, the single term

$$\int\limits_{0}^{\delta} v \frac{\partial u}{\partial y} \, dy = -\int\limits_{0}^{\delta} \left\{ \frac{\partial u}{\partial y} \int\limits_{0}^{y} \frac{\partial u}{\partial x} \, dy \right\} dy = - \left[U \int\limits_{0}^{\delta} \frac{\partial u}{\partial x} \, dy - \int\limits_{0}^{\delta} u \frac{\partial u}{\partial x} \, dy \right]$$

so that we obtain

$$\int\limits_{0}^{\delta} \left(2u \frac{\partial u}{\partial x} - U \frac{\partial u}{\partial x} - U \frac{dU}{dx} \right) dy = -\frac{\tau_w}{\rho} \tag{2.45a}$$

or

$$\int\limits_{0}^{\delta} \frac{\partial}{\partial x} [u(U - u)] \, dy + \frac{dU}{dx} \int\limits_{0}^{\delta} (U - u) \, dy = \frac{\tau_w}{\rho} \tag{2.45b}$$

Note
Since the integrand of both integrals vanishes outside the boundary layer, it is permissible to replace $\delta(x)$ by ∞. Thus, employing the definitions (2.42a,b), Eq. (2.45b) can be rewritten as

$$\frac{\tau_w}{\rho} = \frac{d}{dx}(U^2 \delta_2) + \delta_1 U \frac{dU}{dx} \tag{2.46}$$

A third approach using the stream function formulation is deferred to a problem assignment (cf. Sect. 2.4).

Approach (ii)
Transient, two-dimensional laminar or turbulent boundary layer flow with wall injection/suction

Addition of Equation (2.41b) to Equation (2.39) and ordering terms yield

$$\frac{\partial}{\partial t}(u - U) + \frac{\partial}{\partial x}(u^2 - uU) + (u - U)\frac{\partial U}{\partial x} + \frac{\partial}{\partial y}(uv - vU) = \frac{1}{\rho}\frac{\partial \tau_{yx}}{\partial y} \quad (2.47)$$

Setting $v = \pm v_{\text{wall}}(x)$, that is, injection–suction, is allowed and noting that for $y \to \infty$: $v \approx 0$, we obtain

$$\frac{\tau_{\text{wall}}}{\rho} = \frac{\partial}{\partial t}\int_0^\infty (U - u)dy + \frac{\partial}{\partial x}\int_0^\infty u(U - u)\,dy + \frac{\partial U}{\partial x}\int_0^\infty (U - u)\,dy - v_{\text{wall}}U$$

or

$$\blacklozenge \quad \frac{\tau_{\text{wall}}}{\rho U^2} \equiv \frac{1}{2}c_f = \frac{1}{U^2}\frac{\partial}{\partial t}(U\delta_1) + \frac{\partial \delta_2}{\partial x} + (2\delta_2 + \delta_1)\frac{1}{U}\frac{\partial U}{\partial x} - \frac{v_{\text{wall}}}{U} \quad (2.48)$$

The mechanical-energy integral relation is derived by multiplying the continuity equation by $(u^2 - U^2)$ and the momentum equation by $(2u)$ and then subtracting them. Integration across the boundary layer yields

$$\frac{2}{\rho}\int_0^\infty \tau\frac{\partial u}{\partial y}\,dy = \frac{\partial}{\partial t}\int_0^\infty u(U - u)\,dy + U^2\frac{\partial}{\partial t}\int_0^\infty \left(1 - \frac{u}{U}\right)dy$$

$$+ \frac{\partial}{\partial x}\int_0^\infty u(U^2 - u^2)\,dy - U^2 v_{\text{wall}} \quad (2.49)$$

$$\text{where} \quad \int_0^\infty \tau\frac{\partial u}{\partial y}\,dy \equiv D$$

is the dissipation integral.

Employing the definition for the energy-dissipation thickness, Equation (2.42c), we obtain

$$C_D \equiv \frac{2D}{\rho U^3} = \frac{1}{U}\frac{\partial}{\partial t}(\delta_2 + \delta_1) + \frac{2\delta_2}{U^2}\frac{\partial U}{\partial t} + \frac{1}{U^3}\frac{\partial}{\partial x}(U^3\delta_3) - \frac{v_{\text{wall}}}{U} \quad (2.50)$$

For a given profile $u(y, x; t)$ and $U(x, t)$ correlations between the δ_i's can be found from Equations (2.42a–c). For example, for $U = $ const and $u(x, y) = Uy/\delta(x)$

$$\delta_1 = \frac{1}{2}\delta; \qquad \delta_2 = \frac{1}{6}\delta; \quad \text{and} \quad \delta_3 = \frac{1}{4}\delta$$

Note:
In order to solve Equations (2.45b) or (2.48) the outer velocity field $U(x, t)$ has to be known from, for example, potential flow theory, a solution of the Euler equation, or measurements. In addition, some functional relationships for the δ_i's (often based on measured data) have

to be postulated. Alternatively, a suitable velocity profile for $u[y, \delta(x)]$, which matches the boundary conditions, has to be postulated in order to reduce a particular momentum integral relation to an ODE for $\delta(x)$. An illustrative example is given here as well as in Sections 3.2.3C, 3.2.3D, 3.2.5B, and 3.3.

Problem (4)

Consider steady laminar high-Reynolds-number flow past a horizontal flat plate with zero pressure gradient. Given $u(x, y) = a + by + cy^2$, employ the momentum integral relation (MIR) to find the boundary-layer thickness $\delta(x)$ and the velocity field $\vec{v} = (u, v, 0)$.

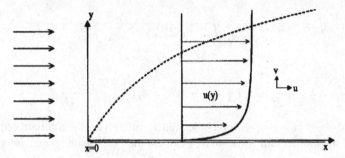

Assumptions

- Steady laminar incompressible 2-D flow with $\partial p / \partial x = 0$
- $U = u_\infty = $ const

Solution

With $U = u_\infty = $ const, Equation (2.45b) reduces to

$$\rho \frac{d}{dx} \int_0^\delta u(u_\infty - u)\, dy = \tau_w = \mu \left. \frac{\partial u}{\partial y} \right|_{y=0} \tag{P2.34}$$

The postulate $u(x, y) = a + by + cy^2$ is a quadratic velocity profile that approximates the (unknown) actual profile. The coefficients are functions of x and have to obey the boundary conditions, that is,

$$u(y = 0) = 0 \qquad u(y = \delta) = u_\infty \quad \text{and} \quad \left. \frac{\partial u}{\partial y} \right|_{y=\delta} = 0 \tag{P2.35a–c}$$

where $\delta = \delta(x)$. Thus,

$$u(x, y) = u_\infty \left[2 \left(\frac{y}{\delta(x)} \right) - \left(\frac{y}{\delta(x)} \right)^2 \right] \tag{P2.36}$$

Inserting Eq. (P2.36) into the MIR (P2.34), carrying out the integrations, and evaluating τ_w, yield

$$0.133 \rho u_\infty^2 \frac{d\delta}{dx} = 2\mu \frac{u_\infty}{\delta}$$

subject to $\delta(x = 0) = 0$. Integration yields

$$\frac{\delta^2}{2} = \frac{\mu x}{0.0665 \rho u_\infty}$$

or

$\blacklozenge$ $$\delta(x) = 5.48\sqrt{\frac{\mu x}{\rho u_\infty}}$$ (P2.37)

which is about 9.6 percent over the exact value. Now, $v(x, y)$ is obtained from the continuity equation, using Eqs. (P2.36) and (P2.37).

2.4   Sample-Problem Solutions

Note
In accordance with the material of Chapter 2, the material in this section emphasizes derivations and tansformations, rather than basic flow problem solutions, which are discussed in Chapter 3. Nevertheless, the classical format of system sketch, problem assumptions, postulates, and solution is adhered to whenever possible.

2.1   Derive the continuity equation in cylindrical coordinates for compressible flow using the control volume approach for a differential fluid element.

Approach
The differential control volume (cf. sketch) is a cylindrical representative elementary volume (REV) given next. The principle of conservation of mass states

$$\left\{ \begin{array}{l} \text{Rate of mass accumulation} \\ \text{per unit volume} \end{array} \right\} = \left\{ \begin{array}{l} \text{net change in mass flow} \\ \text{rate per unit volume} \end{array} \right\}$$

$$\frac{\partial m}{\partial t} = \dot{m}_{\text{in}} - \dot{m}_{\text{out}} \equiv \dot{m}_{\text{net}}$$ (S2.1)

Notes
- $\Delta \forall = \Delta r(r\Delta\theta)\Delta z$ is the REV or control volume.
- $\dot{m}|_{x+\Delta x} = \dot{m}|_x + (\partial m/\partial x)\Delta x$ is the first-order term in Taylor's series expansion.

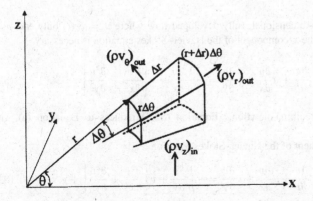

Solution

On the basis of the stated mass balance and in accordance with the sketch we have for 3-D fluid flow

$$\frac{\partial}{\partial t}(\rho \, \Delta V) = \left[(\rho v_z)_z - (\rho v_z)_z - \frac{\partial}{\partial z}(\rho v_z) \, \Delta z\right] \Delta r \cdot r \, \Delta \theta$$

$$+ \left[(\rho v_\theta)_\theta - (\rho v_\theta)_\theta - \frac{\partial}{\partial \theta}(\rho v_\theta) \, \Delta \theta\right] \Delta z \cdot \Delta r + [(\rho v_r)_r] r \, \Delta \theta \, \Delta z$$

$$- \left[(\rho v_r)_r + \frac{\partial}{\partial r}(\rho v_r) \, \Delta r\right] (r + \Delta r) \Delta \theta \, \Delta z$$

Dissolving the brackets, dividing the equation through by $\Delta V = \Delta r(r \Delta \theta)\Delta z$, and neglecting higher-order terms, we obtain

$$\frac{\partial \rho}{\partial t} = -\frac{\partial}{\partial_z}(\rho v_z) - \frac{1}{r}\frac{\partial}{\partial \theta}(\rho v_\theta) - \frac{1}{r}(\rho v_r) - \frac{\partial}{\partial r}(\rho v_r)$$

or

$$\frac{\partial \rho}{\partial t} + \frac{1}{r}\frac{\partial}{\partial r}(r\rho v_r) + \frac{1}{r}\frac{\partial}{\partial \theta}(\rho v_\theta) + \frac{\partial}{\partial z}(\rho v_z) = 0 \tag{S2.2}$$

□

2.2 Starting with the Navier–Stokes equation, state the assumptions and derive each of the following equations.

(a) $u_{,t} + v_0(t)u_{,y} = -\dfrac{1}{\rho}p_{,x} + \nu u_{,yy}$

(b) $\dfrac{d}{dt}v_0(t) = -\dfrac{1}{\rho}p_{,y} + g_y$

(c) $\dfrac{\partial \vec{\zeta}}{\partial t} + (\vec{v} \cdot \nabla)\vec{\zeta} = \dfrac{1}{\text{Re}}\nabla^2\vec{\zeta}$

(d) $\nabla^2\psi = -\zeta$

where $\text{Re}_L = v_0 L/\nu$, ψ is the stream function, and $\zeta = \zeta_z$ is the normalized vorticity for 2-D flow.

Solution

(a) Transient two-dimensional, fully developed flow where $v = v_0(t)$ only and $p_{,x} = $ const; hence, only the x-component of the Navier–Stokes equation is necessary.

$(x\text{-momentum})\quad \dfrac{\partial u}{\partial t} + u\dfrac{\partial u}{\partial x} + v\dfrac{\partial u}{\partial y} = -\dfrac{1}{\rho}\dfrac{\partial p}{\partial x} + \nu\left[\dfrac{\partial^2 u}{\partial x^2} + \dfrac{\partial^2 u}{\partial y^2}\right] \tag{S2.3}$

Thus, employing "comma notation," Equation (S2.3) reduces to Equation (a), since $\partial u/\partial x \equiv 0$.

(b) The y-component of the Navier–Stokes equation reads

$(y\text{-momentum})\quad \dfrac{\partial v}{\partial t} + u\dfrac{\partial v}{\partial x} + v\dfrac{\partial v}{\partial y} = -\dfrac{1}{\rho}\dfrac{\partial p}{\partial y} + \nu\left[\dfrac{\partial^2 v}{\partial x^2} + \dfrac{\partial^2 v}{\partial y^2}\right] + g_y \tag{S2.4}$

Again, since $v = v_0(t)$ only, $\partial v/\partial x = \partial v/\partial y = \partial^2 v/\partial y^2 = \partial^2 v/\partial x^2 \equiv 0$ and Equation (S2.4) reduces to Equation (b).

(c) This vorticity transport equation represents a transient two-dimensional *rotational* flow field. Cross-differentiation of Equations (S2.3) and (S2.4) and subtraction yields, using

$$\tilde{\xi}_z = \frac{\partial \tilde{v}}{\partial \tilde{x}} - \frac{\partial \tilde{u}}{\partial \tilde{y}},$$

$$\frac{\partial \tilde{\xi}_z}{\partial \tilde{t}} + \tilde{u}\frac{\partial \tilde{\xi}_z}{\partial \tilde{x}} + \tilde{v}\frac{\partial \tilde{\xi}_z}{\partial \tilde{y}} = \nu\left[\frac{\partial^2 \tilde{\xi}_z}{\partial \tilde{x}^2} + \frac{\partial^2 \tilde{\xi}_z}{\partial \tilde{y}^2}\right]$$

Note:
The pressure gradients and the body force term dropped out; $\tilde{\xi}_z$ is the only nonzero component of the vorticity vector. Hence, the vorticity equation reads

$$\frac{\partial \tilde{\xi}}{\partial \tilde{t}} + (\vec{v}\cdot\tilde{\nabla})\tilde{\xi} = \nu\tilde{\nabla}^2\tilde{\xi} \qquad \text{where } \vec{v} = (\tilde{u}, \tilde{v}, 0)^{\text{tr}}$$

Introducing the dimensionless variables $\zeta = L\tilde{\xi}/v_0$, $v = \tilde{v}/v_0$, and $\nabla = \tilde{\nabla}/L$, we obtain

$$\left(\frac{v_0}{L}\right)^2\frac{\partial\vec{\zeta}}{\partial t} + \left(\frac{v_0}{L}\right)^2(\vec{v}\cdot\nabla)\vec{\zeta} = \frac{1}{L^2}\frac{v_0}{L}\nu\nabla^2\vec{\zeta}$$

or

$$\frac{\partial\vec{\zeta}}{\partial t} + (\vec{v}\cdot\nabla)\vec{\zeta} = \frac{1}{\text{Re}}\nabla^2\vec{\zeta} \tag{S2.5}$$

where $\quad \text{Re}_L = \dfrac{v_0 L}{\nu}.$

(d) Equation (S2.5) contains two unknowns. A second equation can be generated by using the stream function, which can be obtained for planar flow from the definition

$$u \equiv \frac{\partial\psi}{\partial y} \quad \text{and} \quad v \equiv -\frac{\partial\psi}{\partial x}$$

Hence,

$$\xi_z = \xi = \frac{\partial v}{\partial x} - \frac{\partial u}{\partial y} = \frac{\partial}{\partial x}\left[-\frac{\partial\psi}{\partial x}\right] - \frac{\partial}{\partial y}\left[\frac{\partial\psi}{\partial y}\right] = -\frac{\partial^2\psi}{\partial x^2} - \frac{\partial^2\psi}{\partial y^2}$$

or

$$\zeta = -\nabla^2\psi \tag{S2.6}$$

2.3 Two half-immersed counterrotating rollers produce a liquid film that separates (cf. sketch). State the conditions for velocity and pressure at point $P(x = 0, y = y_1)$.

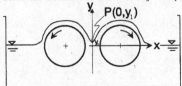

Solution

Assuming that the rollers are the same in diameter, speed, and surface, the separation point $P(0, y_1)$ is similar to a stagnation point representing a mathematical singularity. The conditions are

$$p = p_{atm} \quad \Rightarrow \quad \frac{\partial p}{\partial y} = \frac{\partial p}{\partial x} = 0$$

$$y = y_1 = \text{const} \quad \Rightarrow \quad \frac{\partial v}{\partial y} = 0,$$

and symmetry applies, that is, $\dfrac{\partial v}{\partial x} = 0$. $\qquad\qquad\qquad\qquad\qquad\qquad\square$

2.4 A very viscous liquid in laminar flow moves downward out of a long round tube. After the fluid exits the tube, viscous forces smooth the parabolic exit velocity profile to a uniform value. This happens in a short distance from the exit, and gravity forces are assumed to be negligible. Apply the momentum equation to find the area of the jet where the uniform flow is first established (cf. sketch).

Concept

Free circular jet formation from a vertical tube

Assumptions

- Laminar, steady axisymmetric flow with $z_1 \ll 1$
- Constant fluid properties
- Negligible gravitational forces
- Negligible liquid–air interfacial forces
- Uniform jet profile at $z = z_1$

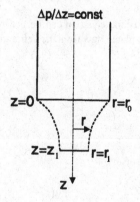

Solution

$$\frac{D(m\vec{v})}{Dt} = \vec{F}_{surf} + \vec{F}_{body} \approx 0 \tag{S2.7}$$

or

$$(m\vec{v})_{\substack{tube \\ exit}} = (m\vec{v})_{jet} = \text{const} \tag{S2.8a}$$

which implies

$$\int_A \rho v_z^2 \, dA = \text{const} \tag{S2.8b}$$

In addition, with $\rho = \text{const}$

$$Q|_{z=0} = Q|_{z=z_1} \tag{S2.8c}$$

The tube exit profile is taken to be

$$v_z(r) = -\left(\frac{\Delta p}{\Delta z}\right)\frac{r_0^2}{4\mu}\left[1 - \left(\frac{r}{r_0}\right)^2\right]$$

From (S2.8c) with the Poiseuille profile given at $z = 0$ and $v_z = v_1 = \text{const}$ at $z = z_1$, we have

$$\left(-\frac{\Delta p}{\Delta z}\right)\frac{\pi\rho r_0^4}{8\mu} = \pi\rho v_1 r_1^2 \tag{S2.9}$$

where r_1 is the radius of the unknown jet area. From (S2.8a)

$$2\pi\rho\int_0^{r_0} v_z^2 r \, dr = 2\pi\rho\int_0^{r_1} v_1^2 r \, dr$$

or

$$\left(-\frac{\Delta p}{\Delta z}\right)^2\frac{\pi\rho r_0^6}{48\mu^2} = \pi\rho v_1^2 r_1^2 \tag{S2.10}$$

Eliminating v_1 using Equation (S2.9) yields with Equation (S2.10)

$$\left(\frac{\Delta p}{\Delta z}\right)^2\frac{r_0^8}{64\mu^2 r_1^4} = \left(\frac{\Delta p}{\Delta z}\right)^2\frac{r_0^6}{48\mu^2 r_1^2}$$

or

$$r_1^2 = \frac{3}{4}r_0^2, \quad \text{i.e.,}$$

$$A_{\text{jet}} = 0.75 A_{\text{tube}}$$

□

2.5 Derive the linear momentum equation in differential form from the Reynolds Transport Theorem.

Solution

$$\iint_{c.s.}\left(-p\vec{\delta} + \vec{\vec{\tau}}\right)d\vec{S} + \iiint_{c.v.}\vec{f}_B \, d\forall = \frac{\partial}{\partial t}\iiint_{c.v.}\vec{v}\rho \, d\forall + \iint_{c.s.}\vec{v}\rho\vec{v}\cdot d\vec{S} \tag{S2.11}$$

Assuming $\rho = \text{const}$ and $\nabla\cdot\vec{\vec{\tau}} = \mu\nabla^2\vec{v}$, applying the divergence theorem

$$\iint\vec{A}\cdot\hat{n}\,dS \equiv \iiint\nabla\cdot\vec{A}\,d\forall \tag{S2.12}$$

to the surface integral terms of Equation (S2.11) yields

$$\iiint_{c.v.} \left\{ \frac{1}{\rho} \nabla \cdot (-p\vec{\delta} + \vec{\vec{\tau}}) + \vec{f}_B - \frac{\partial \vec{v}}{\partial t} - (\vec{v} \cdot \nabla \vec{v}) \right\} d\forall = 0$$

Since equation holds for any REV including $d\forall$, the integrand has to be equal to zero so that

$$\frac{\partial \vec{v}}{\partial t} + (\vec{v} \cdot \nabla)\vec{v} = -\frac{1}{\rho}\nabla p + \nu \nabla^2 \vec{v} + \vec{f}_B \qquad (S2.13)$$

2.6 State the kinematic boundary conditions for free surface flow (cf. sketch).

Concept

$$\left\{ \begin{array}{l} \text{Total displacement rate} \\ \text{of fluid particle at free surface} \end{array} \right\} = \left\{ \begin{array}{l} \text{particle's} \\ \text{upward} \\ \text{velocity} \end{array} \right\}$$

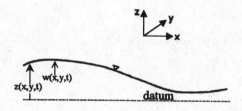

Thus,

$$\frac{Dz}{Dt} = \frac{\partial z}{\partial t} + (\vec{v} \cdot \nabla)z = \frac{\partial z}{\partial t} + u\frac{\partial z}{\partial x} + v\frac{\partial z}{\partial y} = w(x, y, z) \qquad (S2.14)$$

□

2.7 Develop an expression for the wall shear stress, τ_w, in uniform open channel flow (cf. sketch).

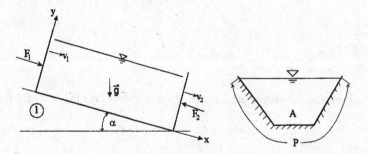

Assumptions

Steady uniform flow in a channel with a small slope

$S = \tan \alpha,$

where $\tan \alpha \approx \sin \alpha; \alpha \ll 1.$

The general force balance, in terms of the momentum equation in integral form,

$$\vec{F}_B + \vec{F}_S = \frac{\partial}{\partial t} \iiint_V \vec{v}\rho\, dV + \iint_V \vec{v}\rho\vec{v} \cdot d\vec{A}$$

reduces to (cf. sketch):

$$F_1 - F_2 + W \sin\alpha - \tau_w P \Delta x = 0 + \dot{m}(v_1 - v_2)$$

where F_1 and F_2 are the hydrostatic pressure forces, W is the fluid weight, and P is the wetted perimeter.
With $v_2 = v_1$, $F_2 = F_1$, and $W = A\Delta x\rho g$,

$$\tau_W = \rho g S \frac{A}{P} \qquad\qquad\qquad\qquad\qquad\qquad\qquad\qquad (S2.15)$$

$\square$

2.8 Discuss the physical meaning of the boundary condition $u''(y = 0) \approx 0$ for steady laminar wall-bounded flow.

Solution
Assuming steady two-dimensional laminar flow, the fluid motion very close to the wall can be approximated by (cf. x-momentum equation)

$$\frac{\partial p}{\partial x} = \nu \frac{\partial^2 u}{\partial y^2}$$

where $\partial p/\partial x \lesseqgtr 0$. Thus $u''(y = 0) \propto \partial\tau_w/\partial y = 0$ implies a zero pressure gradient. $\square$

2.9 State the special condition for internal flow systems to obtain, for example, a pressure relation.

Solution
Conservation of mass in integral form states

$$\dot{m} = \rho\bar{v}A = \int_A \rho u\, dA = \text{const} \qquad\qquad\qquad\qquad\qquad (S2.16a)$$

where $\bar{v}$ is the area-averaged velocity. Since $u = u(p)$, a relation of the form

$$\nabla p = f(\dot{m}) \qquad\qquad\qquad\qquad\qquad\qquad\qquad\qquad (S2.16b)$$

can be established. $\square$

2.10 Derive the kinetic energy equation in the form

$$\frac{D}{Dt}\left(\frac{1}{2}v^2\right) = \dot{W}_{\text{pressure}} + \dot{W}_{\text{shear}} + \dot{W}_{\text{body force}}$$

where $(1/2)\bar{v}^2 = 1/2(u^2 + v^2)$ is the kinetic energy along a streamline and $\dot{W}$ is the rate at which work is done on a unit mass of fluid by the pressure, viscous/turbulent stresses, and body forces per unit volume, respectively.

Solution

The equation that expresses conservation of momentum written in tensor notation is

$$\frac{\partial u_j}{\partial t} + u_k \frac{\partial u_j}{\partial x_k} = \frac{1}{\rho} \frac{\partial \pi_{ij}}{\partial x_i} + f_j \tag{S2.17}$$

If we multiply Equation (S2.17) by u_j, we obtain

$$\frac{1}{2}\left(\frac{\partial u_j^2}{\partial t} + u_k \frac{\partial u_j^2}{\partial x_k}\right) = \frac{u_j}{\rho} \frac{\partial \pi_{ij}}{\partial x_i} + u_j f_j$$

Or, since $\pi_{ij} = -p\delta_{ij} + \tau_{ij}$

$$\frac{D}{Dt}\left(\frac{1}{2}u_j^2\right) = -\frac{u_j}{\rho} \frac{\partial p}{\partial x_j} + \frac{u_j}{\rho} \frac{\partial \tau_{ij}}{\partial x_i} + u_j f_j \tag{S2.18}$$

For incompressible flow of a Newtonian fluid

$$\frac{\partial \tau_{ij}}{\partial x_i} = \mu \frac{\partial^2 u_j}{\partial x_i^2}$$

If we write the kinetic energy equation in the form

$$\frac{D}{Dt}\left(\frac{1}{2}\vec{v}^2\right) = \dot{W}_p + \dot{W}_s + \dot{W}_f \tag{S2.19}$$

then in comparison with (S2.18) we have

$$\dot{W}_{\text{pressure}} = -\frac{u_j}{\rho} \frac{\partial p}{\partial x_j}; \quad \dot{W}_{\text{shear}} = \frac{u_j}{\rho} \frac{\partial \tau_{ij}}{\partial x_i} \quad \text{and} \quad \dot{W}_{\text{body force}} = u_j f_i \qquad \square$$

2.11   Since the governing equations for two-dimensional and axisymmetric boundary-layer flows differ from each other only by the radial distance $r(x, y)$ as sketched below, the axisymmetric flow equations can be placed in a nearly two-dimensional form by using a transformation known as the Mangler transformation. In the case of flow over a body of revolution of radius r_0, where $r_0 = r_0(x)$, we find that if the boundary-layer thickness is small compared with r_0, which implies $r(x, y) \approx r_0(x)$, this transformation puts them exactly into two-dimensional form. Following the problem statement given in Cebeci and Bradshaw (1988), we define the Mangler transformation by

$$d\bar{x} = \left(\frac{r_0}{L}\right)^{2K} dx, \qquad d\bar{y} = \left(\frac{r}{L}\right)^{K} dy \tag{S2.20}$$

to transform an axisymmetric flow with coordinates (x, y) into a two-dimensional flow with coordinates $(\bar{x}, \bar{y})$. In Equation (S2.20) L is an arbitrary reference length and K is the flow index where $K = 0$ indicates 2-D flow and $K = 1$ denotes axisymmetric flow. If a stream function in Mangler variables $(\bar{x}, \bar{y})$ is related to a stream function ψ in (x, y) variables by

$$\bar{\psi}(\bar{x}, \bar{y}) = \left(\frac{1}{L}\right)^{K} \psi(x, y) \tag{S2.21}$$

then

(a)   Show that the relation between the Mangler transformed velocity components $\bar{u}$ and $\bar{v}$ in $(\bar{x}, \bar{y})$ variables and the velocity components u and v in (x, y) variables are

$$u = \bar{u} \qquad\qquad\qquad (S2.22a)$$

and

$$v = \left(\frac{L}{r}\right)^{K}\left[\left(\frac{r_0}{L}\right)^{2K}\bar{v} - \frac{\partial \bar{y}}{\partial x}\bar{u}\right] \qquad\qquad (S2.22b)$$

(b)   By substituting (S2.22) into the axisymmetric flow equations, show that for laminar flows the Mangler-transformed continuity, momentum, and energy equations are

$$\frac{\partial \bar{u}}{\partial \bar{x}} + \frac{\partial \bar{v}}{\partial \bar{y}} = 0$$

$$\bar{u}\frac{\partial \bar{u}}{\partial \bar{x}} + \bar{v}\frac{\partial \bar{u}}{\partial \bar{y}} = -\frac{1}{\rho}\frac{dp}{d\bar{x}} + v\frac{\partial}{\partial \bar{y}}\left[(1+t)^{2K}\frac{\partial \bar{u}}{\partial \bar{y}}\right]$$

$$\bar{u}\frac{\partial T}{\partial \bar{x}} + \bar{v}\frac{\partial T}{\partial \bar{y}} = \frac{k}{\rho c_p}\frac{\partial}{\partial \bar{y}}\left[(1+t)^{2K}\frac{\partial T}{\partial \bar{t}}\right]$$

where $\qquad t = -1 + \left(1 + \dfrac{2L\cos\phi}{r_0^2}\right)^{\frac{1}{2}}$

Note that for $t = 0$, the transformed equations in the $(\bar{x}, \bar{y})$ plane are in exactly the same form as those for two-dimensional flows in the (x, y)-plane (cf. sketch).

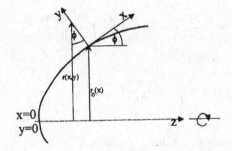

Note

$$r(x, y) = r_0(x) + y\cos\phi$$

where $\qquad \phi = \tan^{-1}\left(\dfrac{dr_0}{dx}\right)$

Solution
If the flow index K is zero, then the transformation becomes trivial. Thus, we consider $K = 1$.
(a)   Recall that $u = (1/r)\partial\psi/\partial y$ and $v = -(1/r)\partial\psi/\partial x$. Hence,

$$ru = \frac{\partial\psi}{\partial y} = \frac{\partial\bar{\psi}}{\partial \bar{y}}\frac{\partial \bar{y}}{\partial y} = L\frac{\partial\bar{\psi}}{\partial \bar{y}}\frac{r}{L} = r\bar{u} \quad\Rightarrow\quad u = \bar{u}$$

$$rv = -\frac{\partial \psi}{\partial x} = -\frac{\partial \psi}{\partial \bar{x}}\frac{\partial \bar{x}}{\partial x} - \frac{\partial \bar{\psi}}{\partial \bar{y}}\frac{\partial \bar{y}}{\partial x} = -L\frac{\partial \bar{\psi}}{\partial \bar{x}}\left(\frac{r_0}{L}\right)^2 - L\frac{\partial \psi}{\partial \bar{y}}\frac{\partial \bar{y}}{\partial x}$$

$$= \bar{v}\frac{r_0^2}{L} - \bar{u}L\frac{\partial \bar{y}}{\partial x} \Rightarrow v = \frac{L}{r}\left[\left(\frac{r_0}{L}\right)^2 \bar{v} - \frac{\partial \bar{y}}{\partial x}\bar{u}\right]$$

(b)   The axisymmetric thermal flow equations are

$$\frac{\partial(ru)}{\partial x} + \frac{\partial(rv)}{\partial y} = 0 \tag{S2.23}$$

$$u\frac{\partial u}{\partial x} + v\frac{\partial u}{\partial y} = -\frac{1}{\rho}\frac{dp}{dx} + \frac{\nu}{r}\frac{\partial}{\partial y}\left(r\frac{\partial u}{\partial y}\right) \tag{S2.24}$$

$$u\frac{\partial T}{\partial x} + v\frac{\partial T}{\partial y} = \frac{\alpha}{r}\frac{\partial}{\partial y}\left(r\frac{\partial T}{\partial y}\right) \tag{S2.25}$$

The term-by-term transformations are

$$\frac{\partial r}{\partial x} = \frac{\partial}{\partial x}(r_0 + y\cos\phi) = \frac{dr_0}{dx} + y\frac{d(\cos\phi)}{dx}$$

$$\frac{du}{dx} = \frac{\partial \bar{u}}{\partial \bar{x}}\frac{d\bar{x}}{dx} + \frac{\partial \bar{u}}{\partial \bar{y}}\frac{\partial \bar{y}}{\partial x} = \left(\frac{r_0}{L}\right)^2\frac{\partial \bar{u}}{\partial \bar{x}} + \frac{\partial \bar{y}}{\partial x}\frac{\partial \bar{u}}{\partial \bar{y}}$$

$$\frac{\partial(ru)}{\partial x} = \bar{u}\left[\frac{dr_0}{dx} + y\frac{d(\cos\phi)}{dx}\right] + r\left[\left(\frac{r_0}{L}\right)^2\frac{\partial \bar{u}}{\partial \bar{x}} + \frac{\partial \bar{y}}{\partial x}\frac{\partial \bar{u}}{\partial \bar{y}}\right]$$

$$\frac{\partial(rv)}{\partial y} = L\frac{\partial}{\partial y}\left[\left(\frac{r_0}{L}\right)^2\bar{v} - \frac{\partial \bar{y}}{\partial x}\bar{u}\right] = \frac{r_0^2}{L}\frac{\partial \bar{y}}{\partial y}\frac{\partial \bar{v}}{\partial \bar{y}} - L\bar{u}\frac{\partial}{\partial y}\left(\frac{\partial \bar{y}}{\partial x}\right) - L\frac{\partial \bar{y}}{\partial y}\frac{\partial \bar{u}}{\partial \bar{y}}\frac{\partial \bar{y}}{\partial y}$$

Now,

$$d\bar{y} = \frac{r}{L}dy = \left(\frac{r_0}{L} + \frac{y\cos\phi}{L}\right)dy \Rightarrow \bar{y} = \frac{r_0}{L}y + \frac{\cos\phi}{2L}y^2$$

so that

$$\frac{\partial}{\partial y}\left(\frac{\partial \bar{y}}{\partial x}\right) = \frac{\partial}{\partial y}\left(\frac{y}{L}\frac{dr_0}{dx} + \frac{y^2}{L}\frac{d\cos\phi}{dx}\right) = \frac{1}{L}\frac{dr_0}{dx} + \frac{y}{L}\frac{d\cos\phi}{dx}$$

$$\therefore \quad \frac{\partial(rv)}{\partial y} = \frac{r_0^2}{L}\frac{r}{L}\frac{\partial \bar{v}}{\partial \bar{y}} - \bar{u}\left[\frac{dr_0}{dx} + y\frac{d(\cos\phi)}{dx}\right] - r\frac{\partial \bar{y}}{\partial x}\frac{\partial \bar{u}}{\partial \bar{y}}$$

Thus,

$$\frac{\partial(ru)}{\partial x} + \frac{\partial(rv)}{\partial y} = r\left(\frac{r_0}{L}\right)^2\left(\frac{\partial \bar{u}}{\partial \bar{x}} + \frac{\partial \bar{v}}{\partial \bar{y}}\right) = 0$$

or

$$\frac{\partial \bar{u}}{\partial \bar{x}} + \frac{\partial \bar{v}}{\partial \bar{y}} = 0 \tag{S2.26}$$

Now,

$$\frac{\partial u}{\partial y} = \frac{r}{L}\frac{\partial \bar{u}}{\partial \bar{y}}$$

$$\frac{\partial}{\partial y}\left(r\frac{\partial u}{\partial y}\right) = \frac{\partial r}{\partial y}\frac{\partial u}{\partial y} + r\frac{\partial^2 u}{\partial y^2} = \frac{r}{L}\frac{\partial \bar{u}}{\partial \bar{y}}\cos\phi + r\left[\frac{\cos\phi}{L}\frac{\partial \bar{u}}{\partial \bar{y}} + \left(\frac{r}{L}\right)^2\frac{\partial^2 \bar{u}}{\partial \bar{y}^2}\right]$$

and Equation (S2.24) becomes

$$\bar{u}\left[\left(\frac{r_0}{L}\right)^2\frac{\partial \bar{u}}{\partial \bar{x}} + \frac{\partial \bar{y}}{\partial x}\frac{\partial \bar{u}}{\partial \bar{y}}\right] + \frac{L}{r}\left[\left(\frac{r_0}{L}\right)^2\bar{v} - \frac{\partial \bar{y}}{\partial x}\bar{u}\right]\frac{r}{L}\frac{\partial \bar{u}}{\partial \bar{u}}$$

$$= -\frac{1}{\rho}\frac{dp}{d\bar{x}}\left(\frac{r_0}{L}\right)^2 + \frac{v}{r}\left[\frac{r}{L}\frac{\partial \bar{u}}{\partial \bar{y}}\cos\phi + \frac{r\cos\phi}{L}\frac{\partial \bar{u}}{\partial \bar{y}} + r\left(\frac{r}{L}\right)^2\frac{\partial^2 \bar{u}}{\partial \bar{y}^2}\right]$$

Note

$$\frac{\partial p}{\partial y} = 0 \quad \Rightarrow \quad \frac{\partial p}{\partial \bar{y}}\frac{r}{L} = 0 \quad \Rightarrow \quad \frac{\partial p}{\partial \bar{y}} = 0$$

$$\therefore \quad \bar{u}\frac{\partial \bar{u}}{\partial \bar{x}} + \bar{v}\frac{\partial \bar{u}}{\partial \bar{y}} = -\frac{1}{\rho}\frac{dp}{d\bar{x}} + v\left[\left(\frac{r}{r_0}\right)^2\frac{\partial^2 \bar{u}}{\partial \bar{y}^2} + \frac{2L\cos\phi}{r_0^2}\frac{\partial \bar{u}}{\partial \bar{y}}\right]$$

If we let

$$(1+t)^2 = 1 + \frac{2L\cos\phi}{r_0^2}\left(\frac{r_0}{L}y + \frac{\cos\phi}{2L}y^2\right) = 1 + 2\frac{y\cos\phi}{r_0} + \frac{y^2\cos^2\phi}{r_0^2} = \left(\frac{r}{r_0}\right)^2$$

then we may write the momentum equation (S2.24) as

$$\bar{u}\frac{\partial \bar{u}}{\partial \bar{x}} + \bar{v}\frac{\partial \bar{u}}{\partial \bar{y}} = -\frac{1}{\rho}\frac{dp}{dx} + v\frac{\partial}{\partial \bar{y}}\left[(1+t)^2\frac{\partial \bar{u}}{\partial \bar{y}}\right] \qquad\qquad \text{(S2.27)}$$

With

$$\frac{\partial T}{\partial x} = \left(\frac{r_0}{L}\right)^2\frac{\partial T}{\partial \bar{x}} + \frac{\partial \bar{y}}{\partial x}\frac{\partial T}{\partial \bar{y}}$$

$$\frac{\partial T}{\partial y} = \frac{r}{L}\frac{\partial T}{\partial \bar{y}}$$

$$\frac{\partial}{\partial y}\left(r\frac{\partial T}{\partial y}\right) = \cos\phi\frac{\partial T}{\partial y}r\left(\frac{\cos\phi}{L}\frac{\partial T}{\partial \bar{y}} + \frac{r^2}{L^2}\frac{\partial^2 T}{\partial \bar{y}^2}\right)$$

the energy equation (S2.25) becomes

$$\bar{u}\left[\left(\frac{r_0}{L}\right)^2\frac{\partial T}{\partial \bar{x}} + \frac{\partial y}{\partial x}\frac{\partial T}{\partial \bar{y}}\right] + \frac{L}{r}\left[\left(\frac{r_0}{L}\right)^2\bar{v} - \frac{\partial \bar{y}}{\partial x}\bar{u}\right]\frac{r}{L}\frac{\partial T}{\partial \bar{y}} = \alpha\left[\frac{2\cos\phi}{L}\frac{\partial T}{\partial \bar{y}} + \frac{r^2}{L^2}\frac{\partial^2 T}{\partial \bar{y}^2}\right]$$

$$\bar{u}\frac{\partial T}{\partial \bar{x}} + \bar{v}\frac{\partial T}{\partial \bar{y}} = \alpha\left[\frac{2\cos\phi}{r_0^2}\frac{\partial T}{\partial \bar{y}}\frac{r^2}{r_0^2} + \frac{\partial^2 T}{\partial \bar{y}^2}\right]$$

or

$$\bar{u}\frac{\partial T}{\partial \bar{x}} + \bar{v}\frac{\partial T}{\partial \bar{y}} = \alpha\frac{\partial}{\partial \bar{y}}\left[(1+t)^2\frac{\partial T}{\partial \bar{y}}\right] \qquad\qquad \text{(S2.28)}$$

where $\quad t = -1 + \left[1 + \frac{2L\cos\phi}{r_0^2}\bar{y}\right]^{1/2}$ $\qquad\qquad\qquad\qquad$ $\square$

2.12   Derive on physical grounds the steady two-dimensional momentum and thermal boundary-layer equations using the control volume approach (cf. sketch) and assuming that $\partial^2/\partial y^2 \gg \partial^2/\partial x^2$.

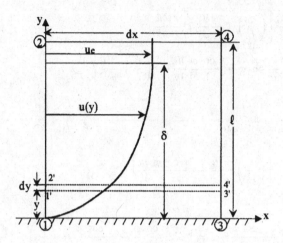

Solution

(i)   Assuming steady two-dimensional flow and constant properties, the *mass flow* rate through surface $1'\text{–}2'$ is $(\rho u\,dy)$. This rate changes along the length dx by $(\partial/\partial x(\rho u\,dy)\,dx)$, so that the mass flow rate leaving surface $3'\text{–}4'$ is $(\rho u\,dy + \partial/\partial x(\rho u)\,dx\,dy)$. The excess of mass flow rate leaving surface $3'\text{–}4'$ over the one entering surface $1'\text{–}2'$ is therefore $(\rho(\partial u/\partial x)\,dx\,dy)$. In the same way, the excess of mass flow rate leaving surface $2'\text{–}4'$ over that entering surface $1'\text{–}3'$ is $(\rho(\partial v/\partial y)\,dx\,dy)$. Since no mass is created or destroyed within the differential control, volume element $1'\text{–}2'\text{–}3'\text{–}4'$, the mass balance yields $(-\rho(\partial u/\partial x)\,dx\,dy) - (\rho(\partial v/\partial y)\,dx\,dy) = 0$. That is, the continuity equation for incompressible fluid flow is

$$\frac{\partial u}{\partial x} + \frac{\partial v}{\partial y} = 0 \tag{S2.29}$$

(ii)   We now consider the *force balance* on the volume element in the x-direction. The momentum flux through $1'\text{–}2'$ is $(\rho u^2\,dy)$. The momentum flux through $3'\text{–}4'$ is then $(\rho u^2\,dy + \partial/\partial x(\rho u^2)\,dx\,dy)$. The excess of the momentum leaving surface $3'\text{–}4'$ over that entering $1'\text{–}2'$ for constant density is thus $(\rho\partial u^2/\partial x\,dx\,dy)$. There is also a transport of x-momentum through $1'\text{–}3'$ which is $(\rho uv\,dx)$, because the mass flow rate $\rho v dx$ carries x-momentum when the velocity has a component u. The corresponding momentum transport through $2'\text{–}4'$ is then again $(\rho uv\,dx + \partial/\partial y(\rho uv)\,dx\,dy)$, and the excess of x-momentum leaving through $2'\text{–}4'$ over that entering through $1'\text{–}3'$ is $(\rho\partial/\partial y(uv)\,dx\,dy)$. These excesses of momentum fluxes have now to be balanced with the forces acting on the volume element. In the x-direction there are pressure forces acting on $1'\text{–}2'$ and $3'\text{–}4'$. The pressure force on $1'\text{–}2'$ is $p\,dy$, and on $3'\text{–}4'$ is $(p dy + \partial p/\partial x\,dy\,dx)$. The difference between the forces on $1'\text{–}2'$ and $3'\text{–}4'$ is $(-\partial p/\partial x\,dy\,dx)$. This is the net pressure force in the x-direction. Similarly, the net shear stress force is $\partial\tau/\partial y dx dy = \mu\partial^2 u/\partial y^2\,dx\,dy$, acting in the positive x-direction. Equating the excess of momentum flux with the forces in the position x-direction results in

$$\rho\left(\frac{\partial u^2}{\partial x} + \frac{\partial uv}{\partial y}\right) = -\frac{\partial p}{\partial x} + \mu\frac{\partial^2 u}{\partial y^2}$$

Considering that

$$\frac{\partial u^2}{\partial x} + \frac{\partial uv}{\partial y} = u\frac{\partial u}{\partial x} + v\frac{\partial u}{\partial y} + \underbrace{\left(\frac{\partial u}{\partial x} + \frac{\partial v}{\partial y}\right)}_{\equiv 0}$$

The final form of the momentum boundary-layer equation is

$$u\frac{\partial u}{\partial x} + v\frac{\partial u}{\partial y} = -\frac{1}{\rho}\frac{\partial p}{\partial x} + v\frac{\partial^2 u}{\partial y^2} \qquad (S2.30)$$

(iii) Considering an *energy balance* for the control volume, the flow entering the volume element through $1'$–$2'$ carries with it heat in the form of internal energy. This energy transport per unit time is $(\rho c u T dy)$.

Note
For a perfect gas $d\hat{u} = c_v\,dT$ and $d\hat{h} = c_p\,dT$, whereas for an incompressible fluid $d\hat{u} = c\,dT$ and $d\hat{h} = c\,dT + (1/\rho)dp$.
   The energy transport convected out of the volume element through $3'$–$4'$ is $(\rho c(uT)dy + \rho c\partial/\partial x(uT)\,dx\,dy)$. This represents the difference of the heat convection leaving through $3'$–$4'$ over that entering through $1'$–$2'$ as $(\rho c\partial/\partial x(uT)\,dx\,dy)$. In the same way, the excess of the heat transport leaving through $2'$–$4'$ over that entering through $1'$–$3'$ is $(\rho c\partial/\partial y(uT)\,dy\,dx)$. The excess conductive heat leaving through $2'$–$4'$ over that entering through $1'$–$3'$ is $(-k\partial^2 T/\partial y^2\,dx\,dy)$. In general, there will also be heat conduction through $1'$–$2'$ and $3'$–$4'$.
   However, in the boundary layer, these fluxes are of much smaller order of magnitude because the temperature variation in the x-direction is much less than that in the y-direction. Correspondingly, these flux terms are neglected in the thermal equation. For the energy balance, we also need the amount of work that is converted in the volume element into heat. The mechanism by which this is accomplished is internal friction. The work performed per unit time by the shear stress acting on $1'$–$3'$ is $(\tau u\,dx)$. Correspondingly, the work performed by the shear stress on $2'$–$4'$ is $(\tau u\,dx + \partial/\partial y(\tau u)\,dx\,dy)$. The difference of both terms is the net work performed by shear stresses, one acting in the positive and the other in the negative x-direction. This work term can be written as

$$\frac{\partial(\tau u)}{\partial y}dx\,dy = u\frac{\partial\tau}{\partial y}dx\,dy + \tau\frac{\partial u}{\partial y}dx\,dy$$

The first of the two terms describes frictional work converted, together with work done by pressure forces, into kinetic energy, as can be seen when the momentum equation is multiplied by u. The second term then expresses energy that is converted within the volume element into heat. This term must therefore be considered in the heat balance. Equating the excesses of the convective and conductive heat flux leaving the volume element over that entering it with the internal heat generation term results in

$$\rho c\left(\frac{\partial(uT)}{\partial x} + \frac{\partial(vT)}{\partial y}\right) = k\frac{\partial^2 T}{\partial y^2} + \tau\frac{\partial u}{\partial y}$$

But,

$$\frac{\partial}{\partial x}(uT) + \frac{\partial}{\partial y}(vT) = u\frac{\partial T}{\partial x} + v\frac{\partial T}{\partial y} + T\left(\frac{\partial u}{\partial x} + \frac{\partial v}{\partial y}\right) \quad \text{and} \quad \tau = \mu\frac{\partial u}{\partial y}$$

Therefore the thermal-boundary-layer equation can be written as

$$\rho c\left(u\frac{\partial T}{\partial x} + v\frac{\partial T}{\partial y}\right) = k\frac{\partial^2 T}{\partial y^2} + \tau\left(\frac{\partial u}{\partial y}\right)^2 \tag{S2.31}$$

□

2.13 Show that for an incompressible zero-pressure gradient flow over a wall at uniform temperature, the integral energy relation can be expressed as

$$\frac{d\theta_T}{dx} = \text{St} \tag{S2.32a}$$

where $\quad \theta_T = \int\limits_0^\infty \frac{u}{u_e}\left(\frac{(T - T_e)}{T_x - T_e}\right)dy \quad$ and $\quad \text{St} = \frac{q_w}{\rho c_p(T_w - T_e)u_e}$

with constant edge quantities u_e and T_e (cf. sketch).

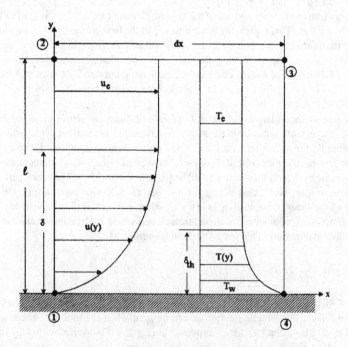

Assumptions

- Steady, two-dimensional flow with constant fluid properties
- Isothermal wall and constant boundary-layer edge conditions $l > \delta > \delta_{th}$ (cf. sketch)

Fluxes

- Rate of energy input through 1–2

$$\rho c \int_0^l uT\,dy$$

- Rate of energy output through 3–4

$$\rho c \int_0^l uT\,dy + \rho c \frac{d}{dx}\left[\int_0^l uT\,dy\right]dx$$

- Rate of energy input through the wall 1–4

$$q_w\,dx$$

- Flow rate of mass entering through 2–3

$$\rho \frac{d}{dx}\left[\int_0^l u\,dy\right]dx$$

- Rate of energy input through 2–3

$$\rho c \frac{d}{dx}\left[\int_0^l uT_e\,dy\right]dx$$

Energy Balance

$$\rho c \frac{d}{dx}\left[\int_0^l uT_e\,dy\right]dx + q_w\,dx = \rho c \frac{d}{dx}\left[\int_0^l uT\,dy\right]dx$$

$$\frac{d}{dx}\left[\int_0^l u(T - T_e)\,dy\right] = \frac{q_w}{\rho c}$$

Since u_e, T_e, and T_w are constant

$$\frac{d}{dx}\left[\int_0^l \frac{u}{u_e}\frac{T - T_e}{T_w - T_e}\,dy\right] = \frac{q_w}{\rho c u_e(T_w - T_e)} \quad\Rightarrow\quad \frac{d\theta_T}{dx} = \mathrm{St} \qquad \text{(S2.32b)}$$

$\square$

2.14 Derive the governing equations for the following two problems:

(a) Derive the transient mechanical energy equation considering a macroscopic balance for kinetic, potential, and (Helmholtz) free energy.

Solution

To obtain the macroscopic mechanical energy balance, that is, Bernoulli's equation, we could start

(i)   with the mechanical energy equation in differential form and integrate it over the volume of the system at hand; or

(ii)   with a control volume of (the) system and balance macroscopical energy fluxes considering potential, kinetic, and free energy only.

For approach (i) see Bird (1957). In approach (ii) use total energy balance for control volume (cf. sketch).

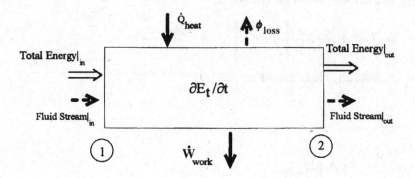

Notation

$$E_t = E_{kin} + E_{pot} + E_{H\,free}$$

where $$E_{kin} = \int \frac{\rho}{2} v^2 d\forall, \qquad E_{pot} = \int \rho g d\forall$$

and

$$E_{Hf} = \int \rho (\hat{u} - Ts) d\forall$$

Now, when $\dot{W}$ represents the rate of work performed by the system, ϕ represents the irreversible friction loss, $\dot{m}$ represents the mass flow rate, and $G = H - TS$, we can write the global mechanical energy balance for isothermal laminar flow in the form

$$\frac{\partial}{\partial t} E_t = \Delta \left[\left(\frac{v^2}{2} + \frac{E_{pot}}{m} + \frac{G}{m} \right) \dot{m} \right] - \dot{W} - \phi \qquad (S2.33)$$

□

(b)   Consider two horizontal parallel disks of radius r_0, a small distance $2h_0$ apart. The gap is completely filled with a non-Newtonian (power-law) fluid, which is very slowly squeezed out of the gap (cf. sketch). Postulating that $v_r = v_r(r, z)$, $v_z = v_z(z)$, and $p = p(r)$ only, set up the problem-oriented equations. This problem was first suggested by Bird et al. (1987).

Assumptions

For quasi-steady, two-dimensional flow of a power-law fluid, we assume that inertia forces and normal stresses are negligible.

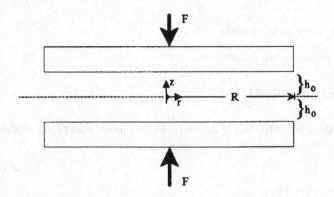

Solution

With the given postulates, the continuity equation and the momentum equation in terms of τ_{ij} (cf. App. B) reduce to cylindrical coordinates

$$\frac{1}{r}\frac{\partial}{\partial r}(rv_r) + \frac{dv_z}{dz} = 0 \tag{S2.34}$$

$$0 = -\frac{dp}{dr} - \frac{d\tau_{rz}}{\partial z} \tag{S2.35}$$

and

$$0 = -\frac{1}{r}\frac{\partial}{\partial r}(r\tau_{rz}) \tag{S2.36}$$

where the shear stress is *nonlinearly* related to the velocity gradients. The power-law model reads (cf. Sect. 1.3.1, Eq. (1.20b,c))

$$\tau_{rz} = \eta\dot{\gamma}_{rz} := m(\dot{\gamma}_{rz})^{(n-1)}\dot{\gamma}_{rz} = m\dot{\gamma}_{rz}^n \tag{S2.37a}$$

$$\text{where} \qquad \dot{\gamma}_{rz} = \frac{\partial v_r}{\partial z} + \underbrace{\frac{\partial v_z}{\partial r}}_{\approx 0} \tag{S2.37b}$$

$\square$

2.15 Consider the one-dimensional transient diffusion equation representing heat conduction (thermal diffusion) with source term or mass diffusion with chemical reaction. Focusing on the latter,

$$\frac{\partial c}{\partial t} = \frac{\partial}{\partial x}\left[D(c)\frac{\partial c}{\partial x}\right] + R(c) \qquad 0 < x < \infty \quad \text{and} \quad t > 0 \tag{S2.38}$$

IC $\quad c(x,0) = h(x);$ $\qquad$ BC $\quad c(0,t) = g(t)$ $\quad$ and $\quad c(\infty,t) = c_\infty$

Now, convert the nonlinear PDE to an ODE and suggest a method of solution.

Solution

The semiinfinite domain suggests *similar* profiles for $c(x,t)$ as a solution of the parabolic PDE (S2.38).

Postulates

$\hat{t} = a^{\alpha}t, \quad \hat{x} = a^{\beta}x \quad$ and $\quad \hat{c} = a^{\gamma}c$

Inserting the postulates into (S2.38) yields

$$a^{\alpha-\gamma}\frac{\partial\hat{c}}{\partial\hat{t}} = a^{2\beta-\gamma}\frac{\partial}{\partial\hat{x}}\left[D\left(a^{-\gamma}\hat{c}\right)\frac{\partial\hat{c}}{\partial\hat{x}}\right] + R(a^{-\gamma}\hat{c}) \tag{S2.39}$$

For this system to be conformally invariant, the transformation must be independent of a, that is, implies

$$\alpha - \gamma = 2\beta - \gamma \tag{S2.40}$$

where we set $\gamma = 0$. Thus

$$\alpha = 2\beta$$

Note:

In general, $D(c)$ and $R(c)$ are not compatible. We recall from Section 2.3.1 that

$$\eta = \frac{1}{2}x(D_0 t)^{-\beta/\alpha} \quad \text{and} \quad f(\eta) = \frac{c(x,t)}{t^{\gamma/\alpha}}$$

Hence

$$\eta = \frac{1}{2}x(D_0 t)^{-1/2} \quad \text{and} \quad f(\eta) = c(x,t) \tag{S2.41a,b}$$

Transformation of BC:

$c(\hat{x}a^{-\beta}, 0) = h(\hat{x}a^{-\beta}), \quad$ i.e., function of x only

$c(\infty, a^{-\alpha}\hat{t}) = c_{\infty} \quad$ i.e., generally fct (t) only

As a result

$$c(\hat{x}a^{-\beta}, 0) = c(\infty, a^{-\alpha}\hat{t}) \equiv c_{\infty} = \text{const}$$

and

$$c(0, \hat{t}a^{-\alpha}) = g(\hat{t}a^{-\alpha}) = g(t) = c_1$$

Note

The initial condition $h(x)$ is actually suppressed since $c(t = 0)$ and $c(x = 0)$ are incompatible. The reaction term is dropped since $R(c^n)$ would require $D(c) = \text{const}$.

Finally, with

$$C = \frac{c - c_{\infty}}{c_1 - c_{\infty}}, \qquad \eta = \frac{x}{\sqrt{4D_0 t}} \quad \text{and} \quad K(C) = \frac{D(C)}{D_0} \tag{S2.42a-c}$$

we obtain

$$\blacklozenge \qquad \frac{d}{d\eta}\left[K(C)\frac{dC}{d\eta}\right] + 2\eta\frac{dC}{d\eta} = 0 \tag{S2.43}$$

subject to

$$C(\infty) = 0 \quad \text{and} \quad C(0) = 1$$

In order to simplify the solution of (S2.43), we introduce the Kirchhoff transformation

$$\psi = \int_0^C K(C')dC'$$

$$\therefore \quad \frac{d\psi}{d\eta} = K(C)\frac{dC}{d\eta} \quad \text{and} \quad \frac{d^2\psi}{d\eta^2} = \frac{d}{d\eta}\left[K(C)\frac{dC}{d\eta}\right]$$

If we can write $K(\psi) := K(C)$, that is, invert the transformation for a given $K(C)$, we have

◆ $$K(\psi)\frac{d^2\psi}{d\eta^2} + 2\eta\frac{d\psi}{d\eta} = 0 \qquad\qquad\qquad\qquad\qquad (S2.44)$$

□

2.16 Consider two-dimensional similar flow past a flat porous plate (Blasius flow with wall mass transfer, i.e., fluid injection or suction). Using the stream function approach, $\psi(x, y) = \sqrt{U\nu x}\, f(x, \eta)$, implies that $u(x, y) = Uf'$ and $v(x, y) \sim f$. One relevant boundary condition is

$$f(x, 0) = f_w = -\frac{1}{\sqrt{U\nu x}}\int_0^x v_w\, dx$$

Find the functional dependence for the wall flux $v_w(x)$ so that *similarity* is preserved.

Solution
Similarity requires that $f_w = $ const; otherwise an explicit x-dependence would appear in the governing equation (cf. Sect. 2.3.3) and the $v_w(x)$-function would create nonsimilar velocity profiles. Thus

$$f_w \sim \frac{1}{\sqrt{x}}\int_0^x x^n dx = \text{const} \qquad \text{where } x^n \sim v_w(x)$$

Hence,

$$\frac{x^{n+1}}{\sqrt{x}} = c \quad \Rightarrow \quad n = -\frac{1}{2}$$

or

◆ $$v_w \sim x^{-1/2}$$

□

2.17 The velocity distribution of fully developed flow of an incompressible, viscous fluid moving in the (streamwise) x-direction through a horizontal conduit, was found to be $u = a[1 - (y/b)^2 - (z/c)^2]$ where a, b, and c are constants.
(a)   Determine mathematically the shape of the conduit.
(b)   Derive $u(y, z)$ from the Navier–Stokes equations and evaluate a in terms of the "driving force."
(c)   Make a reasonable assumption and calculate the pressure gradient.

Solution
The shape of the conduit is revealed by imposing the "no-slip" boundary condition.

$$u_{wall} = 0 \quad \text{implies} \left(\frac{y}{b}\right)^2 + \left(\frac{z}{c}\right)^2 = 1$$

This is the equation of an ellipse. It can be shown (cf. Problem Solution 2.18) that the x-momentum equation for fully developed duct flow ($\partial u/\partial x \equiv 0$) reads

$$\frac{\partial^2 u}{\partial y^2} + \frac{\partial^2 u}{\partial z^2} = \frac{1}{\mu}\frac{d\hat{p}}{dx} = \text{const} \tag{S2.45}$$

Note: This is Poisson's equation, which can be normalized by using

$$\tilde{y} = \frac{y}{l}, \qquad \tilde{z} = \frac{z}{l} \quad \text{and} \quad \tilde{u} = \mu u \left(\frac{d\hat{p}}{dx}\right)$$

so that (S2.45) reads

$$\tilde{\nabla}^2 \tilde{u} = 1$$

subject to

$$\tilde{u}|_{wall} = 0 \tag{S2.46}$$

Inserting the given solution $u(y, t)$ into Equation (S2.45) yields by comparison

$$a = -\frac{1}{2\mu}\left(\frac{d\hat{p}}{dx}\right)\frac{(bc)^2}{b^2 + c^2}$$

where the constant axial pressure gradient is the driving force. For incompressible duct flow, mass conservation can be expressed as

$$Q = \bar{u}A = \text{const}$$

Thus,

$$\frac{d\hat{p}}{dx} = -\frac{4\pi Q}{\pi}\frac{(b^2 + c^2)}{(bc)^2} \tag{S2.47}$$

□

2.18 Consider laminar flows in horizontal conduits such as circular pipes and slits (semi-infinite channels, porous tubes, etc.).

(a) Of interest is steady laminar developing flow in a circular, porous wall duct with constant wall mass flux (cf. sketch). Develop the problem-oriented equations and transform the resulting PDEs into a single ODE using a new variable $\zeta = (r/r_0)^2$.

Assumptions

- Steady, axisymmetric laminar flow with constant fluid properties
- Wall Reynolds number $\text{Re}_w = v_w r_0/\nu \leq 1.0$

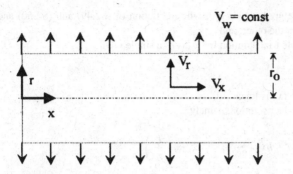

Postulates

Physical insight, suitable boundary conditions, and the continuity equation are utilized to postulate the mathematical dependence of the velocity field and the pressure on the independent variables.

(i)   Boundary conditions:

at $r = r_0$ $v_x = 0$ and $v_r = \pm v_w$

at $r = 0$ $\dfrac{\partial v_x}{\partial r} = 0$ and $v_r = 0$

(ii)   Axisymmetric flow

$v_\theta = 0$ and $\dfrac{\partial}{\partial \theta} \equiv 0$

Note: Considering the boundary conditions, axisymmetric flow condition and the fact that v_w is constant, we postulate that $v_r = v_r(r)$ only.

(iii)   Continuity equation:

$$\frac{1}{r}\frac{\partial}{\partial r}(r v_r) + \frac{\partial v_x}{\partial x} = 0 \tag{S2.48}$$

Since $v_r = v_r(r)$ only, v_x has to change *linearly* with x in order to conserve mass (cf. Eq. (S2.48)).

$$\therefore v_x = v_x(r, x); \qquad v_r = v_r(r) \quad \text{and} \quad v_\theta = 0; \qquad p = p(r, x)$$

Hence, the Navier–Stokes equations in cylindrical coordinates (cf. App. B.2) can be reduced to

$$\rho v_r \frac{\partial v_r}{\partial r} = \mu \frac{\partial}{\partial r}\left(\frac{1}{r}\frac{\partial}{\partial r}(r v_r)\right) - \frac{\partial p}{\partial r} \tag{S2.49}$$

and

$$\rho\left(v_r \frac{\partial v_x}{\partial r} + v_x \frac{\partial v_x}{\partial x}\right) = \mu \frac{1}{r}\frac{\partial}{\partial r}\left(r \frac{\partial v_x}{\partial r}\right) - \frac{\partial p}{\partial x} \tag{S2.50}$$

Transformation

Since it is a steady axisymmetric problem, we take the stream function approach in order to

• satisfy the continuity equation (S2.48) automatically;

- eliminate the pressure gradient via cross-differentiation of (S2.49) and (S2.50) and subtracting one PDE from the other; and
- generate with a suitable trial solution for $\psi(x, r)$ a single ODE.

Notes
- $\psi(x)$ is linear since $v_x(x)$ is linear;
- (xr) is split into a product solution, namely,

$$\psi = A(x) \cdot B(r) = (a + bx) \cdot g(\zeta); \qquad \zeta = \left(\frac{r}{r_0}\right)^2$$

Recall
A volumetric fluid flow balance between pipe entrance $(x = 0)$ and any station x allows us to determine the constants a and b since lines of constant ψ are streamlines, that is,

$$Q_{1\to2} = \int_1^2 (\vec{v} \cdot \hat{n}) \, dA = 2\pi(\psi_2 - \psi_1)$$

or

$$\psi(x, r) = \left[r_0^2 \bar{u}(0) - 2r_0 x v_w\right] g(\zeta) \tag{S2.51}$$

where the factor π has been absorbed in the g-function and $\bar{u}(0)$ is the average inlet velocity. From the definition of the stream function in cylindrical coordinates

$$v_r = -\frac{1}{r}\frac{\partial\psi}{\partial x} \quad \text{and} \quad v_x = \frac{1}{r}\frac{\partial\psi}{\partial r}$$

we have

$$v_r = z v_w \frac{g}{\sqrt{\zeta}} \tag{S2.52a}$$

and

$$v_x = 2\left[\bar{u}(0) - \frac{2x v_w}{r_0}\right] g' \tag{S2.52b}$$

where $'$ denotes $d/d\zeta$.
    Inserting (S2.51) into the stream function equation (cf. App. D)

$$\frac{\partial\psi}{\partial r}\frac{\partial\psi}{\partial x}\nabla^2\psi - \frac{\partial\psi}{\partial x}\frac{\partial}{\partial x}\nabla\psi = \nu\nabla^4\psi \tag{S2.53}$$

yields

$$2\zeta g^{IV} + 4g''' + \text{Re}_w(g'g'' - gg''') = 0 \tag{S2.54a}$$

subject to

$$g(1) = 1 \quad \text{and} \quad g'(1) = 0 \tag{S2.54b,c}$$

Solution

This initial value problem (cf. Sect. 2.3.3) can be solved with a standard Runge–Kutta routine (cf. App. F.1). □

(b) Of interest is transient, fully developed slit flow (cf. sketch). Determine the long-time solution of the flow under a pressure gradient that varies harmonically with time.

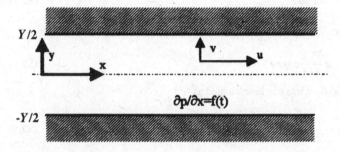

Assumptions
- Transient, fully developed, 1-D laminar flow
- Constant fluid properties

Postulates

Since $v = 0$ at the walls and at the centerline, v is zero everywhere in this symmetric flow field. From continuity, $\partial u/\partial x + \partial v/\partial y = 0$, it follows that $\partial u/\partial x = 0$, which implies fully developed flow, that is, no entrance effects; hence $u = u(y, t)$. From the given pressure gradient we deduce that

$$\frac{\partial p}{\partial x} \sim \cos \omega t \quad \Rightarrow \quad p(x, t) = (P_0 \cos \omega t)x \tag{S2.55}$$

Solution

$$(x\text{-}\,momentum) \quad \frac{\partial}{\partial t}u(y, t) = -\frac{1}{\rho}\frac{\partial}{\partial x}p(x, t) + v\frac{\partial^2}{\partial y^2}u(y, t) \tag{S2.56}$$

where $\quad u\left(y = \left\{ \begin{array}{c} Y/2 \\ -Y/2 \end{array} \right. \right) = 0 \quad$ for all t.

Nondimensionalization

$$\tilde{y} = \frac{y}{Y}; \qquad \tilde{u} = \frac{u}{U}; \qquad \tilde{t} = \frac{t}{\tau}; \quad \text{and} \quad \tilde{p} = \frac{p}{\rho U^2}$$

where Y is the channel width, U is a characteristic velocity to be determined, and τ is the large observation time, that is, $\tau \gg 1$. Insertion into (S2.56) yields

$$\frac{Y}{U\tau}\frac{\partial \tilde{u}}{\partial \tilde{t}} = -\frac{Y P_0}{\rho U^2}\cos \omega t + \frac{v}{UY}\frac{\partial^2 \tilde{u}}{\partial \tilde{y}^2} \tag{S2.57}$$

Selecting $U \equiv Y^2 P_0/\mu$ brings (S2.57) into the form

$$\frac{\mu}{\tau P_0 Y}\frac{\partial \tilde{u}}{\partial \tilde{t}} = -\cos \omega t + \frac{\partial^2 \tilde{u}}{\partial \tilde{y}^2}$$

which reduces for $\tau \gg 1$ to

$$\frac{\partial^2 \tilde{u}}{\partial \tilde{y}^2} = \cos \omega t \tag{S2.58a}$$

subject to

$$\tilde{u}\begin{pmatrix} +1/2 \\ -1/2 \end{pmatrix} = 0 \quad \text{for all } t \tag{S2.58b}$$

Now, separating variables and invoking the boundary conditions yields

$$\tilde{u}(\tilde{y}, t) = \frac{\tilde{y}^2}{2} \cos \omega t - \frac{1}{8} \cos \omega t \tag{S2.58c}$$

The dimensionless flow rate can be evaluated as

$$2 \int_0^{1/2} \tilde{u}(\tilde{y}, t) \, d\tilde{y} = -\frac{1}{2} \cos \omega t = \frac{\bar{u}}{U} \tag{S2.59a}$$

where the mean velocity at any time t is

$$\bar{u} = \frac{1}{12} \frac{Y^2 P_0}{\mu} \cos \omega t = \frac{P_0}{6\rho\omega} \left(\frac{Y}{\delta}\right)^2 \cos \omega t \quad \text{with} \quad \delta = \sqrt{\frac{2\nu}{\omega}} \tag{S2.59b}$$

Interpretations

(i) $y/\delta \to 0$, that is, $\nu \to \infty$ or $\omega \to 0 \Rightarrow \bar{u}(t) \to 0$: viscous forces damp the oscillating forcing function; and

(ii) $y/\delta \to \infty$, that is, $\nu \to 0$ or $\omega \to \infty \Rightarrow \bar{u}(t) \to \pm\infty$: the pressure force dominates. $\quad \square$

2.19 Describe the flow of a film that results when a cylindrical jet of liquid impinges on the vertex of a right circular cone as depicted in the sketch. If the s-coordinate is in the direction of flow, and n is the (normal) distance into the film from the liquid–air interface, compute (a) the liquid velocity distribution $v_s(n)$, that is, $v_s = \text{fct}(\rho, g, \mu, \beta, \delta, n)$; (b) the volumetric flow rate Q; and (c) the dependence of the film thickness δ on the distance s from the cone apex (cf. Bird et al. 1960).

Assumptions

• Steady, axisymmetric laminar flow
• Thin, fully developed liquid film
• Constant fluid properties; no end effects

Postulates
The thin film approximation implies $\partial p / \partial n \approx 0$ and at the interface

$$\tau_{ns}\big|_{n=0} = -\mu \frac{\partial v_s}{\partial n}\bigg|_{n=0} \approx 0$$

Because of symmetry, $v_\theta = 0$ and $\partial/\partial\theta = 0$. With $v_n(n = \delta) = v_n(n = 0) = 0$, we have $v_n = 0$ everywhere, which implies fully developed flow and $v_s = v_s(n)$ only.

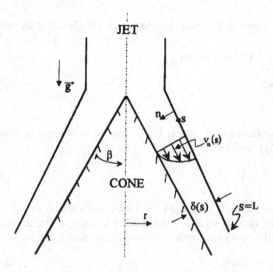

Governing Equations

On the basis of the postulates, the Navier–Stokes equations reduce to

$$0 = \mu \frac{d^2 v_s}{dn^2} + \rho g \cos \beta \tag{S2.60}$$

subject to

$$v_s(n = \delta) = 0 \quad \text{and} \quad \mu \frac{dv_s}{dn}\bigg|_{n=0} = 0$$

Integrating and involving the boundary conditions yield

$$\blacklozenge \qquad v_s = \frac{\rho g \delta^2 \cos \beta}{2\mu} \left[1 - \left(\frac{n}{\delta} \right)^2 \right] \tag{S2.61}$$

The volumetric flow rate $Q = \bar{v} A = \text{const}$ is

$$Q = \frac{1}{\delta} \int_0^\delta v_s (r \pi \delta)\, dn \tag{S2.62a}$$

where $r = s \sin \beta$ and $\sin \beta \cdot \cos \beta = 1/2 \sin 2\beta$, so that

$$Q = \frac{\pi \rho g (\delta^2 s) \sin \beta \cos \beta}{\mu} \int_0^1 \left[1 - \left(\frac{n^2}{\delta} \right) \right] d\left(\frac{n}{\delta} \right) = \frac{\pi \rho g L \sin 2\beta}{3\mu} \delta_L^3 \tag{S2.62b}$$

where $\delta_L = \delta(s = L)$. Note that $(\delta^2 s)$ is constant!

$$\blacklozenge \qquad \delta(s) = \left[\frac{3\mu Q}{\pi \rho g L \sin 2\beta} \left(\frac{L}{s} \right) \right]^{1/3} \tag{S2.63}$$

□

2.20 Show the invariance of the following PDEs and transform them to ODEs using the similarity method.

(a) Consider the transient 1-D diffusion equation

$$\frac{\partial}{\partial t}u(x,t) - \frac{\partial^2}{\partial x^2}u(x,t) = 0 \tag{S2.64}$$

Solution
Using the (stretching) transformation $\tilde{x} = \varepsilon x$, $\tilde{t} = \varepsilon^2 t$, $\tilde{u} = u$, where ε is a constant parameter, Equation (S2.64) takes on the form

$$\frac{\partial \tilde{u}}{\partial \tilde{t}} - \frac{\partial^2 \tilde{u}}{\partial \tilde{x}^2} = \varepsilon^{-2}\left(\frac{\partial u}{\partial t} - \frac{\partial^2 u}{\partial x^2}\right) \tag{S2.65}$$

It is apparent that Equation (S2.65) is unchanged under the transformation up to ε^{-2}. This factor can be eliminated by introducing a new independent variable, η, which *combines,* the coordinates x and t and is also an invariant of the stretching transformation. We take

$$\eta = \tilde{\eta} = \frac{x}{\sqrt{t}} \quad \text{and} \quad f(\eta) = u(x,t) \tag{S2.66a,b}$$

Inserting (S2.66) into (S2.64) yields

$$f'' + \frac{\eta}{2}f' = 0 \tag{S2.67}$$

where the primes denote differentiation with respect to (w.r.t.) η. Integration and back substitution yield the similarity solution

$$u(x,t) = C_1 \int_0^{x/\sqrt{t}} \exp(-\eta^2)\,d\eta + C_2 \tag{S2.68}$$

$\square$

(b) Consider Burger's equation, which describes transient, one-dimensional convection and axial momentum diffusion

$$\frac{\partial u}{\partial t} + u\frac{\partial u}{\partial x} = \nu\frac{\partial^2 u}{\partial x^2} \tag{S2.69}$$

Typically, the viscous term is small when compared with the convective term. Equation (S2.69) describes traveling waves, among other phenomena, and hence we assume a similarity variable of the form $\eta = x - ct$ and a similarity function $f(\eta) = u(x,t)$. This reduces Equation (S2.69) to an ODE

$$-cf'(\eta) + f(\eta)f'(\eta) = \nu f''(\eta) \tag{S2.70}$$

for which an analytic solution exists (cf. Ames 1968 or Zwillinger 1992). $\square$

2.21 Consider a boiler into which cold liquid is continuously fed and partially evaporated by superheated steam as illustrated (cf. sketch). Derive rate equations for the amount of vapor produced and the heat required. This two-phase mass transfer process is accompanied by heat transfer; all necessary constant system parameters are available. It is required to compute $\Psi_{\text{liquid}}(t)$, $m_{\text{gas}}(t)$, $T(t)$, and $q_{\text{steam}}(t)$.

Concept
Employ the Reynolds Transport Theorem to compute global system parameters.

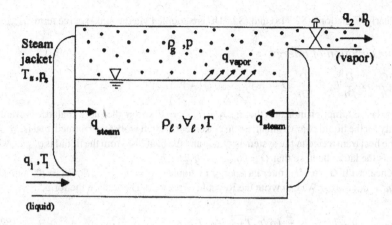

(liquid)

Assumptions

- Both phases, that is, the vapor and the liquid compartments, are completely mixed (no spatial gradients) at all times.
- Densities, velocities, and fluxes are independent of geometric parameters (i.e., volumes and cross-sectional areas).
- Thermodynamic equilibrium exists at all times between the two phases.

Basic Equations (cf. Sect. 2.1.1)

$$\frac{D}{Dt} G_{syst.} = \frac{\partial}{\partial t} \int_{c.\forall.} \rho \eta \, d\forall + \int_{c.s.} \rho \eta \vec{v} \cdot d\vec{A}$$

Conservation of Mass

$$\frac{DG_s}{Dt} \equiv 0$$

For liquid phase $G_s \equiv \rho_\ell \quad \therefore \quad \eta \equiv 1$

With $\rho_\ell = \rho = $ constant

$$0 = \frac{d}{dt}(\rho \forall)_l - \rho q_1 + \dot{m}_{evap}$$

or

$$\left. \frac{d\forall}{dt} \right|_{liq} = q_1 - q_{vapor} \tag{S2.71a}$$

For vapor phase $G_s = \rho_g \therefore \eta \equiv 1$

$$0 = \frac{d}{dt} m_{gas} - q_{vapor}\rho_g + q_2\rho_g$$

or

$$\frac{dm_g}{dt} = (q_{vapor} - q_2)\rho_{gas} \tag{S2.71b}$$

Clearly, Equations (S2.71a) and (S2.71b) are coupled via the sink–source term, q_{vapor}.

Conservation of Energy

$$\frac{DG_s}{Dt} \equiv \frac{D}{Dt} E_s = \frac{D}{Dt} (Q - W)|_{syst.} = (\dot{Q} - \dot{W})$$

The last assumption indicates that $T_{liquid} = T_{vapor} = T$ so that the energy balance is evaluated only for the liquid compartment; no displacement or shaft work is performed, that is, $\dot{W} = 0$. The heat transferred to the system is Q_{steam} and the heat loss from the liquid is $q_{vap}h_v$, where h_v is the latent heat, so that $\dot{Q} = (q_{steam} - q_{vap}h_v)\rho_\ell$.

Hence, with $G_s \equiv U_l$ (internal energy of liquid), $\eta = \hat{u} = c_p T$, $\dot{W} = 0$, and $\dot{Q} = (q_{st} - q_{vap}h_v)\rho_\ell$. We can write the Reynolds Transport Theorem in the form

$$(q_{steam} - q_{vap.}h_v)\rho_\ell = \frac{d}{dt}(\rho \forall c_p T)_{liq.} - c_p T_1 q_1 \rho_1 \tag{S2.72}$$

Note that $\int_{c.s.} \rho \eta \vec{v} \cdot d\vec{A} \equiv -\rho_l c_p T_1 q_1$, the influx of thermal energy due to the flowing stream of (cold) liquid.

With constant liquid density the third tailored equation reads

$$\blacklozenge \quad c_p \frac{d}{dt}(T\forall) = c_p T_1 q_1 + q_{st} - q_{vap}h_v \tag{S2.73}$$

Now, the coupled first-order rate equations (S2.71a), (S2.71b), and (S2.73) have to be integrated (cf. App. F.3) to obtain the required results. $\square$

Problem Assignments

Follow the basic format for problem solutions: sketch, assumptions, postulates, derivations, problem-oriented equations, plus boundary conditions; solutions and graphs, if required.

2.1 Consider a cylindrical element in r–θ space. Derive the continuity equation for steady 2-D incompressible flow in polar coordinates.

2.2 Consider steady laminar flow through a vertical circular tube in r–z coordinates. Set up a momentum balance for a control volume, that is, a cylindrical fluid shell of radius r, thickness Δr, and length $\Delta z = \ell$. Derive and plot the momentum flux τ_{rz} and the velocity v_z. Calculate the volumetric flow rate Q and the force of the fluid on the wetted surface of the tube, that is, the drag force F_z.

2.3 Consider a smooth steady liquid film of constant thickness on an inclined plate without end effects. Focusing on a thin slice of the moving fluid parallel to the plate, perform a momentum balance to find the shear stress and the velocity profile. In case the fluid viscosity is a function of position, for example, $\mu = \mu_0 \exp[-\alpha(y/\delta)]$ where α is a constant, δ is the film thickness, and y is the coordinate normal to the plate, compute the velocity profile and the average velocity.

2.4 Reduce the Navier–Stokes equations for steady laminar fully developed flow through an annulus of inner radius κR and outer radius R where $\kappa < 1$.

(a) Compute the shear stress, the axial velocity profile, the drag force, and the location, that is, the surface of zero shear stress.

(b) Plot the velocity profile and the shear stress for $\kappa R \leq r \leq R$.

2.5 The momentum equation in integral form for a stationary control volume around fixed deflectors or hydraulic structures can be written as

$$\sum \vec{F}_{pressure} + \sum \vec{F}_{stress} + \vec{F}_{gravity} = \iint_{c.s.} \rho \vec{v} (\vec{v} \cdot d\vec{A})$$

Alternatively, with uniform velocity and density distributions over both the inlet and outlet sections with indexes (1) and (2), respectively, we can write

$$\vec{P}_1 + \vec{P}_2 + \vec{R} + \vec{F}_G = m(\vec{v}_2 - \vec{v}_1)$$

where $\vec{P}_i = p_i \vec{A}_i$ and $\vec{R}$ is a resultant force due to wall shear stress and the pressure normal to the wall.

(a) Consider a bent diffusion that changes both the velocity and the flow direction in the x–y plane. Find the resultant (or dynamic) force, $R = \sqrt{R_x^2 + R_y^2}$, exerted by the moving fluid on the bend.

(b) Consider water flowing over a dam-spillway structure. Compute the horizontal force, R_x, the dam has to withstand.

2.6 Set up and solve the transient 1-D diffusion equation for $c(x, t)$ where at time $t = 0$ a slug of dye is concentrated at $x = 0$ in the middle of a tube filled with a stagnant carrier fluid. Plot $c(x)$ for different times $0 \leq t \leq \infty$.

2.7 Consider the mass transfer equation for atmospheric diffusion of a pollutant c_i from a smokestack or an area source (cf. Seinfeld 1984; Eskinazi 1975)

$$\frac{\partial c_i}{\partial t} + \bar{u}_j \frac{\partial c_i}{\partial x_j} = \frac{\partial}{\partial x_j} \left(K_{jj} \frac{\partial c_i}{\partial x_j} \right) + S_i$$

where the K's are the three diagonal elements of the eddy diffusivity tensor.

(a) Set up your tailored equation and boundary conditions and find the concentration distribution $c(x, y, z)$ due to a continuous point source S_p in grams/sec and (g/s) at the origin $x = y = 0$ and $z = H$ (effective stack height); at what distance x does the plume touch the ground $y = z = 0$?

(b) Set up the governing equation and find the pollutant concentration $c(x, y, z, t)$ for a ground-level area source $S(x, y, t)$ [g/m²-s] when a constant mean wind $\bar{u} = U$ is blowing in the x-direction.

2.8 Consider the surface wave problem for irrotational, incompressible flow depicted (cf. LeMehaute 1976):

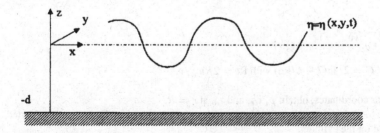

Show that this problem can be formulated in terms of the dependent variables ϕ, that is, a velocity potential where $\vec{v} = \operatorname{grad}\phi$, and the free surface elevation η. Note that

$$\nabla^2\phi = 0 \qquad \text{continuity}$$

$$\frac{\partial\phi}{\partial t} = \frac{1}{2}\left[\left(\frac{\partial\phi}{\partial x}\right)^2 + \left(\frac{\partial\phi}{\partial y}\right)^2 + \left(\frac{\partial\phi}{\partial z}\right)^2\right] + g\eta \qquad \left(\begin{array}{c}\text{dynamic}\\\text{free surface}\\\text{condition}\end{array}\right)$$

$$\frac{\partial\phi}{\partial z}\bigg|_{z=\eta} = -\frac{\partial\eta}{\partial t} + \frac{\partial\phi}{\partial x}\bigg|_{z=\eta}\frac{\partial\eta}{\partial x} + \frac{\partial\phi}{\partial y}\bigg|_{z=\eta}\frac{\partial\eta}{\partial y} \qquad \left(\begin{array}{c}\text{kinematic}\\\text{free surface}\\\text{condition}\end{array}\right)$$

and

$$\frac{\partial\phi}{\partial z}\bigg|_{z=-d} = 0 \qquad \text{(boundary condition)}$$

2.9 Derive the momentum integral equation for flat plate boundary-layer flow.
(a) Start with the definition of the stream function $u = \partial\phi/\partial y$ and $v = -\partial\psi/\partial x$.
(b) Start by multiplying the continuity equation with $(u - U_{\text{outer}})$ and subtract this from the momentum equation to obtain

$$\frac{\partial}{\partial t}(U_0 - u) + \frac{\partial}{\partial x}(uU_0 - u^2) + (U_0 - u)\frac{\partial U_0}{\partial x}\frac{\partial}{\partial y}(vU_0 - uv) = -\frac{1}{\rho}\frac{\partial\tau}{\partial y}$$

2.10 For laminar free convection on a vertical flat plate, find a local Nusselt number correlation using integral relations.

2.11 Consider a cylindrical solid slug of mass M, radius R, and length L falling at constant velocity v_s through an oil-filled long, narrow tube of radius R_2. This "falling cylinder viscometer" is used to find the viscosity of the oil as a function of v_s and the geometric parameters. Note that the axes of the slug and that of the tube are identical at all times while the oil is forced upward through the very small, annular space. Find an expression for μ.

2.12 Consider a long horizontal cylinder of radius R in cross-flow of low-Reynolds-number approach velocity $\vec{v}_\infty \equiv u_\infty$.
(a) Show that the velocity and pressure fields near the cylinder surface can be described by

$$\vec{v} = C\vec{v}_\infty\left[\frac{1}{2}\ln\left(\frac{r}{R}\right) + \frac{1}{4} - \frac{1}{4}\left(\frac{R}{r}\right)^2\right] - C\vec{v}_\infty\frac{(\vec{v}_\infty\cdot\vec{r})}{2r^2}\left[1 - \left(\frac{R}{r}\right)^2\right]$$

and

$$p = p_0 - C\mu\frac{(\vec{v}_\infty\cdot\vec{r})}{r^2} + \rho(\vec{g}\cdot\vec{r})$$

where $C = 2/\ln(7 - 4/\text{Re})$ with $\text{Re} = 2Ru_\infty/\nu$.

(b) Using polar coordinates, obtain p, τ_{rr}, and $\tau_{r\theta}$ at $r = R$.
(c) Obtain the total drag force, due to pressure and wall shear stress, in the x-direction exerted by the fluid on the cylinder.

A Typical Second Homework Set

1 Consider the equation of motion in tensor notation

$$\frac{\partial}{\partial t}(\rho u_j) + \frac{\partial}{\partial x_k}(\rho u_j u_k) = \frac{\partial \pi_{ij}}{\partial x_i} + \rho f_j \tag{1}$$

Show that Eq. (1) is equivalent to

$$\rho \frac{\partial u_j}{\partial t} + (\rho u_k)\frac{\partial u_j}{\partial x_k} = \frac{\partial \pi_{ij}}{\partial x_i} + \rho f_j \tag{2}$$

and that for constant property flows, Eq. (2) reduces to

$$\frac{\partial u_j}{\partial t} + u_k \frac{\partial u_j}{\partial x_k} = -\frac{1}{\rho}\frac{\partial p}{\partial x_j} + \nu \frac{\partial^2 u_j}{\partial x_i \partial x_i} + g_j \tag{3}$$

Comment!

2 Derive Equation (2.36c).

3 Evaluate the radial component of the inertia term $(\vec{v} \cdot \nabla)\vec{v}$ in cylindrical coordinates by using rules and definitions from Appendix A.

4 In cylindrical coordinates, the velocity components for ideal fluid flow around a circular cylinder are

$$v_r = U\left(1 - \frac{a^2}{r^2}\right)\cos\theta$$

$$v_\theta = -U\left(1 + \frac{a^2}{r^2}\right)\sin\theta$$

Here U is the constant magnitude of the velocity approaching the cylinder and a is the radius of the cylinder. If compressible and viscous effects are negligible, determine the pressure $p(r, \theta)$ at any point in the fluid in the absence of any body forces. Take the pressure far from the cylinder to be constant and equal to p_0.

Specialize the result obtained to obtain an expression for the pressure $p(a, \theta)$ on the surface of the cylinder.

5–7 Do text Problems 2.2, 2.3, 2.4.

8 Rework Stokes's first problem, that is, the suddenly accelerated plate in a viscous fluid, from the point of view of vorticity generation.

$$\xi_z = -\frac{\partial u}{\partial y} := -U_0 f' \frac{\partial \eta}{\partial y}$$

is generated by virtue of the shear stress created at the plate surface and transmitted/attenuated in the fluid by means of molecular diffusion/viscosity.

(a) Find $\xi_z(y, t)$ and calculate $\xi_{z_{max}} = \xi_{wall}$ to form ξ_z/ξ_w.

(b) Consider that the fluid is water at $T = 20°C$ and that the plate has moved for 1 hour; at what point from the surface has the vorticity dropped to 50% of the maximum value?

(c) Introducing $Re = U_0 x/\nu$, where $x = U_0 t$, fluid $y(x, \xi_z; \xi_w, Re)$, and calculate the normal distance from the plate that marks the points where the velocity is only 1% of its surface value.

References and Further Reading Material

Al-Khafayi, A.-W., and J. R. Tooley. 1986. *Numerical Methods in Engineering Practice*. Holt,
       Rinehart & Winston, New York.
Ames, W. F. 1965. *Nonlinear Partial Differential Equations in Engineering*. Academic Press, New
       York.
Ames, W. F. 1968. *Nonlinear Ordinary Differential Equations in Transport Phenomena*. Academic
       Press, Boston.
Ames, W. F. 1979. *Numerical Methods for Partial Differential Equations*, 2nd ed. Academic Press,
       New York.
Anderson, D. A., J. C. Tannehill, and R. H. Pletcher. 1984. *Computational Fluid Mechanics and
       Heat Transfer*. McGraw-Hill, New York.
Arparci, V. S., and P. S. Larsen. 1984. *Convection Heat Transfer*. Prentice-Hall, Englewood Cliffs,
       NJ.
Aziz, A., and T. Y. Na. 1984. *Perturbation Methods in Heat Transfer*. Hemisphere, Washington, DC.
Batchelor, G. K. 1967. *An Introduction to Fluid Dynamics*. Cambridge University Press, Cambridge.
Bejan, A. 1984. *Convection Heat Transfer*. Wiley, New York.
Bender, C. M., and S. A. Orszag. 1979. *Advanced Mathematical Methods for Scientists and
       Engineers*. McGraw-Hill, New York.
Bird, R. B. 1957. *Chem. Eng. Sci.*, **6**, 123–131.
Bird, R. B., R. C. Armstrong, and O. Hassager. 1987. *Dynamics of Polymeric Liquids*. Vol. I. *Fluid
       Mechanics*, 2nd ed. Wiley-Interscience, New York.
Bird, R. B., W. E. Stewart, and E. N. Lightfoot. 1960. *Transport Phenomena*. John Wiley, New York.
Boyce, W. E., and R. C. DiPrima. 1977. *Elementary Differential Equations and Boundary Value
       Problems*. John Wiley, New York.
Braun, M. 1978. *Differential Equations and Their Applications*, 2nd ed. Springer-Verlag, New York.
Burden, R. L, J. D. Faires, and A. C. Reynolds. 1978. *Numerical Analysis*. Prindle, Weber &
       Schmidt, Boston.
Carslaw, H. S., and J. C. Jaeger. 1959. *Conduction of Heat in Solids*. Oxford University Press,
       Oxford.
Cebeci, T., and P. Bradshaw. 1977. *Momentum Transfer in Boundary Layers*. McGraw-Hill, New
       York.
Cebeci, T., and P. Bradshaw. 1988. *Physical and Computational Aspects of Convection Heat
       Transfer*. Springer-Verlag, New York.
Crank, J. 1956. *The Mathematics of Diffusion*. Oxford University Press, Oxford.
Cuvelier, C., A. Segal, and A. A. van Steenhoven. 1986. *Finite Element Methods and Navier–Stokes
       Equations*. D. Reidel, Dordrecht, Holland.
Dorf, R. C., ed. 1995. *The Engineering Handbook*. CRC Press, and IEEE Press, Lewis Publishers,
       Boca Raton, FL.
Eskinazi, S. 1975. *Fluid Mechanics and Thermodynamics of Our Environment*. Academic Press,
       New York.
Finlayson, B. A. 1978. *Nonlinear Analysis in Chemical Engineering*. McGraw-Hill, New York.
Gebhart, B., Y. Jaluria, R. L. Mahajan, and B. Sammakia. 1988. *Buoyance-Induced Flows and
       Transport*. Hemisphere, New York.
Habib, I. S. 1975. *Engineering Analysis Methods*. Lexington Books/D. C. Heath, Lexington, MA.
Hansen, A. G. 1964. *Similarity Analyses of Boundary Value Problems in Engineering*.
       Prentice-Hall, Englewood Cliffs, NJ.

Hirsch, C. 1988. *Numerical Computation of Internal and External Flows*. John Wiley & Sons, New York.

Kamke, E. 1959. *Differentialgleichungen: Lösungsmethoden und Lösungen*, Chelsea, New York.

LeMehaute, B. 1976. *An Introduction to Hydrodynamics and Water Waves*. Springer-Verlag, New York.

Minkowycz, W. J., E. H. Sparrow, G. E. Schneider, and R. H. Pletcher. 1988. *Handbook of Numerical Heat Transfer*. Wiley-Interscience, New York.

Morse, P. M., and H. Feshbach. 1953. *Methods of Theoretical Physics*. McGraw-Hill, New York.

Na, T. Y. 1979. *Computational Methods in Engineering Boundary Value Problems*. Academic Press, New York.

Nayfeh, A. H. 1981. *Introduction to Perturbation Techniques*. Wiley-Interscience, New York.

Özişik, M. N. 1993. *Heat Conduction*. Wiley-Interscience, New York.

Patankar, S. V. 1980. *Numerical Heat Transfer and Fluid Flow*. McGraw-Hill, New York.

Ralston, A., and P. Rabinowitz. 1978. *A First Course in Numerical Analysis*, 2nd ed. McGraw-Hill, New York.

Reider, W. G., and H. R. Bushy. 1986. *Introductory Engineering Modeling*. John Wiley, New York.

Roache, P. J. 1976. *Computational Fluid Dynamics*. Hermosa, Albuquerque, NM.

Schlichting, H. 1979. *Boundary-Layer Theory*, McGraw-Hill, New York.

Seinfeld, J. H. 1983. *Air Pollution*, 2nd ed. McGraw-Hill, New York.

Shih, T. M. 1984. *Numerical Heat Transfer*. Hemisphere, Washington, DC.

Vandergraft, J. S. 1983. *Introduction to Numerical Computations*. Academic Press, New York.

White, F. M. 1988. *Heat and Mass Transfer*. Addison-Wesley, New York.

White, F. M. 1991. *Viscous Fluid Flow*. McGraw-Hill, New York.

Zwillinger, D. 1992. *Handbook of Differential Equations*. Academic Press, Boston.

CHAPTER 3

Analyses of Basic Fluid Flow Problems

3.1 Introduction and Review

In this chapter, the basic concepts, equations, and solution methods discussed previously are utilized to understand and solve standard fluid mechanics problems. The focus is clearly on *incompressible viscous* fluid flow; however, a few introductory aspects of compressible flow and inviscid flow are discussed.

3.1.1 Compressible Flow

In *compressible* flow, the fluid density is not constant. For example, in natural or free convection, buoyancy forces caused by local temperature or concentration differences set up flow recirculation or induce upward flow against gravity. In acoustics, sound waves propagate because of extremely small changes in density, pressure, and temperature. At high-speed fluid flow past a body, density effects become important, for example, in the range $0.3 < \mathrm{Ma} < 0.8$; at even higher velocities (i.e., $\mathrm{Ma} > 1.0$) shock waves dominate the flow field. Here, $\mathrm{Ma} = u/a$ is the Mach number, where a is the speed of sound defined as

$$a = \left(\left. \frac{\partial p}{\partial \rho} \right|_S \right)^{1/2} = \left(\gamma \left. \frac{\partial p}{\partial \rho} \right|_T \right)^{1/2} \tag{3.1}$$

where S is the entropy and $\gamma = c_p/c_v$ the ratio of specific heats of a perfect gas. Typical values for the speed of sound are $a_{\text{air}} = 317$ m/s, $a_{\text{water}} = 1{,}490$ m/s, and $a_{\text{steel}} = 5{,}060$ m/s.

With the fluid density varying because of changes in flow field parameters, an equation of state for $\rho = \rho(p, T, c)$ is needed in addition to the continuity, momentum, and energy equations for a complete description of a compressible flow system. High-speed flow, in which compressibility effects become very important, typically deals with the dynamics of gases. Natural convection deals with incompressible subsonic flow, which is buoyancy driven (cf. Boussinesq assumption introduced in Chapter 2 and applied in Chapter 4).

For steady, *quasi-one-dimensional flow* the flow properties are *uniform* across any given stream tube and the integral form of the steady-state conservation equations (cf. Sect. 2.1.2) can be reduced to a set of algebraic equations. Specifically, mass conservation in integral form

$$\iint_{c.s.} \rho \vec{v} \cdot d\vec{A} = 0 \tag{3.2a}$$

reduces to

$$\rho u A = \text{const}$$

or

$$d(\rho u A) = 0 \quad \text{which implies} \quad \rho u^2 \, dA + \rho u A \, du + A u^2 d\rho = 0 \tag{3.2b}$$

Similarly, the integral momentum equation (cf. Sect. 2.1.2)

$$\iint_{c.s.} \vec{v}\rho\vec{v} \cdot d\vec{A} - \iint_{c.s.} \left(-p\vec{\delta} + \vec{\tau}\right) \cdot d\vec{A} = 0 \tag{3.3a}$$

reduces to

$$A\,dp + Au^2 d\rho + \rho u^2 \, dA + 2\rho u A\,du = 0 \tag{3.3b}$$

Subtracting (3.2b) from (3.3b) yields a reduced form of Euler's equation

$$dp = -\rho u \, du \tag{3.4}$$

Now, the continuity equation (3.2b) can be rewritten as

$$\frac{du}{u} + \frac{dA}{A} = -\frac{d\rho}{\rho}$$

and combined with the momentum equation (3.4) in the form

$$\frac{dp}{\rho} = \frac{dp}{d\rho}\frac{d\rho}{\rho} = -u\,du$$

or

$$\frac{d\rho}{\rho} = -\frac{u\,du}{a^2} = -\text{Ma}^2 \frac{du}{u}$$

to

$$\frac{dA}{A} = (\text{Ma}^2 - 1)\frac{du}{u} \tag{3.5}$$

where $\quad \text{Ma}^2 = \dfrac{u^2}{a^2} \quad$ and $\quad a^2 = \dfrac{dp}{d\rho} := \left(\dfrac{\partial p}{\partial \rho}\right)_S$

For Ma $\to 0$, Equation (3.5) reduces to

$$\frac{dA}{A} = \frac{du}{u} \tag{3.6a}$$

or

$$uA = \text{const} \tag{3.6b}$$

which is the continuity equation in integrated form for incompressible flow.

Notes

- For subsonic flow $(0 < \text{Ma} < 1)$, u increases with decreasing A, and vice versa, as for incompressible flow.

- For sonic flow (Ma = 1), $dA/A = 0$, which implies a local minimum in the cross-sectional area of a stream tube as in a convergent–divergent duct. This minimum area where Ma = 1 is called a *throat*.
- For *supersonic* flow (Ma > 1), the velocity *increases* in diverging ducts and *decreases* in converging ducts.

More elaborate analyses and discussions can be found in Anderson (1990), Nunn (1989), Schreier (1982), or Landau and Lifshitz (1987), among other texts. A few illustrative examples of 1-D compressible flow are given in Section 3.3.

3.1.2 Incompressible Inviscid Flow

Fluids in motion for which viscous effects and thermal conductivity are unimportant are said to be *ideal*. The dynamics of ideal fluids is described by Euler's equation (cf. Sect. 2.2.1)

$$\frac{\partial \vec{v}}{\partial t} + (\vec{v} \cdot \nabla)\vec{v} = -\frac{1}{\rho}\nabla p + \vec{g} \qquad (3.7)$$

which is the Navier–Stokes equation without the viscous term ($\nu \nabla^2 \vec{v}$). For incompressible flow, ρ = const and the continuity equation, $\partial \rho / \partial t + \nabla \cdot (\rho \vec{v}) = 0$, reduces to

$$\nabla \cdot \vec{v} = 0 \qquad (3.8)$$

Thus, Euler's equation could be written as

$$\frac{\partial \vec{v}}{\partial t} + (\vec{v} \cdot \nabla)\vec{v} = -\nabla\left(\frac{p}{\rho}\right) + \vec{g} \qquad (3.9)$$

Since no thermal diffusion occurs within the fluid, the fluid motion is adiabatic, and if the entropy is constant at all times the flow is called *isentropic*; that is, S = const or

$$\frac{DS}{Dt} = \frac{\partial S}{\partial t} + (\vec{v} \cdot \nabla)S = 0 \qquad (3.10)$$

The isentropic flow condition (3.10) has a bearing on the pressure term in Equation (3.9) since the thermodynamic relation $dh = T\,ds + \hat{v}\,dp$ can be used to replace $\nabla p/\rho$ with ∇h (cf. Sects. 1.2.3 and 2.2.3). Using the vector identity $(\vec{v} \cdot \nabla)\vec{v} \equiv \nabla(v^2/2) + \vec{v} \times (\nabla \times \vec{v})$ Equation (3.9) now reads

$$\frac{\partial \vec{v}}{\partial t} + \vec{v} \times (\nabla \times \vec{v}) = -\nabla\left(h + \frac{1}{2}v^2\right) + \vec{g} \qquad (3.11)$$

or

$$\frac{\partial \vec{v}}{\partial t} + \nabla\left(\frac{1}{2}v^2 + \frac{p}{\rho}\right) - \vec{g} + \vec{v} \times \vec{\zeta} = 0 \qquad (3.12)$$

where $\nabla \times \vec{v} \equiv \vec{\zeta}$ is the vorticity vector. When $\vec{\zeta} = 0$, the flow field is *irrotational*, and the velocity vector can be expressed as the gradient of a potential function, that is, $\vec{v} = \nabla \phi$. This implies that for incompressible flow

$$\nabla^2 \phi = 0 \tag{3.13}$$

where ϕ is the velocity potential, evaluated later.

Multiplying Equation (3.12) by a displacement vector, $d\vec{r}$, that is parallel to $\vec{v}$ and hence measured along a streamline yields

$$\frac{\partial \vec{v}}{\partial t} \cdot d\vec{r} + d\left(\frac{1}{2}v^2\right) + \frac{dp}{\rho} + g\,dz = 0 \tag{3.14}$$

since $(\vec{\zeta} \times \vec{v}) \cdot d\vec{r} \equiv 0$. Integration along a particular streamline for inviscid flow results in *Bernoulli's equation*

$$\int_{①}^{②} \frac{\partial v}{\partial t}\,ds + \int_{1}^{2} \frac{dp}{\rho} + \frac{1}{2}\left(v_2^2 - v_1^2\right) + g(z_2 - z_1) = 0 \tag{3.15}$$

where ① and ② are two points on a representative streamline and ds is the arc length.

Introducing the velocity potential function in Equation (3.14), in order to replace the velocity terms, we obtain for

$$\frac{\partial \vec{v}}{\partial t} \cdot d\vec{r} = \frac{\partial}{\partial t}(\nabla \phi) \cdot d\vec{r} = d\left(\frac{\partial \phi}{\partial t}\right) \quad \text{and} \quad v^2 = |\nabla \phi|^2$$

and after integration

$$\frac{\partial \phi}{\partial t} + \int \frac{dp}{\rho} + \frac{1}{2}|\nabla \phi|^2 + gz = \text{const} \tag{3.16}$$

Thus, for *incompressible* flow, Equation (3.13) can be solved for ϕ and Equation (3.16) can be used to calculate p. For steady incompressible flow, Bernoulli's equation reduces to

$$\frac{p}{\rho} + \frac{1}{2}v^2 + gz = \text{const} \tag{3.17}$$

Equation (3.17) applied to a streamline connecting infinity (i.e., free stream conditions p_∞, u_∞) with the stagnation point (i.e., nose of a submerged body where $p = p_{\max}$ and $\vec{v} \equiv 0$) yields the stagnation point pressure

$$p_{\max} = p_\infty + \frac{1}{2}u_\infty^2 \tag{3.18}$$

A few illustrative examples are given in Section 3.3.

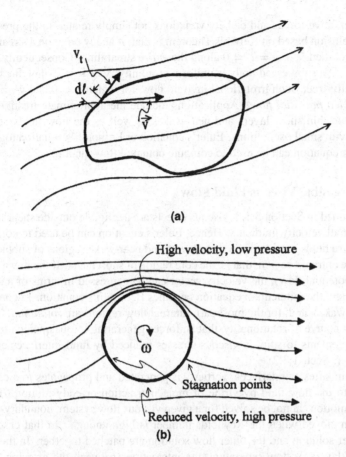

v_t

$d\ell$

$\vec{v}$

(a)

High velocity, low pressure

ω

Stagnation points

Reduced velocity, high pressure

(b)

Fig. 3.1. Schematics and application for induced circulation: (a) Concept of circulation; (b) Uniform flow and circulation about a rotating cylinder.

3.1.3 Potential Flow

Potential flow is irrotational, that is, vorticity-free flow described by Equation (3.13) and best explained with Kelvin's Theorem (cf. Sect. 1.3.2)

$$\frac{D\Gamma}{Dt} = 0 \quad \text{or} \quad \Gamma = \zeta_n A = \text{const} \tag{3.19a,b}$$

Here $\Gamma = \oint \vec{v} \cdot dl$ is the velocity circulation for a closed "fluid contour" (Fig. 3.1). Rewriting Equation (3.19) as

$$\Gamma = \oint \vec{v} \cdot dl = \iint_A \vec{\zeta} \cdot d\vec{A} = \text{const}$$

where A is the fluid surface spanning the closed contour, the following physical interpretation is possible: If we follow a fluid element in motion, the vorticity associated with that particle will not change. On the other hand, vorticity may be changed as a result of viscous, nonconservative forces and density variations not simply related to the pressure. An important conclusion based on Kelvin's Theorem is that, if at any point on a streamline the vorticity is zero, then $\nabla \times \vec{v} = \vec{\zeta} = 0$ along the *entire* streamline. Consequently, if an approach stream past a submerged body is uniform at infinity, $\nabla \times \vec{v} \equiv 0$, then the entire flow field is vorticity-free. Such irrotational inviscid flow, for which $\vec{\zeta} = 0$ and $\nu \nabla^2 \vec{v} \approx 0$ everywhere, is called *potential flow*. Applications include the approximate treatment of flow regions outside thin shear layers and near-wakes as well as the flow fields around immersed bodies with small oscillations. Euler's equation or Bernoulli's equation together with the continuity equation can be used to compute potential flow quantities.

3.2   Incompressible Viscous Fluid Flow

As mentioned in Section 3.1.3, viscous effects are negligible outside shear layers because of very small velocity gradients. Hence, Euler's equation can be used to compute the velocity/pressure fields outside these shear layers and near-wake regions of submerged bodies or inside the entrance region, that is, the core region of short ducts. If the flow is also irrotational (i.e., potential flow), the velocity vector can be expressed in terms of a scalar potential function and the momentum equation becomes Bernoulli's equation. In contrast, shear layers, near-wakes, and, for the most part, internal flow regions are *rotational*. Thus, the most important source of rotationality, that is, forces generating torques, which in turn are causing fluid elements to spin, is viscous stresses induced by fluid interfaces and/or solid boundaries (cf. Sect. 1.3.2).

Along such boundaries or interfaces, vorticity is generated and propagates to a certain extent outward into the flow field by virtue of molecular action or eddy motion. If the entire equation of motion cannot be solved for a given viscous flow system, boundary-layer theory could be employed when the Reynolds number is high enough. In that case, the thin boundary layer solution and the outer flow solution are patched together. In the case of thick boundary layers or flow separation, the interactions between the growing shear layers or separation bubbles and the outer flow have to be recalculated until convergence has been obtained. *It is within these shear layers where steep velocity, temperature, and concentration gradients occur, leading to transport of momentum, energy, and mass.* Thus a detailed knowledge of viscous fluid flow with a special emphasis on boundary-layer-type flows, separating flows, jets, and wakes is most important. On the other hand, a number of computational fluid dynamics (CFD) softwares that solve the Navier–Stokes equations for a lot of practical problems on affordable workstations are now available and hence deemphasize the exclusive study of unique fluid flow patterns.

Considering incompressible, single-phase flow of a Newtonian fluid, the Navier–Stokes equations (Sect. 2.2) are applicable. They are, together with the continuity equation, the parent equations of numerous solved and unresolved flow problems. Hence, fluid mechanics problems are traditionally classified as those that are *exact solutions* of the Navier–Stokes equations and those that are *approximate solutions*. Alternatively, flows are categorized according to the Reynolds number into low-, moderate-, and high-Reynolds number flows. Occasionally, flow systems are divided into internal (or pipe) flows and external (or boundary-layer) flows.

Flow systems that are *exactly* represented by *a reduced form of* the Navier–Stokes equations and solvable by analytic or numerical methods fall into the first category. Examples include flow problems solved by Poiseuille, Couette, Hiemenz, Jeffrey-Hamel, von Karman,

Stokes, and Rayleigh. Typically, the flow is laminar, possibly transient, one-dimensional, that is, quasiparallel, nonaccelerating, or fully developed; the fluid properties are constant, the system geometry is simple; and the wall surfaces are idealized. Since the Navier–Stokes equations are known to produce multiple solutions for the same boundary conditions, there is no guarantee that a new exact solution will actually occur in reality.

In the second category, that is, for approximate solutions, the Navier–Stokes equations have to be simplified (or approximated) on the basis of physically sound assumptions. Examples include (i) very-low-Reynolds-number or "creeping" flows, that is, slow fluid motion around submerged bodies or in conduits of complicated geometry (e.g., Oseen, Reynolds, Hele–Shaw flows), and (ii) high-Reynolds-number flows (cf. Prandtl's boundary-layer theory for laminar and turbulent thin shear layers).

Surely, one day such a classification will become obsolete when a new breed of affordable workstations is capable of solving the complete Navier–Stokes equations accurately, subject to any given set of initial and boundary conditions, and when applied mathematicians have resolved the uniqueness problem mentioned. In fact, for today's numerical analysis purposes, it is most convenient to classify a given fluid flow problem according to its flow *field characteristics* and *degree of complexity*. Thus one should primarily consider whether

(i)   the flow regime is (a) laminar or (b) turbulent
(ii)   the flow's time dependence is (a) steady-state or (b) transient
(iii)   the flow field dimensionality is (a) 1-D fully developed, (b) 2-D accelerating, or (c) fully 3-D
(iv)   the "nature" of the flow field is (a) parabolic or (b) elliptic
(v)   the fluid is (a) Newtonian or (b) non-Newtonian
(vi)   the flow field is (a) isothermal or (b) temperature-dependent
(vii)   the flow is (a) single-phase or (b) multi-phase

In any case, the availability of exact and approximate solutions discussed later is very valuable

- for learning more about fluid dynamics,
- as a starting point to analyze complex real-world problems, and
- for comparisons with new numerical solutions.

3.2.1 Low-Reynolds-Number Flows

Of interest are internal or external incompressible viscous flows that are exact or approximate solutions of the Navier–Stokes equations. "Low Reynolds number" implies that the inertia terms are either identically zero or approximately zero, or they can be linearized. In any case, the key *nonlinearity* is removed from the equation of motion, the principle of superposition can be applied, and analytic or simple approximate solutions of two distinct families of problems are available. Highly viscous laminar flows are examples of *approximate* solutions of the Navier–Stokes equations and are characterized by Re $\leq$ 1.0, also known as *creeping flows*: (i) flow past a sphere (Stokes or Oseen); (ii) flow in lubrication films (Reynolds); (iii) flow past a cylinder in a slit (Hele–Shaw); (iv) flow through porous media (Darcy); and (v) flow of certain non-Newtonian fluids such as foodstuff, thick polymers, plastics, and slurries. When the inertia terms, $(\vec{v} \cdot \nabla)\vec{v}$, are equal to zero or

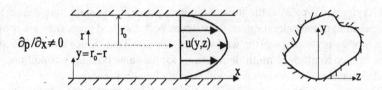

Fig. 3.2. Coordinate system for parallel flows.

negligible, the Navier–Stokes equation reduces to (cf. Sect. 2.2)

♦ $$\rho \frac{\partial \vec{v}}{\partial t} + \nabla p = \mu \nabla^2 \vec{v} + \rho \vec{g}$$ (3.20)

The continuity equation remains

♦ $$\nabla \cdot \vec{v} = 0$$ (3.21)

The second family of flows are those for which *exact solutions* of the Navier–Stokes equations have been found. Actually, the so-called exact solutions are exact *representations* of a given flow problem by the reduced Navier–Stokes equations that may be solved analytically or numerically. Such flows may exhibit higher Reynolds numbers than creeping flows. Either they are fully developed and one-dimensional – which implies that the convective acceleration is equal to zero everywhere – or the inertia terms have to be retained.

A   Parallel Flow Solutions of the Navier–Stokes Equations

Since the fully developed flow of interest is only in one direction, $(\vec{v} \cdot \nabla)\vec{v} \equiv 0$ and $\vec{v}$ collapses to $u\hat{i}$. We postulate this $u = u(y, z, t)$, so that Equation (3.20) reduces to

$$\frac{\partial u}{\partial t} = -\frac{1}{\rho} \frac{\partial p}{\partial x} + v \left(\frac{\partial^2 u}{\partial y^2} + \frac{\partial^2 u}{\partial z^2} \right) + g_y$$ (3.22)

in rectangular coordinates (Fig. 3.2). An initial condition would prescribe $u(t = 0) = u_0(y, z)$ and the boundary condition is $u|_{\text{wall}} = 0$.

Example (1)
Consider Poiseuille flow, that is, steady laminar fully developed flow in a conduit of constant cross section. The driving forces for the fluid motion are, in general, the pressure gradient and the gravitational force. The left-hand side (LHS) of Equation (3.22) is zero and the pressure term can be extended to absorb the constant gravitational acceleration. As a result we obtain Poisson's equation

$$\nabla^2 u = f(x)$$ (E3.1)

For simple duct geometries, analytic solutions are available. For example, Poiseuille flow in a horizontal circular pipe of radius, R, is described by (cf. Sect. 2.2 or App. B):

$$\frac{1}{r} \frac{d}{dr} \left(r \frac{du}{dr} \right) = \frac{1}{\mu} \frac{dp}{dx} = \text{const}$$ (E3.2)

because $u = u(r)$ only. This postulate is based on the assumptions, continuity equation, and boundary conditions outlined in Section 2.2.4. Equation (E3.2) has the solution

$$u(r) = -\frac{1}{4\mu}\left(\frac{dp}{dx}\right)(R^2 - r^2) \tag{E3.3}$$

which implies that $u(r = R) = 0$ and $du/dr = 0$ at $r = 0$. Note that the condition $u(r = 0) = U_{max}$ is often not known a priori. For fully developed pipe flow, the pressure gradient can be approximated by

$$-\frac{dp}{dx} \approx -\frac{p_L - p_0}{L} = \text{const} \tag{E3.4}$$

where L is the length of the pipe section of interest. Using the known pressure drop $\Delta p/L$, the volumetric flow rate can be computed as

$$Q = \frac{\pi R^4}{8\mu}\left(\frac{\Delta p}{L}\right) = u_{av}A_{pipe} \tag{E3.5}$$

from which the friction factor, f, for laminar flow in smooth pipes can be obtained, where $f = f[\tau_{wall} \text{ or } (\Delta p/L)]$ (cf. Sect. 3.3). □

Example (2)

Consider Couette flow, that is, steady laminar nonaccelerating flow in a small gap, h, formed by a semiinfinite fixed wall and a parallel moving plate, or by long concentric cylinders, of which one rotates. Pressure and/or viscous forces keep the incompressible fluid in motion. Ignoring end effects, we postulate for the parallel-plate case that $u = u(y)$, $dp/dx = \text{const}$, and body forces are not applicable. Thus Equation (3.22) reduces to (cf. Fig. 3.3)

$$\frac{dp}{dx} = \mu\frac{d^2u}{dy^2} = \text{const} \tag{E3.6}$$

subject to the boundary conditions

$$u(y = 0) = 0 \tag{E3.7}$$

and

$$u(y = h) = U \tag{E3.8}$$

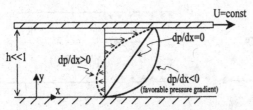

Fig. 3.3. Coordinate system and three velocity profiles for Couette flow with different constant pressure gradients.

Introducing the pressure parameter

$$P = -\frac{h^2}{2\mu U}\frac{dp}{dx} = \text{const} \qquad\qquad\qquad\text{(E3.9)}$$

the solution can be written as

$$\frac{u(y)}{U} = \frac{y}{h} + P\frac{y}{h}\left(1 - \frac{y}{h}\right) \qquad\qquad\qquad\text{(E3.10)}$$

Depending upon the magnitude of P, we have *simple shear flow* for $P = 0$ (i.e., zero pressure gradient), locally *reverse flow* for $P < 0$ (i.e., positive or adverse pressure gradient), and extended *parabolic flow* for $P > 0$ (i.e., negative or favorable pressure gradient). An interesting extension to Couette's flow is a rotating and translating shaft within a fixed or counterrotating cylinder (cf. Sect. 3.3). □

It should be noted that this class of steady parallel or quasiparallel flows (Couette, Poiseuille, slit flows, thin film flows on inclined surfaces, etc.) can be described by a simple second-order ODE of the form

♦ $$u'' = C \qquad\qquad\qquad\text{(3.23)}$$

where u is the streamwise velocity, the derivatives of u are w.r.t. the normal coordinate, and the constant C represents the pressure gradient and/or a gravitational component.

Example (3)

Consider exact solutions to *transient* problems. Examples include flows near suddenly accelerated or oscillating plates (i.e., Stokes's problems I and II) and flows in a pipe with pulsating pressure gradient, that is, $\partial p/\partial x = (\Delta p/l)\cos\omega t$. For those situations,

$$u = u(y, t) \quad \text{and} \quad v = v(y, t)$$

which implies that the velocity components are independent of x (Fig. 3.4). Again, the postulates for u and v are based on the system assumptions and inspection of the continuity equation and the boundary conditions.

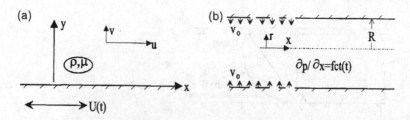

Fig. 3.4. Coordinate system for transient flow problems: (a) Planar flow near oscillating wall; (b) Transient pipe flow with possible fluid injection.

With these postulates for u and v, the Navier–Stokes equations and the continuity equation reduce to

$$\text{(x-momentum)} \qquad \frac{\partial u}{\partial t} + v\frac{\partial u}{\partial y} = -\frac{1}{\rho}\frac{\partial p}{\partial x} + v\frac{\partial^2 u}{\partial y^2} \tag{E3.11}$$

$$\text{(y-momentum)} \qquad \frac{\partial v}{\partial t} = -\frac{1}{\rho}\frac{\partial p}{\partial y} \tag{E3.12}$$

$$\text{(continuity)} \qquad \frac{\partial v}{\partial y} = 0 \tag{E3.13}$$

For a given set of arbitrary initial and boundary conditions, this is not an easy problem to solve. However, solutions are known when $v = 0$ or $v = v_0 = \text{const.}$ In either case, mass conservation is satisfied (Eq. (E3.13)) and the pressure becomes independent of y (Eq. (E3.12)). The resulting governing equation is then

$$\blacklozenge \qquad \frac{\partial u}{\partial t} + v_0\frac{\partial u}{\partial y} = -\frac{1}{\rho}\frac{dp}{dx} + v\frac{\partial^2 u}{\partial y^2} \tag{E3.14}$$

The pressure gradient is often related to the wall motion $U(t)$ or to a free stream far away from the wall with a velocity of the form $U(t) = U_0[1 + g(t)]$. Thus, employing Euler's equation

$$-\frac{1}{\rho}\frac{\partial p}{\partial x} = \frac{dU}{dt} := U_0\frac{dg}{dt} \tag{E3.15}$$

The function $g(t)$ can be constructed so that $U(t)$ is a constant in the form of a step function or a linear function or an oscillatory function with or without damping. A classical transient flow problem (Stokes I) is revisited next. Additional transient flow problems are discussed in Sect. 3.3. and in Ch. 5.   □

Stokes's First Problem

The simplest application of Equations (E3.14) and (E3.15) is the suddenly accelerated impermeable plane wall:

$$\begin{array}{lll}
\text{when } t \leq 0 & u = 0 & \text{for all } y \\
\text{when } t > 0 & u = U_0 & \text{for } y = 0 \quad \text{and} \\
& u = 0 & \text{for } y \to \infty
\end{array} \tag{E3.16a–c}$$

It can be seen from Equation (E3.15) that in this case the pressure in the flow domain is constant. Equation (E3.14) can be further reduced with $v_0 \equiv 0$ to (cf. Fig. 3.5)

$$\blacklozenge \qquad \frac{\partial u}{\partial t} = v\frac{\partial^2 u}{\partial y^2} \tag{E3.17}$$

Equation (E3.17) is known, in general, as the *transient one-dimensional diffusion equation*, where momentum, heat, or species mass starts to diffuse at $t > 0$ into a space $y > 0$ because of a step change in velocity, temperature, or concentration at $y = 0$. Combining the two

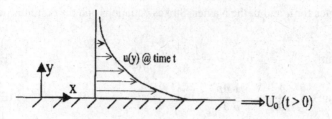

Fig. 3.5. Velocity profile caused by a suddenly accelerated wall (Stokes's first problem).

independent variables into one (cf. Sect. 2.3.3A) we obtain the dimensionless similarity variable

$$\eta = \frac{y}{2\sqrt{vt}} \tag{E3.18}$$

We recall, that Equation (E3.18) is based on the postulate $\eta \cong y^a v^b t^c$, where the exponents a to c are such that η is dimensionless. It is apparent that $u(y, t) \sim f[\eta(y, t)]$, where $f(\eta)$ is an unknown similarity function. Because one boundary condition requires that $u(y = 0, t > 0) = U_0$, we select the wall velocity as a dimensionally correct proportionality factor. Thus,

$$u = U_0 f(\eta) \tag{E3.19}$$

and the governing equation (E3.17) can be transformed to an ODE

$$\blacklozenge \qquad f'' + 2\eta f' = 0 \tag{E3.20}$$

subject to

$$f(\eta = 0) = 1 \quad \text{and} \quad f(\eta \to \infty) \to 0 \tag{E3.21a,b}$$

The solution is $f(\eta) = 1 - \text{erf}(\eta)$ so that (cf. Fig. 3.5)

$$u = U_0 \text{erfc}(\eta) = U_0 \left[1 - \frac{2}{\sqrt{\pi}} \int_0^{\eta} \exp(-\eta^2) \, d\eta \right] \tag{E3.22}$$

If u/U_0 is plotted versus η, it turns out that $u/U_0 \approx 0.01$ at $\eta = 2.0$. This means that the disturbance created by the sudden motion of the wall sets the adjacent fluid into motion, but it dies out (as a result of viscous dissipation) a certain distance from the wall. Hence a penetration depth (or boundary-layer thickness δ) could be computed from this condition, that is,

$$y(\eta = 2.0) = \delta \qquad \text{when } u \approx 0.01 U_0$$

or

$$\blacklozenge \qquad \delta \approx 4\sqrt{vt} \tag{E3.23}$$

The essence of Equation (E3.23) can also be evaluated from the point of view of vorticity generation at the surface and vorticity transport/attenuation normal to the wall. With

$$\vec{\zeta} \ \Rightarrow \ \zeta_z := -\frac{\partial u}{\partial y} = -U_0 f' \frac{\partial \eta}{\partial y}$$

we obtain from Equations (E3.18) and (E3.22)

$$\zeta_z = \frac{U_0}{\sqrt{\pi \nu t}} \exp\left[-\left(\frac{y}{2\sqrt{\nu t}}\right)^2\right] \tag{E3.24}$$

or with $\zeta_z(y = 0) = \zeta_w = \frac{U_0}{\sqrt{\pi \nu t}}$

$$\frac{\zeta_z}{\zeta_w} = \exp\left[-\left(\frac{y}{2\sqrt{\nu t}}\right)^2\right] \tag{E3.25}$$

Introducing $\mathrm{Re}_x = U_0 x / \nu$, where $x = U_0 t$, Equation (E3.25) can be rewritten as

$$\frac{\zeta_z}{\zeta_w} = \exp\left[-\left(\frac{y\sqrt{U_0}}{2\sqrt{\nu x}}\right)^2\right]$$

or

$$\frac{y}{x} = \frac{2}{\sqrt{\mathrm{Re}_x}} \left(\ln \frac{\zeta_w}{\zeta_z}\right)^{1/2} \tag{E3.26}$$

Clearly, when for example $\zeta_z = 1$ percent of ζ_w, $y \approx \delta$, and Equation (E3.26) indicates that

$$\blacklozenge \quad \frac{\delta}{x} \approx \frac{4.3}{\sqrt{\mathrm{Re}_x}} \tag{E3.27}$$

B   Nonparallel Flow Solutions of the Navier–Stokes Equations

For the unidirectional flow previously discussed, the inertia terms are identical to zero. There are a few real flows *with inertial effects* that can be exactly described by the Navier–Stokes equations and for which analytic solutions are known. In each case, the convective acceleration terms have to be retained and the continuity equation has to be satisfied since $\partial u / \partial x \neq 0$. Examples of such flows include (cf. Schlichting 1979): (i) steady two-dimensional stagnation flow (Hiemenz); (ii) planar stagnation flow normal to an oscillating wall (Rott); (iii) flow near a rotating disk in a fluid reservoir (Von Karman); and (iv) other boundary-layer-type flows past bodies with simple geometry (Prandtl). For these flow systems, suitable transformations based on physical grounds that convert the governing PDEs into a set of coupled ODEs for which analytic or approximate solutions are known (cf. Sect. 2.3.3 and App. F.3/F.4) have been found.

Transient two-dimensional flows in rectangular coordinates can be described as (cf. Sect. 2.2.3 or App. C):

$$\frac{\partial u}{\partial t} + u\frac{\partial u}{\partial x} + v\frac{\partial u}{\partial y} = -\frac{1}{\rho}\frac{\partial p}{\partial x} + v\left(\frac{\partial^2 u}{\partial x^2} + \frac{\partial^2 u}{\partial y^2}\right) \tag{3.24}$$

$$\frac{\partial v}{\partial t} + u\frac{\partial v}{\partial x} + v\frac{\partial v}{\partial y} = -\frac{1}{\rho}\frac{\partial p}{\partial y} + v\left(\frac{\partial^2 v}{\partial x^2} + \frac{\partial^2 v}{\partial y^2}\right) \tag{3.25}$$

$$\frac{\partial u}{\partial x} + \frac{\partial v}{\partial y} = 0 \tag{3.26}$$

Special applications allow a *reduction* of these momentum equations as shown later. The objective is to demonstrate how postulates for u, v and p can be constructed and to elaborate on the physical meaning of the resulting equations.

Problem (1) Two-Dimensional Stagnation Point Flow

Consider a planar uniform stream approaching perpendicularly a solid stationary wall or very large cylinder. This problem contains components of (i) the *inviscid* stagnation point flow analysis (cf. Sect. 3.1.2) and (ii) the viscous flow effects (cf. Fig. 3.6).

The *inviscid* flow solution in terms of the stream function

$$\psi = \alpha x y \tag{P3.1a}$$

fulfills the continuity equation, $\partial \hat{u}/\partial x + \partial \hat{v}/\partial y = 0$, and can be readily depicted with the streamline equation $\psi = $ const or $y = c/x$. From Bernoulli's equation (cf. Sect. 3.1.2) evaluated at the stagnation point, $p = p_0$ and $\vec{v} = 0$, we obtain

$$p_0 - p = \frac{\rho}{2}(\hat{u}^2 + \hat{v}^2) := \frac{\rho\alpha^2}{2}(x^2 + y^2) \tag{P3.1b}$$

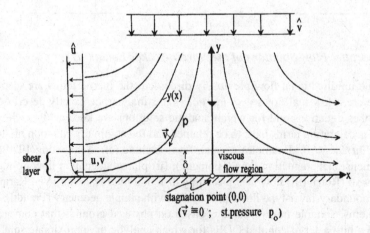

Fig. 3.6. Hiemenz's two-dimensional stagnation point flow.

where $\hat{u} = \alpha x$ and $\hat{v} = -\alpha y$ was employed. In summary, it is observed that the ideal velocity field obeys only *one* wall condition: $\hat{v}(y = 0) = 0$.

For the *viscous* flow solution, the wall presence, which largely determines the near-wall flow field, has to be accounted for in terms of $u = u(x, y)$, $v = v(y)$, and $p = p(x, y)$. A suitable postulate is

$$u = xf'(y) \qquad \text{so that} \qquad v = -f(y) \tag{P3.2a,b}$$

from continuity. Similarly, in an extension of (P3.1b)

$$p_0 - p = \frac{1}{2}\rho\alpha^2[x^2 + F(y)] \tag{P3.3}$$

to reflect the wall influence.

With Eqns. (P3.2a,b) and (P3.3), the reduced x- and y-components of the Navier–Stokes equations can be transformed to

$$f'^2 - ff'' = \alpha^2 + vf''' \tag{P3.4}$$

and

$$ff' = \frac{\alpha^2}{2}F' - vf'' \tag{P3.5}$$

subject to

$$f = f' = F = 0 \qquad \text{at } y = 0 \quad \text{and} \quad f' \to \alpha \qquad \text{when } y \to \infty$$

Equation (P3.4) could be further transformed in order to absorb α^2 and v and then solved independently. The postulate $f(y) = A\phi(y)$ and $\eta = \beta y$ yields with Eq. (P3.4)

$$\eta = \sqrt{\frac{\alpha}{v}}\, y \quad \text{and} \quad \phi(\eta) = \frac{f(y)}{\sqrt{\alpha v}}$$

so that the new x-momentum equation reads

$$\phi''' + \phi\phi'' - (\phi')^2 + 1 = 0 \tag{P3.6a}$$

subject to

$$\phi = 0 \quad \text{and} \quad \phi' = 0 \qquad \text{at } y = 0 \tag{P3.6b,c}$$

and

$$\phi' = 1 \qquad \text{at } y \to \infty \tag{P3.6d}$$

This boundary-value problem (BVP) can be solved with an initial-value solver, such as the Runge–Kutta algorithm in App. F.3, when a second initial condition for $\partial u/\partial y$ at $y = 0$, that is, a value for $f''(0)$ or $\phi''(0)$ is known. By trial and error, $\phi''(0) = 1.2326$, which allows us to match the far-field condition $\phi'(\eta \to \infty) = 1$. In fact, already at $\eta = 2.4$, $\phi' = 0.99$, so that the wall-influenced region or shear layer thickness, δ, is

$$\blacklozenge \qquad \delta \cong 2.4\sqrt{\frac{v}{\alpha}}$$

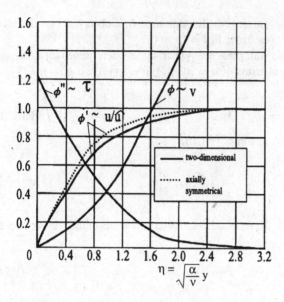

Fig. 3.7. Stagnation point flow profiles.

Figure 3.7 shows $u \sim \phi'(\eta)$, $v \sim \phi(\eta)$, and $\tau \sim \phi''(\eta)$. Clearly $u(\eta)$ resembles a laminar boundary-layer profile and $\tau(\eta)$ decreases nonlinearly from its maximum, τ_{wall}, to $\tau(\eta = 2.8) \approx 0$ where the inviscid flow field begins.

Problem (2) Flow near a Rotating Disk (Von Karman's Viscous Pump)

Consider a large smooth disk rotating slowly with $\omega_0 = $ const in a stagnant, semi-infinite fluid reservoir. Neglecting end effects and assuming $p = p(z)$ only, fluid particles near the disk are flung outward as a result of a centrifugal force, which implies that fluid is drawn axially toward the disk for replacement to maintain mass continuity. In addition, a swirling motion is induced by the rotating disk surface. Thus, all velocity components in cylindrical coordinates are nonzero. The first step is to arrive at postulates for v_r, v_θ, and v_z, based on physical insight gained in conjunction with the list of assumptions, the boundary conditions, and the continuity equation (Fig. 3.8).

Assumptions

Steady, axisymmetric laminar incompressible flow without gravitational or end effects.

Postulates

- From the boundary conditions for v_θ we know that

$$v_\theta(z = 0) = r\omega_0 \quad \text{and} \quad v_\theta(z \to \infty) = 0$$

- On the basis of symmetry, it can be stated that

$$v_\theta = v_\theta(r, z) \quad \text{and} \quad v_r = v_r(r, z)$$

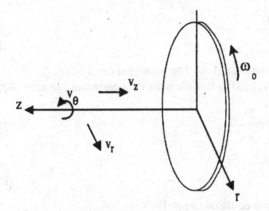

Fig. 3.8. Coordinate system and velocity components for rotating disk in an infinite body of viscous fluid.

- Because it is largely an unconfined flow field and the small disk rotates rather slowly, $p = p(z)$ only.
- Using $r\omega_o$ as a scale factor and recognizing the fact that *walls always produce complicated flow fields*, we postulate that

$$\frac{v_r}{r\omega_0} \text{ is a fct } (z) \text{ only} \quad \text{and} \quad \frac{v_\theta}{r\omega_0} \text{ is a fct } (z) \text{ only}$$

This implies that the velocity components change linearly with the radius.
- Checking the continuity equation and recalling that $\partial/\partial\theta \equiv 0$ yields

$$\frac{1}{r}\frac{\partial}{\partial r}(rv_r) + \frac{\partial v_z}{\partial z} = 0 \tag{P3.7}$$

Inserting the postulate for v_r into (P3.7), we find that $v_z = v_z(z)$ only; this is mathematically satisfactory and physically realistic.

Utilizing the previous result for the penetration depth $\delta \sim \sqrt{\nu t}$ (cf. Stokes I Problem, page 124 in the form $\delta \sim \sqrt{\nu/\omega_0}$, we can construct a dimensionless distance from the wall, $\zeta \sim z/\delta$, as

$$\zeta = z\sqrt{\frac{\omega_0}{\nu}}$$

Consequently, defining

$$f(\zeta) = \frac{v_r}{r\omega_0}, \qquad g(\zeta) = \frac{v_\theta}{r\omega_0}$$

and

$$h(\zeta) \equiv \frac{v_z}{\sqrt{\nu\omega_0}},$$

Equation (P3.7) is transformed to

♦ $2f + h' = 0$ (P3.8)

which relates the radial flow and the axial flow. The prime indicates $d/d\zeta$.

Inserting the postulates for v_r, v_z, and v_θ into the reduced momentum equation (App. B)

$$v_r \frac{\partial v_r}{\partial r} + v_z \frac{\partial v_r}{\partial z} - \frac{v_\theta^2}{r} = v \frac{\partial^2 v_r}{\partial z^2}$$ (P3.9)

we obtain

♦ $f^2 + hf' - g^2 = f''$ (P3.10)

Similarly, the reduced θ-momentum equation (App. B)

$$v_r \frac{\partial v_\theta}{\partial r} + v_z \frac{\partial v_\theta}{\partial z} + \frac{v_r v_\theta}{r} = v \frac{\partial^2 v_\theta}{\partial z^2}$$ (P3.11)

is transformed to

♦ $fg + hg' + fg = g''$ (P3.12)

The pressure $p(z)$ is nondimensionalized as

$$\hat{p}(\zeta) = \frac{p(z)}{\rho v \omega_0}$$

so that the z-momentum equation

$$v_z \frac{\partial v_z}{\partial z} = -\frac{1}{\rho} \frac{\partial p}{\partial z} + v \frac{\partial^2 v_z}{\partial z^2}$$ (P3.13)

now reads

♦ $h'' - hh' = (\hat{p})'$ (P3.14)

The associated boundary conditions for the coupled set of ODEs are

at $\zeta = 0$ $f = 0$, $g = 1$, $h = 0$ and $p = 0$ (P3.15a–d)

as $\zeta \to \infty$ $f = 0$ and $g = 0$ (P3.16a,b)

This constitutes a two-point boundary value problem (BVP), which can be rewritten as six first-order ODEs with f' and g' the additional unknowns. Using the shooting method (App. F.3) the BVP can be treated as an initial value problem (IVP) with $f'(0)$-values and $g'(0)$-values replacing the end conditions at $\zeta = \infty$. Thus, higher-order Runge–Kutta codes are suitable to solve the problem numerically by trial and error. Good values for the new initial conditions are (Panton 1984) $f'(0) = 0.510233$ and $g'(0) = 0.514922$ (cf. Sample-Problem 3.16). It turns out that the penetration depth is $\delta = 5.5\sqrt{v/\omega_0}$ and the disk-induced approach velocity $v_z(\infty) = 0.886\sqrt{v\omega_0}$, which is a volumetric flow rate per unit area, representing the *pumping effect*.

C Creeping Flow Solutions

When the Reynolds number is very small, typically Re $<$ 1 is associated with very slow fluid motion or "creeping flow," the inertia terms may be negligible. Hence, for Newtonian fluids the Navier–Stokes equations reduce to *Stokes's equations*

$$\blacklozenge \qquad \nabla p = \mu \nabla^2 \vec{v} \quad \text{and} \quad \nabla \cdot \vec{v} = 0 \qquad \qquad (3.27\text{a,b})$$

Actually, there are four possibilities for creeping flow:

$$\text{Re}_l = \frac{\rho v l}{\mu} \to 0 \quad \text{implies} \quad \begin{cases} \rho \to 0 & \text{(zero-mass fluid)} \\ v \to 0 & \text{(very slow motion)} \\ l \to 0 & \text{(small length scale)} \\ \mu \to \infty & \text{(highly viscous fluid)} \end{cases}$$

Thus, Stokes's equations are applicable to slow flow around a submerged body ($v \to 0$), highly viscous film lubrication ($\mu \to \infty, l < 1$), flow in porous media and locally small scale flows ($l \to 0$), or very low density fluid flow ($\rho \to 0, v < 1$). The approximation Re $\to$ 0 removes the nonlinearity from the momentum equation and hence a number of analytic solutions for steady two-dimensional creeping flow problems are available. In this section, we discuss flow past a sphere (Stokes), planar lubrication (Reynolds), and flow in very small conduits or passages (Darcy). It is interesting to note that these creeping flows may exhibit instabilities leading to *turbulence* at rather low Reynolds numbers. For example, uniform flow past a sphere becomes transient and three-dimensional as a result of vortex shedding at $\text{Re}_d \geq 140$. Viscous flow in rotating concentric cylinders develops Taylor vortices when $Ta = \omega_0 r_0 \Delta r / v \sqrt{\Delta r / r_0} > 41$. Darcy flow in porous media becomes turbulent when $\text{Re}_{d_m} \geq 100$.

Occasionally, it is convenient to solve Eq. (3.27) for planar flow by employing the stream function approach, where $\psi(x, y)$ is defined by $u = \partial \psi / \partial y$ and $v = -\partial \psi / \partial x$ to satisfy (3.27b). Cross-differentiation and subtraction eliminate the pressure gradient in (3.27a) but generate a fourth-order PDE:

$$\blacklozenge \qquad \nabla^4 \psi = 0 \qquad \qquad (3.28)$$

With the solution of (3.28), the 2-D velocity components can be computed and the pressure can be evaluated from Eq. (3.27a). The pressure field can also be directly obtained from

$$\blacklozenge \qquad \nabla^2 p = 0 \qquad \qquad (3.29)$$

The Laplace equation (3.29) appears when taking the divergence, that is, div or $\nabla \bullet$ operation of both sides of (3.27a).

C.1 Stokes Flow Past a Sphere

The Stokes solution implies very low Reynolds number flow past a rigid sphere of radius a_0, or slow motion of a sphere through a stagnant ambient (cf. sketch). The flow field is symmetric about the r–θ plane, that is, $v_\phi = 0$, and Eq. (3.28) can be expressed in spherical coordinates as

$$L^4 \psi = 0 \qquad \qquad (3.30)$$

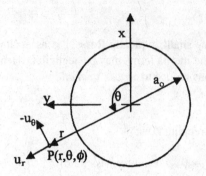

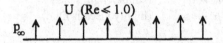

where (App. C and D)

$$L^2 \equiv \frac{\partial^2}{\partial r^2} + \frac{1-\alpha^2}{r^2}\frac{\partial^2}{\partial \alpha^2}; \qquad \alpha \equiv \cos\theta; \quad \text{and} \quad L^2\{L^2\} = L^4$$

By definition

$$u_r = -\frac{1}{r^2}\frac{\partial \psi}{\partial \alpha} = \frac{1}{r^2 \sin\theta}\frac{\partial \psi}{\partial \theta}$$

$$u_\theta = -\frac{1}{r\sqrt{1-\alpha^2}}\frac{\partial \psi}{\partial r} = -\frac{1}{r \sin\theta}\frac{\partial \psi}{\partial r}$$

The boundary conditions are

$$u_r = u_\theta = 0 \qquad \text{at } r = a_0 \qquad \text{whereas}$$

at $r \to \infty$: $u_r = U \cos\theta$ and

$$u_\theta = -U\sin\theta \qquad \text{from which } \psi = \int U r \sin^2\theta \, dr$$

This implies that at $r \gg a_0$

$$\psi(r \to \infty, \theta) = \frac{1}{2}Ur^2 \sin^2\theta \tag{3.31a}$$

Clearly, $\psi(\theta) \sim \sin^2\theta$, but $\psi(r)$ is more complicated than $\psi \sim r^2$ because of the wall presence. Thus, we can postulate

$$\psi(r,\theta) = (\sin^2\theta)f(r) \tag{3.31b}$$

and use $L^4\psi = 0$ to find $f(r)$. Specifically,

$$L^2\psi = L^2[\sin^2\theta f(r)] = \sin^2\theta \frac{d^2 f}{dr^2} + \frac{\sin^2\theta}{r^2}f\frac{d^2}{d\alpha^2}(1-\alpha^2) = \sin^2\theta\left[\frac{d^2 f}{dr^2} - \frac{2f}{r^2}\right]$$

and

$$L^4 \psi = L^2\{L^2 \psi\} = 0 \quad \Rightarrow \quad L^2\{L^2 f\} = 0$$

or

$$\left(\frac{d^2}{dr^2} - \frac{2}{r^2}\right)\left(\frac{d^2 f}{dr^2} - \frac{2}{r^2}f\right) = 0 \tag{3.32}$$

A trial solution of the form

$$f(r) = \frac{A}{r} + Br + Cr^2 + Dr^3 \tag{3.33}$$

fulfills the 4th-order ODE (3.32), so that the coefficients A to D can be determined via the four boundary conditions for u_r and u_θ. Thus, with

$$u_r = 2\left(\frac{A}{r^3} + \frac{B}{r} + C + Dr\right)\cos\theta$$

and

$$u_\theta = -\left(-\frac{A}{r^3} + \frac{B}{r} + 2C + 3Dr\right)\sin\theta$$

$$A = \frac{1}{4}Ua_0^3, \qquad B = -\frac{3}{4}Ua_0, \qquad C = \frac{1}{2}U \quad \text{and} \quad D = 0$$

As a result,

$$\psi(r,\theta) = \frac{r^2}{2}U\left(1 - \frac{3a_0}{2r} + \frac{a_0^3}{2r^3}\right)\sin^2\theta \tag{3.34a}$$

$$u_r = U\left[1 - \frac{3a_0}{2r} + \frac{a_0^3}{2r^3}\right]\cos\theta \tag{3.34b}$$

and

$$u_\theta = -U\left[1 - \frac{3a_0}{4r} - \frac{a_0^3}{4r^3}\right]\sin\theta \tag{3.34c}$$

The fluid flow field can be expressed in rectangular coordinates with x the streamwise coordinate as (cf. Schlichting 1979):

$$u = U\left[\frac{3}{4}\frac{a_0 x^2}{r^3}\left(\frac{a_0^2}{r^2} - 1\right) - \frac{a_0}{4r}\left(3 + \frac{a_0^2}{r^2}\right) + 1\right] \tag{3.35a}$$

$$v = \frac{3}{4}U\frac{a_0 xy}{r^3}\left(\frac{a_0^2}{r^2} - 1\right) \tag{3.35b}$$

and

$$w = \frac{3}{4} U \frac{a_0 xz}{r^3} \left(\frac{a_0^2}{r^2} - 1 \right)$$

(3.35c)

where $r = (x^2 + y^2 + z^2)^{1/2}$. From $\nabla p = \mu \nabla^2 \vec{v}$, we obtain

$$p - p_\infty = -\frac{3}{2} U \frac{\mu a_0 x}{r^3}$$

(3.36a)

For $r = a_0$, the surface pressure appears as a linear distribution (cf. graph).

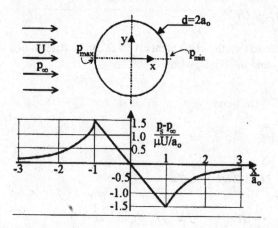

$$p_s = p_\infty - \frac{3}{2} U \frac{\mu x}{a_0^2}$$

(3.36b)

The drag force

$$F_D = a_0^2 \int_0^{2\pi} d\phi \int_0^\pi \tau_{xr}|_{r=a_0} \sin \theta \, d\theta = 6\pi \mu a_0 U$$

(3.37a)

which in dimensionless form, $c_D = \dfrac{F_D}{[(\rho/2)U^2(\pi a_0^2)]}$, yields

$$c_D = \frac{24}{\mathrm{Re}_d} \qquad \text{for } 0 < \mathrm{Re}_D < 1$$

(3.37b)

C.2   Reynolds Flow in Lubrication Systems

The Reynolds one-dimensional solution is based on steady laminar "confined wedge" flow driven by a surface moving with a velocity, $U = $ const, relative to a stationary surface $h(x)$ apart (cf. sketch). Neglecting end effects, a (varying) pressure gradient that contributes to the load capacity and unique velocity field is generated.

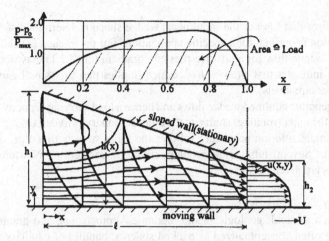

The very small gap size h provides a length scale for the y-direction with which it can be shown that the viscous forces are dominant. Specifically,

$$\frac{F_{\text{inertia}}}{F_{\text{viscous}}} \sim \frac{u\partial u/\partial x}{v\partial^2 u/\partial y^2} \sim \frac{U^2/l}{vU/h^2} = \frac{Ul}{v}\left(\frac{h}{l}\right)^2 < 1$$

Although $\text{Re}_l = Ul/v = O(10 \text{ to } 100)$ or higher in lubrication, $\text{Re}_h = \text{Re}_l(h/l)^2 = O(0.1)$. Scale analysis also shows that

$$v \ll u, \qquad \partial^2 u/\delta x^2 \ll \partial^2 u/\partial y^2 \quad \text{and} \quad \partial p/\partial y \approx 0$$

so that Eq. (3.27a) can be reduced to

♦
$$\frac{dp}{dx} = \mu\frac{\partial^2 u}{\partial y^2} \tag{3.38}$$

subject to

$$u(y = 0) = U, u[y = h(x)] = 0 \quad \text{and} \quad p(x = 0) = p(x = l) = p_0$$

Partial integration yields

$$u(x, y) = U\left(1 - \frac{y}{h(x)}\right) - \frac{[h(x)]^2}{2\mu}\left(\frac{dp}{dx}\right)\frac{y}{h(x)}\left(1 - \frac{y}{h(x)}\right) \tag{3.39}$$

where the pressure gradient is obtained from

$$Q = \int_0^{h(x)} u(x, y)\, dy = \text{const} \tag{3.40}$$

that is,

$$\frac{dp}{dx} = \frac{12\mu}{2h^3}(Uh - 2Q) \tag{3.41}$$

Further analysis requires that $h(x)$ or the (nonlinear) wedge shape is determined such that the pressure distribution is optimal for a specific application. For the linear $h(x)$ sketched, backflow similar to Couette flow for a positive pressure gradient (cf. 3.2.1) may occur. In any case, Eq. (3.41) indicates that $p(x) \sim h^{-3}$, which implies that very small gap sizes may produce high pressure levels.

Journal bearings generate confined wedge flows and hence a load carrying capacity via an *eccentricity* between the inner (rotating) shaft and the outer stationary cylinder, or vice versa. The relevant approximate solution procedure follows the same steps outlined previously (cf. Hamrock 1994). Several lubrication problem solutions considering Newtonian or non-Newtonian fluids are given in Sect. 3.3.

C.3 Darcy Flow in Small Passages

Porous materials, such as sand, geologic media, ceramics, composites, and granular or fibrous materials, are often conceptualized as packed spheres, bundles of capillary tubes, stacks of capillary fissures, and so forth. On the basis of the mean diameter of a representative conduit, the Reynolds number is less than unity. Typically, the porous medium is assumed to be homogeneous, isotropic, saturated, and rigid (cf. sketches of porous media models).

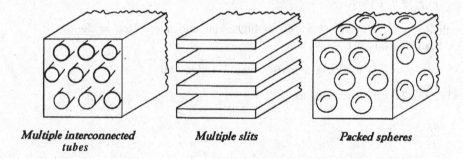

| *Multiple interconnected* | *Multiple slits* | *Packed spheres* |
| *tubes* | | |

Porous Media Models Darcy (1856) postulated an empirical equation for the average velocity in these tiny channels to be

$$u = \frac{Q}{\varepsilon A} = -\frac{k}{\varepsilon \mu}\left(\frac{\partial p}{\partial s} - \rho g \frac{\partial z}{\partial s}\right) \tag{3.42a}$$

or in terms of the Darcy velocity $v = Q/A$ in three-dimensional coordinates

$$\frac{\mu}{k} v_i = -\frac{\partial}{\partial x_i}(p - \rho g z)$$

or

$$\blacklozenge \qquad \vec{v} = -\frac{k}{\mu}(\nabla p - \rho \vec{g}) \tag{3.42b,c}$$

Here Q is the flow rate, ε is the porosity, A is the geometric cross-sectional area, εA is the total effective flow area, k is the permeability, and s is the coordinate in flow direction. Considering horizontal flow, that is, eliminating gravitational effects, and comparing

Eq. (3.42) with Eq. (3.27a) it is transparent that the viscous term has been replaced by a resistance force per unit volume, that is,

$$\mu\frac{\partial^2 v_x}{\partial y^2} \quad \Rightarrow \quad R_f = \frac{\varepsilon\mu}{k}u = \frac{\mu}{k}v$$

The fluid balance takes on the form

$$\frac{\partial}{\partial t}(\varepsilon\bar{\rho}) + \nabla\cdot(\bar{\rho}\vec{v}) = 0 \tag{3.43}$$

where $\bar{\rho}$ is the ensemble averaged or macroscopic density. Taking the divergence of (3.42b) produces again Eq. (3.29) for constant properties. Numerous attempts have been made to enhance Eq. (3.42) by adding inertia and/or viscous terms while reinterpreting R_f as a fluid–matrix interaction force. A derivation of Eq. (3.42) from the Navier–Stokes equations is given by Whitaker (1986).

Direct application of the Navier–Stokes equation to porous media flow is only possible for a *macroscopic* description of the fluid motion after careful implementation of a volume averaging procedure (cf. Cvetkovic 1986). When the flow passages are modeled as symmetric, slightly varying slits, the momentum equation reduces (cf. Eq. (3.27a) and sketch) to

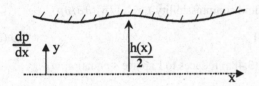

$$\mu\frac{\partial^2 u}{\partial y^2} = \frac{dp}{dx} \tag{3.44}$$

which yields for each unconnected capillary tube

$$u(x, y) = -\frac{[h(x)]^2}{8\mu}\left(\frac{dp}{dx}\right)\left[1 - 4\left(\frac{y}{h}\right)^2\right] \tag{3.45}$$

and

$$Q_{\text{tube}} = -\frac{[h(x)]^3}{12\mu}\left(\frac{dp}{dx}\right) = \text{const} \tag{3.46}$$

The solutions (3.45) and (3.46) have little practical value even for the simplest problems in porous media flow, such as "radial flow near a pumping well" or "gas flow in a porous slab." Solutions to these problems are based on reduced forms of Equations (3.42a) and

(3.43). Specifically, the transient term of (3.43) can be related to the specific fluid mass storage coefficient S_ϕ and (piezometric) head or potential ϕ, that is:

$$\frac{\partial(\varepsilon\bar{\rho})}{\partial t} = S_\phi \frac{\partial\phi}{\partial t}$$

so that the continuity equation reads

$$-\nabla \cdot (\bar{\rho}\vec{v}) = S_\phi \frac{\partial\phi}{\partial t} \quad \text{or} \quad -\bar{\rho}\nabla \cdot \vec{v} = S_\phi \frac{\partial\phi}{\partial t}$$

Expressing now the momentum equation (3.42b) in terms of the potential $\phi = z + p/\bar{\rho}g$ and introducing the hydraulic conductivity $K = \bar{\rho}gk/\mu = \text{const}$, we obtain

$$\vec{v} = -K\nabla\phi \tag{3.47}$$

Inserting (3.47) into the modified continuity equation yields for a homogeneous isotropic medium

♦ $$\bar{\rho}K\nabla^2\phi = S_\phi \frac{\partial\phi}{\partial t} \tag{3.48a}$$

The storage coefficient, that is, specific mass storativity, S_ϕ, is a function of the medium porosity ε, the elastic property of the solid matrix $\alpha = -(1/V_s)(\partial V_s/\partial p) = 1/(1-\varepsilon)(\partial\varepsilon/\partial p)$, and the coefficient of fluid compressibility $\beta = (1/\bar{\rho})(\partial\bar{\rho}/\partial p)$.

$$S_\phi = g\bar{\rho}^2[\alpha(1-\varepsilon) + \varepsilon\beta] \tag{3.48b}$$

For steady-state applications, Eq. (3.48a) reduces to Laplace's equation.

Steady Radial Flow near a Pumping Well   Consider a homogeneous semi-infinte confined aquifer of thickness B and with constant K. When pumping at a constant rate Q_w takes place, a cone of depression is formed with a drawdown $s(r) = \phi_R - \phi(r)$ near the well of radius r_w (cf. sketch). For steady flow in cylindrical coordinates, Eq. (3.48a) reads

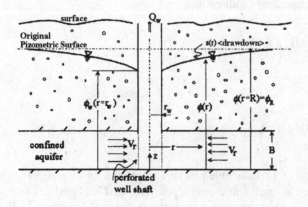

$$\frac{\partial^2 \phi}{\partial r^2} + \frac{1}{r}\frac{\partial \phi}{\partial r} = 0$$

subject to

$$\phi(r = r_w) = \phi_w \quad \text{and} \quad \phi(r = R) = \phi_R$$

Integration yields

$$\phi(r) = \phi_w + \frac{Q_w}{2\pi K B} \ln \frac{r}{r_w}$$

The radial specific flux or Darcy velocity is

$$v_r = K\frac{\partial \phi}{\partial r} = \frac{Q_w}{2\pi r B}$$

Unsteady Pressure-Driven Flow of a Gas   Consider an ideal gas of viscosity μ at pressure p_0 saturating a semiinfinite porous medium of constant porosity ε and permeability k. Gravitational effects can be neglected. At time $t = 0$ the pressure is changed at the surface and maintained at a new level p_1.

Modeling Equations
Equations (3.43) and (3.42b) can be reduced to

$$\varepsilon\frac{\partial \rho}{\partial t} + \frac{\partial}{\partial x}(v_x \rho) = 0 \quad \text{and} \quad v_x = -\frac{k}{\mu}\frac{\partial p}{\partial x}$$

where $p = \rho R T.$

Combining these equations yields

$$\frac{\partial}{\partial x}\left(p\frac{\partial p}{\partial x}\right) = \frac{\varepsilon\mu}{k}\frac{\partial p}{\partial t}$$

subject to

$$p(x, 0) = p_0, \ p(0, t) = p_1 \quad \text{and} \quad p(\infty, t) = p_0; \qquad p_1 > p_0$$

Solution
Seeking similarity, the PDE for $p(x, t)$ is transformed to a two-point BVP for $f(\eta)$. Analogously to the solution of the 1-D transient diffusion equation (E3.17), we try

$$\eta = C\frac{x}{\sqrt{t}}$$

where $C = \sqrt{\mu/kp_0}$ to make the similarity variable dimensionless. Thus, with

$$f = \frac{p - p_0}{p_1 - p_0} = \frac{p - p_0}{\Delta p}$$

each term of the PDE can be transformed as follows:

$$\frac{\partial p}{\partial x} = \Delta p f' \frac{C}{\sqrt{t}}; \qquad p\frac{\partial p}{\partial x} = \frac{\Delta p C}{\sqrt{t}}(\Delta p f f' + p_0 f')$$

$$\frac{\partial}{\partial x}\left(p\frac{\partial p}{\partial x}\right) = \frac{\Delta p C^2}{t}[\Delta p(f'^2 + ff'') + p_0 f'']$$

$$\frac{\partial p}{\partial t} = -\frac{\Delta p C x}{2t^{\frac{3}{2}}}f'$$

As a result, we obtain

$$f'' + \kappa f f'' + \kappa f'^2 + \frac{\varepsilon\eta}{2}f' = 0$$

where $\quad \kappa = \Delta p/p_0 \quad$ and $\quad f(\eta = 0) = 1 \quad$ and $\quad f(\eta = \infty) = 0.$

As outlined in Sect. 3.2.3, a "shooting method" could be applied numerically, where a *second* initial condition $f'(\eta = 0)$ is found by trial and error so that the end condition $f(\eta = \infty) = 0$ can be matched. Thus, the Runge–Kutta IVP-solver of App. F.3 can be employed to find $f(\eta)$ and hence $p(x, t)$.

D Non-Newtonian Fluid Flow Models

In Section 1.3.1, the shear stress tensor $\vec{\vec{\tau}}$ was related *linearly* for Newtonian fluids and related *nonlinearly* for non-Newtonian fluids to the rate-of-deformation tensor $\vec{\vec{\dot{\gamma}}} = \nabla\vec{v} + (\nabla\vec{v})^{\text{tr}}$. Thus, for incompressible flow (cf. Fig. 1.7)

$$\tau_{ij} = \begin{cases} -\mu\dot{\gamma}_{ij} & \text{for Newtonian fluids} \\ -\eta(\dot{\gamma} \text{ or } \tau)\dot{\gamma}_{ij} & \text{for non-Newtonian fluids} \end{cases}$$

Some practical fluid models and their applications are summarized in the following:

- The "power-law" model of Ostwald and deWaele is a two-parameter model for a wide variety of shear-thinning ($n < 1$) or shear-thickening ($n > 1$) aqueous solutions: simple polymeric liquids, and so on.

$$\eta = m\dot{\gamma}^{n-1}, \qquad |\dot{\gamma}| \geq 1s^{-1} \tag{3.49}$$

  For Newtonian fluids, the dimensionless exponent n is equal to 1, which implies that m becomes μ.
- The Carreau–Yasuda model is a five-parameter expression that extends the application of the power law model to concentrated polymer solutions and melts.

$$\frac{\eta - \eta_\infty}{\eta_0 - \eta_\infty} = [1 + (\lambda\dot{\gamma})^a]^{(n-1)/a} \tag{3.50}$$

where η_0 is the zero-shear-rate viscosity; η_∞ is the infinite-shear-rate viscosity (typically $\eta_\infty \to 0$); λ is a time constant that represents the "fading memory" of certain polymers, that is, $\lambda = O(10 \text{ s})$ to $O(100 \text{ s})$; and a (typically $a = 2$) is a

dimensionless parameter describing the transition region between the zero-shear-
stress and power-law regions.
- The Bingham model is a two-parameter formula for pastes and slurries exhibiting
  a threshold or yield stress τ_0 resisting motion.

$$\eta = \begin{cases} \infty & \text{for } \tau \le \tau_0 \\ \mu_p + \dfrac{\tau_0}{\dot{\gamma}} & \tau > \tau_0 \end{cases} \tag{3.51a,b}$$

- The basic Casson model is a nonlinear extension of the Bingham model suitable
  for simulating suspensions of (spherical) particles in polymer solutions.

$$\sqrt{\tau_{ij}} = \begin{cases} 0 & \text{for } \tau \le \tau_0 \\ \sqrt{\tau_0} + \sqrt{\mu_0}\sqrt{\dot{\gamma}_{ij}} & \text{for } \tau > \tau_0 \end{cases} \tag{3.52a,b}$$

- A modified Casson model can be employed to simulate blood rheology, which is
  represented by the stress tensor $\vec{\vec{\tau}}$ with the following relations:

$$\vec{\vec{\tau}} = 2\eta(II_D)\vec{\vec{D}} \tag{3.53a}$$

Here, $\vec{\vec{\varepsilon}} \equiv \vec{\vec{D}} = \frac{1}{2}[\nabla\vec{v} + (\nabla\vec{v})^{\text{tr}}]$ is the rate-of-strain tensor, and the apparent
viscosity η is a function of the shear rate

$$\eta(II_D) = \frac{1}{2\sqrt{II_D}}\left[C_1(Ht) + C_2(Ht)\sqrt{2\sqrt{II_D}}\right]^2 \tag{3.53b}$$

where Ht is the hematocrit and II_D is the second scalar invariant of $\vec{\vec{D}}$, that is,

$$II_D = \frac{1}{2}\left[(\text{trace}\vec{\vec{D}})^2 + \text{trace}(\vec{\vec{D}}^2)\right] \tag{3.53c}$$

Writing out the right-hand side in component form, yields

$$II_D = D_{11}D_{22} + D_{11}D_{33} + D_{22}D_{33} - D_{12}D_{21} - D_{13}D_{31} - D_{23}D_{32} \tag{3.53d}$$

The coefficients C_1 and C_2 were determined for $Ht = 40$ percent as $C_1 = 0.2\,(\text{dyn/cm}^2)^{1/2}$
and $C_2 = 0.18\,(\text{dyn}\cdot\text{s/cm}^2)^{1/2}$, based on Merrill's experimental data (cf. Merrill 1968).
The following graph shows the variation of blood viscosity η (Eq. 3.53b) with shear
rate $\dot{\gamma} = 2\sqrt{II_D}$. For human hemodynamic studies (cf. Kleinstreuer et al. 1996) we used
$\mu = 0.0348\,(\text{dyn}\cdot\text{s/cm}^2)$, that is, $\nu = 0.033\,(\text{cm}^2/\text{s})$, which guarantees a smooth transition from Casson model to Newtonian fluid. At the other end of the curve, since Casson's
model is only suitable for $\dot{\gamma} > 1(s^{-1})$, we take $\eta = \eta|_{\dot{\gamma}=1} = 0.1444\,(\text{dyn}\cdot\text{s/cm}^2)$ when
$\dot{\gamma} < 1(s^{-1})$, which is the "zero-shear rate" condition (cf. graph).

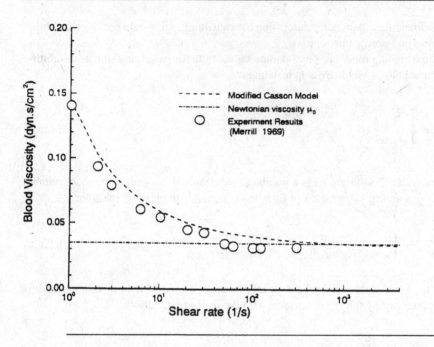

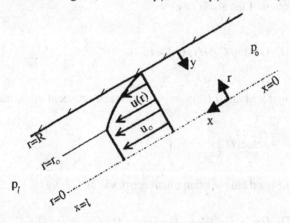

Example

Calculate the volumetric flow rate of a basic Casson fluid through a slanted circular tube of radius R, length l, where $-dp/dx \approx \Delta p/l = $ const (cf. sketch).

Assumptions

- Steady laminar fully developed flow
- Shear-rate dependent viscosity
- Uniform fluid motion when $\tau_{rx} < \tau_0$ (yield stress)
- Gravitational effects absorbed in $\Delta p/l$

Postulates

$v = w = 0$ and $u = u(r)$ only; $\dot{\gamma}_{ij} \Rightarrow \dot{\gamma}_{rx} = \dfrac{du}{dr}$; $\dot{\gamma}_{rx} = 0$ for $0 \leq r \leq r_0$

Governing Equations

The Navier–Stokes equations are *not* applicable. Thus, the reduced equation of motion in the x-direction reads (cf. App. B)

$$0 = -\frac{dp}{dx} - \frac{1}{r}\frac{d}{dr}(r\tau_{rx})$$

Integration yields

$$\tau_{rx} = \frac{\Delta p}{2l}r \qquad \text{where } \tau_{rx}(r=0) = 0$$

Invoking the Casson model, we obtain

$$\text{at } r = r_0 \qquad \tau_{rx}|_{r=r_0} = \tau_0 = \frac{\Delta p}{2l}r_0$$

and

$$\text{at } r = R \qquad \tau_{rx}|_{r=R} = \tau_w = \frac{\Delta p}{2l}R$$

so that $\tau_{rx} = \tau_w\, r/R$ as for Newtonian fluids. Continuity is preserved in terms of

$$Q = 2\pi \int_0^R ur\,dr := 2\pi\left[\int_0^{r_0} u_0 r\,dr + \int_{r_0}^R u(r)r\,dr\right] = \text{const}$$

Solution

From the Casson relation we know that

$$-\dot{\gamma}_{rx} = -\frac{du}{dr} = -\frac{1}{\mu_0}(\sqrt{\tau_{rx}} - \sqrt{\tau_0})^2 \qquad \text{for } \tau > \tau_0$$

Now we have two possibilities:

(i) Inserting $\tau_{rx} = Ar$, where $A \equiv \Delta p/(2l)$, and integrating to find $u(r)$ or
(ii) expressing the equation for $Q = \text{const}$ in terms of τ and integrating to find Q directly.

For the second approach we recall that

$$r = \frac{\tau_{rx}}{\tau_w}R \quad \text{or} \quad dr = \frac{R}{\tau_w}d\tau_{rx}, \qquad r_0 = \frac{2l}{\Delta p}\tau_0, \qquad R = \frac{2l}{\Delta p}\tau_w$$

and, applying integration by parts,

$$Q = 2\pi \int_0^R ur\,dr = \pi\left[ur^2|_0^R - \int_0^R r^2 du\right] = -\pi \int_{r_0}^R r^2\left(\frac{du}{dr}\right)dr$$

After substitution

$$Q = \frac{\pi}{\mu_0}\left(\frac{R}{\tau_0}\right)^3 \int\limits_{\tau_0}^{\tau_w} \tau_{rx}^2 [\tau_{rx} - 2\sqrt{\tau_{rx}}\sqrt{\tau_0} + \tau_0]\, d\tau_{rx}$$

Introducing $\kappa = (2l\tau_0/\Delta p R)$ and using $\tau_w = (\Delta p/2l) \cdot R$ we obtain after integration

$$Q = \frac{\pi R^4}{8\mu_0}\left(\frac{\Delta p}{l}\right)\left(1 - \frac{16}{7}\sqrt{\kappa} + \frac{4}{3}\kappa - \frac{1}{21}\kappa^4\right)$$

Note: For $\tau_0 = 0$, $\kappa = 0$ and with $\mu_0 \equiv \mu$ the well-known Hagen–Poiseuille solution is obtained.   □

Additional information on non-Newtonian fluid models may be found in Tanner (1985), Churchill (1988), Bird et al. (1987), and Macosko (1994), among others. A few analytic problem solutions appear in Sections 2.4 and 3.3.

3.2.2 Moderate-Reynolds-Number Flows

The moderate-Reynolds-number regime falls roughly between creeping flows and boundary-layer flows. As expected, pressure, viscous, and inertia forces are all important in these flows, which may exhibit strong vortices depending upon the given system geometry. The governing equations are spatially elliptic; that is, boundary conditions have to be prescribed all around the flow domain. Examples include laminar flow in cavities, flow past cylinders or spheres, flow along walls with obstructions, and laminar entrance or exit flows where a reservoir feeds a cascade of plates or a sluice gate (cf. Panton 1984). The resulting flow patterns are strongly Reynolds-number-dependent and may abruptly change when the Reynolds number changes slightly. Needless to say, numerical methods are required to solve practical medium-Reynolds-number flow problems. Specific case studies are outlined and discussed in Chapter 5.

3.2.3 High-Reynolds-Number or Boundary-Layer Flows

Flows at high Reynolds numbers are not necessarily turbulent. For example, *laminar* shear layers exist on flat plates up to $\mathrm{Re}_L \approx 5 \times 10^5$. In the entrance region of ducts, in nozzles and diffusers, or at the surface of submerged bodies, boundary layers develop when the free stream Reynolds number is high enough. These shear layers may grow and fill the entire duct of a constant cross-sectional area as in developing *internal* flows, or thicken and ultimately separate from the body (as a result of adverse pressure gradients), as in *external* flows. Typically, the boundary-layer concept becomes applicable for Reynolds numbers, say, $\mathrm{Re}_L = (u_\infty L/\nu) \geq O(10^3)$ for *wall-bounded* flows. A proper selection of the length scale $L := l$, an axial length, or $L := d$, a mean diameter, is important. Transition to turbulence, caused by instabilities, is a highly nonlinear phenomenon discussed elsewhere (cf. Warsi 1993). In any case, no matter what the subsonic free stream Reynolds number is, the boundary layer starts as a laminar thin shear layer. In fact, at the stagnation point, or leading edge, a mathematical singularity exists, as discussed later.

First we consider *external* high-Reynolds-number flows; *internal* flow applications are discussed later. Thus, of interest are

- the concept of boundary-layer theory,
- the derivation of the boundary-layer equations (Prandtl's first-order approximation of the Navier–Stokes equations), and
- the application of Prandtl's theory to steady, laminar, two-dimensional wedge flows and then to *internal* boundary-layer-type flows.

A   Boundary-Layer Concept

A *boundary layer* is a moving-fluid region near a wall or interface, in which there is a single predominant direction of flow with a high level of vorticity. In such a thin region, shear stresses, heat fluxes, and mass fluxes are significant only normal to the predominant or streamwise direction. Examples of momentum (velocity), thermal (temperature), or mass (concentration) boundary layers are the following:

- Shear layers along flat plates, wedges, and airfoils; flows in duct entrance regions; planar stagnation and rotating disk flows; moving-boundary flows
- Thermal boundary layers in ducts (Graetz problem) or along plates or surfaces of enclosures with constant wall temperature or constant wall heat flux condition
- Concentration boundary layers in porous channels (e.g., ultrafiltration) or along external surfaces in convection mass transfer systems (e.g., reactors)

For certain applications (e.g., fuel droplet vaporization or mass transfer on catalytic surfaces) all three boundary layers are coupled and have to be solved simultaneously. Outside these layers of steep gradients, changes in velocity, temperature, or concentration normal to the streamwise direction are negligible. Thus, in momentum boundary-layer problems, Euler's equation or potential flow theory can be used to describe the flow field outside thin shear layers. Some of the features of laminar and turbulent boundary-layer flows are depicted in Figure 3.9.

In order to capture a physical–mathematical understanding of boundary-layer formation, we consider two different points of view: (i) the approach based on the "no-slip condition" at the wall and (ii) the approach of vorticity generation at the wall with outward diffusion.

No-Slip Condition Approach

For fluids with constant properties, the equation of motion is (cf. Sect. 2.2.3)

$$\rho \frac{D\vec{v}}{Dt} = -\nabla \hat{p} + \mu \nabla^2 \vec{v} \tag{3.54a}$$

where $\hat{p} = (p + \phi)$ with ϕ some conservative potential such as gravity. Equation (3.54a) in nondimensional form reads

$$\frac{D\tilde{\vec{v}}}{D\tilde{t}} = -\tilde{\nabla}\tilde{p} + \frac{1}{\text{Re}} \tilde{\nabla}^2 \tilde{\vec{v}} \tag{3.54b}$$

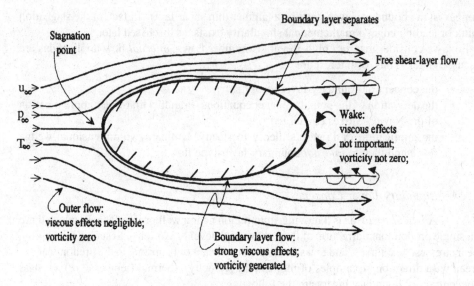

Fig. 3.9. Features of external high-Reynolds-number flow.

where $\tilde{v}_i = v_i/u_\infty$, $\tilde{t} = tu_\infty/L$, $\tilde{\nabla} = \nabla L$, and $\mathrm{Re} = u_\infty L/\nu$. It turns out that $\tilde{p} = (\hat{p} - p_\infty)/(\rho_\infty u_\infty^2)$, which is dimensionally consistent. When $\mathrm{Re} \to \infty$, as in potential flows, viscous forces apparently become negligible in most of the domain and Euler's equation is obtained, that is, in dimensional form

$$\rho \frac{D\vec{v}}{Dt} = -\nabla \hat{p} \tag{3.55}$$

Equation (3.55) is of first order and fulfills *only* the permeability or impermeable-wall condition. The second-order term, $\mu \nabla^2 \vec{v}$, is missing and the "no-slip condition" cannot be invoked. Since the no-slip boundary condition has been experimentally proven to exist for solid walls at all Reynolds numbers, there has to be a thin layer close to the wall surface where the viscous terms, $\mu \nabla^2 \vec{v}$, always remain significant. In addition, within such a layer of thickness δ, the velocity has to increase rapidly from $u(y = 0) = 0$ to $u(y = \delta) = U_{\text{outer}}$ or u_{edge}, which implies that steep velocity changes, $\partial u/\partial y$, can be expected near a solid boundary. Thus, within the thin shear layer of variable thickness δ, inertia forces and viscous forces balance each other; hence, they are of comparable order of magnitude:

$$|(\vec{v} \cdot \nabla)\vec{v}| \sim |\nu \nabla^2 \vec{v}|$$

In terms of an "order-of-magnitude-analysis" or scale analysis (cf. Sect. 2.3.1) in streamwise direction, that is, considering $f_{\text{inertia}} \sim a_x$ versus $f_{\text{viscous}} \sim (\partial \tau_{yx}/\partial y)$ or $u(\partial u/\partial x)$ versus $\nu(\partial^2 u/\partial y^2)$, we have (Fig. 3.10)

$$\frac{U_0^2}{L} \sim \nu \frac{U_0}{\delta^2}$$

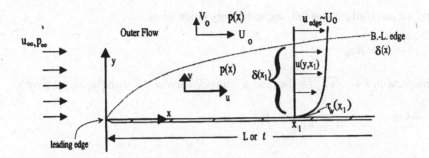

Fig. 3.10. Laminar boundary layer flow past a flat plate.

from which

$$\frac{\delta}{L} \sim \left(\frac{U_0 L}{\nu}\right)^{-1/2} = \frac{1}{\sqrt{\mathrm{Re}_L}} \ll 1 \qquad (3.56)$$

In addition, when for $\mathrm{Re}_x \gg 1$

$$\frac{\delta}{x} \ll 1 \qquad \text{then} \qquad \frac{v}{u} \ll 1 \qquad (3.56c)$$

that is, streamlines in horizontal, flat plate boundary layer flow curve slightly upward.

Wall Vorticity Generation and Diffusion Approach
The equation of motion in terms of vorticity transport can be written as (cf. Sect. 1.3.2) follows:

$$\underbrace{\frac{D\vec{\zeta}}{Dt}}_{\substack{\text{vorticity} \\ \text{time rate} \\ \text{of change}}} = \underbrace{(\vec{\zeta} \cdot \nabla)\,\vec{v}}_{\substack{\text{vorticity} \\ \text{sketching}}} + \underbrace{\nu\nabla^2\vec{\zeta}}_{\substack{\text{vorticity} \\ \text{diffusion}}} \qquad (3.57)$$

where $\vec{\zeta} = \nabla \times \vec{v}$. Consider a free stream of nonrotating "fluid particles" approaching a flat plate as thin as a razor blade. Either the fluid elements reaching the solid surface stick to the surface (no slip), or the ones next to them start rotating as a result of frictional effects. In turn, the next "layer" of fluid particles starts to spin, and so forth. Hence vorticity is generated along a solid boundary, and the vorticity diffuses away from the wall, represented by the term $\nu\nabla^2\vec{\zeta}$ in Equation (3.57). At large Reynolds numbers, the vorticity is predominantly swept downstream, thus leaving the flow a certain distance from the wall, that is, outside the shear layer, essentially irrotational. In summary,

Rate of Vorticity Convection $\to\gg$ *Rate of Vorticity Diffusion* $\uparrow$

or in terms of characteristic times (cf. Eq. (E3.23) of Sect. 3.2.1).

$$\text{Residence Time } \frac{L}{U_0} \gg \text{Diffusion Time } \frac{\delta}{U_0} = \sqrt{\frac{\nu L}{U_0^3}}$$

From this we can confirm that the inequality only holds when

$$\sqrt{\frac{U_0 L}{\nu}} = \mathrm{Re}_L^{1/2} \gg 1$$

Furthermore, with $\delta = \sqrt{\nu L / U_0}$ the distance traveled across the shear layer, we have

$$\Delta t_{\mathrm{diff}} \sim \frac{\delta}{U_0} \sim \frac{L}{U_0 \sqrt{\mathrm{Re}_L}} \ll 1$$

or

♦ $$\frac{\delta}{L} \sim \frac{1}{\sqrt{\mathrm{Re}_L}} \ll 1$$

as before (cf. Eq. (3.56a)).

B Prandtl's Laminar Boundary-Layer Equations

In the previous section we established the condition for applying Prandtl's boundary-layer theory as

♦ $$\frac{\delta}{l} \sim \frac{v}{u} \sim \frac{1}{\sqrt{\mathrm{Re}_l}} \ll 1 \tag{3.58}$$

The criterion (3.58) implies that

- the boundary layer is very thin when compared with a length scale of the submerged body;
- the Reynolds number is large; for example, for laminar flow past a horizontal flat plate, the range is $10^3 \le \mathrm{Re}_l \le 5 \times 10^5$ and for turbulent flow, $\mathrm{Re}_l > 10^7$;
- the streamwise velocity is dominant, that is, $u \gg v$;
- the outer flow may be deflected because of the growing boundary layer, where again $V_0 / U_0 \sim v/u \sim \delta/l \ll 1$; and
- rapid changes in the direction normal to the solid boundary can be expected for all principal variables, except the pressure.

Restricting the *derivation* of the boundary-layer equations to two-dimensional flow with constant fluid properties, we start with the Navier–Stokes equations in rectangular coordinates.

$$(x\text{-momentum}) \qquad \frac{\partial u}{\partial t} + u \frac{\partial u}{\partial x} + v \frac{\partial u}{\partial y} = -\frac{1}{\rho} \frac{\partial p}{\partial x} + v \left(\frac{\partial^2 u}{\partial x^2} + \frac{\partial^2 u}{\partial y^2} \right) \tag{3.59}$$

$$(y\text{-momentum}) \qquad \frac{\partial v}{\partial t} + u \frac{\partial v}{\partial x} + v \frac{\partial v}{\partial y} = -\frac{1}{\rho} \frac{\partial p}{\partial y} + v \left(\frac{\partial^2 v}{\partial x^2} + \frac{\partial^2 v}{\partial y^2} \right) \tag{3.60}$$

$$(\text{continuity}) \qquad \frac{\partial u}{\partial x} + \frac{\partial v}{\partial y} = 0 \tag{3.61}$$

In order to nondimensionalize Equations (3.59) to (3.61), we select the following reference quantities: plate length, l; boundary layer thickness, $\delta(x)$; free stream velocity, u_∞; free stream pressure, p_∞; and the y-component of the outer flow velocity vector, V_0

(cf. Fig. 3.10). Thus, we generate dimensionless variables that are all of the order of 1, in defining

$$\tilde{x} = \frac{x}{l}, \qquad \tilde{y} = \frac{y}{\delta}, \qquad \tilde{u} = \frac{u}{u_\infty}, \qquad \tilde{v} = \frac{v}{V_0}, \qquad \tilde{p} = \frac{p}{p_\infty} \quad \text{and} \quad \tilde{t} = \frac{t u_\infty}{l}$$

This leads to a set of normalized equations with $\text{Re} = u_\infty l / \nu$ where the dimensionless coefficients have a relative order of magnitude as symbolically indicated with $\varepsilon \ll 1$.

(continuity)
$$\underbrace{\frac{u_\infty}{V_0} \frac{\partial \tilde{u}}{\partial \tilde{x}}}_{O(\varepsilon^{-1})} + \underbrace{\frac{l}{\delta} \frac{\partial \tilde{v}}{\partial \tilde{y}}}_{O(\varepsilon^{-1})} = 0 \tag{3.62}$$

(x-momentum)
$$\frac{\partial \tilde{u}}{\partial \tilde{t}} + \tilde{u} \frac{\partial \tilde{u}}{\partial \tilde{x}} + \underbrace{\frac{l}{\delta} \frac{V_0}{u_\infty} \tilde{v} \frac{\partial \tilde{u}}{\partial \tilde{y}}}_{O(1)} = - \underbrace{\frac{p_\infty}{\rho u_\infty^2} \frac{\partial \tilde{p}}{\partial \tilde{x}}}_{O(1)}$$

$$+ \underbrace{\frac{1}{Re}}_{O(\varepsilon^2)} \left(\frac{\partial^2 \tilde{u}}{\partial \tilde{x}^2} + \underbrace{\frac{l^2}{\delta^2} \frac{\partial^2 \tilde{u}}{\partial \tilde{y}^2}}_{O(\varepsilon^{-2})} \right) \tag{3.63}$$

(y-momentum)
$$\frac{\partial \tilde{v}}{\partial \tilde{t}} + \tilde{u} \frac{\partial \tilde{v}}{\partial \tilde{x}} + \underbrace{\frac{l}{\delta} \frac{V_0}{u_\infty} \tilde{v} \frac{\partial \tilde{v}}{\partial \tilde{y}}}_{O(1)} = - \underbrace{\frac{l}{\delta} \frac{u_\infty}{V_0} \frac{p_\infty}{\rho u_\infty^2} \frac{\partial \tilde{p}}{\partial \tilde{y}}}_{O(\varepsilon^{-2})}$$

$$+ \underbrace{\frac{1}{Re}}_{O(\varepsilon^2)} \left(\frac{\partial^2 \tilde{v}}{\partial \tilde{x}^2} + \underbrace{\frac{l^2}{\delta^2} \frac{\partial^2 \tilde{v}}{\partial \tilde{y}^2}}_{O(\varepsilon^{-2})} \right) \tag{3.64}$$

This scaling step produces tilded derivatives of $O(1)$ as well as coefficients of $O(1)$, $O(\varepsilon^2)$, $O(1/\varepsilon)$, and $O(1/\varepsilon^2)$ where $\varepsilon \ll 1$. Hence a comparison of the magnitude of each term in each equation can be performed. This is called a *relative order of magnitude analysis* (ROMA), which allows reduction of a given equation by retaining only significant terms. Obviously, a proper selection of the reference quantities based on physical insight is a very important first step (cf. Sect. 2.3.1). Starting with the continuity equation, it is apparent that the coefficients of both terms are of the same large magnitude so that Equation (3.62) has to be fully preserved. The LHS of the x-momentum equation (3.63) is of $O(1)$ and so is the coefficient of the pressure gradient, as can be deduced from Bernoulli's equation for an outer flow streamline. Now, the coefficient of the second term on the RHS, $(1/\text{Re}) = O(\varepsilon^2)$, which makes the term $[(1/\text{Re})(\partial^2 \tilde{u}/\partial \tilde{x}^2)]$ negligible, whereas the last term, $(1/\text{Re})(l^2/\delta^2)(\partial^2 \tilde{u}/\partial \tilde{y}^2)$, has to be retained since its coefficient $(1/\text{Re})(l^2/\delta^2) = O(1)$. So far, we could dismiss only one term because its coefficient was, relatively speaking, very small. In the y-momentum equation, the coefficient of the normal pressure gradient is of $O(\varepsilon^{-2})$, and hence we multiply Equation (3.64) through by a factor of $O(\varepsilon^2)$ so that the entire LHS is of $O(\varepsilon^2)$, that is, negligible. The pressure gradient term, now of $O(1)$, has to be retained, and the other two terms on the RHS are negligible because they are of $O(\varepsilon^4)$ and $O(\varepsilon^2)$, respectively. In summary, we are left with:

(continuity) $\dfrac{\partial u}{\partial x} + \dfrac{\partial v}{\partial y} = 0$ (3.65)

(x-momentum) $\dfrac{\partial u}{\partial t} + u\dfrac{\partial u}{\partial x} + v\dfrac{\partial u}{\partial y} = -\dfrac{1}{\rho}\dfrac{\partial p}{\partial x} + v\dfrac{\partial^2 u}{\partial y^2}$ (3.66)

(y-momentum) $0 = -\dfrac{1}{\rho}\dfrac{\partial p}{\partial y}$ (3.67)

Applying Euler's equation to the external stream along the boundary-layer edge where $\partial u/\partial y \approx 0$, we can deduce that

$$-\frac{1}{\rho}\frac{\partial p}{\partial x} = \frac{\partial U_0}{\partial t} + U_0\frac{\partial U_0}{\partial x}$$ (3.68)

Equation (3.68) can be used to eliminate the pressure gradient in the x-momentum equation (3.66). At time $t = 0$, the velocity fields (U_0, V_0) and (u, v) have to be known. The boundary conditions are

$$u(y = 0) = 0, \qquad v(y = 0) = \begin{cases} 0 \\ \pm v_w \end{cases}$$ (3.69a–c)

and $u(y \to \infty) = U_0(x, t)$

A number of observations are appropriate:

- The Navier–Stokes equations, our point of departure, are with respect to the co-ordinates, elliptic equations that collapsed basically to Equation (3.66), which is *parabolic*. From a fluid mechanics point of view $(v\partial^2 u/\partial x^2) \ll (u\partial u/\partial x)$; that is, streamwise momentum diffusion is very much smaller than axial convection.
- With regard to Equation (3.67), the pressure does not change normal to the solid boundary that allowed us to compute the streamwise pressure variation along the boundary-layer edge (cf. Eq. (3.68)). Thus, at any station x, the outer pressure is imposed normal to the wall onto the boundary layer. On the other hand, the force balance in the transverse direction is basically wiped out; that implies that a fluid particle in the shear layer, moving in the y-direction, has no mass and experiences no inertia, viscous, or pressure forces. This situation changes for thick boundary-layer flow past *curved* surfaces, where second-order effects such as the centrifugal force on fluid particles have to be accommodated.
- The leading edge or stagnation point is a location with a mathematical singularity where the boundary-layer equations are invalid. In the vicinity of $(x = 0, y = 0)$, $u \ll 1$ and the Reynolds number is certainly not high, as required (cf. Eq. (3.58)).

C   Solution Methods for the Boundary-Layer Equations

On the basis of the condition

$$\frac{\delta}{l} \sim \frac{v}{u} \sim \frac{1}{\sqrt{Re_l}} \ll 1$$ (3.70)

we derived for two-dimensional flow with constant properties the boundary-layer equations in rectangular coordinates as

♦ $$\frac{\partial u}{\partial t} + u\frac{\partial u}{\partial x} + v\frac{\partial u}{\partial y} = \frac{\partial U_0}{\partial t} + U_0\frac{\partial U_0}{\partial x} + v\frac{\partial^2 u}{\partial y^2} \qquad (3.71)$$

and

♦ $$\frac{\partial u}{\partial x} + \frac{\partial v}{\partial y} = 0 \qquad (3.72)$$

where $U_0(x, t)$ is the outer flow or boundary-layer edge velocity and the boundary conditions are given as Eq. (3.69a–c).

Equation (3.71) is *invalid* (a) at the leading edge, (b) near the point of flow separation, (c) when the free stream Reynolds number is too low, and (d) when the attached boundary layer becomes so thick that $\delta/l \approx O(1)$. In general, before boundary-layer-type flow problems can be solved, it is necessary to check whether

(i) the given flow is steady or transient, laminar or turbulent, incompressible or compressible;

(ii) the fluid is Newtonian or non-Newtonian; and

(iii) the outer flow, that is, boundary-layer edge conditions, is known or has to be calculated with Euler's equation or from potential flow theory, typically using an iterative displacement body method (cf. Cebeci and Bradshaw 1977; Bradshaw et al. 1981; Anderson et al. 1984).

An *analytic* solution of Equation (3.71) in conjunction with (3.72) is not known. However, for *self-similar* boundary-layer flows (cf. Sect. 2.3.3A), the explicit x-dependence of Eq. (3.71) can be eliminated and the PDE is reduced to an ODE, that is, a boundary-value problem that can be solved numerically (App. F.3). For more complex boundary-layer flows, approximation methods such as the integral method (cf. Sect. 2.3.3B) have to be employed for solution. Both techniques, similarity theory and the integral method, are illustrated later. There are a number of finite-difference solutions available for the *laminar* boundary-layer equation (cf. Anderson et al. 1984). It has to be noted that *turbulent* boundary layers require submodels for the Reynolds stresses that may change the type of PDE (cf. Sect. 3.2.4). The parabolic equations are solved numerically with so-called marching techniques (cf. App. F.4). The input requirements are (1) an upstream profile $u(0, y)$ for starting; (2) the edge or outer flow conditions $u_e(x) \equiv U_0(x)$ and $T_e(x)$ if applicable; and (3) the velocity wall conditions and, for *thermal* boundary layers, a prescribed T_w or q_w. Instead of solving for the primitive variables u, v, and p or U_0 directly, the vorticity-transport/stream-function approach can be employed (cf. Sect. 3.2.3D and App. D). However, the most popular are coordinate transformations that reduce the complexity of (3.71) as indicated later.

D   Laminar Boundary-Layer Flow Applications

Consider steady laminar boundary-layer flow with arbitrary outer flow and general wall conditions. Ignoring body forces, the governing equations are

$$\frac{\partial u}{\partial x} + \frac{\partial v}{\partial y} = 0$$

and

$$u\frac{\partial u}{\partial x} + v\frac{\partial u}{\partial y} = U_0\frac{dU_0}{dx} + \nu\frac{\partial^2 u}{\partial y^2}$$

subject to

$$u(x, y = 0) = 0, \qquad u(x, y \to \infty) = U_0 \quad \text{and} \quad v(y = 0) = \begin{cases} 0 \\ \pm v_w \end{cases}$$

Using the stream function approach

$$u \equiv \frac{\partial \psi}{\partial y} \quad \text{and} \quad v \equiv -\frac{\partial \psi}{\partial x}$$

the continuity equation is automatically satisfied and the momentum equation becomes

◆ $$\frac{\partial \psi}{\partial y}\frac{\partial^2 \psi}{\partial x \partial y} - \frac{\partial \psi}{\partial x}\frac{\partial^2 \psi}{\partial y^2} = U_0\frac{dU_0}{dx} + \nu\frac{\partial^3 \psi}{\partial y^3} \qquad (3.73a)$$

From similarity theory (cf. Sect. 2.3.3A) we use the "combined" variable

$$\eta = \frac{y}{\delta(x)} \cong y\sqrt{\frac{U_0}{\nu x}} \sim \frac{y}{\sqrt{x}}$$

where $\delta(x) \sim (U_0/\nu x)^{-1/2}$ from boundary-layer scale analysis has been utilized. For the stream function, we postulate that

$$\psi = \sqrt{U_0 \nu x}\, f(x, \eta)$$

where $f(x, \eta)$ is a *dimensionless* stream function, which requires that the dimensions of $\sqrt{U_0 \nu x} \doteq$ (velocity) $\times$ (length) $\doteq L^2 T^{-1}$. Inserting these transformations into Equation (3.73a) and applying the chain rule (cf. Sect. 2.3.3)

$$\frac{\partial}{\partial x} = \left(\frac{\partial}{\partial x}\right)_\eta + \left(\frac{\partial}{\partial \eta}\right)_x\frac{\partial \eta}{\partial x} \quad \text{and} \quad \frac{\partial}{\partial y} = \left(\frac{\partial}{\partial \eta}\right)_x\frac{\partial \eta}{\partial y}$$

we obtain for

$$u = \frac{\partial \psi}{\partial y} = (U_0\nu x)^{1/2} f'\left(\frac{U_0}{\nu x}\right)^{1/2} = U_0 f'$$

$$-v = \frac{\partial \psi}{\partial x} = f\frac{d}{dx}(U_0\nu x)^{1/2} + (U_0\nu x)^{1/2}\frac{\partial f}{\partial x} + (U_0\nu x)^{1/2} f'\frac{\partial \eta}{\partial x}$$

$$\frac{\partial u}{\partial y} = \frac{\partial^2 \psi}{\partial y^2} = U_0\left(\frac{U_0}{\nu x}\right)^{1/2} f''$$

$$\frac{\partial^2 u}{\partial y^2} = \frac{\partial^3 \psi}{\partial y^3} = \frac{U_0^2}{\nu x} f'''$$

$$\frac{\partial u}{\partial x} = \frac{\partial}{\partial x}\left(\frac{\partial \psi}{\partial y}\right) = \frac{\partial}{\partial x}(U_0 f') = f'\frac{dU_0}{dx} + U_0\frac{\partial f'}{\partial x} + U_0 f''\frac{\partial \eta}{\partial x}$$

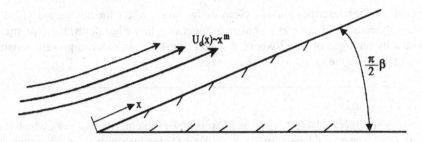

Fig. 3.11. Falkner—Skan wedge flows.

so that Equation (3.72a) yields after some manipulations

$$\blacklozenge \qquad f''' + \frac{m+1}{2} f f'' + m(1 - f'^2) = x\left(f' \frac{\partial f'}{\partial x} - f'' \frac{\partial f}{\partial x} \right) \qquad (3.73b)$$

where a prime denotes $d/d\eta$ and

$$\blacklozenge \qquad m \equiv \frac{x}{U_0} \frac{dU_0}{dx}$$

is related to the pressure gradient

$$\blacklozenge \qquad -\frac{1}{\rho} \frac{\partial p}{\partial x} = U_0 \frac{dU_0}{dx}$$

The transformed boundary conditions are

$$\underbrace{f'(x,0) = 0,}_{\text{(no wall slip)}} \qquad \underbrace{f'(x,\infty) = 1}_{\text{(outer flow matching)}} \quad \text{and} \quad \underbrace{f(x,0) = 0}_{\text{(solid surface)}}$$

or if wall permeability allows for suction or injection of the fluid

$$f(x,0) = -\sqrt{U_0 \nu x} \int_0^x v_w(s)\,ds$$

Application I: Self-Similar Boundary-Layer Flows
For a particular outer flow velocity distribution (cf. Fig. 3.11)

$$\blacklozenge \qquad U_0(x) = Cx^m; \qquad m = \text{const}$$

the pressure distribution $p(x)$ is such that the velocity profiles $u(x, y)$ are similar. The dimensionless stream function reduces to $f = f(\eta)$ only, and without explicit x-dependence Equation (3.73b) collapses to (Falkner and Skan 1931)

$$\blacklozenge \qquad f''' + \frac{m+1}{2} f f'' + m(1 - f'^2) = 0 \qquad \text{(Falkner–Skan)} \qquad (3.73c)$$

The ODE (3.73c) describes *Falkner–Skan wedge flows*, where the unchanged $\eta(x, y)$ is a similarity variable and $f(\eta)$ is a similarity function. In wedge flows, the free stream is deflected by an angle of $\beta\pi/2$ where $\beta = 2m/(m + 1)$. In order to prevent separation, values of m are in the range $-0.09 < m < \infty$.

Example (1)

A special case of (3.73c) is the *Blasius* flqw problem, where $\beta = 0$, which implies $m = 0$, $U_0 = $ const, and hence $\partial p/\partial x = 0$. Thus for laminar high-Reynolds-number flow past a flat horizontal plate with zero pressure gradient, Equation (3.73c) reduces to (Blasius 1908)

♦ $$f''' + \frac{1}{2}ff'' = 0 \qquad \text{(Blasius)}$$

♦ $$f(0) = f'(0) = 0 \quad \text{and} \quad f'(\infty) = 1 \qquad\qquad (3.73d)$$

If we were to have a value for $f''(0)$ to match the end-condition $f'(\infty) = 1$, we could use an IVP solver such as the Runge–Kutta algorithm (cf. App. F.3). Trial and error yielded $f''(0) = 0.33206$. Once $f(\eta)$ is known, any variable or parameter of the boundary layer can be plotted or evaluated. For example, since $f'(\eta = 5.0) = 0.99$, we find the boundary-layer thickness, or edge location, $\delta(x)$, from this condition by inspection, which implies $u = 0.99U_0$. Hence, for the flat-plate Blasius flow

$$\eta \approx 5.0 = \delta\sqrt{\frac{U_0}{\nu x}} \qquad \text{where } \delta \equiv \delta_{99\%}$$

or

♦ $$\frac{\delta(x)}{x} \approx \frac{5.0}{\sqrt{\text{Re}_x}} \qquad\qquad (E3.28)$$

that is, the shear-layer thickness for Blasius flow varies as $\delta \sim x^{1/2}$. The local wall shear stress

$$\tau_w(x) = \mu\frac{\partial u}{\partial y}\bigg|_{y=0} = \mu U_0\sqrt{\frac{U_0}{\nu x}}f''(0) \qquad \text{where } f''(0) = 0.332$$

Hence the local skin friction coefficient

♦ $$c_f \equiv \frac{2\tau_w(x)}{\rho U_0^2} = \frac{0.664}{\sqrt{\text{Re}_x}} \qquad\qquad (E3.29)$$

A *displacement thickness*, $\delta_1(x)$, can be defined as the distance normal to the wall by which the outer flow field is displaced as a result of the decrease in velocity within the shear layer. Thus, the *mass flow defect* due to curved streamlines caused by wall friction is (cf. Fig. 3.12)

$$\rho\int_{y=0}^{\delta}(U_0 - u)\,dy \equiv \rho U_0\,\delta_1$$

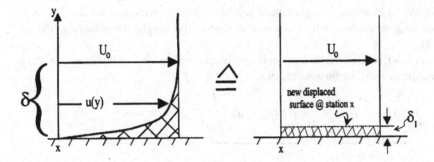

Fig. 3.12. Displacement thickness concept.

or for self-similar flow

$$\delta_1(x) \equiv \int\limits_{y=0}^{\infty} \left(1 - \frac{u}{U_0}\right) dy = \sqrt{\frac{\nu x}{U}} \int\limits_{\eta=0}^{\infty} [1 - f'(\eta)]\, d\eta$$

or applied to Blasius flow

$$\blacklozenge \qquad \frac{\delta_1(x)}{x} = \frac{1.721}{\sqrt{\mathrm{Re}_x}} \qquad\qquad\qquad\qquad\qquad \text{(E3.30)}$$

Similarly, *momentum defect* in relation to the outer flow can be expressed as

$$\rho \int\limits_{y=0}^{\delta} u(U_0 - u)\, dy \equiv \rho U_0^2\, \delta_2$$

or for Blasius flow

$$\blacklozenge \qquad \frac{\delta_2(x)}{x} = \sqrt{\frac{\nu}{x U_0}} \int\limits_{\eta=0}^{\infty} f'(1 - f')\, d\eta := \frac{0.664}{\sqrt{\mathrm{Re}_x}} = c_f \qquad\qquad \text{(E3.31)}$$

These system parameters or integral properties are useful characteristics of a given boundary-layer problem. For example, the displacement thickness, $\delta_1(x)$, added as a variable layer to the body contour, is used to simulate the *true deflection* of the approach stream, that is, a deflection due to the submerged body plus a displacement caused by the thin shear layer. In turn, the momentum thickness, $\delta_2(x)$, is directly related to the wall shear stress.   □

Application II: Nonsimilar Boundary-Layer Flows
Similarity theory applicable to the flat-plate problem (Blasius) or to wedge flows (Falkner–Skan) is inappropriate for *nonsimilar flows* such as separating flows, turbulent flows, general wall suction/injection flows, or curved surface flows. Furthermore, numerical solutions of nonsimilar flow problems might be cumbersome. In such cases, approximate solution techniques such as power series, methods of weighted residuals, or perturbation theory may

be suitable. As an example of approximate solution methods, we apply the Von Karman–Pohlhausen *integral method*, which is a special case of the weighted residual method (cf. Sect. 2.3.3).

In Section 2.3.3, we derived for transient two-dimensional *laminar* or *turbulent* boundary-layer flows, the *momentum integral relation*

$$
\blacklozenge \quad \frac{\partial}{\partial t} \underbrace{\int_0^\infty (U_0 - u)\, dy}_{\sim \delta_1(x)} + \frac{\partial}{\partial x} \underbrace{\int_0^\infty u(U_0 - u)\, dy}_{\sim \delta_2(x)}
$$

$$
+ \frac{\partial U_0}{\partial x} \underbrace{\int_0^\infty (U_0 - u)\, dy}_{\sim \delta_1(x)} - U_0 v_w = \frac{\tau_w}{\rho} \tag{3.74}
$$

This equation allows for fluid injection or withdrawal at the wall. Given an outer flow velocity field $U_0(x, t)$ and a possible wall flux v_w, and recalling that $\tau_w = \mu(\partial u/\partial y)|_{y=0}$, Equation (3.74) can be used to determine $u(x, y; t)$. Typically, a functional form of $u(x, y; t)$, which has to fulfill the boundary conditions, is postulated and Equation (3.74) is reduced to an ODE, usually for $\delta(x)$, the boundary-layer thickness. Alternatively, empirical correlations among the integral parameters related to δ_2, U_0, and so on, are introduced and a resulting ODE, for example, $\delta_2(x)$, is solved (Thwaites 1949). The procedure for flat-plate boundary-layer flow parameter evaluation using the integral method is summarized next.

- Step (la): Make a reasonable assumption about the form of $u(x, y)$ that satisfies the "no-slip" plus "matching" conditions (cf. sketch).

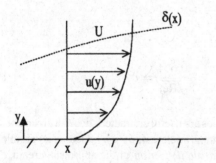

Sample Postulates

$$
u(x, y) \approx U(1 - e^{-\alpha y}) \qquad \text{where} \qquad \alpha = \alpha(x)
$$

$$
u(x, y) \approx U \tanh \frac{y}{a(x)}
$$

$$
u(x, y) = \begin{cases} U \sin\left(\frac{\pi}{2} \frac{y}{\delta(x)}\right) & \text{for } y \le \delta \\ U & \text{for } y > \delta \end{cases}
$$

$$
u(x, y) \approx a_0 + a_1 y^* + a_2 y^{*2} + \cdots; \qquad \text{where} \qquad y^* = \frac{y}{\delta(x)}
$$

- Alternative Step (1b): Introduce empirical correlations among the integral parameters (cf. Example (3)).
- Step (2): Calculate $\delta_2(x)$, $\delta_1(x)$, and $\delta(x) \equiv \delta_{99\%}$ all in terms of the trial solution parameter $\alpha(x)$ or $a(x)$ or $\delta(x)$. Also estimate, for example, $C_D[\delta(x)]$, $\tau_w[\delta(x)]$, and $c_f[\delta(x)]$ and obtain an equation for $\delta(x)$.
- Step (3): Employ problem-specific conditions to fix $\delta(x = x_0) = \delta_0$ (e.g., $\delta(x = 0) = 0$) and solve the ODE obtained in Step (2) for a relationship $\delta(x)$.

Example (2)

Evaluation of thin-shear-layer parameters for steady Blasius flow, that is, $U_0 =$ const. Thus, the momentum-integral relation, Equation (3.74), reduces for steady 2-D incompressible boundary-layer flow to

$$\frac{\tau_w}{\rho} = \frac{d}{dx}(U_0^2 \delta_2) \tag{E3.32}$$

where $\tau_w = \mu(\partial u/\partial y)|_{y=0}$ and $\delta_2 = \int_0^\infty (u/U_0)[1 - (u/U_0)]\,dy$ with $U_0 =$ const. Now we employ the trial solution (cf. Step 1a):

$$u(x, y) \approx U_0 \tanh\left[\frac{y}{a(x)}\right]$$

subject to $u(y = 0) = 0$ and $u(y = 2.65a(x)) = U$. The upper limit for y stems from the observation that $\tanh(y/a(x)) \approx 0.99$ when $(y/a) \approx 2.65$. Evaluating

$$\delta_2 = \int_0^\delta \frac{u}{U_0}\left(1 - \frac{u}{U_0}\right)dy, \qquad \text{where } \delta \approx 2.65a(x)$$

yields

$$\delta_2 \approx a(x)(1 - \ln 2) = 0.307a(x) \tag{E3.33}$$

and

$$\tau_w = \mu \frac{\partial u}{\partial y}\bigg|_{y=0} \approx \frac{\mu U_0}{a(x)} \tag{E3.34}$$

From Eq. (E3.32)

$$\tau_w = \rho U_0^2 \frac{d\delta_2}{dx} \tag{E3.35}$$

so that, combining Eqs. (E3.33) to (E3.35),

$$\frac{\mu}{a(x)} = 0.307\rho U_0 \frac{da}{dx}$$

or

$$a\,da = \frac{\mu}{0.307\rho U_0}\,dx \tag{E3.36}$$

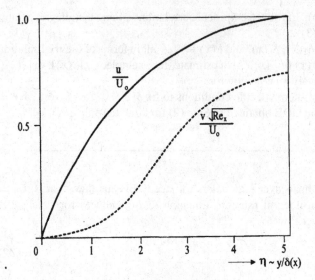

Fig. 3.13. Self-similar velocity profiles in thin shear layer (Blasius Flow).

subject to the leading edge condition $a(x = 0) = 0$. Integration yields

$$\blacklozenge \qquad \frac{a(x)}{x} = \sqrt{\frac{2\mu}{0.307\rho U_0 x}} = \frac{2.553}{\sqrt{Re_x}}, \qquad \text{where} \qquad Re_x = \frac{U_0 x}{\nu} \qquad (E3.37)$$

Substituting $a(x)$ back into the trial solution $u(x, y)$ as well as into the expressions for δ, δ_1, δ_2, τ_w, c_f, and $v(x, y)$ from the continuity equation finalizes the problem solution (cf. Fig. 3.13 and Sect. 3.3).

Example (3)

Consider Thwaites's integral method for determining thin *boundary-layer separation*. For steady two-dimensional laminar boundary-layer flow past an impermeable surface, Equation (3.74) becomes

$$\frac{\tau_w}{\rho} = \frac{d}{dx}(U^2 \delta_2) + \delta_1 U \frac{dU}{dx} \qquad (E3.38a)$$

or

$$\frac{\tau_w}{\rho U^2} \equiv \frac{c_f}{2} = \frac{d\delta_2}{dx} + \left(2 + \frac{\delta_1}{\delta_2}\right) \frac{\delta_2}{U} \frac{dU}{dx} \qquad (E3.38b)$$

Multiplying (E3.38b) by $(U\delta_2/\nu)$ yields

$$\frac{\tau_w \delta_2}{\mu U} = \frac{U \delta_2}{\nu} \frac{d\delta_2}{dx} + \frac{\delta_2^2 U'}{\nu}\left(2 + \frac{\delta_1}{\delta_2}\right) \qquad (E3.39)$$

Table 3.1. *Shear function and shape function*

λ	0.25	0.20	0.14	0.12	0.10	0.080	0.064	0.048	0.032
$H(\lambda)$	2.00	2.07	2.18	2.23	2.28	2.34	2.39	2.44	2.49
$S(\lambda)$	0.500	0.463	0.404	0.382	0.359	0.333	0.313	0.291	0.268
λ	0.016	0.0	−0.016	−0.032	−0.040	−0.048	−0.052	−0.056	−0.060
$H(\lambda)$	2.55	2.61	2.67	2.75	2.81	2.87	2.90	2.94	2.99
$S(\lambda)$	0.244	0.220	0.195	0.168	0.153	0.138	0.130	0.122	0.113
λ	−0.064	−0.068	−0.072	−0.076	−0.080	−0.084	−0.086	−0.088	−0.090
$H(\lambda)$	3.04	3.09	3.15	3.22	3.57	3.66	3.71	3.49	3.55
$S(\lambda)$	0.104	0.095	0.085	0.072	0.056	0.038	0.027	0.015	0.0000[a]

Note: [a]That is, BL separation because $\tau_w \sim S(\lambda) = 0$.
Source: Thwaites 1949.

where the momentum thickness $\delta_2 = \int_0^\infty (u/U)[1-(u/U)]\,dy$, the displacement thickness $\delta_1 = \int_0^\infty [1 - (u/U)]\,dy$, $(\tau_w \delta_2/\mu U) \equiv S$ is a "shear factor," and $(\delta_1/\delta_2) \equiv H$ is a "shape factor," where typically $2 \le H \le 3.5$. Now, the streamwise velocity $u(x, y)$ is approximated by a one-parameter function, $\lambda(x)$, that is,

$$u = U(x) f[\eta, \lambda(x)] \tag{E3.40}$$

where $\eta = y/\delta$ is the dimensionless normal coordinate and $\lambda = \lambda[\delta(x)]$ is a *dimensionless system parameter* that determines boundary-layer integral properties such as δ_1, δ_2, and c_f.

Equation (E3.39) can be rewritten in terms of λ, where

$$\lambda(x) \equiv \frac{\delta_2^2}{\nu}\frac{dU}{dx} \tag{E3.41}$$

as well as the shear correlation $S = S(\lambda)$ and the shape-factor correlation $H = H(\lambda)$. Specifically, with $\delta_2 d\delta_2 \equiv d(\delta_2^2/2)$, Equation (E3.39) can be rewritten and takes on the form

$$\frac{U}{\nu}\frac{d(\delta_2^2)}{dx} = U\frac{d}{dx}\left(\frac{\lambda}{U'}\right) \approx 2[S(\lambda) - \lambda(2 + H)] \equiv F(\lambda) \tag{E3.42}$$

Now, Thwaites (1949) observed that a linear correlation for the boundary-layer function $F(\lambda) = (U/\nu)\,d(\delta_2^2)/dx$ can be established:

$$F(\lambda) \approx 0.45 - 6.0\lambda \tag{E3.43}$$

The solution of (E3.42) in conjunction with (E3.43) is

$$\delta_2^2 \approx \frac{0.45\nu}{U^6}\int_0^x [U(s)]^5\,ds \tag{E3.44}$$

where s is a dummy integration variable. Once $\delta_2(x)$ is known, all other parameters can be calculated by using tables for $S(\lambda)$ and $H(\lambda)$ (cf. Table 3.1).

E   Summary of Solution Approaches for the Navier–Stokes Equations

The Navier–Stokes equations, describing fluid flow with constant fluid properties, are a set of transient, *nonlinear*, second-order inhomogeneous *elliptic* PDEs. Depending on the characteristics of the flow system (system geometry, flow patterns, boundary conditions, etc.) it might be justifiable to simplify the Navier–Stokes equations drastically. Specifically:

- Delete the nonlinear terms: $(\vec{v} \cdot \nabla)\vec{v} \equiv \vec{a}_{conv} \sim \vec{F}_{inertia} \approx 0$ (e.g., for creeping flows, fully developed unidirectional flows).
- Neglect 2nd-order terms in streamwise direction: $\partial^2 u / \partial x^2 \approx 0$ (e.g., boundary-layer flows).
- Assume a constant pressure gradient: $-\nabla p \Rightarrow \Delta p / l$ (e.g., in fully developed pipe flows).
- Assume steady state: $(\partial / \partial t) = 0$ if there are no time variations.
- Neglect gravity effects, developing flow effects, end effects, and so on.
- Check the problem's dimensionality as 1-D or 2-D flows, (i.e., reduce the number of directions in which the velocity can vary).

Summary of Basic Exact Solutions
Frictionless Flow $(\nu \equiv 0$ or $\nabla^2 \vec{v} \approx 0)$ as discussed in Sections 3.1.2 and 3.1.3.
   Euler's equation

$$\blacklozenge \qquad \frac{D\vec{v}}{Dt} = -\frac{1}{\rho}\nabla p + \vec{g} \tag{3.75a}$$

or
   Laplace's equation

$$\blacklozenge \qquad \nabla^2 \phi = 0 \tag{3.75b}$$

   where $\vec{v} = \nabla\phi;$ $\phi \triangleq$ velocity potential.

Solution Notes

- Use the numerical solution of (3.75a) to obtain velocity and pressure fields.
- For Equation (3.75b), obtain pressure field from Bernoulli's equation.
- Solve Laplace Equation analytically or numerically.

Parallel Flows   (Couette, Poiseuille, suddenly accelerated plate (Stokes I or Rayleigh), pipe flow start-up, oscillating plate (Stokes II))
   Note the following:

$$(\vec{v} \cdot \nabla)\vec{v} = 0; \qquad \nabla \cdot \vec{\vec{\tau}} \sim \nabla p \quad \text{and} \quad \nabla \cdot \vec{v} = 0$$

   General postulate $u = u(y, z, t)$

Stokes's Equation

$$\blacklozenge \qquad \frac{\partial u}{\partial t} = -\frac{1}{\rho}\frac{dp}{dx} + \nu\left(\frac{\partial^2 u}{\partial y^2} + \frac{\partial^2 u}{\partial z^2}\right) + g_x \tag{3.76}$$

Typically $u = u(y, t)$ but if $u = u(y)$ only as in Couette and Poiseuille flows

$$\frac{d^2 u}{dy^2} = \frac{1}{\mu}\frac{dp}{dx} - g_x := \text{const} \qquad (3.77\text{a})$$

or

♦ $\qquad u'' = C \qquad\qquad\qquad\qquad\qquad\qquad\qquad\qquad\qquad (3.77\text{b})$

Application schematics

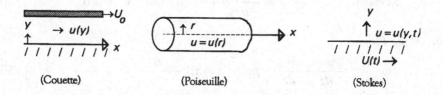

| (Couette) | (Poiseuille) | (Stokes) |

Solution Notes

- Employ transformations, that is, separation of variables or similarity method; solve analytically or numerically. Equation (3.77) can be integrated directly.
- For laminar flow in slightly varying conduits, such as tapered pipes and nonparallel slits, quasiparallel flow may be assumed; however, via the "no-slip" boundary condition, $u = u(x, y)$ and $dp/dx = \text{fct}(x)$, as demonstrated in Sect. 3.3.

Steady Flow with Convective Acceleration   (Flow around bodies of 2-D geometry, flow near rotating disk, flow in conduits)
Note the following:

$$(\vec{v} \cdot \nabla)\vec{v} \sim \nabla p \quad \text{and} \quad \sim \nabla \cdot \vec{\vec{\tau}}$$

$$uu_{,x} + vu_{,y} = -\frac{1}{\rho}p_{,x} + v(u_{,xx} + u_{,yy}) \qquad (3.78\text{a})$$

$$uv_{,x} + vv_{,y} = -\frac{1}{\rho}p_{,y} + v(v_{,xx} + v_{,yy}) \qquad (3.78\text{b})$$

Solution Notes
Seek further reduction of the y-momentum equation; use clever postulates as in the Von Karman solution for the rotating disk (cf. Sect. 3.2.1); solve numerically.

Summary of Basic Approximate Solutions
Very-Low-Reynolds-Number Flows   Very Slow Motion around Bodies and in Conduits of "Nonparallel" Geometry (Creeping Flows): Re $\to$ 0

- Flow past a sphere (Stokes, Oseen)
- Lubrication theory (Reynolds equation)
- 1-D Stretching flows (e.g., fibers, sheets, films, coats)
- Hele–Shaw flow

Note linearization of

$$(\vec{v} \cdot \nabla)\vec{v} \approx \begin{cases} 0 & \text{(Stokes)} \\ U \dfrac{\partial \vec{v}}{\partial x} & \text{(Oseen)} \end{cases}$$

Stokes: $\nabla p = \mu \nabla^2 \vec{v} \rightarrow \nabla^2 p = 0 \quad \text{or} \quad \nu L^4 \psi = 0$ (3.79a–c)

where L is the biharmonic differential operator and ψ is the stream function.

Reynolds: $(\vec{v} \cdot \nabla)\vec{v} = -\dfrac{1}{\rho}\nabla p + \nu \nabla^2 \vec{v} \quad \text{and} \quad \nabla \cdot \vec{v} = 0$ (3.80a,b)

Stretching Flows:

$$\rho w \frac{\partial w}{\partial z} = -\frac{dp}{dz} + \frac{d\tau_{zz}}{dz} + \rho g; \quad p \approx \tau_{rr} - \frac{\sigma}{r(z)}$$ (3.80c,d)

Notes:

- Low-Reynolds-number flow examples are discussed in Sections 3.2.1C and 3.3.4.
- For simple bearings with mildly varying gap size, such as $h(x) \ll L$, the inertia term $(\vec{v} \cdot \nabla)\vec{v}$ in Eq. (3.80a) may be negligible (cf. Sect. 3.3).
- Stretching flows are near-parallel flows with 1-D inertia, surface tension, viscous, and gravitational effects (cf. Problem 3.14c in Sect. 3.3.4).

High-Reynolds-Number Flows   Boundary-Layer Flows near Solid Walls (cf. sketch)

$$\frac{\delta}{l} \sim \frac{v}{u} \sim \frac{1}{\sqrt{\text{Re}_l}} \ll 1$$

Note: Viscosity μ might be small but $(\partial u / \partial y) \gg 1$ near the wall (shear layer)

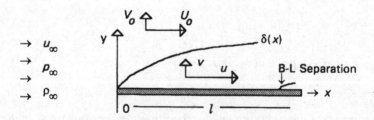

Prandtl's boundary-layer equations

$$\frac{\partial u}{\partial x} + \frac{\partial v}{\partial y} = 0$$ (3.81)

$$\frac{\partial u}{\partial t} + u\frac{\partial u}{\partial x} + v\frac{\partial u}{\partial y} = -\frac{1}{\rho}\frac{dp}{dx} + \nu\frac{\partial^2 u}{\partial y^2}$$ (3.82)

where $-\dfrac{1}{\rho}\dfrac{dp}{dx} = \dfrac{\partial U_0}{\partial t} + U_0\dfrac{\partial U_0}{\partial x} \quad \text{and} \quad p(y) = \text{const}$ (3.83a,b)

Solution Notes

- Use the Falkner–Skan transformation for self-similar boundary-layer flows; solve resulting ODEs numerically (e.g., Runge–Kutta method with two "initial" conditions).
- Use the integral method for nonsimilar boundary-layer flows using suitable velocity profiles.
- Employ numerical parabolic equation solvers (FDM, FEM, CVM, etc.) for general cases.

3.2.4 Turbulent Flow Models

The vast majority of natural and industrial flows are *turbulent*. Thus, the basic knowledge and mathematical skills gained from laminar flow studies have to be significantly extended in order to evaluate, at least qualitatively, *transition* from laminar to turbulent flow and to compute quantitatively *turbulence effects* in a variety of practical applications.

After a brief discussion of laminar flow instabilities and turbulence characteristics, approaches to turbulent flow modeling are summarized. The next section then deals with turbulent flow applications: boundary-layer, duct, and free shear-layer flows. Two case studies of turbulent external and internal flows are discussed in Section 5.2.

A Aspects of Turbulence

Transitions from laminar-to-turbulent and/or turbulent-to-laminar flows are inadvertently of importance in numerous applications including thrombosis in locally occluded blood vessels, drag reduction with laminarized flows, and transition control of fighter jets. For all practical purposes, at least two problems have to be solved: (i) determination of the onset of transition and (ii) modeling of the influence of the transition region on the specific flow field. In general, numerous "irregularities and disturbing sources" exist a priori in free shear flows and on surfaces of wall-bounded flows. At sufficiently large Reynolds numbers, these flow perturbations (e.g., free-stream turbulence, flow instabilities in the form of vortices and wall roughness effects) are not damped out; just the opposite, they amplify and start to interact with the entire flow field. Specifically, in two-dimensional flows, the natural transition begins with the amplification of Tollmien–Schlichting (T–S) waves, that is the linear and nonlinear development of initially small amplitude T–S waves into large amplitude disturbances that break down into a chaotic motion (cf. Fig. 3.14). Free stream fluctuations; Taylor- or Goertler-type vortices; body surface roughness, waviness or vibration; and acoustic or particle environments all can contribute to the onset of transition and the development of (locally) turbulent flow fields. Linear stability theory allows the calculation of characteristics of these T–S waves as eigensolutions of the linearized Navier–Stokes equations (cf. Warsi 1993). A crude evaluation of the transition region may be obtained with an intermittency function γ applied to the turbulent shear stress $\tau^{\text{total}} = \tau_{\text{laminar}} + \gamma \tau_{\text{turb}}$; γ is zero in laminar flow and $\gamma \geq 1$ in turbulent flow, where $\gamma > 1$ near the end of the transition region (cf. Sect. 3.3.5).

The development of a general theory for *turbulent flow* is hampered by the fact that turbulence is generally randomly chaotic, nonlinear, three-dimensional, inhomogeneous, nonstationary, and largely dependent upon its origin and the given flow configuration. However, the detection of various types of spatially *coherent structures* in important engineering flows

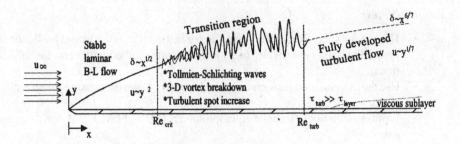

Fig. 3.14. Flat-plate laminar/transitional/turbulent boundary-layer characteristics.

has spawned a new class of turbulence research. Coherent structures, associated with the large scale motion, are basically recognizable patterns with characteristic orientations that recur throughout the flow field. A Karman vortex street behind a cylinder at high Reynolds number may serve as an example. On a smaller scale, coherent substructure rules exist such as longitudinal vortices, hairpin structures, and horseshoe vortices of random position and variable shape and size. These coherent structures contain a significant fraction of the turbulence kinetic energy, and they are associated with the transport of scalar quantities in the flow field. Thus, they are believed to be responsible for the apparent stresses or Reynolds stresses and for most of the energy containing motion. In wall turbulence, for example, there exists a hierarchy of eddy scales with the smaller ones proportional to ν/u_τ, where $u_\tau \equiv (\tau_w/\rho)^{1/2}$ is the friction velocity. In contrast, there are also detached *incoherent* motions, such as eddies of the Kolmogorov length scale, $(\nu^3/\varepsilon)^{1/4}$, which are responsible for the energy dissipation, expressed in terms of the eddy dissipation rate ε, as discussed later.

   The traditional approach for modeling turbulence is the use of the Reynolds-averaged Navier–Stokes equations (cf. Eq. (3.86)). Because of the nonlinear terms, the time-averaging or time-smoothing process of these equations produces dominant, apparent stresses that make turbulence closure models necessary. Since averaging occurs over all the turbulence dynamics simultaneously, it is necessary to model all structures requiring flow-system-specific rather than universal relationships. This closure problem, that is, the necessary specification of some auxiliary equations in order to match the number of new unknowns, is usually solved with eddy-viscosity models or, now more frequently, with Reynolds-stress transport models. In any case, the classical point of departure is Reynolds's decomposition of the instantaneous, dependent variable v,

♦ $$v = \bar{v} + v' \tag{3.84a}$$

into a time-averaged $\bar{v}$, and a fluctuating component v', where v could be a tensor, a vector, or a scalar. Substituting the decomposition into the basic transport equations (cf. Sect. 2.2.2) yields after averaging (cf. Warsi 1993), for example,

$$\bar{v} = \frac{1}{\Delta t} \int_{t_1}^{t_2} v\, dt \tag{3.84b}$$

the turbulent transport equations. Basic averaging axioms include $\overline{f+g} = \bar{f} + \bar{g}$, $\overline{cf} = c\bar{f}$, and $\overline{\bar{f}g} = \bar{f}\bar{g}$, and $\bar{\bar{f}} = \bar{f}$, where f and g are random functions and c is a constant.

For example, the Navier–Stokes equations can be transformed as follows:

(i) Continuity: $\text{div } \bar{\mathbf{v}} + \text{div } \mathbf{v}' = 0$

With $\overline{\text{div } \bar{\mathbf{v}}} = \text{div } \bar{\mathbf{v}}$, we obtain

♦ $\text{div } \bar{\mathbf{v}} = 0$ (3.85)

which, of course, implies that $\overline{\text{div } \mathbf{v}'} = \text{div } \bar{\mathbf{v}}'$ because $\bar{\mathbf{v}}' \equiv 0$.

(ii) Momentum: $\dfrac{\partial \bar{\mathbf{v}}}{\partial t} + \dfrac{\partial \mathbf{v}'}{\partial t} + \nabla \cdot (\bar{\mathbf{v}}\bar{\mathbf{v}}) + \nabla \cdot (\bar{\mathbf{v}}\mathbf{v}') + \nabla \cdot (\mathbf{v}'\bar{\mathbf{v}}) + \nabla \cdot (\mathbf{v}'\mathbf{v}')$

$$= -\nabla\left(\frac{\bar{p}}{\rho}\right) - \nabla\left(\frac{p'}{\rho}\right) + \nu\nabla^2\bar{\mathbf{v}} + \nu\nabla^2\mathbf{v}'$$

After each term is averaged or filtered (cf. Warsi 1993), the Reynolds-averaged Navier–Stokes equations appear:

♦ $\dfrac{\partial \bar{\mathbf{v}}}{\partial t} + \nabla \cdot (\bar{\mathbf{v}}\bar{\mathbf{v}}) = -\nabla\left(\dfrac{\bar{p}}{\rho}\right) + \nu\nabla^2\bar{\mathbf{v}} - \nabla \cdot (\overline{\mathbf{v}'\mathbf{v}'})$ (3.86)

Note the following:

- Subtracting the last two equations yields an equation for the flow perturbations.
- The dyad, $-\rho\overline{\mathbf{v}'\mathbf{v}'}$, is called the *Reynolds stress tensor.* It contains nine, actually six different, so-called apparent or turbulent stress components. For example, in rectangular coordinates

$$\rho\overline{\mathbf{v}'\mathbf{v}'} = \begin{vmatrix} \rho\overline{u'^2} & \rho\overline{u'v'} & \rho\overline{u'w'} \\ \rho\overline{u'v'} & \rho\overline{v'^2} & \rho\overline{v'w'} \\ \rho\overline{u'w'} & \rho\overline{v'w'} & \rho\overline{w'^2} \end{vmatrix} \qquad (3.87)$$

- The sum of the diagonal in the form

$$\frac{1}{2}\left(\overline{u'^2} + \overline{v'^2} + \overline{w'^2}\right) = k \qquad (3.88)$$

is the *turbulence kinetic energy*, an important measure of turbulent flows.
- In cartesian coordinates, using the convention of summation over repeated indices, the turbulent incompressible flow equations read

♦ $\dfrac{\partial \bar{v}_i}{\partial x_i} = 0$ (3.89)

♦ $\dfrac{\partial \bar{v}_i}{\partial t} + \bar{v}_j\dfrac{\partial \bar{v}_i}{\partial x_j} = -\dfrac{1}{\rho}\dfrac{\partial \bar{p}}{\partial x_i} + \underbrace{\nu\dfrac{\partial^2 v_i}{\partial x_j \partial x_j}}_{\sim\nabla\cdot\bar{\bar{\tau}}_{\text{laminar}}} - \underbrace{\dfrac{\partial}{\partial x_j}\left(\overline{v_i'v_j'}\right)}_{\sim\nabla\cdot\bar{\bar{\tau}}_{\text{turbulent}}}$ (3.90)

- Whereas $\tau_{ij}^{\text{total}} = \tau_{ij}^{\text{lam}} + \tau_{ij}^{\text{turb}}$ throughout a flow field,

$$\tau_{ij}^{\text{turb}} \gg \tau_{ij}^{\text{lam}}$$

everywhere in turbulent shear flow except in the near-wall region, where the local Reynolds number approaches zero and a "viscous sublayer" is formed (cf. Fig. 3.14).
- For two-dimensional incompressible turbulent *boundary-layer* flows

♦ $$\frac{\partial \bar{u}}{\partial x} + \frac{\partial \bar{v}}{\partial y} = 0 \tag{3.91}$$

and

♦ $$\frac{\partial \bar{u}}{\partial t} + \bar{u}\frac{\partial \bar{u}}{\partial x} + \bar{v}\frac{\partial \bar{u}}{\partial y} = -\frac{1}{\rho}\frac{\partial \bar{p}}{\partial x} + \frac{\partial}{\partial y}\left(\nu\frac{\partial \bar{u}}{\partial y} - \overline{u'v'}\right) \tag{3.92}$$

Turbulence modeling, that is, a solution to the closure problem, revolves around the turbulent stresses (cf. Eq. (3.90))

♦ $$\tau_{ij}^{\text{turb}} = -\rho\overline{v_i'v_j'} \tag{3.93a}$$

One prevailing idea is the *Boussinesq concept*, that is, the eddy viscosity model (EVM), where in an extension to Stokes's hypothesis (cf. Sect. 1.3.1)

♦ $$\frac{\tau_{ij}^{\text{turb}}}{\rho} = \nu_{\text{turb}}\frac{\partial \bar{v}_i}{\partial x_j} \tag{3.93b}$$

where the turbulent eddy viscosity, ν_{turb}, is a rather complex function. In general, $\nu_t = \nu_t$ (eddy size or length scale, turbulence kinetic energy, turbulence energy dissipation, turbulence time scale, velocity field gradients, system geometry, wall roughness, fluid viscosity, etc.).

Depending upon the level of turbulence complexity, which is typically associated with the type of flow geometry, we may need several ensemble-averaged PDEs, in addition to the mean flow equations, to model turbulence effects. For example, Prandtl's mixing length hypothesis (MLH) is an *algebraic* equation for the eddy viscosity of the form $\nu_t[l(y)]$. It is called a "zero-equation" model because no additional PDE is required for the solution of a particular turbulent flow. On the other hand, if two PDEs for the turbulence kinetic energy

$$k = 1/2\overline{(v_i')^2}$$

and the dissipation function

$$\varepsilon_{ij} = 2\nu\overline{(\partial v_i'/\partial x_k)(\partial v_j'/\partial x_k)}$$

are solved to form, for example,

$$\nu_t = C\left(k^2/\varepsilon\right)$$

the approach is called a "two-equation" model, and so on.

Alternatively, the cumbersome solution of *any* differential equation can be avoided altogether, when streamwise *velocity profiles for specific turbulent flows* are directly *postulated*.

Examples include the log-profiles and power-law profiles. Other recent advances in turbulence modeling have not matured enough and/or are too costly to be considered in complex turbulent systems analyses. Examples include the theory of dynamical systems where chaotic yet deterministic behavior of solution trajectories, controlled by sets of strange attractors, is being investigated; or direct turbulence simulation on parallel supercomputers, where the spatial and temporal resolutions have to be fine enough to represent the smallest and fastest turbulent flow phenomena. Since the exact simulation of the turbulence dynamics for complex flows is unattainable, traditional closure concepts will be used for today's and tomorrow's engineering computations. It is also evident that there is not one turbulence model that can compute a whole range of flows to acceptable engineering accuracy, and each particular model requires "fine tuning" with empirical data that depend on the flow type and system geometry.

B Turbulence Scales

Before a specific turbulence model is discussed, it may be of interest to consider first characteristic time and length scales of turbulent flows. Such scales are usually constructed from representative flow system parameters, such as approach velocity, wall friction velocity, or mean velocity; body length, shear layer thickness, or duct radius; fluid properties (i.e., v, ρ, k, and c_p); and mean temperature.

Time scales:

- Convection time scale $t_c = L/U$, where L is a typical system dimension and U is a mean flow velocity
- Viscous diffusion time scale $t_v = L^2/v$
- Kolmogorov time scale $t_k = (v/\varepsilon)^{1/2}$, where the eddy dissipation rate $\varepsilon \propto U^3/L$ or $\varepsilon \propto v(u'/l')^2$ with u' and l' are eddy velocity and size scales (here, viscous diffusion time is about equal to the eddy convection time l'/u')
- Heat diffusion time scale $t_h = L^2/\alpha$ where $\alpha = k/(\rho c_p)$

Length scales:

- Viscous wall length $l_\tau = v/u_\tau = O(l')$, where $u_\tau \equiv \sqrt{\tau_w/\rho}$ is the friction velocity
- Mixing length $l_m = \kappa y$ or $l_m = \kappa y[1 - \exp(-y/A)]$, where $\kappa = 0.41$ and A is the Van Driest damping length
- Kolmogorov length scale $l_k = (v^3/\varepsilon)^{1/4}$ or with $\mathrm{Re}_L = UL/v$, $l_k/L \sim \mathrm{Re}_L^{-3/4}$

Landahl and Mollo-Christensen (1992) pointed out an interesting effect of small-scale mixing. Consider 1 liter of water in a 10-cm-diameter household mixer with a net power input of 10 W. This implies that at equilibrium conditions $\varepsilon = 10$ W/kg and with $v_{H_2O} = 10^{-6}$ m^2/s, $l_k = 0.02$ mm and $l_k/L = 2 \times 10^{-4}$. Now, because $l_k \ll 1$ is desirable but $l_k \sim \varepsilon^{-1/4}$, one has to pay a high price in power consumption for good mixing; a halving of the smallest eddy size would require a power increase by a factor of 16. Additional examples and physical insight may be found in Hinze (1975), George and Arndt (1989), and Lesieur (1990), among others.

The brief discussion of the complexities of turbulence should imply that *turbulence modeling* is a very challenging, never-ending task. Numerous volumes have been written on this subject matter, and hence it is only appropriate to provide here a brief *summary* of some viable engineering approaches.

C   Summary of Turbulence Modeling

Despite some frustrating failures (in the past), computational fluid dynamics (CFD) modeling of turbulence is advantageous because of lower turn-around time, lower cost, more flexibility, and occasionally higher accuracy when compared to physical modeling. In a *decreasing order of complexity*, which implies reduced needs for computational resources and measured "constants," *deterministic* turbulence modeling approaches could be grouped as follows:

1. *Direct numerical simulation (DNS)* of turbulence, which requires time and length scales small enough to resolve turbulent fluctuations and tiny, near-wall eddies. In other words, DNS does not require any turbulence *model*.
2. *Large eddy simulation (LES)* with subgrid scale (SGS) modeling in which the coherent, large-scale structures are directly computed with the filtered Navier–Stokes equations while the small-scale eddies are modeled on the basis of the Smagorinsky eddy viscosity concept.
3. *Reynolds stress modeling (RSM)*, in which transport equations (PDEs) for each of the important components of the turbulence stress tensor, $\rho \overline{u'_i u'_j} = \tau_{ij}^{\text{turb}}$, plus the turbulence energy, $k = 1/2\overline{u_i'^2}$, have to be solved. Numerous empirical coefficients have to be tuned to match system-specific turbulent flow patterns.
4. *Eddy viscosity modeling (EVM)*, based on the postulates, or constitutive equations, by Boussinesq (1877) and Kolmogorov (1942):

$$\overline{u'_i u'_j} = \frac{\tau_{ij}^t}{\rho} = \frac{2}{3}k\delta_{ij} - \nu_t\left(\frac{\partial \bar{u}_i}{\partial x_j} + \frac{\partial \bar{u}_j}{\partial x_i}\right) \tag{3.94}$$

   It requires the calculation of one or more turbulence quantities (e.g., length scale l, turbulence kinetic energy k, dissipation function ε, time scale τ), which are then combined and directly related to the "turbulent eddy viscosity,"

$$\nu_t = \frac{\mu_t}{\rho} = \text{fct}\,(l, k, \varepsilon, \tau, \text{etc.})$$

   Zero-equation models, such as Prandtl's mixing length hypothesis (MLH) and the Van Driest extension for wall damping, require the solution of an *algebraic* equation, for example, an eddy length scale, where $\nu_t \sim l_m^2$. One-equation models require the solution of *one* transport equations (PDE), for example, $k = 1/2\overline{u_i'^2}$, where $\nu_t \sim \sqrt{k}$. Two-equation models require the solution of *two* transport equations (PDEs), for example, k and ε, where $\nu_t \sim k^2/\varepsilon$ as mentioned earlier.
5. *Empirical correlations*: The use of *empiricism* for fully developed pipe flows and flat-plate boundary-layer flows allows bypassing the equations of motion and hence eliminates the need for basic turbulence modeling. For example, turbulent pipe flow problems are solved with the extended Bernoulli equation, which relates pressure drops to losses in terms of the friction factor f.

$$\Delta z + \frac{\Delta p}{\rho g} = h_f = f\frac{L}{d}\frac{v^2}{2g} \qquad \text{where } f = f\left(\text{Re}_d, \frac{e}{d}\right)$$

$$\text{from} \begin{cases} \text{Moody chart } or \\ \text{formulas by Colebrook,} \\ \text{Blasius, Von Karman, etc.} \end{cases}$$

Pipe sizing, flow rates, and pump requirements are directly related to the friction loss h_f, where $h_f = (8Q^2 Lf/\pi^2 gd^5)$, with $f = 4c_f = (8\tau_w/\rho v^2)$, power $P = F_D \cdot v$, with $F_D = \tau_w A_{\text{surf}}$.

Alternatively, on a more differential basis, semiempirical *turbulent velocity profiles* could be employed:

$$\bar{u} \sim \begin{cases} \left(\dfrac{y}{r_0}\right)^{1/7} \quad \text{or} \quad \left(\dfrac{y}{\delta}\right)^{1/7} & \text{power-law} \\[2mm] \ln\left(\dfrac{y}{r_0}\right) \quad \text{or} \quad \ln y & \text{log-law} \end{cases}$$

Most turbulent velocity profiles $\bar{u}(x, y)$ require knowledge of the friction velocity $u_\tau = (\tau_w/\rho)^{1/2}$.

Notes to 1 and 2 above

DNS is useful for studying the detailed physics of turbulent "building block" flows. This understanding may lead to new, or at least improved, turbulence models. Current DNS examples include wall-bounded flows such as channel and boundary-layer flows as well as free shear flows, such as mixing layers and plane jets. However, all flow examples are presently restricted to homogeneous turbulence of relatively *low* Reynolds numbers and *simple* geometries.

Incorporation of length and time *scales* in turbulent flow calculations is as follows:

- Kolmogorov length or dissipation scale

$$l_k = \left(v^3/\varepsilon\right)^{1/4},$$

$$\varepsilon \sim \frac{U^3}{L} \tag{3.95a,b}$$

where ε is the energy dissipation function. Thus

$$\blacklozenge \qquad \frac{l_k}{L} \sim \text{Re}_L^{-3/4}, \qquad \text{Re}_L = \frac{UL}{v} \tag{3.96a,b}$$

where L is a reference length (e.g., BL thickness or largest eddy size) and U is a reference velocity (e.g., $u_{\max}$ or u_∞). In DNS, the Navier–Stokes equations have to "cover" all length scales as schematically illustrated (cf. sketch, p. 170).

- Kolmogorov time scale $\qquad t_k = \left(\dfrac{v}{\varepsilon}\right)^{1/2}, \qquad \varepsilon \sim \dfrac{U^3}{L} \tag{3.97a,b}$

$$\blacklozenge \qquad \therefore \ \frac{t_k}{t_{\text{ref}}} \sim \text{Re}_L^{1/2}, \qquad t_{\text{ref}} = \frac{L}{U} \tag{3.98a,b}$$

Note: Time step $\Delta t \ll t_k, t_{\text{ref}}$.

In conclusion, DNS is presently too costly for simple flows and prohibitive for most engineering flows. LES, although presently capable of depicting more complex flows than DNS, is also very cost-intensive and perhaps not suitable for engineering problem solutions.

Navier-Stokes Equations

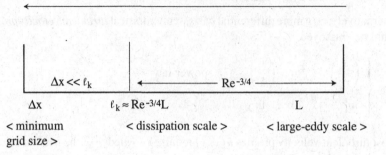

$$\Delta x \ll \ell_k \qquad\qquad \longleftarrow \text{— Re}^{-3/4} \text{—} \longrightarrow$$

Δx $\qquad\qquad\qquad \ell_k \approx \text{Re}^{-3/4}\text{L}$ $\qquad\qquad\qquad\qquad$ L

< minimum $\qquad\qquad$ < dissipation scale > $\qquad$ < large-eddy scale >
grid size >

Note to 3
RSM accounts for the history and nonlocal effects of the mean velocity gradients, for example, those due to relaxational effects, wall curvature, nonparallel flows, and counter gradient transport.

Mean Flow Equations:

$$u_i = \bar{u}_i + u'_i \qquad \text{where} \qquad \bar{u}_i = \frac{1}{\Delta t} \int_{t_1}^{t_2} u_i \, dt \tag{3.99a,b}$$

$$\therefore \ \bar{u}_{i,i} = 0 \quad \text{and} \quad \frac{D\bar{u}_i}{Dt} = -\frac{1}{\rho}\bar{p}_{,i} - (\overline{u'_i u'_j})_{,j} + \nu \bar{u}_{i,jj} \tag{3.100a,b}$$

Obviously, the apparent stresses $-\rho\overline{u'_i u'_j}$ have to be modeled in order to gain closure.
Reynolds Stress Transport Equations (RST Equations):

(i) Differential closures:

$$\frac{D}{Dt}(\overline{u'_i u'_j}) = D_{ij} + P_{ij} + T_{ij} + \prod_{ij} - \varepsilon_{ij} \tag{3.101}$$

$D_{ij} = \nu(\overline{u'_i u'_j})_{,kk} \hat{=} \text{diffusive transport} \qquad$ (i.e., on the viscous laminar level)

$P_{ij} = -\overline{u'_i u'_k} u_{j,k} - \overline{u'_j u'_k} u_{i,k} \hat{=} \text{stress production}$

$T_{ij} = -[\overline{u'_i u'_j u'_k} + \frac{1}{\rho}(\overline{u'_i p'}\delta_{ik} + \overline{u'_j p'}\delta_{ik})]_{,k} \hat{=} \text{turbulent diffusion transport}$
$\qquad\qquad\qquad\qquad\qquad\qquad\qquad\qquad\qquad$ (to be modeled)

$\prod_{ij} = \frac{1}{\rho}\overline{p'(u'_{i,j} + u'_{j,i})} \hat{=} \text{presure-strain correlation (to be modeled)}$

$\varepsilon_{ij} = 2\nu\overline{u'_{i,k} u'_{j,k}} \hat{=} \text{viscous dissipation (to be modeled)}$

(ii) Algebraic closures:
Assuming the transport of $\overline{u'_i u'_j}$ to be proportional to the transport of turbulence energy, $k = 1/2\overline{u'^2_i}$, the LHS of the RST equation is approximated as (Rodi 1980):

$$\frac{D}{Dt}(\overline{u'_i u'_j}) \approx \frac{\overline{u'_i u'_j}}{k} \frac{Dk}{Dt} \tag{3.102}$$

resulting in algebraic stress equations for

$$\overline{u'_i u'_j} = \text{fct}\left(k, \frac{Dk}{Dt}, \varepsilon, \text{etc.}\right)$$

where simultaneously solutions to auxiliary PDEs for k and ε have to be supplied.

In conclusion, although very computer- and data-intensive, RSM is now being incorporated into fluid dynamics softwares such as PHOENICS, FLUENT, FIDAP, CFX, FLOW-3D, FLOTRAN, and CFD 2000. However, *wall effects*, such as rapid near-wall variations, especially in the $\prod_{ij}$-terms, may still cause substantial errors. Furthermore, RSM relies typically on a single turbulence time scale to characterize rate processes in turbulence. However, at least two independent time scales accounting for the different response mechanisms of the large-scale and the small-scale motions are needed in some applications.

Note to 4

EVM, typically based on zero-equation models or two-equation models, works reasonably well for nonseparating, *near-parallel* shear flows. However, model inclusion of correction factors and extension terms, representing effects due to streamline curvature, flow separation, flow rotation, fluid compressibility, and other factors, have kept EVM the first choice for a large variety of engineering problems. Virtually all fluid dynamics software packages entertain MLH (i.e., zero-equation) and k–ε (i.e., two-equation) closures. Again, proper near-wall shear-layer modeling is crucial for successful turbulent flow simulations.

The underlying concept is Boussinesq's analogy with molecular transport of momentum. Thus, in simplified form

♦ $$\tau_t = \rho \nu_t \frac{\partial \bar{u}}{\partial y} \qquad \text{where } \rho \nu_t \equiv \mu_t \tag{3.103a,b}$$

Now, for wall-bounded shear layers, based on Prandtl's mixing length hypothesis (MLH):

$$\nu_t = l^2 \left|\frac{\partial \bar{u}}{\partial y}\right|, \qquad l = \kappa y \qquad \text{(Prandtl and Von Karman)}$$

or

$$\nu_t = l^2 \left|\frac{\partial \bar{u}}{\partial y}\right|,$$
$$l = \begin{cases} \kappa y & \text{for } 0 \leq y/\delta \leq \lambda/\kappa \\ \lambda \delta & \text{for } \lambda/\kappa \leq y/\delta \leq 1 \end{cases} \quad \text{(Cebeci and Bradshaw)} \tag{3.103c–e}$$
$$\kappa = 0.435 \quad \text{and} \quad \lambda = 0.09$$

or

$$\nu_t = \begin{cases} \nu_t^{\text{inner}} = \text{fct}(l^2, |\frac{\partial u}{\partial y}|, \gamma^i) \text{ for } 0 \leq y \leq y_c & \text{(Cebeci and Smith)} \\ \nu_t^{\text{outer}} = \text{fct}(U_0, \delta_1, \gamma^0) \text{ for } y_c \leq y \leq \delta \end{cases}$$

where $\quad l = \kappa y[1 - \exp(-y/A)]$, A is the Van Driest damping length, γ's are intermittency factors, U_0 is the outer flow, and δ is the BL thickness; $\tau_t^{\text{inner}} = \tau_t^{\text{outer}}$ at $y = y_c$.

Table 3.2. *Exponent n and velocity ratio u_{av}/u_{max} as a function of Reynolds number*

Reynolds number	4×10^3	23×10^4	1.1×10^2	1.1×10^6	2.0×10^6	3.2×10^6
n	6.0	6.6	7.0	8.8	10.0	10.0
u_{av}/u_{max}	0.791	0.806	0.817	0.849	0.865	0.865

Note: cf. Schlichting (1979).

For a *complete* eddy viscosity turbulence model, at least two turbulence quantities have to be specified: typically the turbulence kinetic energy k plus a length scale L, energy dissipation rate ε, or dissipation time scale τ. For example,

$$\nu_t = c\sqrt{k}L \qquad \text{(Kolmogorov and Prandtl)}$$

$$\nu_t = C_\mu \frac{k^2}{\varepsilon}, \qquad C_\mu = 0.09; \qquad \varepsilon \propto \frac{k^{3/2}}{L} \qquad \text{(Jones and Launder)} \qquad (3.104a,b)$$

where

$$\frac{Dk}{Dt} = P_{ij} - \varepsilon + \frac{\partial}{\partial x_j}\left(\frac{\nu_t}{\sigma_k}\frac{\partial k}{\partial x_j}\right) + \nu\nabla^2 k \quad \text{and} \qquad (3.105)$$

$$\frac{D\varepsilon}{Dt} = G_{ij} - C_\varepsilon\frac{\varepsilon^2}{k} + \frac{\partial}{\partial x_i}\left(\frac{\nu_t}{\sigma_\varepsilon}\frac{\partial\varepsilon}{\partial x_i}\right) + \nu\nabla^2\varepsilon \qquad (3.106)$$

with $\sigma_k \approx 1.0$, $\sigma_\varepsilon \approx 1.92$, $C_\varepsilon = 1.44 - 1.96$, $P_{ij} \triangleq k$-production terms, and $G_{ij} \triangleq \varepsilon$-generation terms.

Note to 5
The turbulent boundary-layer velocity profile

$$\blacklozenge \qquad u^+ \equiv \frac{\bar{u}}{u_\tau} = \frac{1}{\kappa}\ln y^+ + B; \qquad y^+ = u_\tau\frac{y}{\nu} \qquad (3.107a,b)$$

can be *derived* on the basis of Prandtl's near-wall shear stress hypotheses as shown in Sect. 3.2.5. The empirical constants are $\kappa = 0.41$ and $B = 5.24$. The friction velocity $u_\tau \equiv \sqrt{\tau_w/\rho}$ is either indirectly measured or obtained from appropriate skin friction or friction factor correlations $f(\tau_w) = \text{fct}(\text{Re}, e)$ where e is the surface roughness. Terms representing laminar sublayer, buffer zone, and near-wake effects have all been added to the basic log-law profile (cf. Spalding 1961, among others).

An alternative turbulent velocity profile, typically applied to pipe flow, is the power law

$$\frac{\bar{u}}{u_{max}} = \left(1 - \frac{r}{r_0}\right)^{1/n} \qquad (3.108)$$

which can be constructed where n is a weak function of the Reynolds number $\text{Re} = u_{av}D/\nu$ with $D = 2r_0$ and $u_{av} = Q/\pi r_0^2$ (cf. Table 3.2).

It has to be noted that the power law profile has a nonzero gradient at the centerline, that is, $(\partial u/\partial r)|_{r=0} \neq 0$. Equation (3.108) can be rewritten in terms of the friction velocity

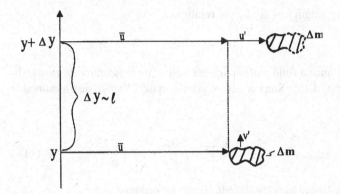

Fig. 3.15. Mixing length model.

$u_\tau = \sqrt{\tau_w/\rho}$ and the wall coordinate $y = r_0 - r$, with $n = 7$, as

$$\quad \frac{u}{u_\tau} = 8.74 \left(\frac{y u_\tau}{\nu}\right)^{1/7} \tag{3.109}$$

which is known as the 1/7-law. It has to be noted that neither the log-law nor the power-law fulfills all boundary conditions at $y = 0$ and $y = \begin{cases} \delta \\ r_0 \end{cases}$.

3.2.5   Turbulent Flow Calculations

For turbulent flat-plate shear layers, the stress tensor τ_{ij}^{turb} reduces to

$$\tau_{yx} = -\rho \overline{u'v'} := \rho \nu_t \frac{\partial \bar{u}_i}{\partial x_j} \equiv \rho \varepsilon_m \frac{\partial \bar{u}_i}{\partial x_j} \tag{3.110a,b}$$

following Boussinesq's eddy viscosity concept (cf. Sect. 3.2.4A).

Clearly, the eddy viscosity ν_t plays a central role in turbulence modeling. With an increasing degree of complexity we express $\nu_t = \nu_t(l)$, $\nu_t = \nu_t(\sqrt{k})$, $\nu_t = \nu_t(k^2, \varepsilon^{-1})$, and so on. In order to obtain some physical understanding of the Reynolds stresses, we consider a fluid mass Δm, large in comparison with a molecule, at level y (cf. Fig. 3.15). It moves upward with the incremental velocity v' through the distance Δy. The mass Δm, which had the velocity $\bar{u}$, is now in a region where the flow velocity is $\bar{u} + u'$, where $u' \cong (d\bar{u}/dy)\Delta y$. The lump of fluid arrives at the new position with a deficit of x-momentum equal to $\rho v'(-u')$, where $(\rho v')$ is the fluctuating mass flux. By the same token, it is this excess of momentum relative to the fluid at level y, that is, $\rho(-v')u'$, that gives rise to a turbulent stress. It is apparent that $+v'$ is usually associated with $-u'$, and vice versa, so that $(\rho u'v') < 0$. In summary, the Reynolds stress is equal to the time-averaged momentum deficiency $(-\rho \overline{u'v'})$. This retardation is in the same direction as the laminar viscous stress.

Now, Prandtl (1921) suggested that the eddy viscosity depends on only one length scale: $\nu_t = \nu_t(l)$. This is known as a "zero-equation" model since no transport equations (PDEs) for the turbulence quantities have to be employed.

Prandtl defined the mixing length $l \sim \Delta y$ by the relation

$$u' \cong l\frac{d\bar{u}}{dy}$$

where l is the transverse distance a fluid particle travels before its momentum is changed and it loses its identity (cf. Fig. 3.15). Since u' and v' are correlated, Prandtl also assumed

$$v' \cong l\frac{d\bar{u}}{dy}$$

$$\therefore \tau_{\text{total}} = \tau_{\text{viscous}} + \tau_{\text{turbulent}} = \mu\frac{d\bar{u}}{dy} - \rho\overline{u'v'} = \mu\frac{d\bar{u}}{dy} + \rho l^2\left(\frac{d\bar{u}}{dy}\right)^2 \qquad (3.111)$$

from which Prandtl's mixing length hypothesis (MLH) can be deduced.

$$\nu_t = l_m^2\left|\frac{\partial\bar{u}}{\partial y}\right| \qquad (3.112)$$

where $l_m = \kappa y$, $\kappa = 0.41$, and y is measured from the wall. Clearly, turbulence modeling introduces an additional nonlinearity into the equation of motion. Either Equation (3.112) and its extensions (cf. Van Driest 1951; Rodi 1984; Granville 1990) can be directly used to solve the turbulent transport equations or Prandtl's MLH can be employed to derive turbulent velocity profiles suitable for use in the integral method, as shown below.

A Turbulent Velocity Profiles for Thin Shear Layers

The *simplest* representation of a turbulent flow velocity profile is the power law (cf. Summary of Sect. 3.2.4C).

$$\blacklozenge \qquad \frac{\bar{u}}{U_0} = \left[\frac{y}{\delta_t(x)}\right]^{1/n} \qquad (3.113)$$

where $U_0 = u_{\text{edge}}$ is the constant outer flow velocity and $n = n(\text{Re}_l)$, typically, $n = 7$; the local turbulent boundary-layer thickness $\delta_t = \delta_t(\text{Re}_x)$.

Alternatively, Prandtl's MLH can be employed to derive a *logarithmic* velocity profile that represents a wide range of internal and external turbulent flows that are fully developed, steady, two-dimensional, and homogeneous (cf. sketch). Measurements in turbulent open channel flow and boundary-layer flow showed the existence of basically two layers: (1) the inner region in the vicinity of the wall and (2) the outer, fully turbulent region, which makes up over 90 percent of a boundary layer.

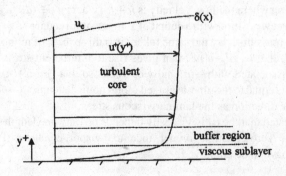

1. At the wall, that is, in the inner region where viscous shear is dominant, called the *viscous sublayer*, $u = u(\tau_{\text{wall}}, \nu_{\text{lam}}, y)$. Thus, Prandtl postulated a "law of the wall" in dimensionless form

$$u^+ = \frac{u}{u_\tau} = \phi(y^+)$$

where $\tau_w = \text{const}$ and $y^+ = u_\tau y/\nu$ is a wall Reynolds number based on the friction velocity $u_\tau = \sqrt{\tau_w/\rho}$.

At the wall, $0 < y^+ < 5$, $\nu_t = 0$, and $\tau_{\text{total}} = \tau_{\text{wall}}$, so that Equation (3.111) reduces to

$$\frac{\tau_w}{\rho} \approx \text{const} = \nu \frac{\partial u}{\partial y} \qquad \text{subject to } u(y = 0) = 0$$

Integration yields

◆ $$u^+ = y^+ \qquad \text{for } 0 < y^+ < 5 \tag{3.114}$$

that is, a linear velocity profile within the laminar sublayer.

Near the wall, $30 < y^+ < 400$, $\tau_l \ll \tau_t$, and $\tau_{\text{total}} = \tau_{\text{turb}} \approx \tau_{\text{wall}} = \text{const}$. The mixing length is taken after Prandtl as $l_m = \kappa y$, that is, a linear increase in eddy size away from the wall so that Equation (3.110b) becomes

$$\tau_w = \rho(\kappa y)^2 \left(\frac{\partial u}{\partial y}\right)^2 = \text{const}$$

After integration

$$u = \frac{1}{\kappa}\sqrt{\frac{\tau_w}{\rho}} \ln y + C$$

or with $\sqrt{\tau_w/\rho} = u_\tau$,

◆ $$u^+ = \frac{1}{\kappa}\ln(y^+) + B \qquad \text{for } 30 < y^+ < 400 \tag{3.115}$$

where $\kappa = 0.41$ is a universal or Von Karman constant and $B \approx 5.0\text{--}5.5$ is an empirical (integration) constant reflecting possible surface roughness and the type of system considered.

In the *buffer region*, $5 < y^+ < 30$, a semiempirical log-law is appropriate

$$u^+ = 5.0\ln(y^+) - 3.0 \qquad \text{for } 5 < y^+ < 30 \tag{3.116}$$

2. The outer region, $0.2 \le y/\delta \le 1.0$ or $y^+ > 400$, which makes up most of the turbulent shear layer, is characterized by

$$\frac{u_e - u}{u_\tau} = f\left(\frac{y}{\delta(x)}, \delta_1, \tau_w, \frac{dp}{dx}, \frac{d\delta}{dx}\right) \tag{3.117}$$

An expression of the form (3.117) is known as the *velocity defect law*, where u_e is the edge velocity. Since the outer region of the composite boundary-layer model

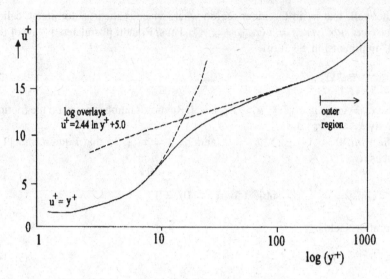

Fig. 3.16. Distinct regions of turbulent boundary layer (u_{edge} = const).

has to match the inner region, that is, Equation (3.115), an empirically extended log-law has been postulated

$$\blacklozenge \qquad u^+ = \frac{1}{\kappa} \ln y^+ + C + \frac{\Pi}{\kappa} W \tag{3.118}$$

where the added function depends on a pressure gradient parameter (Clauser 1954)

$$\beta = \frac{\delta_1}{\tau_w} \frac{dp}{dx} = \text{const}$$

and the far-wake behavior. On the basis of curve fitting

$$\Pi \approx 0.8(\beta + 0.5)^{0.75} \qquad \text{(wake parameter)} \tag{3.119a}$$

and

$$W \approx 2 \sin^2\left(\frac{\pi}{2} \frac{y}{\delta}\right) \qquad \text{(wake function)} \tag{3.119b}$$

For many applications wake effects are negligible and Equation (3.115) prevails. Instead of patching several profiles together, that is, Equations (3.114) to (3.118) as shown in Figure 3.16, Spalding (1961) suggested an implicit relationship for $u^+(y^+)$ that holds for the entire boundary layer or duct cross-sectional area:

$$y^+ = u^+ + 0.1108\left[\exp\left(0.4u^+ - 1 - 0.4u^+ - \frac{1}{2}(0.4u^+)^2 - \frac{1}{6}(0.4u^+)^3\right)\right] \tag{3.120}$$

Example (8)

Both velocity profiles (Eq. 3.113 and Eq. 3.115) are now used in the evaluation of steady incompressible turbulent boundary layers with constant outer flow field. The momentum integral relation for this class of boundary-layer flows reads (cf. Sects. 2.3.3B and 3.2.3D).

$$\frac{d}{dx}\left[\delta\int_0^1 \frac{\bar{u}}{U}\left(1-\frac{\bar{u}}{U}\right)d\left(\frac{y}{\delta}\right)\right] = \frac{\tau_w}{\rho U^2} \qquad \text{(E3.1)}$$

Starting with the power-law profile

$$\frac{\bar{u}}{U} = \left(\frac{y}{\delta}\right)^{1/n} \qquad \text{(E3.2)}$$

where $n \approx 7$ for $\mathrm{Re} = O(10^7)$, we first evaluate the wall shear stress, for example, from

$$\tau_w \equiv \frac{f}{8}(\rho v^2) \qquad \text{(E3.3a)}$$

It has to be noted that neither the power law nor the log law is valid near the wall. The friction factor f could be obtained from the Colebrook formula or with the Blasius correlation for smooth plates and relatively low turbulent Reynolds numbers (cf. derivation in Sect. 3.2.5B).

$$f \approx 0.3164\,\mathrm{Re}^{-1/4} \qquad \text{(E3.3b)}$$

where $\mathrm{Re} = v(2\delta)/v$ and the mean velocity v is

$$v = \frac{1}{A}\int \bar{u}(y, \delta(x), n, U)\,dA := \frac{2n^2 U}{(n+1)(2n+1)}$$

Hence, from Equation (3.123)

$$\tau_w \approx 0.03325\rho v^{7/4}\left(\frac{v}{\delta}\right)^{1/4} \qquad \text{(E3.4)}$$

Inserting (3.122) and (3.124) into Eq. (3.121) yields with $y/\delta \equiv \eta$

$$\frac{d}{dx}\left[\delta\int_0^1 \eta^{1/7}(1-\eta^{1/7})d\eta\right] = 0.03325\left(\frac{v}{U\delta}\right)^{1/4}$$

or

$$\delta^{1/4}\frac{d\delta}{dx} = 0.240\left(\frac{v}{U}\right)^{1/4}$$

subject to $\delta(x=0) = 0$. Integration yields

$$\delta = 0.380\left(\frac{v}{U}\right)^{1/5}x^{4/5}$$

or

$$\blacklozenge \qquad \frac{\delta}{x} \approx \frac{0.380}{\mathrm{Re}_x^{1/5}} \qquad \text{(E3.5)}$$

where $5\times 10^5 < \mathrm{Re}_x = (Ux/v) < 10^7$. Substituting (3.125) into Eq. (3.124) yields an expression for the skin friction coefficient.

Using the usually more accurate log law in the form

$$\frac{u}{u_\tau} = \frac{1}{0.41} \ln\left(\frac{y u_\tau}{\nu}\right) + 5.0$$

where $u_\tau \equiv \sqrt{\tau_w/\rho} = \sqrt{c_f U^2/2}$, one can show (cf. White 1991) that

$$\frac{\delta}{x} \approx \frac{0.140}{\text{Re}_x^{1/7}} \tag{E3.6}$$

It is of interest to note that whereas

$$\delta_{\text{laminar}} \sim x^{1/2} \tag{E3.7a}$$

$$\delta_{\text{turbulent}} \sim x^{4/5} \text{ to } x^{6/7} \tag{E3.7b}$$

depending upon the type of trial solution for $\bar{u}(x, y)$ employed.   □

B   Turbulent Duct Flow

In order to generate some appreciation for the structure of turbulence in fully developed pipe flow, Figures 3.17a–f depicts key turbulence parameters across the pipe or near the wall for $\text{Re}_d = \bar{u}_{\max} d/\nu = 5 \times 10^5$. Additional profiles can be found in Laufer (1954), Hinze (1975), Tennekes and Lumley (1972), and Schlichting (1979), among others.

As in laminar internal flows, turbulent flows are subdivided into a developing entrance region and a fully developed region. The hydrodynamic entrance lengths can be estimated for circular smooth pipes as

$$\frac{l_e}{d} \approx \begin{cases} 0.06\,\text{Re} & \text{for laminar flow} \\ 4.4\,\text{Re}^{1/6} & \text{for turbulent flow} \end{cases} \tag{3.128a,b}$$

Actually, the entrance region consists of boundary-layer-type flow near the inlet, where the core velocity increases continuously and the pressure drop is nonlinear. Further downstream, the core vanishes and the thick shear layers meet and merge at the centerline. Representative profiles for high-Reynolds-number pipe flow are depicted in Fig. 3.18. Note that the profile areas are the same for incompressible flow because of mass conservation. Two entrance lengths are shown: one, $l_{e,p}$, is characterized by a variable pressure gradient, and the other reflects turbulent flow development, $l_{e,d}$. As derived later, the total shear stress varies linearly with the radius in fully developed flow, that is, $\bar{\tau}_{\text{total}}(r) = \bar{\tau}_{\text{turb}}(r) + \tau_{\text{lam}}(r) = (\Delta p/2L)r$, which implies that $\tau_w = (\Delta p/2L)r_0$. As always for internal flow, the condition

$$\dot{m} = \text{const} \tag{3.129}$$

can be used to compute the axial pressure gradient $\partial p/\partial x$. The centerline velocity is employed as an "outer flow" velocity when evaluating the turbulent inlet boundary layers.

In the order of increasing complexity, there are three basic approaches for solving *fully developed* turbulent duct flow problems such as finding the pump power, duct size (i.e., hydraulic diameter), or flow rate:

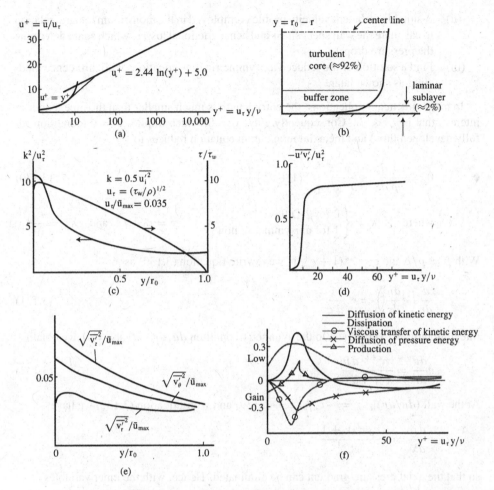

Fig. 3.17. Turbulence data for fully developed flow in a smooth pipe (after Laufer 1954): (a) velocity profile in inner wall region; (b) turbulent velocity profile; (c) turbulence kinetic energy and shear stress; (d) turbulence shear stress in the wall region; (e) turbulence intensities; (f) near wall energy budget. Note: All terms are relatively small in the core.

(i)   Compute the energy loss $gh_f = f(l/d)(v^2/2)$, where the friction factor $f = f(\mathrm{Re}_d, e/d)$ is obtained from the Moody chart or other sources; the extended Bernoulli equation (or mechanical energy equation) then relates the losses to the pressure drop.

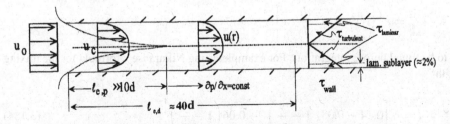

Fig. 3.18. Velocity and shear stress profiles in turbulent smooth pipe flow.

(ii)   A suitable turbulent velocity profile is employed in the momentum integral relation to calculate the wall shear stress and hence frictional losses, which again determine the pressure drop.

(iii) Find a solution of the reduced axisymmetric flow equations with turbulence modeling as shown later.

In many practical applications, the entry length is much smaller than the pipe length of interest, that is, $l_e \ll L$. Consequently, $\partial u / \partial x = 0$, which implies $v = 0$ (condition for fully developed flow) and the axial momentum equation reduces to

$$\blacklozenge \qquad 0 = -\frac{1}{\rho}\frac{dp}{dx} + \frac{v}{r^\kappa}\frac{d}{dy}\left[r^\kappa\left(1 + \varepsilon_m^+\right)\frac{du}{dy}\right] \qquad (3.130)$$

$$\text{where} \qquad \kappa = \begin{cases} 0 \text{ for 2-D flow} \\ 1 \text{ for axisymmetric flow} \end{cases}, \qquad y = r_0 - r \quad \text{and} \quad \varepsilon_m^+ = \frac{\varepsilon_m}{v}$$

With $\hat{p} = p/\rho$ and $c = r^\kappa(1 + \varepsilon_m^+)v$, we rewrite Equation (3.130) as

$$\frac{d}{dr}\left(c\frac{du}{dr}\right) = r^\kappa\frac{d\hat{p}}{dx} \qquad (3.131)$$

Integration of (3.131) subject to the symmetry condition $du/dr = 0$ at $r = 0$, we obtain

$$c\frac{du}{dr} = \frac{r^{\kappa+1}}{\kappa+1}\frac{d\hat{p}}{dx} \qquad (3.132)$$

At the wall $(du/dr)|_{r=r_0} = -\tau_w/\mu \equiv -(u_\tau^2/v)$ and with Equation (3.132) we have

$$v\frac{d\hat{p}}{dx} = -\frac{c_0 u_\tau^2(\kappa+1)}{r_0^{\kappa+1}}$$

so that the axial pressure gradient can be eliminated. Hence, with the inner variables

$$u^+ = \frac{u}{u_\tau}, \qquad y^+ = \frac{u_\tau y}{v} \quad \text{and} \quad r_0^+ = \frac{r_0 u_\tau}{v}$$

Equation (3.132) can be written as

$$\blacklozenge \qquad \left(1 - \frac{y^+}{r_0^+}\right) = \left(1 + \varepsilon_m^+\right)\frac{du^+}{dy^+} \qquad (3.133)$$

For laminar duct flow, $\varepsilon_m^+ \equiv 0$ and for fully developed turbulent duct flow an expression for

$$\varepsilon_m^+ = \varepsilon_m^+\left[l(y), u_{\max}, \frac{du}{dy}; v\right]$$

has to be found before integration. For example, using Nikuradse's formula for the mixing length

$$l = r_0\left[0.14 - 0.08\left(1 - \frac{y}{r_0}\right)^2 - 0.06\left(1 - \frac{y}{r_0}\right)^4\right] \qquad (3.134)$$

and Van Driest's damping factor, $(1 - e^{-y}/A_D)$, we obtain

$$\blacklozenge \qquad \varepsilon_m^+ = \left\{ l^+[1 - \exp(-y^+/A_D^+)] \right\}^2 \frac{du^+}{dy^+} \qquad (3.135)$$

where $l^+ = lu_\tau/\nu$ and $A_D^+ = 26$. Clearly, an expression for $u_\tau \sim f$ has to be found before Equation (3.133) can be solved.

Pipe Friction Factors

In order to calculate $u(y)$ after inserting Eq. (3.135) in Eq. (3.133) and integrating once with $u^+(y^+ = 0) = 0$, we need an expression for the friction velocity $u_\tau = \sqrt{\tau_w/\rho}$. We use the log-law for turbulent pipe flow, evaluated at the centerline where $y = r_0$ and $u = u_{\max}$, that is,

$$\frac{u_{\max}}{u_\tau} = 0.25 \ln\left(\frac{r_0 u_\tau}{\nu}\right) + 5.5 \qquad (3.136)$$

and use the empirical relation for the average velocity

$$\blacklozenge \qquad u_{av} = u_{\max} - 4.07 u_\tau \qquad (3.137)$$

to obtain Prandtl's friction law for smooth pipes (cf. Section 3.3)

$$\frac{1}{\sqrt{f}} = 0.87 \ln\left(\text{Re}\sqrt{f}\right) - 0.8 \qquad (3.138)$$

where the pipe friction factor f is discussed later in conjunction with the derivation of the Blasius friction factor correlation. Given a volumetric flow rate, $Q = u_{av} A$, the Reynolds number and hence the friction factor can be computed and u_τ is known. Since empiricism was used, mass conservation might not be exactly fulfilled, that is,

$$\rho \iint_A u(y)\, dA \neq \dot{m} \qquad (3.139)$$

In such a case, the formula for f or ε_m^+ has to be adjusted in order to conserve mass 100 percent.

To derive the friction factor for fully developed turbulent pipe flow, we start with the *integrated* momentum equation for a finite control volume (cf. Reynolds Transport Theorem in Sect. 2.1.1)

$$\frac{\partial}{\partial t} \iiint_{c.\forall.} \rho\vec{v}d\forall + \iint_{c.s.} \vec{v}\rho\vec{v} \cdot d\vec{S} = \sum \vec{F}_{ext} = \iint_{c.s.} -p\, dS + \iint_{c.s.} \overset{=}{\tau} \cdot d\vec{S} \qquad (3.140)$$

For steady, fully developed duct flow, Eq. (3.140) reduces to

$$0 = -\iint p\, dS + \iint \overset{=}{\tau} \cdot d\vec{S} \qquad (3.141)$$

where p and τ are time-smoothed variables. For an arbitrary cylindrical control volume in a circular pipe (Fig. 3.19), Eq. (3.141) yields

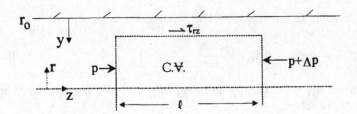

Fig. 3.19. Control volume in a pipe.

$$0 = -\Delta p r^2 \pi + \tau_{rz} 2r\pi l \tag{3.142a}$$

which evaluated at the wall, $r = r_0$, reads

$$0 = -\Delta p r_0^2 \pi + \tau_w 2r_0\pi l \tag{3.142b}$$

Dividing the last two equations yields

$$\tau = \tau_w \frac{r}{r_0} \tag{3.143}$$

With $\Delta p = (2l/r_0)\tau_w$ from Equation (3.142b) and recalling the "frictional pipe loss" expression (White 1986)

$$h_f = \frac{\Delta p}{\rho g} = f \frac{l}{2r_0} \frac{u_{av}^2}{2g} \tag{3.144}$$

we obtain

♦ $$\tau_w = \frac{1}{4} f \frac{\rho u_{av}^2}{2} \quad \text{or} \quad f = \frac{8\tau_w}{\rho u_{av}^2} \tag{3.145a,b}$$

In order to derive a correlation for $f = f(\text{Re})$, we start with an empirical turbulent velocity profile for circular pipes, the 1/7-law.

♦ $$u^+ = \frac{u}{u_\tau} = 8.7 \left(\frac{y u_\tau}{\nu} \right)^{1/7} \quad \text{for } 2 \times 10^2 < \text{Re} < 10^5 \tag{3.146}$$

The approach is to find

$$f \sim u_{av}, \quad \text{i.e., } f \sim \text{Re}$$

From Equation (3.145b)

$$f \equiv \frac{8\tau_w}{\rho u_{av}^2}$$

and by definition

$$\tau_w = \rho u_\tau^2$$

so that

$$f = 8 \left(\frac{u_\tau}{u_{av}} \right)^2$$

With $u^+(y^+)$ evaluated at the centerline, $y = r_0$, we have from (3.146)

$$\frac{u_{max}}{u_\tau} = 8.7\left(\frac{r_0 u_\tau}{\nu}\right)^{1/7}$$

or

$$u_\tau = \left(\frac{u_{max}}{8.7}\right)^{7/8}\left(\frac{\nu}{r_0}\right)^{1/8}$$

Also by definition

$$u_{av} = \frac{1}{A}\int u \, dA = \frac{2}{r_0^2}\int_0^{r_0} ur \, dr$$

$$= \frac{2}{r_0^2}(8.7)u_\tau\left(\frac{u_\tau}{\nu}\right)^{1/7}\int_0^{r_0}(r_0 - r)^{1/7}r \, dr$$

$$= 7.105 u_\tau\left(\frac{r_0 u_\tau}{\nu}\right)^{1/7}$$

or

$$u_{av} = 0.817 u_{max} \tag{3.147}$$

Now, substituting u_τ with $u_{max} = (u_{av}/0.817)$ into Equation (3.145b), we obtain

$\blacklozenge$ $\qquad f = 0.3164\,\mathrm{Re}^{-1/4}$ $\qquad\qquad\qquad\qquad\qquad\qquad\qquad$ (3.148)

Equation (3.148) is the explicit Blasius correlation $f(\mathrm{Re})$ *for smooth pipes* where $\mathrm{Re} = u_{av}(2r_0)/\nu$. In general, as can be deduced from the Moody chart (cf. Fig. 3.20), there are three regimes where wall roughness, e, may be important or not:

$$\frac{e}{d}\mathrm{Re}_d < 10 \qquad\qquad \text{wall roughness is unimportant}$$

$$10 \leq \left[\frac{e}{d}\mathrm{Re}_d\right] \leq 1{,}000 \qquad f = f\left(\frac{e}{d}, \mathrm{Re}_d\right)$$

$$\frac{e}{d}\mathrm{Re}_d > 10^3 \qquad\qquad f = f\left(\frac{e}{d}\right) \text{ only}$$

The following example (cf. Gerhart et al. 1992) may illustrate the use of the friction factor correlation and the different turbulent pipe flow velocity profiles. Only *smooth* pipe walls are considered in this problem.

Problem

Water flows at $\mathrm{Re}_d = 2.3 \times 10^4$ through a horizontal smooth pipe with $u_{av} = 0.807$ m/s. Compare $\bar{u}(y^+ = 5, 100,$ and $500)$ from (a) the log law and (b) the power law to the measured data, that is, the solid line in Fig. 3.16.

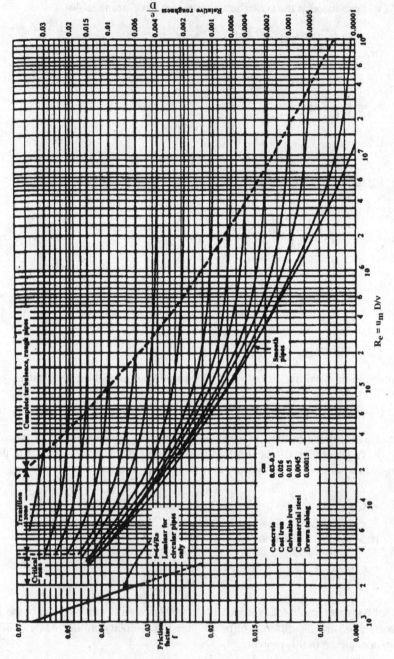

Fig. 3.20. Friction factor for flow inside circular pipes (the Moody Chart after Colebrook 1938 and Moody 1944).

Table 3.3. *Data comparison for turbulent velocity profiles*

y^+	$(y/r_0) \times 100\%$	Log law	Power law	Measurements
5	0.8%	NA (0.229 from $u^+ = y^+$)	0.478	0.229
100	15.3%	0.741	0.752	0.754
500	76.4%	0.921	0.960	0.983
	100%	0.969	1.00	

Solution

(a) The log law $u^+ = 2.44 \ln y^+ + 5.0$, $40 \leq y^+ \leq 600$, has to be augmented for $y^+ = 5$, which is located in the laminar sublayer for which $u^+ = y^+$. Again, $u^+ \equiv \bar{u}/u_\tau$ and $y^+ \equiv yu_\tau/\nu$, where $y = r_0 - r$. First we find $u_\tau \equiv (\tau_w/\rho)^{1/2}$ where

$$\tau_w = \rho u_{av}^2 f/8 \quad \text{with} \quad f \approx 0.3164 \, \text{Re}_d^{-1/4} \qquad \text{(cf. Eq.(3.148))} \qquad \text{(P3.17)}$$

Hence, with $\rho = 998$ kg/m^3 and $\nu = 1 \times 10^{-6}$ m^2/s,

$$u_\tau = 0.0457 \text{ m/s}$$

so that $\bar{u}$ at all three locations can be calculated.

(b) The power law for pipe flow can be written as (cf. Sect. 3.2.4)

$$\bar{u} = u_{max} \left(\frac{y}{r_0} \right)^{1/n} \qquad \text{where } n \approx 6.6 \qquad \text{for } \text{Re}_d = 2.3 \times 10^4$$

Now, from

$$u_{av} = \frac{1}{A} \int \bar{u} \, dA$$

the centerline velocity $u_{max} \equiv u_{\mathfrak{t}}$ can be derived as

$$u_{max} = \frac{(n+1)(2n+1)}{2n^2} u_{av} := 1.0 \text{ m/s}$$

Summary of Results

In Table 3.3, the results are compared with experimental findings of Fig. 3.16. The power law performs quite well near the wall.

C Turbulent Free Shear Flows

Free shear flows, such as jets, separated boundary layers, wakes, and mixing layers (cf. Fig. 3.21a), possess velocity gradients caused by some *upstream* mechanism, for example, nozzle walls, splitter plate, submerged body, or multistream interfaces. Typically, Re $\gg 1$ and $\partial p/\partial x \approx 0$. Downstream of the disturbance, free shear flows decelerate, entrain ambient fluid, or mix layers and spread. Viscous effects smooth the velocity field, which may become *self-similar* farther downstream (known as *far-field similarity*).

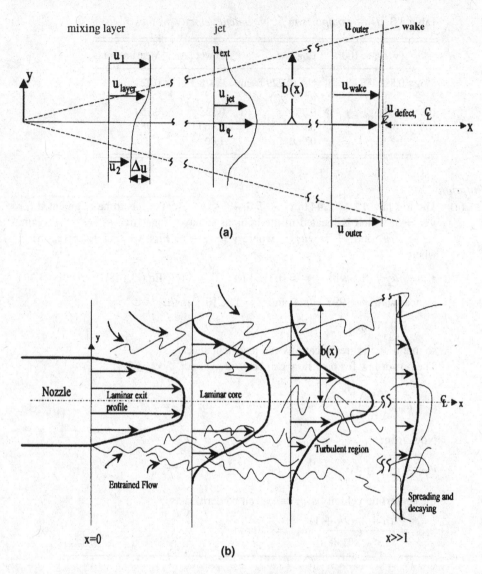

Fig. 3.21. (a) Schematics of three different free shear flows. (b) Laminar/turbulent jet discharging into a reservoir (after Panton 1984).

In general, free shear flows are highly unstable and hence become *turbulent* quite close to their origin. Large-scale coherent vortex structures are the dominant component of such turbulent flows. Specifically in mixing layers, pairing of these vortices, as they move downstream, contributes largely to the lateral growth. For turbulent round jets, $Re_d > 30$, the potential core disappears at about $x/d \approx 4$. For $x/d > 30$, the time-averaged velocity profiles are self-preserving, that is, $u_{jet}(x, y)/\Delta u(x) = f[y/b(x)]$. In between, $4 \leq x/d \leq 30$, the flow field including the Reynolds stresses are strongly developing.

Laminar/Turbulent Jets

Consider an axisymmetric or two-dimensional jet emerging from a nozzle and mixing with the same surrounding, stagnant fluid (cf. Fig. 3.21b).

With $b/l \sim v/u \ll 1$ and very small pressure variations throughout the jet, the steady-state boundary-layer equations can be written as (cf. Sect. 3.2.3.B)

$$\frac{\partial u}{\partial x} + \frac{\partial v}{\partial y} = 0 \tag{3.149}$$

and

$$u\frac{\partial u}{\partial x} + v\frac{\partial u}{\partial y} = \frac{1}{\rho}\frac{\partial \tau_{yx}}{\partial y} \tag{3.150}$$

where, in general,

$$\tau_{yx} = \mu\frac{\partial u}{\partial y} - \rho\overline{u'v'} \tag{3.151}$$

The auxiliary conditions are

$$v = 0 \quad \text{and} \quad \frac{\partial u}{\partial y} = 0 \quad \text{at } y = 0 \quad \text{(symmetry condition)}, \tag{3.152a,b}$$

and

$$u = 0 \quad \text{at } y = \infty \quad \text{(boundary condition)} \tag{3.152c}$$

Neglecting net pressure and interfacial drag forces, a 1-D momentum balance, $\sum F_x = \dot{M}_{\text{out}} - \dot{M}_{\text{in}} = 0$, yields

$$(\rho\bar{u}A)\bar{u}|_{x+\Delta x} - (\rho\bar{u}A)\bar{u}|_x = 0$$

which implies that the momentum flux

$$\blacklozenge \qquad J = 2\int_0^\infty \rho u^2 dy = \text{const} \tag{3.153}$$

Thus, J, being independent of x, is a measure of the "jet strengths," rather than $(\rho\bar{u}_{\text{exit}})$. There are two points of departure for the development of suitable postulates for jet spreading (cf. Fig. 3.21a and Problem (1) of Sect. 2.3.3A).

(i) Stream function approach:

$$\psi \sim x^s f\left(\frac{y}{b}\right) \qquad \text{where } b(x) \sim x^p \quad \text{and} \quad \frac{y}{b(x)} \equiv \eta$$

With $u \equiv \partial\psi/\partial y$, $u \sim x^{s-p} f'(\eta)$, that is, the axial velocity profiles have the same form, with different scales, at all x-stations. Now, employing Eqs. (3.153) and (3.150), one can show that $s = 1/3$ and $p = 2/3$.

(ii) Empirical approach:
One can deduce from the velocity profiles (cf. Fig. 3.21a) that $u(x, y) = u_{\max}(x)g(y/b)$; say, $u_{\max} \sim x^m$ and $b \sim x^n$. From Eq. (3.150),

$$u\frac{\partial u}{\partial x} \sim F_{\text{inertia}} \sim x^{2m-1} \quad \text{and} \quad v\frac{\partial^2 u}{\partial y^2} \sim F_{\text{viscous}} \sim x^{m-2n}$$

both must vary similarly with x, so that

$$2m - 1 = m - 2n$$

In addition, Eq. (3.153) indicates that with $d\eta = d(y/b)$

$$J = 2\rho u_{\max}^2 b \int_0^\infty g^2 \, d\eta \sim x^{2m+n} := C \equiv 1$$

which implies

$$2m + n = 0$$

or

$$m = -\frac{1}{3} \quad \text{and} \quad n = \frac{2}{3}$$

Either approach can be employed to postulate expressions for the stream function and the similarity variable in far-field *laminar* jet flow as

$$\psi(x, y) = v^{1/2}x^{1/3} f(\eta) \qquad \text{where } \eta = \frac{y}{3v^{1/2}x^{2/3}}$$

so that

$$u = \frac{1}{3x^{1/3}} f'(\eta) \quad \text{and} \quad v = -\frac{v^{1/2}}{3x^{2/3}}(f - 2\eta f')$$

While (3.149) is automatically satisfied, Eqs. (3.150) and (3.152) transform to

$$\blacklozenge \qquad f''' + ff'' + f'^2 = 0 \quad \text{subject to } f(0) = f''(0) = 0 \quad \text{and} \quad f'(\infty) = 0 \quad (3.154)$$

Schlichting (1979) provides an analytic solution for Eq. (3.154) given in Sect. 3.3.

In *nonsimilar* free shear flows the dimensionless stream function $f = f(\eta \text{ and } \xi)$, where the dimensionless streamwise distance is $\xi = x/L$. Now, with the new reference length L and an average discharge velocity u_0, we postulate, on the basis of dimensional analysis (cf. Sect. 2.3)

$$\psi = \sqrt{u_0 v L}\, \xi^{1/3} f(\xi, \eta) \qquad \text{where } \eta = \sqrt{\frac{u_0}{vL}} \frac{y}{3\xi^{2/3}} \qquad \xi = \frac{x}{L}$$

so that Eq. (3.150) is transformed to

$$\blacklozenge \qquad ff'' + f'' + f'^2 = 3\xi \left(f' \frac{df'}{d\xi} - f'' \frac{\partial f}{\partial \xi} \right) \qquad\qquad (3.155)$$

For *turbulent* jets (cf. Eq. (3.151))

$$\frac{\tau_{yx}^{\text{turb}}}{\rho} = -\overline{u'v'} = \nu_t \frac{\partial u}{\partial y}$$

has to be defined. If ν_t is assumed to be independent of η and if, following Schlichting (1979),

$$\nu_t = 0.037 u_c b$$

where $b(x) = 0.115x$ and $u_c \sim x^{-1/2}$, one can show that for similar turbulent 2-D jet flow

$$\blacklozenge \qquad ff'' + c(f'' + f'^2) = 0 \qquad\qquad\qquad (3.156)$$

The constant $c = 1.5523$ is determined from the condition $f' \equiv u/u_c = 0.5$ at $y = b$. The boundary conditions (3.152) are the same for turbulent jet flows.

In summary, the analytic solutions for Eqs. (3.154) and (3.156) are

- laminar self-similar jets: $f = 1.414 \tanh(0.707\eta)$ and $b(x) \sim x^{2/3}$
- turbulent similar jets: $f = 1.135 \tanh(0.881\eta)$ and $b(x) \sim x$

Indeed, with all the assumptions made, the axial velocity profiles are the same for both flow regimes; however, the rates of jet spreading differ (cf. Abramovich 1963).

Mixing Layers

Quasiuniform streams of different velocities, densities, temperatures, and/or viscosities form a growing mixing layer at the stream interface (cf. Fig. 3.22). For $U_2 = 0$ and $\kappa \equiv (\rho_1 \mu_1)/(\rho_2 \mu_2) \to 0$, the Blasius problem emerges. At the interface, $\tau_1 = \tau_2$; however, fluid transfer between the decelerating streams is usually neglected, that is, the normal components $v_1(0) = v_2(0) = 0$. Applications of this idealized system appear in the food sciences, oceanography, combustion, chemical engineering, and aerodynamics.

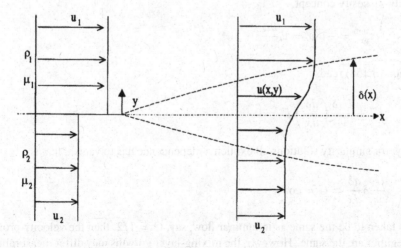

Fig. 3.22. Schematics of mixing layer.

The boundary-layer equations (3.149)–(3.151) are still valid; however, the boundary conditions are now

$$u = \begin{cases} U_1 & \text{for } y = \infty \\ U_2 & \text{for } y = -\infty \end{cases}$$

Seeking similarity solutions for the laminar or turbulent mixing layer, we define

$$\frac{u(x, y)}{U_1} = f'(\eta) \qquad \text{where } \eta = \frac{y}{\delta(x)}$$

The dimensionless stream function $f(\eta)$ is an integral part of the stream function $\psi(x, y)$, which has the dimensions $[L^2 T^{-1}]$. Hence, for dimensional consistency, we postulate

$$\psi(x, y) = [U, \delta(x)] f(\eta) \tag{3.157}$$

which implies

$$v = U_1 \left(\frac{d\delta}{dx} \right) (f'\eta - f)$$

The x-momentum equation (3.150) can be transformed to

$$U_1^2 \left(\frac{d\delta}{dx} \right) ff'' + \frac{1}{\rho} \frac{\partial \tau_{yx}}{\partial y} = 0 \tag{3.158}$$

subject to

$$f' = 1 @ \eta = \eta_\infty; \qquad \text{if } f' = \frac{U_2}{U_1} @ \eta = -\eta_\infty \quad \text{and} \quad f = 0 @ \eta = 0$$

In turbulent mixing layers, the laminar shear stress contributions can be neglected and with the eddy-viscosity concept

$$\frac{1}{\rho} \tau_{yx}^{\text{turb}} = -\overline{u'v'} = v_t \frac{\partial u}{\partial y}$$

Equation (3.158) reads

$$f''' + \frac{U_1 \delta}{v_t} \left(\frac{d\delta}{dx} \right) f'' = 0 \tag{3.159}$$

Clearly, for similarity solutions all explicit x-dependence has to vanish, that is,

$$\frac{U_1}{v_t} \delta \frac{d\delta}{dx} = C = \text{const} \tag{3.160}$$

If C is taken to be the same as in laminar flow, say, $C = 1/2$, then the velocity profiles in both regimes are the same. However, the mixing-layer growths may differ measurably, that is, $\delta(x)^{\text{turb}} \neq \delta(x)^{\text{lam}}$ depending upon the turbulent viscosity expression v_t in Eq. (3.160).

3.3 Sample-Problem Solutions

After a few basic *compressible* flow and incompressible *inviscid* flow problem solutions (cf. Sect. 3.1), laminar *incompressible viscous* flows are discussed (cf. Sects. 2.2.3 and 2.3.3 as well as Sects. 3.2.1–3.2.3). A few simple turbulent flow problem solutions complete this section. Problem solutions deal with Newtonian as well as non-Newtonian fluids, internal flows, and thin shear layer flows. In studying the sample solutions, the emphasis should be on the *understanding* of

- the problem *conceptualization*;
- *postulates* for the dependent variable(s) of more complicated problems, based on system assumptions, boundary conditions, and the continuity equation;
- coordinate transformations and/or efficient solution method; and
- *interpretation* of the results.

3.3.1 One-Dimensional Compressible Flows

3.1 Steady air flow through a nozzle has the inlet conditions $p_1 = 350$ kPa (abs), $v_1 = 183$ m/s, and $T_1 = 60°C$ (cf. sketch).

The flow accelerates to a Mach number at the outlet of $M_2 = 1.3$ where the local isentropic stagnation conditions are $p_{02} = 385$ kPa (abs) and $T_{02} = 350$ K. Find p_{01}, T_{01}; p_2; and T_2. Assume that air is an ideal gas.

Background
Local isentropic stagnation properties are those that would be obtained at any point in a flow field if the fluid at that point were decelerated from local conditions to zero velocity after a frictionless adiabatic, that is, isentropic, process. From Bernoulli's equation we know that

$$\frac{dp}{\rho} + d\left(\frac{v^2}{2}\right) = 0 \tag{S3.1}$$

and for ideal gas with constant specific heats, the first law of thermodynamics, $\delta Q_{rev} - \delta W_{rev} = dU$ or $TdS - pd\forall = \rho c_v dT$, can be expressed on a per-unit-mass basis as

$$ds = c_v \frac{dT}{T} + R \frac{d\hat{v}}{\hat{v}}, \qquad \text{where } \hat{v} = \rho^{-1} \text{ is the specific volume.}$$

Now, for isentropic processes, $ds \equiv 0$, and integration yields

$$\ln \frac{T_2}{T_1} = \frac{R}{c_v} \ln \frac{\hat{v}_2}{\hat{v}_1} \quad \text{or} \quad \ln \frac{T_2}{T_1} = \frac{R}{c_p} \ln \frac{p_2}{p_1}$$

With $\gamma \equiv c_p/c_v$ and $R = c_p - c_v$ we obtain

$$\frac{p}{\rho^\gamma} = C = \text{const} \tag{S3.2a}$$

where $\qquad p = \rho RT \tag{S3.2b}$

Eliminating the density in (S3.1) using (S3.2a) and integrating between the initial state (v, p) and the corresponding stagnation state $(0, p_0)$, we obtain

$$-\int_v^0 d\left(\frac{v^2}{2}\right) = C^{1/\gamma} \int_p^{p_0} p^{-1/\gamma} dp$$

After some manipulations and using (S3.2b) we have

$$\frac{p_0}{p} = \left[1 + \frac{\gamma - 1}{2} \frac{v^2}{\gamma RT}\right]^{\gamma/(\gamma-1)} \tag{S3.3}$$

With the speed of sound $a = \sqrt{\gamma RT}$ for an ideal gas, and (S3.2a) rewritten as

$$\frac{p_0}{p} = \left(\frac{\rho_0}{\rho}\right)^\gamma \quad \text{or} \quad \frac{\rho_0}{\rho} = \left(\frac{p_0}{p}\right)^{1/\gamma} \quad \text{and} \quad \frac{T_0}{T} = \left(\frac{p_0}{p}\right)^{(\gamma-1)/\gamma} \tag{S3.4a–c}$$

we obtain the equations for determining the isentropic stagnation properties of an ideal gas as (cf. Sect. 3.1)

$$\frac{p_0}{p} = \left[1 + \frac{\gamma - 1}{2} \text{Ma}^2\right]^{\gamma/(\gamma-1)} \tag{S3.5}$$

where $\text{Ma} = v/a$ is the Mach number and $\gamma = c_p/c_v$ for air.

Solution
Using Equations (S3.3) to (S3.5) with $a_1 = \sqrt{\gamma RT_1} = 366$ m/s and $\text{Ma}_1 = v_1/a_1 = 0.5$

$$\frac{p_{01}}{p_1} = 1.186 \quad \text{or} \quad p_{01} = 415 \text{ kPa}$$

$$\frac{T_{01}}{T_1} = 1.05 \quad \text{or} \quad T_{01} = 350 \text{ K}$$

$$\frac{p_{02}}{p_2} = 2.77 \quad \text{or} \quad p_2 = 139 \text{ kPa} \qquad \qquad \square$$

3.2   Consider isentropic air flow in a converging–diverging nozzle with exit area $A_e = 0.001$ m^2 and a throat Mach number $\text{Ma}_{th} = 0.68$ (see sketch on left). The upstream stagnation conditions are $T_0 = 350$ K and $p_0 = 1.0$ MPa. The back pressure $p_b = 954$ kPa. Find the flow properties at the throat and the exit Mach number.

Background
When the valve is opened, $p_b < p_0$ (see diagram on right), flow through the nozzle occurs, and the pressure distribution is shown with curve A, that is, the flow is subsonic and basically incompressible. When the flow rate is increased, that is, via a lower back pressure, compressibility effects become important and the flow velocity in the throat may

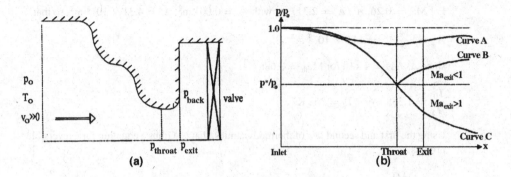

(a) (b)

reach $v_{th} = a_{th}$, that is, $Ma_{th} = 1$ and the nozzle is choked: the flow rate reached is the maximum possible for the given nozzle and stagnation conditions

$$\dot{m} = \rho^* v^* A^* \qquad \text{where } A^* = A_{throat}$$

In order to obtain supersonic flow at the exit, the back pressure has to be further decreased and the flow accelerates isentropically in the diverging section from $Ma_{th} = 1$ to $Ma_e > 1$ as indicated with curve C, where $p_C = p_b$.

Solution

From the isentropic flow relations (S3.3) to (S3.5), we obtain, with T_0 and p_0 constant, the following throat conditions:

$$\frac{T_0}{T_{th}} = 1 + \frac{\gamma - 1}{2} Ma_{th}^2 := 1.094 \quad \text{or} \quad T_{th} = 320 \text{ K}$$

$$\frac{p_0}{p_{th}} = \left(\frac{T_0}{T_{th}}\right)^{\gamma/(\gamma-1)} \quad \text{so that } p_{th} = 734 \text{ kPa,}$$

$$\rho_{th} = \frac{p_{th}}{R T_{th}} = 7.99 \text{ kg/m}^3$$

and

$$v_{th} = Ma_{th} \sqrt{\gamma R T_{th}} = 244 \text{ m/s}$$

Since $Ma_{throat} < 1$ (cf. curve A), the exit flow is subsonic and $p_{exit} = p_{back}$. Thus, from (S3.5) in the form

$$\frac{p_0}{p_e} = \left(1 + \frac{\gamma - 1}{2} Ma_e^2\right)^{\gamma/(\gamma-1)}$$

we obtain $Ma_e = 0.26$. The throat cross-sectional area is computed from the continuity equation in integral form

$$\rho A v = \rho^* A^* v^* = \text{const} \tag{S3.6}$$

or for an ideal gas (cf. Sect. 3.1)

$$\frac{A}{A^*} = \frac{1}{Ma} \frac{\rho^*}{\rho} \sqrt{\frac{T^*}{T}} = \frac{1}{Ma} \left[\frac{1 + (\gamma - 1)/2Ma^2}{1 + (\gamma - 1)/2}\right]^{(\gamma+1)/2(\gamma-1)} \tag{S3.7}$$

For $Ma_e = 0.26$, $A_e/A^* = 2.317$, or with $A_e = 0.001$ m^2, $A^* = 4.31 \times 10^{-4}$ m^2, so that

$$A_{th} = \frac{A_{th}}{A^*} A^* := 4.80 \times 10^{-4} \text{ m}^2$$

since $A_{th}/A^* = 1.110$ for $Ma_{th} = 0.68$;

$$\frac{T_{02}}{T_2} = 1.338 \quad \text{or} \quad T_2 = 262 \text{ K}$$

Using the first and second law of thermodynamics, that is, Gibbs's equation for a reversible process

$$T\,dS = \frac{\delta Q}{dT} = dU + p\,d\forall \tag{S3.8}$$

or

$$T\,ds = dh - \hat{v}\,dp \qquad \text{where } \hat{v} = 1/\rho \text{ is the specific volume.}$$

We can calculate the entropy change $s_2 - s_1 = s_{02} - s_{01}$ as

$$\Delta s = s_2 - s_1 = \int_{T_1}^{T_2} c_p \frac{dT}{T} - \int_{p_1}^{p_2} R \frac{dp}{p} = c_p \ln \frac{T_2}{T_1} - R \ln \frac{p_2}{p_1}$$

or

$$\Delta s = 0.0252 \text{ kJ/kg} \cdot \text{K}$$

Note, in the temperature–entropy (T-S) diagram, $T_{01} = T_{02}$, that is, an isotherm for isentropic flow and $s_{01} - s_1 = s_{02} - s_2 = 0$ because $\delta Q = 0$ (cf. Eq. (S3.8)).

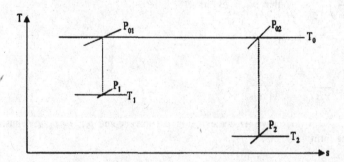

3.3 Determine the fluid pressure rise in a finite section of a rigid pipe after sudden closure of a valve.

Concept

At the suddenly closed valve, the fluid is brought to rest and its kinetic energy is converted to strain energy of compression, that is, change in fluid volume times mean pressure change.

$$\frac{v^2}{2} = \frac{(\Delta p)^2}{2K\rho} \tag{S3.9}$$

where $\Delta p/(\rho K)$ is the change in volume with K the bulk modulus. Taking an average velocity of $v = 1.5$ m/s and water ($K = 2.14 \times 10^4$ N/m^2 and $\rho = 10^3$ kg/m^3) we obtain a pressure rise of

$$\Delta p = v\sqrt{K\rho} := 2194 \times 10^3 \text{ N/m}^2$$

Needless to say, this pressure wave moves back and forth with the speed of sound, $a = \sqrt{K/\rho}$, from one end of the closed pipe to the other, neglecting damping effects.   □

3.4   Free and forced vortex flows: Consider flow fields with purely tangential motion (circular streamlines): $V_r = 0$ and $V_\theta = f(r)$. Evaluate the rotation, vorticity, and circulation for rigid-body rotation, that is, a *forced vortex* in viscous flow. Show that it is possible to choose $f(r)$ so that flow is irrotational, that is, to produce a free vortex in inviscid flow.

Assumptions/Postulates

- Purely tangential motion
- $V_r = 0$
- $V_\theta = f(r)$

Solution
Basic equation $\vec{\zeta} = 2\vec{\omega} = \nabla \times \vec{v}$

For motion in the r–θ plane, the only components of rotation and vorticity are in the z-direction.

$$\zeta_z = 2\omega_z = \frac{1}{r}\frac{\partial(rV_\theta)}{\partial r} - \frac{1}{r}\frac{\partial V_r}{\partial \theta}$$

Because $V_r = 0$ everywhere in this field, this reduces to

$$\zeta_z = 2\omega_z = \frac{1}{r}\frac{\partial(rV_\theta)}{\partial r}$$

(a)   For rigid-body rotation $V_\theta = r\omega$. Then

$$\omega_z = \frac{1}{2r}\frac{\partial(rV_\theta)}{\partial r} = \frac{1}{2r}\frac{\partial}{\partial r}\omega r^2 = \frac{1}{2r}(2\omega r) = \omega$$

and

$$\zeta_z = 2\omega$$

The circulation is

$$\Gamma = \oint_c \vec{v} \cdot d\vec{s} = \int_a 2\omega_z\, dA$$

Since $\omega_z = \omega = $ const, the circulation about any closed contour is given by $\Gamma = 2\omega A$, where A is the area enclosed by the contour. Thus for rigid-body motion (a forced vortex), the rotation and vorticity are constants; the circulation depends on the area enclosed by the contour.

(b)   For irrotational flow, $(1/r)\partial/\partial r\,(rv_\theta) = 0$. Integrating, we find

$$rv_\theta = \text{const}   \text{ or }   v_\theta = f(r) = C/r$$

For this flow, the origin is a singular point where $v_\theta = \infty$. The circulation for any contour enclosing the origin is

$$\Gamma = \int_c \vec{v} \cdot d\vec{s} = \int_0^{2\pi} \frac{C}{r} r \, d\theta = 2\pi C$$

The circulation around any contour not enclosing the singular point at the origin is zero. $\Box$

3.3.2   *Incompressible Inviscid Flows*

3.5   Consider fluid flow start-up in a pipe suddenly fed by a large reservoir (cf. sketch).

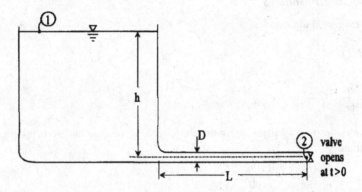

Assumptions

- Incompressible frictionless flow
- Representative streamline for flow

Background

Euler's equation is integrated along a streamline or stream tube. To begin with, the equation of ideal fluid motion is multiplied by a streamline arc length, $d\vec{s}$.

$$\frac{D\vec{v}}{Dt} \cdot d\vec{s} = -\frac{1}{\rho} \nabla p \cdot d\vec{s} + \vec{g} \cdot d\vec{s} \tag{S3.10}$$

Thus, along a streamline

$$v_s \, dv_s + \frac{\partial v_s}{\partial t} \, ds = -\frac{1}{\rho} dp - g \, dz$$

Integration with $\rho = $ const yields the unsteady Bernoulli equation

$$\frac{p_1}{\rho} + \frac{v_1^2}{2} + g z_1 = \frac{p_2}{\rho} + \frac{v_2^2}{2} + g z_w + \int_1^2 \frac{\partial v_s}{\partial t} \, ds \tag{S3.11}$$

Solution

In our case, (S3.11) reduces to

$$gz_1 = \frac{v_2^2}{2} + \int_1^2 \frac{\partial v_s}{\partial t}\,ds$$

With

$$z_1 \equiv h \quad \text{and} \quad \int_1^2 \frac{\partial v_s}{\partial t}\,ds \approx \int_0^L \frac{dv_2}{dt}\,ds = L\frac{dv_2}{dt}$$

we obtain

$$gh = \frac{v_2^2}{2} + L\frac{dv_2}{dt} \tag{S3.12}$$

or

$$\int_0^{v_2} \frac{dv}{2gh - v^2} = \frac{t}{2L}$$

or

$$(2gh)^{-1/2}\tanh^{-1}\left(\frac{v_2}{\sqrt{2gh}}\right) = \frac{t}{2L}$$

and, finally (cf. graph),

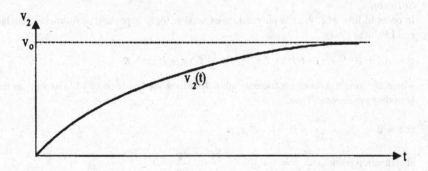

$$\frac{v_2}{v_0} = \tanh\left(\frac{t}{2L}v_0\right) \qquad \text{where } v_0 = \sqrt{2gh} \qquad \square$$

3.6 Calculate the potential flow near a stagnation point (cf. sketch below).

Assumptions

- Steady potential flow
- Planar body surface at stagnation point

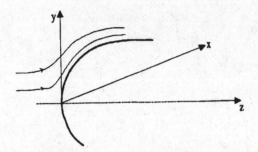

Background

From continuity, $\nabla \cdot \vec{v} = 0$, with $\vec{v} = \nabla\phi$ where ϕ is a potential function, we obtain Laplace's equation

$$\nabla^2\phi = 0 \tag{S3.13}$$

The associated boundary conditions are

$$\text{(normal velocity)} \qquad v_n = \frac{\partial\phi}{\partial n} = \begin{cases} 0 & \text{for stationary solid wall} \\ f(t) & \text{for moving surface} \end{cases}$$

Note that Equation (S3.13) is time-independent and that transient effects may only enter the picture via the boundary condition of an accelerating body. The velocity is related to the pressure via Bernoulli's equation

$$\frac{\partial\phi}{\partial t} + \frac{v^2}{2} + \frac{p}{\rho} = g(t) \tag{S3.14}$$

Solution

In order to find $\phi(x, y, z)$, a polynomial expansion for ϕ is postulated following Landau and Lifshitz (1987):

$$\phi = Ax + By + Cz + Dx^2 + Ey^2 + Fz^2 + Gxy + Hyz + Kzx$$

where the nine unknown coefficients are determined via Equation (S3.13) as well as the boundary conditions. Thus,

$$\text{at } z = 0 \qquad v_z = \frac{\partial\phi}{\partial z} = 0; \qquad \text{all } x, y$$

$$\text{at stagnation point } (x = y = z = 0) \qquad \frac{\partial\phi}{\partial x} = \frac{\partial\phi}{\partial y} = 0$$

As a result, $A = B = C = 0$, $F = -D - E$, and $G = H = K = 0$, so that

$$\phi = Dx^2 + Ey^2 - (D + E)z^2$$

For axisymmetric flow $D = E$ and we obtain

$$\phi = D(x^2 + y^2 - 2z^2)$$

so that

$$v_x = 2Dx, \qquad v_y = 2Dy \quad \text{and} \quad v_z = -4Dz$$

The streamlines are determined by the system of differential equations

$$\frac{dx}{v_x} = \frac{dy}{v_y} = \frac{dz}{v_z} \tag{S3.15}$$

In our case, we have $x^2 z = c_1$ and $y^2 z = c_2$, representing the streamlines near a stagnation point. □

3.7 Describe the fluid motion created by a solid sphere moving with a constant velocity U in the x-direction through an infinite reservoir of an ideal liquid at rest (cf. sketch).

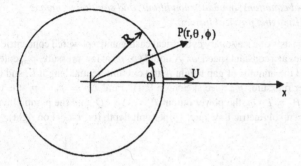

Approach
A velocity potential, that is, a ϕ-function, has to be constructed such that

- Laplace's equation, $\nabla^2 \phi = 0$, is fulfilled everywhere in the axisymmetric flow field;
- the spatial derivatives of ϕ must vanish at infinity (i.e., no fluid motion); and
- at the surface, $r = R$, $v_n = \partial\phi/\partial r = U \cos\theta$, which implies spherical symmetry.

Solution
In order for $\phi(r)$ and its derivatives to satisfy the conditions at infinity, we require $\phi \sim 1/r$. For ϕ to be a scalar function involving only the x-component U of the velocity vector $\vec{v}$ and the derivatives of $1/r$, we set

$$\phi = \vec{A} \cdot \nabla\left(\frac{1}{r}\right) := A\frac{\partial}{\partial x}\left(\frac{1}{r}\right) = -A\frac{\cos\theta}{x^2} \qquad \text{where } \cos\theta = \frac{x}{r}$$

From the condition that $\nabla^2\phi = 0$ everywhere, we obtain $A = (1/2)UR^3\cos^2\theta$, so that

$$\phi = -\frac{1}{2}UR\left(\frac{R}{r}\right)^2\cos\theta \tag{S3.16}$$

The pressure distribution along the sphere's surface can be obtained from (cf. Eq. (S3.14))

$$p = p_\infty - \frac{1}{2}\rho v^2 - \rho\frac{\partial\phi}{\partial t}$$

as

$$p = p_\infty + \frac{1}{8}\rho U^2(9\cos^2\theta - 5) \tag{S3.17}$$

□

3.3.3 Laminar Incompressible Viscous Flows

Note

Whenever possible, the standard format *for (fluid flow) problem solutions is executed:*

   (i)   *System sketch and/or conceptional model*
  (ii)   *Assumptions w.r.t. time-dependence, dimensionality, fluid properties, and so on*
 (iii)   *Concept/approach needed for complex problem solutions*
 (iv)   *Postulates for dependent variables based on boundary conditions plus continuity equation in addition to assumptions*
  (v)   *Solution with reduced equations, boundary conditions, and solution technique*
 (vi)   *Graphs/comments displaying physical insight*

3.8 A viscosity pump consists of a stationary cylindrical casing and a powered concentric drum of radius R that rotates at a constant speed, ω. A liquid (ρ, μ) at low pressure, p_{in}, enters the pump, flows through the annulus of gap height h and circumferential length L, and leaves the pump at a higher pressure, p_{out} (see sketch on left). Find $\Delta p = p_o - p_i$, the torque T, the power input $P_i = T \cdot \omega$, the power output $P_o = \Delta p \cdot Q$, and the pump efficiency, $\eta = P_o/P_i$, for a given volumetric flow rate, Q, per unit depth (cf. model on right).

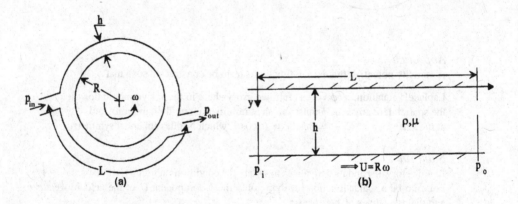

Assumptions

- Steady laminar fully developed flow; $h \ll 1$
- Constant fluid properties
- Flat plate approximation: $R \gg 1$
- $\partial p/\partial x \approx \Delta p/L = \text{const}$

Concept

The problem conceptualization (cf. sketch (b)) and the assumptions indicate Couette-type flow.

Postulates

$v = 0$, $u = u(y)$ only   and $-\partial p/\partial x \approx \Delta p/L = \text{const}$

Solution

The Navier–Stokes equations reduce to the x-component of the form

$$\frac{\partial^2 u}{\partial y^2} = \frac{\Delta p}{\mu L} = \text{const} \tag{S3.18}$$

subject to $u(y = 0) = 0$ and $u(y = h) = \omega R$. Integration yields

$$u(y) = \frac{\Delta p}{2\mu L} y^2 + \left[\frac{R\omega}{h} - \frac{\Delta p}{2\mu L} h\right] y \tag{S3.19}$$

Using the volumetric flow rate,

$$Q = \int_0^h u(y)\, dy = -\frac{\Delta p}{6\mu L} h^3 + \frac{R\omega}{2} h \tag{S3.20}$$

we obtain

$$\Delta p = \frac{6\mu L}{h^3}(R\omega h - 2Q) \tag{S3.21}$$

The torque exerted on the shaft is

$$T = \iint_A r\, dF; \qquad dF = \tau_{\text{wall}}\, dA \tag{S3.22a}$$

so that

$$T = R\tau_w \cdot A \tag{S3.22b}$$

where $\tau_w = \mu(du/dy)|_{y=h} = \text{const}$ and $A = L \cdot W; W \equiv 1$.
Thus,

$$T = \frac{\Delta p R h}{2} + \frac{\mu R^2 \omega L}{h} \tag{S3.22c}$$

The system efficiency is defined as

$$\eta \equiv \frac{P_{\text{out}}}{P_{\text{in}}} = \frac{\Delta p \cdot Q}{\omega \cdot T} \tag{S3.23a}$$

so that

$$\eta = \frac{3Q(hR\omega - 2Q)}{hR\omega(2hR\omega - 3Q)} \tag{S3.23b}$$

Recommendation

Sketch $u(y, \Delta p/L)$ and comment on Eqs. (S3.22c) and (S3.23b) $\square$

3.9 Stream Function for Flow in a Corner.
Given the velocity field for steady, incompressible flow $\vec{v} = Ax\hat{i} - Ay\hat{j}$, with $A = 2\text{ s}^{-1}$, determine the stream function that will yield this velocity field (cf. sketch).

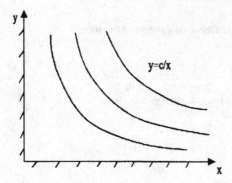

Assumptions

- Steady flow
- Incompressible flow

Solution

The 2-D flow is incompressible, so the stream function satisfies the continuity equation from which $u = \partial\psi/\partial y$ and $v = -\partial\psi/\partial x$. With

$$u = Ax = \frac{\partial\psi}{\partial y}$$

integrating with respect to y gives

$$\psi = Axy + f(x)$$

where $f(x)$ is arbitrary. The function $f(x)$ may be evaluated using the equation for v. Thus,

$$v = -\frac{\partial\psi}{\partial x} = -Ay - \frac{df}{dx}$$

Comparing this with the preceding equation shows that $df/dx = 0$, or $f(x) = $ const. Therefore,

$$\psi = Axy + C$$

Lines of constant ψ represent streamlines in the flow field. The constant C may be chosen as any convenient value for plotting purposes. If the constant is chosen as zero, the stream-line through the origin may be designated as $\psi = \psi_1 = 0$. Then the value for any other streamline represents the flow from left to right between the origin and that streamline. So, with $C = 0$ and $A = 2$ s^{-1}

$$\psi = 2xy \quad \text{in } (\text{m}^3/\text{s/m}) \qquad \square$$

3.10 Consider an incompressible boundary-layer flow with $U_\infty = $ const (cf. sketch). Assumming a velocity distibution $u = U_\infty(2y/\delta - y^2/\delta^2)$, determine the wall shearing stress $\tau_0(x)$ and the boundary-layer thickness $\delta(x)$. (The given parabolic velocity profile is an approximation to a laminar boundary-layer profile.)

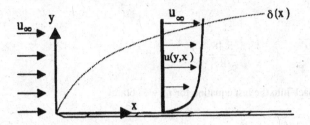

Assumptions

- Steady laminar thin shear-layer flow
- Constant fluid properties
- $U_\infty = u_{\text{edge}} = \text{const}$
- Given $u(x, y)$ fulfills BCs

Postulates

$$u = u[y, \delta(x)]; \qquad \partial p/\partial x = 0 \qquad \text{because } U_\infty = \text{const}$$

Solution

In this case, the momentum integral relation yields (Sect. 3.2.3)

$$\tau_0 = \rho \frac{d}{dx} \int_0^\delta u(U_\infty - u)\, dy$$

$$= \rho \frac{d}{dx} \int_0^\delta U_\infty^2 \left(\frac{2y}{\delta} - \frac{y^2}{\delta^2} \right) \left(1 - \frac{2y}{\delta} + \frac{y^2}{\delta^2} \right) dy$$

$$= \rho U_\infty^2 \frac{d}{dx} \left(\frac{2\delta}{15} \right)$$

$$= 0.133 \rho U_\infty^2 \frac{d\delta}{dx}$$

Now, the shear stress is also equal to $\mu(\partial u/\partial y)_{\text{wall}}$, resulting in

$$\mu U_\infty \left(\frac{2}{\delta} \right) = 0.133 \rho U_\infty^2 \frac{d\delta}{dx}$$

or

$$\frac{\mu\, dx}{0.0665 \rho U_\infty} = \delta\, d\delta$$

Using $\delta = 0$ at $x = 0$, this integrates to

$$\frac{\delta^2}{2} = \frac{\mu x}{0.0665 \rho U_\infty}$$

so that

$$\delta(x) = 5.48\sqrt{\frac{\mu x}{\rho U_\infty}} \quad \text{or} \quad \frac{\delta}{x} = \frac{5.48}{\sqrt{\text{Re}_x}}$$

Substituting this back into the last equation for τ_0, we obtain

$$\tau_0 = 0.365\rho U_\infty^2 \sqrt{\frac{\mu}{\rho U_\infty x}} \qquad\qquad \square$$

3.11 Consider Couette-type flow between two concentric, rotating, porous cylinders allowing radial flow, that is, $v_r = c/r$, where c is a given constant (cf. sketch).

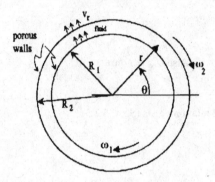

Assumptions

- Steady state
- Laminar flow
- Constant fluid properties
- No end effects or body forces

Postulates
Taking $v_r = v_r(r) = c/r$, it follows from the continuity equation

$$\frac{\partial}{\partial r}(rv_r) + \frac{\partial}{\partial \theta}v_\theta = 0$$

that $v_\theta = v_\theta(r)$ only. In addition, $p = p(r)$ only as shown later.

Tailored Equations
Substituting $v_r(r)$ and invoking the postulate for v_θ in the Navier–Stokes equations yield

$$(r\text{-component}) \qquad \frac{\partial p}{\partial r} = \rho c^2 \frac{1}{r^3} + \rho\frac{v_\theta^2}{r} \qquad\qquad \text{(S3.24a)}$$

$$(\theta\text{-component}) \qquad \frac{\partial p}{\partial \theta} = \rho r f(r) \qquad\qquad \text{(S3.24b)}$$

Equation (S3.24b) together with the condition $p(\theta = 0) = p(\theta = 2\pi)$ indicates that $p = p(r)$ only. Now, writing out Eq. (S3.24b) we obtain

$$\frac{d^2 v_\theta}{dr^2} + \left(1 - \frac{c}{\nu}\right)\frac{1}{r}\frac{dv_\theta}{dr} - \left(1 + \frac{c}{\nu}\right)\frac{v_\theta}{r^2} = 0 \tag{S3.24c}$$

Solution
Equation (S3.24c) can be transformed, using $r \equiv e^x$ to obtain a second-order homogeneous ODE with constant coefficients

$$v_\theta'' + A v_\theta' + B v_\theta = 0 \tag{S3.24d}$$

Alternatively, Cauchy's differential equation (S3.24c) allows the trial solution

$$v_\theta(r) = r^m$$

leading to the characteristic equation

$$m^2 - m\frac{c}{\nu} - \left(1 + \frac{c}{\nu}\right) = 0 \qquad \text{with } m = \begin{cases} 1 + c/\nu \\ -1 \end{cases}$$

so that

$$v_\theta = C_1 r^{1+c/\nu} + C_2 r^{-1} \tag{S3.24e}$$

subject to

$$v_\theta(r = R_1) = R_1\omega_1 \quad \text{and} \quad v_\theta(r = R_2) = R_2\omega_2$$

Recommendation
Program the solution (S3.24e), determine the constants, and plot $v_\theta(r)$ for different c's.

$\square$

3.12 Find an approximate solution for drainage of a very viscous liquid out of a cylindrical tank (cf. sketch).

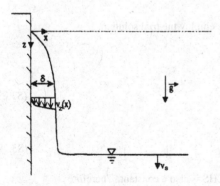

Assumptions

- Creeping flow
- Thin film approximation: $\nabla p \approx 0$ and $\tau_{\text{interface}} \approx 0$
- Constant fluid properties
- Fully developed flow (i.e., no end effects)

Postulates

$v_x = 0, \qquad v_z = v_z(x) \text{only}; \qquad \delta = \delta(z, t), \qquad \nabla p \approx 0$

Governing Equations

The Generalized Transport Equation (GTE) in differential form (cf. Sect 2.2.2)

♦
$$\frac{\partial \psi}{\partial t} + \nabla \cdot (\vec{v}\psi) = \nabla \cdot \vec{\vec{\Omega}} + \dot{\Sigma}$$

is applied to mass conservation: $\psi \equiv m/A = \rho \Psi/A$ or $\psi = \rho\delta$; $\Omega = \Sigma = 0$. Thus with $\rho = \text{const}$ and using the postulates, the GTE can be reduced to

$$\frac{\partial \delta}{\partial t} + \frac{\partial}{\partial z}(\delta v_z) \overset{\cdot}{=} 0 \tag{S3.25}$$

The z-momentum equation reduces to

$$0 = v\frac{\partial^2 v_z}{\partial x^2} + g \tag{S3.26a}$$

and can be directly integrated subject to $v_z(x = 0) = 0$ and $dv_z/dx \approx 0$ at $x = \delta$.

$$v_z = \frac{g\delta^2}{2v}\left[2\left(\frac{x}{\delta}\right) - \left(\frac{x}{\delta}\right)^2\right] \tag{S3.26b}$$

In order to eliminate the x-dependence of the vertical velocity before solving (S3.25), we take the average of (S3.26b), that is,

$$\bar{v}_z = \frac{\int v_z\, dA}{\int dA} = \frac{g\delta^2}{3v} \tag{S3.27}$$

Inserting (S3.27) into (S3.25) yields

♦
$$\frac{\partial \delta}{\partial t} + \frac{g}{v}\delta^2\frac{\partial \delta}{\partial z} = 0 \tag{S3.28}$$

Solution

With the combined variable $\eta = z/t$, we employ the trial solution

$$\delta(t, z) = a\eta^n$$

Equation (S3.28) now reads

$$-naz^n t^{-(n+1)} + n\frac{g}{v}a^2\eta^{2n}at^{-n}z^{n-1} = 0 \tag{S3.29a}$$

This can be reduced to

$$\frac{g}{v}a^2\eta^{2n-1} = 1 \tag{S3.29b}$$

which implies that $n \equiv 1/2$ so that the LHS is also a constant. Therefore

$$a = \sqrt{\frac{\mu}{\rho g}}$$

or

♦
$$\delta(z, t) = \sqrt{\frac{\mu}{\rho g}\frac{z}{t}} \tag{S3.29c}$$

Notes

- How can we assume parallel flow, that is, $v_z = v_z(x)$ only, when $\delta = \delta(z, t)$?
- Plot $v_z(x, \delta)$ and indicate under what conditions Equations (S3.26b) and (S3.29c) are approximately correct. ☐

3.13 Consider a lubrication squeeze film between two parallel disks a distance $2h$ apart (cf. sketch). Calculate the equilibrium velocity, pressure and force fields while the disks are slowly pressed together with a speed of $\dot{h} = dh/dt = $ const. The problem solution expands on the results given in Bird et al. (1987).

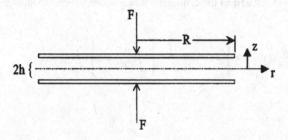

Assumptions

- Creeping flow: inertia terms in r-momentum equation are negligible
- Steady, laminar, axisymmetric flow without body forces
- Constant fluid properties
- No end effects although fluid may leave gap.

Concept
The system evaluation is taken at a "frozen" time frame.

Boundary Conditions

$$v_z(h) = \dot{h} = \text{const} \quad \text{and} \quad v_z(z = 0) = 0; \qquad v_r(z = h) = 0$$

$$\text{and} \quad \partial v_r / \partial z = 0 \quad \text{at } z = 0$$

Postulates
With the stated assumptions, we postulate that at any given time t

$$v_\theta \equiv 0, \qquad v_z = v_z(z) \text{ only}, \qquad v_r = v_r(r, z); \quad \text{and} \quad p = p(r, z) \approx p(r)$$

Governing Equations

(continuity) $\qquad \dfrac{1}{r}\dfrac{\partial}{\partial r}(rv_r) + \dfrac{dv_z}{dz} = 0$ \hfill (S3.30a)

or

$$\dfrac{dv_z}{dz} = -\dfrac{1}{r}\dfrac{\partial}{\partial r}(rv_r) \hfill \text{(S3.30b)}$$

Note, since the LHS of (S3.30b) is a function of z only, the RHS has to be some function of z, too, in order to satisfy mass conservation. Hence, we obtain with

$$v_r(r, z) = r \cdot g(z) \tag{S3.31}$$

(r-momentum) $$0 = -\frac{\partial p}{\partial r} + \mu r \frac{d^2}{dz^2} g(z) \tag{S3.32}$$

(z-momentum) $$\rho v_z \underbrace{\frac{dv_z}{dz}}_{\substack{\text{inertia} \\ \text{force}}} = \underbrace{\mu \frac{d^2 v_z}{dz^2}}_{\substack{\text{normal} \\ \text{stress force}}} \tag{S3.33}$$

Solution

Integration of Equation (S3.30a) over the cylindrical flow domain, in order to utilize $v_z(z = h) = \dot{h}$, yields

$$0 = \int_{z=0}^{z=h} \left\{ 2\pi \int_0^r r \left[\frac{1}{r} \frac{\partial}{\partial r}(rv_r) \right] dr \right\} dz + 2\pi \int_0^r r \left\{ \int_0^h \frac{dv_z}{dz} dz \right\} dr$$

so that

$$2 \int_0^{z=h} (rv_r) \, dz + r^2 \underbrace{v_z(h)}_{=\dot{h}} = 0$$

or

$$2 \int_0^h v_r(r, z) \, dz + r\dot{h} = 0 \tag{S3.34}$$

Note, assuming that basically $p = p(r)$ only, Equation (S3.32) can be directly integrated and Equation (S3.33) reflects a balance between the inertia force per unit area and normal stress. Integrating (S3.32) with respect to z twice, where $\partial p/\partial r \triangleq dp/dr$, we obtain

$$C_1 z + C_2 = -\frac{dp}{dr} \frac{z^2}{2} + \mu r g(z)$$

or with $g(z = h) = 0$ and $dg/dz = 0$ at $z = 0$,

$$-\frac{dp}{dr} \frac{h^2}{2} = -\frac{dp}{dr} \frac{z^2}{2} + \mu v_r$$

so that

$$v_r = r \cdot g(z) = \frac{h^2 r}{2} \left(-\frac{1}{\mu r} \frac{dp}{dr} \right) \left[1 - \left(\frac{z}{h} \right)^2 \right] \tag{S3.35}$$

Substituting (S3.35) into (S3.34) yields

$$-\frac{1}{\mu r} \frac{dp}{dr} = -\frac{3\dot{h}}{2h^3} \tag{S3.36}$$

Combining (S3.36) and (S3.35) to eliminate the pressure gradient gives

$$v_r(r, z) = \frac{3}{4} \left(\frac{-\dot{h}}{h} \right) r \left[1 - \left(\frac{z}{h} \right)^2 \right] \tag{S3.37}$$

The radial pressure distribution from (S3.36) is

$$p - p_{atm} = \frac{3(-\dot{h})\mu R^2}{4h^3}\left[1 - \left(\frac{r}{R}\right)^2\right] \tag{S3.38}$$

Note $v_z(z)$ can be obtained from Equation (S3.30). The force F is in a dynamic equilibrium with pressure and (normal) stress. Thus the force exerted on one plate is:

$$F(t) = \int_0^{2\pi}\int_0^R (p - p_{atm} + \tau_{zz})|_{z=h} r\,dr\,d\theta$$

where

$$\tau_{zz} = -2\mu\frac{\partial v_z}{\partial z} := 2\mu\left(\frac{1}{r}\frac{\partial}{\partial r}(rv_r)\right) := 3\mu\frac{(-\dot{h})}{h}\left[1 - \left(\frac{z}{h}\right)^2\right] \tag{S3.39}$$

so that

$$F(t) = \frac{3\pi R^4\mu(-\dot{h})}{8h^3} \tag{S3.40}$$

For a constant force F, Equation (S3.40) has to be integrated w.r.t. time:

$$-\frac{dh}{dt} = \frac{8F}{3\pi R^4\mu}h^3$$

which gives the disk gap as a function of time

$$\frac{1}{h^2} - \frac{1}{h_0^2} = \frac{16F}{3\mu\pi R^4}t$$

or

$$h(t) = \left(\frac{16F}{3\mu\pi R^4 t} + \frac{1}{h_0^2}\right)^{-1/2} \tag{S3.41}$$

Recommendation
Solve this problem for a power-law fluid and compare. □

3.3.4 Lubrication Flows and Stretching Flows

Some slider bearing problems or slowly rotating journal bearings are examples of creeping flow where Re ≪ 1 and the inertia terms can be neglected. As a result, the momentum equation reduces to Stokes's equation, a special form of the Reynolds equation that is the cornerstone in classical lubrication theory (cf. Hamrock 1994). In contrast to these flows where inertia terms are often negligible, axial convection, that is, the $(u\partial u/\partial x)$-term, is very important in fiber spinning or film casting as a result of the stretching force applied (cf. Middleman 1977; Tanner 1985; Churchill 1988; and Papanastasiou 1994; among others).

3.14a Determine the location and magnitude of the maximum pressure in a slider bearing and hence evaluate the "confined wedge-flow" effect (cf. sketch).

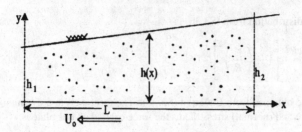

Assumptions

- Steady laminar flow in a small gap
- Negligible inertia terms and no body force
- Constant fluid properties
- No end effects
- $h_2 - h_1 \ll 1$

Basic Equations

$$\frac{\partial u}{\partial x} + \frac{\partial v}{\partial y} = 0 \tag{S3.42}$$

$$u\frac{\partial u}{\partial x} + v\frac{\partial u}{\partial y} = -\frac{1}{\rho}\frac{\partial p}{\partial x} + v\left(\frac{\partial^2 u}{\partial x^2} + \frac{\partial^2 u}{\partial y^2}\right) \tag{S3.43}$$

$$u\frac{\partial v}{\partial x} + v\frac{\partial v}{\partial y} = -\frac{1}{\rho}\frac{\partial p}{\partial y} + v\left(\frac{\partial^2 v}{\partial x^2} + \frac{\partial^2 v}{\partial y^2}\right) \tag{S3.44}$$

Postulates

Nondimensionalization and relative order of magnitude analysis (ROMA) yield with

$$\tilde{x} = \frac{x}{L}; \qquad \tilde{y} = \frac{y}{h_1}; \qquad \tilde{u} = \frac{u}{U_0}; \qquad \tilde{v} = \frac{v}{U_0}$$

the following simplifications

(continuity equation) $\qquad \left(\frac{h_1}{L}\right)\frac{\partial \tilde{u}}{\partial \tilde{x}} + \frac{\partial \tilde{v}}{\partial \tilde{y}} = 0 \quad$ or $\quad$ since $\dfrac{h_1}{L} \ll 1$

$\partial \tilde{v}/\partial \tilde{y} = 0$, that is, $\tilde{v} = \tilde{v}(\tilde{x})$ so far. Along the walls, that is, $y = 0$ and $y = h(x)$, $v = 0$, and since $(h_2 - h_1) \ll 1$, we treat the flow in the gap as fully developed ($\partial u/\partial x \approx 0$) and quasiparallel ($v \approx 0$). As a result, $\partial \tilde{p}/\partial \tilde{y} = 0$ and the basic equations (S3.42 to S3.44) can be simplified significantly (cf. Sample Problem 3.8). Note, the $h(x)$-effect will appear in $u(x, y)$ and is also buried in the expression for dp/dx.

Tailored Equation

$$\frac{\partial \tilde{p}}{\partial x} \approx \mu\frac{\partial^2 \tilde{u}}{\partial \tilde{y}^2} \tag{S3.45a}$$

or

$$\frac{\partial^2 u}{\partial y^2} \approx \frac{1}{\mu}\left(\frac{dp}{dx}\right) = \text{fct}(x) \tag{S3.45b}$$

subject to $u(y = 0) = -U_0$, $u[y = h(x)] = 0$ and $u(y)$ is prescribed at $x = 0$ for $0 < y < h_1$.

Solution

Integrating partially w.r.t. y twice and invoking the boundary conditions yield

$$u(x, y) = -U_0\left(1 - \frac{y}{h}\right) - \frac{h^2}{2\mu}\left(\frac{dp}{dx}\right)\frac{y}{h}\left(1 - \frac{y}{h}\right) \tag{S3.46a}$$

where the x-dependence appeared via the BC $u[y = h(x)] = 0$. For example,

$$h = h(x) = h_1 + \frac{h_2 - h_1}{L}x \tag{S3.46b}$$

Using the volumetric flow rate, $Q = \int_0^{h(x)} u\,dy = \text{const}$, yields an expression for the pressure gradient

$$\frac{dp}{dx} = -\frac{12\mu}{h^3}\left(\frac{U_0 h}{2} + Q\right) \tag{S3.47a}$$

or

$$p_L - p_0 = -12\mu\left(\frac{U_0}{2}\int_0^L \frac{dx}{h^2} + Q\int_0^L \frac{dx}{h^3}\right) \tag{S3.47b}$$

Note

For Couette flow, the gap size h was constant and so was dp/dx, resulting in $u = u(y)$ only (cf. Eqs. (S3.46a) and (S3.47a). In case the pressure is atmospheric at both ends, that is, $p_L = p_0$, we obtain

$$Q = \frac{-(U_0/2)\int_0^L (dx/h^2)}{\int_0^L (dx/h^3)}$$

or

$$Q = -\frac{h_1 h_2}{h_1 + h_2}U_0 \tag{S3.48}$$

The maximum pressure may occur at the location x_{0p}, where $dp/dx = 0$, which implies from (S3.47a) that

$$\frac{U_0 h_{0p}}{2} + Q = 0$$

or

$$h_{0p} = -\frac{2Q}{U_0} := \frac{2h_1 2h_2}{h_1 + h_2} = h_1 + \frac{h_2 - h_1}{L}x_{0p}$$

Thus

$$x_{0p} \equiv x|_{p=p_{max}} = \frac{h_1 L}{h_1 + h_2} \tag{S3.49}$$

The load carrying capacity of the slider bearing is (cf. graphs):

$$F = \int_0^L (p - p_0)\,dx$$

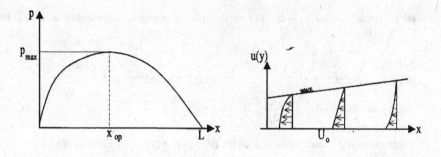

3.14b Consider a stepped bearing with a linear pressure distribution as shown. Calculate the flow rate, the maximum pressure, the load, and the drag force exerted by the oil film on the moving surface.

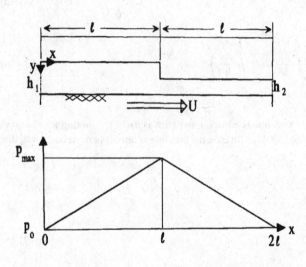

Assumptions

- Steady laminar, fully developed, quasi-parallel flow
- Constant fluid properties
- No end effects
- Quasisymmetric system
- Triangular symmetric pressure distribution as shown

Postulates

In contrast to Sample Problem 3.14a, $u = u(y)$ only because $h = \text{const}$ and $\partial p/\partial x = \text{const}$.

Tailored Equations

(x-momentum) $0 = -\dfrac{1}{\rho}\dfrac{\partial p}{\partial x} + \nu\dfrac{\partial^2 u}{\partial y^2}$

or, concentrating on the left half of the bearing (cf. S3.45b),

$$\frac{d^2u}{dy^2} = \frac{p_{max} - p_0}{\mu l} = P \tag{S3.50a}$$

subject to

$$u(h = h_1) = U \quad \text{and} \quad u(y = 0) = 0$$

Solution

$$u(y) = P\frac{y^2}{2} + \left(\frac{U}{h_1} - P\frac{h_1}{2}\right)y = A\eta^2 + B\eta \tag{S3.50b}$$

where $\eta = y/h_1$, $A = Ph_1^2/2$, and $B = U - A$. A similar solution is obtained for the right half where $h = h_2$ and $P_{right} = -P_{left}$. The volumetric flow rate per unit width is

$$Q = h_1 \int_0^1 u(\eta)\, d\eta \tag{S3.51a}$$

Thus,

$$Q = h_1\left(\frac{A}{3} + \frac{B}{2}\right) \tag{S3.51b}$$

Since no change in mass flow rate occurs across the step we have

$$Q_{left} = Q_{right}$$

from which

$$\Delta p = p_{max} - p_0 = 6\mu l U \frac{h_1 - h_2}{h_1^3 + h_2^3} \tag{S3.52}$$

The load per unit width is

$$F_L = 2\int_0^l \Delta p\, dx = (p_{max} - p_0)l \tag{S3.53a}$$

The drag force per unit width

$$F_D = \int_0^l \tau_w|_{left}\, dx + \int_l^{2l} \tau_w|_{right}\, dx$$

$$= \mu \int_0^l \left(\frac{du}{dy}\Big|_{y=0}^{left} + \frac{du}{dy}\Big|_{y=0}^{right}\right) dx \tag{S3.53b}$$

$$= \mu\left(\frac{U}{h_1}x - P_l\frac{h_1}{2}x + \frac{U}{h_2}x - P_r\frac{h_2}{2}x\right)\Big|_0^l$$

$$= \mu U\left(\frac{l}{h_1} + \frac{l}{h_2}\right) - \frac{\Delta p h_1}{2}\left(1 - \frac{h_2}{h_1}\right)$$

The ratio of drag to load is

$$\frac{F_D}{F_L} = \frac{\mu U}{\Delta p l}\left(\frac{l}{h_1} + \frac{l}{h_2}\right) - \frac{1}{2}\frac{h_1}{l}\left(1 - \frac{h_2}{h_1}\right) \qquad (S3.53c)$$

□

3.14c Derive the governing equations for *stretching flows* of cylindrical fibers (cf. sketch) and thin films under tensile forces (cf. Papanastasiou 1994).

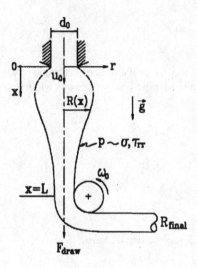

Assumptions

- u is dominant; $v \approx w \ll 1$
- u is (nearly) uniform
- $u(x = L) > u_0 = u(x = 0)$
- $\tau_{ii} = 0$ for $i \neq j$
- τ_{ij} is uniform
- $\tau_{\text{normal}} = p + f(\sigma)$
- $\Sigma F_{\text{fluid}} = F_{\text{inert}} + F_{\text{visc}} + F_{\text{grav}} + F_{\text{surf}}$
- $F_{\text{viscous}} \gg F_{\text{surface tension}}$ in some cases
- neglect extrudate swell region

Background Information

A polymer melt is drawn vertically from a nozzle (diameter d_0, uniform velocity u_0) or a slit (h_0, u_0) and thinned out, that is, $R = R(x)$ for fibers and $h = h(x)$ for films, over a distance L. Although assuming steady laminar quasi-unidirectional flow, there are three to four fluid forces that have to be considered in the draw-down region $0 \leq x \leq L$, where the fiber/sheet slowly hardens.

Solution

Based on the assumptions, the continuity equation reduces to

$$\frac{\partial u}{\partial x} + \frac{1}{r}\frac{\partial}{\partial r}(rv) = 0 \qquad \text{for fiber stretching}$$

and $\qquad$ (S3.54 a,b)

◆ $\dfrac{\partial u}{\partial x} + \dfrac{\partial v}{\partial y} = 0 \qquad$ for sheet stretching

With the *postulates* that $u = u(x)$ and only τ_{ii} and $(\partial/\partial x)$-terms are nonzero, the equations of motion reduce for both fibers and films to

◆ $\rho u \dfrac{\partial u}{\partial x} = -\left(\dfrac{dp}{dx}\right) + \dfrac{d\tau_{xx}}{dx} + \rho g \qquad$ (S3.55a)

Note, as in previous lubrication and thin film problems, we "violate" continuity on the differential scale but conserve mass on the integral scale (cf. *Transformation*).

In order to eliminate the pressure gradient, we employ the free surface condition

◆ $p = \tau_{rr} - \sigma\left(\dfrac{1}{R} + \dfrac{d^2 R}{dx^2}\right) \approx \tau_{rr} - \dfrac{\sigma}{R(x)} \qquad$ for fiber stretching

or $\qquad$ (S3.55b,c)

◆ $p \approx \tau_{yy} - \sigma \dfrac{d^2 h}{dx^2} \qquad$ for sheet stretching

where σ [N/m] is the surface tension to be found in (chemical) engineering handbooks (cf. Perry et al. 1984). Other auxiliary conditions include

$$u(x = 0) = u_0, \qquad Q = u_0 A_0 = u(x)\,A(x) = u_L A_L, \quad \text{and} \quad R(x = L) = R_{\text{final}}.$$

Transformation
Focusing on cylindrical fiber-stretching, we set $u\partial u/\partial x \hateq (1/2)du^2/dx$ and average the x-momentum equation over the fiber cross-sectional area πR^2, namely,

$$\dfrac{2\pi}{\pi R^2} \int\limits_0^{R(x)} \dfrac{d}{dx}\left(\dfrac{\rho}{2}u^2 + p - \tau_{xx}\right) r\, dr = \rho g$$

$$\dfrac{d}{dx} \int\limits_0^R \left(\dfrac{\rho}{2}u^2 + p - \tau_{xx}\right) r\, dr - \dfrac{dR}{dx}\left(\dfrac{\rho}{2}u^2 + p - \tau_{xx}\right) R = \dfrac{\rho}{2}R^2 g$$

Now performing the term-by-term averaging that is, $\bar{v} = [2\pi \int_0^R vr\, dr]/(\pi R^2)$, and so on and using $Q = \pi R^2 \bar{u}$ as well as $p = \tau_{rr} - \sigma/R$, we obtain after integration w.r.t. x:

◆ $\underbrace{\dfrac{\rho}{2}\bar{u}x}_{\sim F_{\text{inertia}}} + \underbrace{\dfrac{\bar{\tau}_{rr} - \bar{\tau}_{xx}}{\bar{u}x}}_{\sim F_{\text{viscous}}} - \underbrace{\sigma\sqrt{\dfrac{\pi}{\bar{u}(x)l}}}_{\sim F_{\text{surf. tens.}}} - \underbrace{\int_0^x \dfrac{\rho g}{\bar{u}x}dx}_{\sim F_{\text{grav}}} = \underbrace{\text{const}}_{\sim F_{\text{net}}} \qquad$ (S3.56)

where $\bar{\tau}_{xx} - \bar{\tau}_{rr} \approx (3\mu)d\bar{u}/dx$ for Newtonian fluids (cf. Middleman 1977). $\qquad\square$

3.15 Calculate the power required to pump oil ($\dot{m} = 50$ tons/hr, $\rho = 915$ kg/m^3, $\nu = 0.00186$ m^2/s) through a pipeline ($d = 10$ cm, $L = 1.6$ km).

Concept
The pump has to overcome frictional head loss to maintain a steady flow of oil.

System Parameter Values

Volumetric flow rate $\quad Q = \dfrac{\dot{m}}{\rho} := \dfrac{50 \times 10^3}{915 \times 3,600} = 0.0152 \text{ m}^2\text{/s}$ $\qquad$ (S3.57a)

Average velocity $\quad \bar{v} = \dfrac{Q}{A} := \dfrac{0.0152}{\pi/4 \times (0.1)^2} = 1.94 \text{ m/s}$ $\qquad$ (S3.57b)

Reynolds number $\quad \text{Re} = \dfrac{\bar{v}d}{\nu} := 104 \quad \Rightarrow \quad \text{laminar flow}$ $\qquad$ (S3.57c)

Friction Factor Derivation

Axial velocity $\quad v(r) = -\dfrac{\Delta p d^2}{16\mu L}\left[4\left(\dfrac{r}{d}\right)^2 - 1\right] \qquad \text{(Poiseuille flow)}$ $\qquad$ (S3.58a)

Volumetric flow rate $\quad Q = \dfrac{1}{A}\int_A v\,dA = \dfrac{\pi \Delta p d^4}{128\mu L} = \bar{v}A$ $\qquad$ (S3.58b)

Mean velocity $\quad \bar{v} = \dfrac{Q}{A} = \dfrac{\Delta p d^2}{32\mu L} \quad \text{or} \quad \Delta p = \dfrac{32\mu L\bar{v}}{d^2}$ $\qquad$ (S3.59a,b)

Darcy's formula $\quad h_f = \dfrac{fL\bar{v}^2}{8gR_h} = \dfrac{\Delta p}{\rho g} \qquad \text{(head loss)}$ $\qquad$ (S3.60)

Here, f is the friction factor; $R_h = d/4$ for a pipe is the hydraulic radius. Eliminating Δp yields Darcy's friction factor $f = 64/\text{Re}_d$ for laminar flow.

Numerical Results

$$f = \dfrac{64}{\text{Re}_d} = \dfrac{64}{104} = 0.615$$

$$h_f = \dfrac{fL\bar{v}^2}{2gd} = 1,891 \text{ m of oil}$$

$$P = \dot{m}g \cdot h_f = \dfrac{50 \times 10^3 \times 9.81}{3,600} \times 1,891$$

or

$$P = 258 \text{ kW} \qquad \text{of power required to pump the oil} \qquad\qquad \square$$

3.16 Consider steady laminar flow of a *power-law fluid* (cf. Sect. 3.2.1D, p. 140) in a slightly tapered tube of length L (cf. sketch). The tapering effect is small enough that the radial velocity is negligible, that is, $v_r \ll v_z$, but significant enough that the axial velocity $v_z = $ fct$[R(z)]$ where the tube radius $R(z)$ increases linearly from $R_0 = R(z = 0)$ to $R_L = R(z = L)$. Given the tube geometry (R_0, R_L, L), the volumetric flow rate (Q), and the non-Newtonian fluid properties (ρ, K, n), find expressions for v_z and Δp.

Note: The problem discussion is based on the solution given by Bird et al. (1987).

Assumptions

- Steady, quasi 1-D, laminar incompressible flow of a power-law fluid
- $v_r \ll v_z$; $\qquad \partial p/\partial z \approx$ const
- Axisymmetric flow

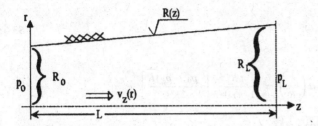

Approach

Neglect "tapering effect" in the fluid flow analysis but recover the true geometry when introducing $R = R(z)$ via a boundary condition.

Postulates

From the stated assumptions, $v_r \approx 0$, $v_\theta \equiv 0$; and from continuity, $v_z = v_z(r)$ only. Also, $\partial p/\partial z \approx \Delta p/L = \text{const.}$

Equations

The (tailored) equations have to be written in terms of τ (cf. Sect. 2.2.1 and App. B) since the Navier–Stokes equations are not valid for *non-Newtonian fluid* flows.

♦   (z-component) $$0 = -\frac{1}{r}\frac{\partial}{\partial r}(r\tau_{rz}) - \frac{\partial p}{\partial z} \qquad (S3.61)$$

Integration yields

$$\tau_{rz} = \frac{p_0 - p_L}{2L}r + \frac{C_1}{r} \qquad (S3.62)$$

Since $\tau_{rz}(r = 0)$ is finite, $C_1 \equiv 0$.

The appropriate power-law viscosity model is (cf. Sect. 1.3.1 or Sect. 3.2.1)

♦ $$\tau_{rz} = -\eta\frac{dv_z}{dr} = -K\dot{\gamma}_{rz}^{n-1}\frac{dv_z}{dr} = K\left(-\frac{dv_z}{dr}\right)^n \qquad (S3.63)$$

Note: In general, $\dot{\gamma}_{rz} = \partial v_r/\partial z + \partial v_z/\partial r$; the minus sign is taken into the parentheses to assure that shear stresses are acting opposite to the flow direction.

Solution

Combining (S3.63) and (S3.62), taking the nth root of both sides, and integrating once yield

$$v_z = -\left(\frac{p_0 - p_L}{2KL}\right)^{1/n}\frac{r^{(1/n)+1}}{(1/n)+1} + C_2 \qquad (S3.64a)$$

Using the "no-slip" condition $v_z[r = R(z)] = 0$ and wall shear stress $\tau_w = \tau_{rz}(r = R)$ (cf. Equation (S3.62)) yields

♦ $$v_z = \left(\frac{\tau_w}{K}\right)^{1/n}\frac{R}{(1/n)+1}\left[1 - \left(\frac{r}{R}\right)^{(1/n)+1}\right] \qquad (S3.64b)$$

The volumetric flow rate is

$$Q = \int_0^{2\pi} \int_0^{R(z)} v_z \, r \, dr \, d\theta \tag{S3.65a}$$

$$Q = 2\pi R^2 \int_0^1 v_z \frac{r}{R} d\left(\frac{r}{R}\right) = \frac{\pi R^3}{(1/n) + 3} \left[\frac{(p_0 - p_L)R}{2KL}\right]^{1/n} \tag{S3.65b}$$

Note: Equation (S3.65b) cannot yet be used to calculate the pressure drop, $\Delta p = p_L - p_0$, since $R = R(z)$.

Recall

$$R(z) = R_0 + \frac{R_L - R_0}{L}z \quad \text{or} \quad dR = \frac{R_L - R_0}{L}dz \quad \text{and} \quad \frac{(p_0 - p_L)}{L} \approx -\frac{dp}{dz}$$

so that Equation (S3.65b) can be rewritten as

◆ $$Q = \frac{\pi[R(z)]^3}{(1/n) + 3} \left[\frac{R(z)}{2K}\right]^{1/n} \left(-\frac{dp}{dz}\right)^{1/n} \tag{S3.65c}$$

Noting that Q is constant for all z and hence all $R(z)$, integration yields

◆ $$\Delta p = \frac{2KL}{3n} \left[\frac{Q}{\pi}\left(\frac{1}{n} + 3\right)\right]^n \frac{R_L^{-3n} - R_0^{-3n}}{R_0 - R_L} \tag{S3.66}$$

Recommendation

Plot $(v_z/\bar{v}_z)$ versus (r/R) to show the evolution from plug flow ($n = 0$) via pseudoplastic flow, that is, $n < 1$, to Poiseuille flow of a Newtonian fluid ($n = 1$). $\qquad\square$

3.17 A plate moves with $U = At^n$ in a reservoir of viscous fluid. Using the trial solution $u(t, y) = At^n f(\eta)$ where $\eta = y/\delta = y/\sqrt{4\nu t}$, calculate the velocity distribution $u(y, t)$ and determine

$$\lim_{t \to 0} u(y, t) \quad \text{for } y = \begin{cases} \text{zero} \\ \text{nonzero} \end{cases}$$

Note: This problem was suggested by Lu (1977).

Concept

Consider a suddenly accelerated plate similar to Stokes's first problem when $n \equiv 0$ (cf. Sect. 3.2.1). For this type of transient *parallel* flow, the inertia or convective acceleration terms vanish identically.

Tailored Equations

Assuming transient one-dimensional flow, the problem-oriented equation is

$$\frac{\partial u}{\partial t} = \nu \frac{\partial^2 u}{\partial y^2} \tag{S3.67a}$$

subject to the initial and boundary conditions

$$u(t = 0, y) = 0; \quad u(y = 0, t > 0) = U = At^n \quad \text{and} \quad u(y \to \infty) = 0 \tag{S3.67b}$$

Transformations

Using $u(y, t) = At^n f(\eta)$ where the similarly variable $\eta = y/\sqrt{4\nu t}$, Equation (S3.67) takes on the form

$$f'' + 2\eta f' - 2(2n) f = 0 \qquad \text{(S3.68a)}$$

subject to

$$f(\eta = 0) = 1 \quad \text{and} \quad f(\eta \to \infty) = 0 \qquad \text{(S3.68b)}$$

Solution

Typically, n is an integer and the solution of (S3.68) is

$$f(\eta) = \text{erfc}(\eta) \qquad \text{so that} \quad u(y, t) = At^n \text{erfc}(\eta) \qquad \text{(S3.69)}$$

When $n = -1/2$ ($n < 0$ implies that the surface starts with an infinite velocity $U = A/\sqrt{t}$), Equation (S3.68) can be written as

$$(f' + 2\eta f)' = 0$$

or

$$f' + 2\eta f = C_1 \qquad \text{(S3.70)}$$

It turns out that the nonhomogeneous term, C_1, makes it difficult to find a closed-form solution. However, for the special case of $C_1 \equiv 0$, Equation (S3.70) has the solution

$$f = A \exp\left(-\frac{\eta^2}{4}\right) \quad \text{or} \quad u = At^n \exp\left(-\frac{\eta^2}{4}\right) \qquad \text{(S3.71a,b)}$$

Recommendation

Plot $u(y)$ for different n's and comment. □

3.18 Consider flow from a point source into a diverging channel or flow toward a point sink (cf. sketch for Jeffery–Hamel flow).

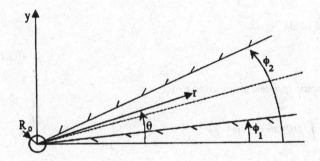

Assumptions

- Laminar steady flow
- Area of radius R_0 with singularity excluded
- Exact radial flow
- Constant fluid properties
- No end effects

Postulates

On the basis of the assumptions, the impermeable wall condition, $u_\theta = 0$, and from the continuity equation, $\partial/\partial r(ru_r) = 0$, it follows that

$$u_r = \frac{f(\theta)}{r} \qquad \text{where } r \geq R_0 \quad \text{and} \quad \phi_1 \leq \theta \leq \phi_2 \tag{S3.72}$$

Tailored Equations

With these postulates, the Navier–Stokes equations can be reduced to

$$(r\text{-momentum}) \qquad -\frac{f^2}{r^3} = -\frac{1}{\rho}\frac{\partial p}{\partial r} + \frac{\nu}{r^3}\frac{d^2 f}{d\theta^2} \tag{S3.73}$$

$$(\theta\text{-momentum}) \qquad 0 = -\frac{1}{\rho r}\frac{\partial p}{\partial \theta} + \frac{2\nu}{r^3}\frac{df}{d\theta} \tag{S3.74}$$

Solution

Partial integration of (S3.74) yields

$$\frac{p}{\rho} = \frac{2\nu}{r^2}f + F(r) \tag{S3.75}$$

or after partial differentiation with respect to r

$$\frac{1}{\rho}\frac{\partial p}{\partial r} = -\frac{4\nu}{r^3}f + \frac{dF}{dr} \tag{S3.76}$$

Using (S3.75) in (S3.73) yields

$$-f^2 = 4\nu f - r^3\frac{dF}{dr} + \nu\frac{d^2 f}{d\theta^2}$$

or after separation

$$\nu\frac{d^2 f}{d\theta^2} + 4\nu f + f^2 = r^3\frac{dF}{dr} = \lambda = \text{const}$$

The RHS integrated yields

$$F(r) = -\frac{\lambda}{2r^2} + C_1 \tag{S3.77}$$

Knowing the volumetric flow rate at any given r

$$Q = \int_{\phi_1}^{\phi_2} u_r\, r\, d\theta = \int_{\phi_1}^{\phi_2} f(\theta)\, d\theta = \text{const}$$

and the pressure at any point in the flow field (cf. Equation (S3.76))

$$p = \frac{2\rho\nu}{r^2}f - \frac{\lambda\rho}{2r^2} + C_1\rho$$

the constants λ and C_1 can be calculated.
The LHS reads

$$\nu\frac{d^2 f}{d\theta^2} + 4\nu f + f^2 = \lambda$$

which multiplied by $(df/d\theta)$ and integrated yields

$$\frac{v}{2}\left(\frac{df}{d\theta}\right)^2 + 2vf^2 + \frac{1}{3}f^3 = \lambda f + C_2$$

or

$$\left(\frac{df}{d\theta}\right)^2 = -\frac{2}{3v}f^3 - 4f^2 + \frac{2\lambda f}{v} + C_1 := -\frac{2}{3v}G(f) \qquad (S3.78)$$

The algebraic equation $G(f)$ reads

$$G(f) = f^3 + 6vf^2 - 3\lambda f + C_3 \qquad (S3.79)$$

This cubic function $G(f)$ should be negative for any real θ; that implies that only a limited range of f-values that include the point $f = 0$ is part of the permissible solution. The implicit solution of (S3.78) is an (indefinite) elliptic integral that is tabulated in mathematical handbooks

$$\theta = \pm \int \frac{df}{\sqrt{-(2/3v)G(f)}} \qquad (S3.80)$$

Needless to say, the behavior of $G(f)$ and the location of its roots for (given) values of λ and C_3 determine the solution of (S3.80) and hence the channel flow field.

Recommendation
Find $G(f)$, evaluate (S3.80) numerically, and plot $u_r(\theta)$ at different distances r. $\qquad\qquad\square$

3.19 Consider similar boundary-layer flows, that is, the Falkner–Skan wedge flows (cf. sketch and Sect. 3.2.3D). Derive the transformed ODE that constitutes a boundary-value problem (BVP).

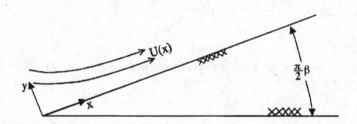

Assumptions

- Steady laminar 2-D flows
- Similar, thin shear layers
- Constant fluid properties

Governing Equations and Transformation

(continuity) $\qquad \dfrac{\partial u}{\partial x} + \dfrac{\partial v}{\partial y} = 0 \qquad\qquad\qquad (S3.81)$

Partial integration of the continuity equation, using the "no-slip" condition, yields

$$v = -\frac{\partial}{\partial x} \int_0^y u \, dy \tag{S3.82}$$

Now, the x-momentum equation reads

$$u\frac{\partial u}{\partial x} - \frac{\partial u}{\partial y}\left(\frac{\partial}{\partial x}\int_0^y u \, dy\right) = U\frac{dU}{dx} + v\frac{\partial^2 u}{\partial y^2} \tag{S3.83}$$

The goal is to combine x and y to a (similarity) variable, η, so that the momentum equation becomes an ODE for certain $U(x)$ distributions (cf. Sect. 2.3.3). The facts that $u(x, y)$ must be integrated once and has to approach $U(x)$ at the boundary-layer edge suggest

$$u(x, y) = U(x)f'(\eta) \tag{S3.84}$$

where $\eta = \eta(x, y)$ and prime denotes $d/d\eta$. If $u = Uf'$, then

$$\frac{\partial u}{\partial x} = \frac{dU}{dx}f' + Uf''\frac{\partial \eta}{\partial x}, \qquad \frac{\partial u}{\partial y} = Uf''\frac{\partial \eta}{\partial y}, \text{ etc.}$$

The integral term in (S3.83) is evaluated using integration by parts and Leibniz's rule. The transformation yields

$$f''' + f'\underbrace{\frac{\partial^2\eta/\partial y^2}{(\partial\eta/\partial y)^2}}_{g_1(\eta)} = ff''\underbrace{\frac{U(\partial^2\eta/\partial x\partial y)}{v(\partial\eta/\partial y)^3}}_{g_2(\eta)} + (f'^2 - ff'' - 1)\underbrace{\frac{dU/dx}{v(\partial\eta/\partial y)^2}}_{g_3(\eta)} \tag{S3.85}$$

We now postulate $\eta = Cyx^a$ so that

$$g_1 \equiv 0, \qquad g_2 = \frac{Ua}{vC^2x^{2a+1}} \quad \text{and} \quad g_3 = \frac{dU/dx}{vC^2x^{2a}}$$

Note

• Similarity requires that the explicit x-dependence in the coefficients $g_i(\eta), i = 1, 2, 3$, disappear; that is, we set

$$U(x) = Kx^{2a+1} = Kx^m \tag{S3.86}$$

• The arbitrary constant C is determined so that (S3.85) reduces to the ODE for flat-plate (or Blasius) flow: $U = \text{const}$ or $m = 0$. This implies that $a = -1/2$ and $g_2 = -1$. In more general terms, we force $g_3 - g_2 = 1$, so that

$$C = \sqrt{\frac{K(1+m)}{2v}} \quad \text{and} \quad \eta = y\sqrt{\frac{m+1}{2}\frac{U(x)}{vx}} \tag{S3.87a,b}$$

As a result, Equation (S3.85) with (S3.86) and (S3.87) reads

$$f''' + ff'' + \beta(1 - f'^2) = 0 \tag{S3.88}$$

where $\beta = 2m/(1+m)$ and the boundary conditions are

$$f(0) = f'(0) = 0 \quad \text{and} \quad f'(\infty) = 1 \tag{S3.89}$$

It has to be noted that the Falkner–Skan equation (S3.88) constitutes a BVP; the special case $\beta = 0$ is the Blasius problem (cf. Section 3.2.3D). Applications are discussed in Sample-Problem 3.20. □

3.20 Consider the Blasius problem of steady laminar flow past a flat horizontal plate with $U(x) = u_\infty = u_e = $ const.

Approach A
The Blasius equation, to be obtained from (S3.89) with $U = $ const, which implies that $m = \beta = 0$, reads

$$f''' + ff'' = 0 \tag{S3.90}$$

It is solved numerically as discussed in Section 3.2.3D or approximately, using the integral method discussed later. Note that the factor 1/2 in Equation (3.73d) is a result of a slightly different definition of the similarity variable, that is, $\eta = y\sqrt{U_0/(vx)}$ previously versus the more general form, $\eta = y\sqrt{((m+1)/2)U(x)/(vx)}$ in Sample-Problem 3.19.

Approach B
The momentum integral equation for the Blasius problem is (cf. Sect. 2.3.3)

$$\rho \frac{d}{dx}\left[\int_0^\delta u(u_\infty - u)\,dy\right] = \tau_w \qquad \text{in } 0 \le y < \infty \tag{S3.91}$$

A function $u(x, y)$ that fits the actual velocity profile well and obeys the boundary and edge conditions has to be postulated and inserted into (S3.91). Suitable candidates are polynomials and trigonometric and exponential functions.

Example (1)
A cubic polynomial to approximate $u(x, y)$ reads

$$u(x, y) = a_0 + a_1 y + a_2 y^2 + a_3 y^3$$

where the coefficients a_i, $i = 0, 1, 2, 3$, are functions of x and have to be determined by using the following conditions:

$$u(y = 0) = 0, \qquad u(y = \delta) = u_\infty, \qquad \left.\frac{\partial u}{\partial y}\right|_{y=\delta} = 0 \quad \text{and} \quad \left.\frac{\partial^2 u}{\partial y^2}\right|_{y=0} = 0$$

The latter condition can be obtained for $U = $ const by evaluating the x-momentum equation near $y = 0$. As a result, the postulated velocity profile can be written as

$$\frac{u(x, y)}{u_\infty} = \frac{3}{2}\left(\frac{y}{\delta(x)}\right) - \frac{1}{2}\left(\frac{y}{\delta(x)}\right)^3 \tag{S3.92}$$

where the boundary-layer thickness $\delta(x)$ is the new unknown. Inserting (S3.92) into (S3.91) yields an ODE for $\delta(x)$

$$u_\infty^2 \frac{d}{dx}\left\{\int_0^\delta \left[\frac{3}{2}\left(\frac{y}{\delta}\right) - \frac{1}{2}\left(\frac{y}{\delta}\right)^3\right]\left[1 - \frac{3}{2}\left(\frac{y}{\delta}\right) + \frac{1}{2}\left(\frac{y}{\delta}\right)^3\right]dy\right\} = vu_\infty \frac{3}{2\delta}$$

or

$$\frac{d}{dx}\left[\frac{39}{280}\delta(x)\right] = \frac{3v}{2u_\infty\delta(x)}$$

or

$$\delta d\delta = \frac{140}{13}\frac{v}{u_\infty}dx \qquad \text{subject to } \delta(x=0)=0 \qquad\qquad\qquad \text{(S3.93a)}$$

Integration yields

$$\delta^2(x) = \frac{280vx}{13u_\infty} \quad \text{or} \quad \delta(x) = \sqrt{\frac{280}{13}\frac{vx}{u_\infty}}$$

In dimensionless form

$$\frac{\delta(x)}{x} = \frac{4.64}{\sqrt{Re_x}} \qquad\qquad\qquad\qquad\qquad\qquad\qquad\qquad\qquad \text{(S3.93b)}$$

while $\left(\dfrac{\delta}{x}\right)_{\text{exact}} = 4.96\,Re_x^{-1/2}$.

Note

Knowing the boundary-layer growth, $\delta \sim x^{1/2}$, other flat-plate boundary-layer parameters can be readily evaluated, for example:

- Displacement thickness $\qquad \delta_1(x) = \displaystyle\int_0^\infty \left(1 - \frac{u}{u_\infty}\right) dy := \frac{3}{8}\delta$

- Momentum thickness $\qquad \delta_2(x) = \displaystyle\int_0^\infty \frac{u}{u_\infty}\left(1 - \frac{u}{u_\infty}\right) dy := \frac{39}{280}\delta$

- Wall shear stress $\qquad \tau_w(x) = \mu\left.\dfrac{du}{dy}\right|_{y=0} := \frac{3\mu u_\infty}{2\delta}$

- Skin friction coefficient $\qquad c_f(x) = \dfrac{2\tau_w}{\rho u_\infty^2} := \dfrac{0.6464}{\sqrt{Re_x}}, \qquad$ i.e., $c_f \sim x^{-1/2}$

- Drag coefficient $\qquad c_D(x) = 2c_f(x) := \dfrac{1.293}{\sqrt{Re_x}}$

Questions

Why are $\delta_2 < \delta_1 < \delta$; $\quad c_f = \delta_2/x$; and $c_D = 2c_f$?

Example (2)

Select a trigonometric function for $u(x, y)$ that fulfills the boundary conditions plus wall compatibility, that is, $\partial^2 u/\partial y^2 = 0$ at $y = 0$, and "smoothness" conditions for $u(y) \rightarrow U_0 = $ const as $y \rightarrow \delta$ for all x.

Postulate

$$\frac{u}{U_0} = \sin\left(\frac{\pi}{2}\frac{y}{\delta(x)}\right) = f(\eta) \qquad \text{where } \eta = \frac{y}{\delta} \qquad\qquad (S3.94)$$

System Parameters

$$\frac{\delta_1}{\delta} = \int_0^1 (1 - f)\, d\eta = \int_0^1 \left(1 - \sin\left(\frac{\pi}{2}\eta\right)\right) d\eta = \frac{\pi - 2}{\pi}$$

$$\frac{\delta_2}{\delta} = \int_0^1 f(1 - f)\, d\eta = \int_0^1 \sin\left(\frac{\pi}{2}\eta\right)\left(1 - \sin\left(\frac{\pi}{2}\eta\right)\right) d\eta = \frac{4 - \pi}{2\pi}$$

Recall

In this case,

$$c_f = 2\, d\delta_2/dx \quad \text{or} \quad \delta^2(x) = [2f'(0)/(\delta_2/\delta)](\nu x/U_0)$$

Hence,

$$\left(\frac{\delta}{x}\right)_{\text{approx}} \approx \frac{3.399}{\sqrt{\text{Re}_x}} \qquad \text{whereas} \qquad \left(\frac{\delta}{x}\right)_{\text{exact}} = \frac{4.96}{\sqrt{\text{Re}_x}}$$

and

$$c_f|_{\text{approx}} = \frac{2\tau_w}{\rho U_0} \approx \frac{0.654}{\sqrt{\text{Re}_x}} \qquad \text{whereas } c_f|_{\text{exact}} = \frac{0.664}{\sqrt{\text{Re}_x}}$$

Questions

Why is (δ/x), using the sine function, less accurate than (δ/x) based on the third-order polynomial and why is the opposite the case for c_f? The best results using the integral method can be achieved with $u = \tanh(y/\delta)$. Why?

Example (3)

Use of an exponential function to approximate the streamwise velocity profile

$$\frac{u}{U} = 1 - \exp[-\alpha(x)y] \qquad\qquad (S3.95)$$

System Parameters

$$\delta_1 = \int_0^\infty \left(1 - \frac{u}{U}\right) dy = \int_0^\infty e^{-\alpha(x)y}\, dy = \frac{1}{\alpha(x)}$$

$$\delta_2 = \int_0^\infty \frac{u}{U}\left(1 - \frac{u}{U}\right) dy = \int_0^\infty (e^{-\alpha y} - e^{-2\alpha y})\, dy = \frac{1}{2\alpha(x)}$$

$$\tau_w = \mu \frac{\partial u}{\partial y}\bigg|_{y=0} = \mu U \left[\alpha(x)e^{-\alpha(x)y}\bigg|_{y=0}\right] = \mu U \alpha(x)$$

With $U(x) = \text{const}$, $c_f = (2\tau_w/\rho U^2) = 2(d\delta_2/dx)$ and inserting $\tau_w(x)$ and $\delta_2(x)$ yields

$$\frac{2\mu}{\rho U} = -\frac{1}{\alpha^3}\frac{d\alpha}{dx} \quad \text{or} \quad \frac{2\mu x}{\rho U} = \frac{1}{2\alpha^2}$$

Thus,

$$\alpha(x) = \frac{\sqrt{\mathrm{Re}_x}}{2x} \tag{S3.96}$$

The system parameters are now, with Equation (S3.96)

$$\frac{\delta_1 \sqrt{\mathrm{Re}_x}}{x} = 2; \quad \frac{\delta_2 \sqrt{\mathrm{Re}_x}}{x} = 1; \quad \frac{\delta \sqrt{\mathrm{Re}_x}}{x} = 9.21; \quad c_f \sqrt{\mathrm{Re}_x} = 1 \text{ and } c_D \sqrt{\mathrm{Re}_x} = 2$$

Note: Except for (δ_1/x), the system parameters are way off (50.6 percent and 84.2 percent) when compared with the exact Blasius solution. The results are poor because actual boundary-layer profiles $u(y)$ for any given x, are inaccurately matched and because the compatibility condition at the wall is not satisfied, that is, here $\partial^2 u/\partial y^2 = 0$ at $y = 0$.   □

3.21 Given the outer flow distribution $U(x) = U_0(1 - x^6)$ for laminar flow past a flat plate. Find the point of flow separation, x_{sep}.

Approach
Use Thwaites's integral method as discussed in Section 3.2.3D.

Solution
The approximated momentum thickness is

$$\delta_2^2 \approx 0.45\nu U^{-6} \int_0^x U^5\, dx \tag{S3.97a}$$

where the integral is evaluated here as

$$\int_0^x U^5\, dx = U_0^5 \int_0^x \left(\frac{U}{U_0}\right)^5 dx = U_0^5 \int_0^x (1 - x^6)^5 dx$$

so that

$$\delta_2^2 = 0.645 \frac{\nu x}{U_0}[6(1 - x^6)^{-6} + 6(1 - x^6)^{-5}] \tag{S3.97b}$$

The dimensionless system parameter

$$\lambda \equiv \frac{\delta^2}{\nu}\frac{dU}{dx} := \frac{\delta^2}{\nu}(-6U_0 x^5) \tag{S3.98a}$$

or

$$\lambda = -0.3858x^6[6(1 - x^6)^{-6} + (1 - x^6)^{-5}] \tag{S3.98b}$$

Now, at the point of flow separation, the shear correlation $(\tau_w \delta_2/\mu U) \approx S(\lambda)$ is equal to zero (cf. Table 3.1):

$$S(\lambda = -0.09) = 0 \tag{S3.99}$$

so that Equation (S3.81b) becomes

$$0.09 = 0.3858x_{\text{sep}}^6[6(1 - x_{\text{sep}}^6)^{-6} + (1 - x_{\text{sep}}^6)^{-5}]$$

or

$$0.4287x_{\text{sep}}^{12} - 4x_{\text{sep}}^6 + 1 = 0$$

Letting $x_{\text{sep}}^6 \equiv \zeta$, we have to solve

$$0.4287\zeta^2 - 4\zeta + 1 = 0 \tag{S3.100}$$

which has the solutions

$$\zeta = \begin{cases} 9.0734 \\ 0.2571 \end{cases}$$

Thus,

$$x_{\text{sep}} = 0.7974 \qquad\qquad\qquad \square$$

3.22 Find an approximate solution to the Blasius problem by integrating

$$f''' + ff'' = 0 \qquad \text{subject to } f(0) = f'(0) = f'(\infty) - 1 = 0$$

where prime denotes $d/d\eta$ with $\eta = y\sqrt{u_\infty/2\nu x}$.

Solution
Rewriting the governing equation as

$$\frac{df''}{d\eta} = -ff'' \quad \text{or} \quad \frac{df''}{f''} = -f\,d\eta \tag{S3.101a,b}$$

allows integration so that

$$f'' = A\exp\left[-\int f\,d\eta\right]$$

Note, at the wall $f''(0) = A$, where A is the wall shear stress.
Integrating twice yields

$$f = A\iint \exp\left[-\int f\,d\eta\right] d\eta\,d\eta + C_1\eta + C_2 \tag{S3.102a}$$

The boundary conditions $f(0) = f'(0) = 0$ imply $C_1 = C_2 = 0$. Thus a numerical solution of

$$f = A\iint \exp\left[-\int f\,d\eta\right] d\eta\,d\eta \tag{S3.102b}$$

should yield the similarity function.

Alternatively, we find an approximate solution near the wall when $f'' = A$ so that $f' = A\eta$ and $f = A\eta^2/2$. Substituting these expressions into (S3.102b)

$$f = A \iint \exp\left[-\int f\, d\eta\right] d\eta\, d\eta$$

yields

$$f' = A \int \exp\left(-\frac{A\eta^3}{3!}\right) d\eta = A^{2/3}(3!)^{1/3} \int_0^\eta \exp(-\eta^3)\, d\eta \qquad (S3.103)$$

The integral $I = \int_0^\zeta \exp(-\zeta^3)\, d\zeta$ is tabulated in mathematical handbooks.
For example

ζ	0.0	0.1	0.2	0.3	0.4	0.5	0.6
I	0.00	0.10	0.20	0.298	0.394	0.485	0.570

ζ	0.8	1.0	1.2	1.4	1.6	1.8	1.9
I	0.711	0.808	0.861	0.884	0.891	0.893	0.897

The factor A (wall shear stress) can be evaluated by using the edge condition, $f'(\infty) - 1 = 0$:

$$A(x) = \left[(3!)^{1/3} \int_0^\delta e^{-\eta^3}\, d\eta\right]^{-3/2} \qquad (S3.104)$$

Recommendation
Program Equation (S3.103) and plot $u(y)$ for different x-stations.   □

3.3.5   Turbulent Boundary-Layer Flows

Note:
Because of the complexity of turbulence and our limited mathematical tools, we have to resort occasionally to a "grab-bag" approach in solving turbulent flow problems.

3.23 Find an expression for the turbulent velocity profile near the wall in terms of the inner variables $u^+(y^+)$.

Solution
Prandtl suggested that near the wall, $\tau_{\text{total}} \approx \tau_w$. Hence,

$$\frac{1}{\rho}\tau_{\text{total}} = \nu\frac{\partial\bar{u}}{\partial y} - \overline{u'v'} := \frac{\tau_w}{\rho} \equiv u_\tau^2 \qquad (S3.105)$$

where u_τ is called (the) friction velocity and $-\overline{u'v'} = \nu_t\partial\bar{u}/\partial y$ following Boussinesq's eddy viscosity concept. Employing Prandtl's mixing length hypothesis $\nu_t \cong l^2|\partial\bar{u}/\partial y|$, extended by Van Driest (1956), $l = l(y, A)$, we obtain

$$\nu\frac{d\bar{u}}{dy} + l^2\left|\frac{d\bar{u}}{dy}\right|^2 = u_\tau^2 \qquad (S3.106a)$$

where $\quad l = \kappa y \left[1 - \exp\left(\dfrac{-y}{A} \right) \right]$ (S3.106b)

with A a "damping length."

In terms of dimensionless quantities $u^+ = \bar{u}/u_\tau$ and $y^+ = y u_\tau/\nu$, Eq. (S3.106a,b) yields

$$a(y^+) \left(\frac{du^+}{dy^+} \right)^2 + \left(\frac{du^+}{dy^+} \right) - 1 = 0 \tag{S3.107}$$

subject to $u^+(y^+ = 0) = 0$ where $a = (\kappa y^+)^2 [1 - \exp(-y^+/A^+)]^2$ with $A^+ = 24.4$. Integration yields

$$u^+ = \int\limits_0^{y+} \frac{2 dy^+}{1 + \sqrt{1 + 4a(y^+)}} \qquad \text{for } 0 \le y^+ \le 350 \tag{S3.108}$$

Numerical integration is required to evaluate $u^+(y^+)$. Note that $y^+ = 350$ corresponds to $y/\delta = 0.20$. Furthermore, once $u^+(y^+)$ is known, an expression for $u_\tau = \sqrt{\tau_w/\rho}$ has to be provided (cf. Sect. 3.2.5A) to obtain $\bar{u}(y)$.   □

3.24 Transform the steady 2-D turbulent boundary-layer equations using the Falkner–Skan transformation and Boussinesq's eddy viscosity model where eddy viscosity expressions are defined for the inner and for the outer boundary layer, for example, as in the Cebeci–Smith turbulence model (cf. Bradshaw et al. 1981).

Governing Equations

$$\frac{\partial \bar{u}}{\partial x} + \frac{\partial \bar{v}}{\partial y} = 0 \tag{S3.109}$$

$$\bar{u} \frac{\partial \bar{u}}{\partial x} + \bar{v} \frac{\partial \bar{u}}{\partial y} = U \frac{dU}{dx} + \frac{1}{\rho} \frac{\partial \tau_{yx}}{\partial y} \tag{S3.110}$$

where $\quad \tau_{yx} = \mu \dfrac{\partial \bar{u}}{\partial y} - \rho \overline{u'v'} = \rho \nu (1 + \varepsilon^+) \dfrac{\partial \bar{u}}{\partial y} \quad$ with $\varepsilon^+ = \dfrac{\varepsilon}{\nu}$ (S3.111)

Falkner–Skan Transformation

Using the stream function approach, we postulate (cf. Sect. 3.2.3D)

$$\psi = (U \nu x)^{1/2} f(\eta) \tag{S3.112}$$

where f is a dimensionless stream function, $\eta = \sqrt{(U/\nu)}(y/\sqrt{x})$; $u = \partial\psi/\partial y$ and $v = -\partial\psi/\partial x$ with $B = 1 + \varepsilon^+$ and $m = \beta/(2 - \beta)$, where $(\pi/2)\beta$ is the wedge angle with $0 \le \beta < 2$ (cf. Sample-Problem 3.10), Equation (S3.110) using (S3.109) reduces to

$$(Bf'')' + \frac{m+1}{2} ff'' + m[1 - f'^2] = 0 \tag{S3.113}$$

A submodel for ε^+ has to be provided in order to gain closure (cf. sketch).

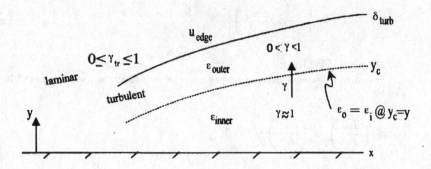

The Inner/Outer Eddy Viscosity Distributions

Following Cebeci and Smith (1974), the turbulent boundary layer is assumed to consist of an inner layer where

$$\varepsilon_{\text{inner}} = l^2 \left| \frac{\partial \bar{u}}{\partial y} \right| \gamma_{\text{tr}} \gamma \qquad \text{for } 0 \leq y \leq y_c \tag{S3.114}$$

and an "outer" layer

$$\varepsilon_{\text{outer}} = \alpha \left| \int_0^\delta (u_e - u)\, dy \right| \gamma_{\text{tr}} \gamma \qquad \text{for } y_c \leq y \leq \delta \tag{S3.115a}$$

or

$$\varepsilon_{\text{outer}} = 0.0168 u_e \delta_1 \gamma_{\text{tr}} \gamma \tag{S3.115b}$$

where $u_e \equiv U(x)$ and γ_{tr} and γ are intermittency factors. The mixing length l is defined as

$$l = \kappa y[1 - \exp(-y/A)]$$

where the Van Driest damping length $A \approx 26(\nu/u_\tau)$ with $u_\tau = \sqrt{\tau_w/\rho}$; γ_{tr} is an intermittency factor for the transition region from laminar to turbulent flow, that is,

$$\gamma_{\text{tr}} = 1 - \exp[-G(x - x_{\text{tr}})] \int_{x_{\text{tr}}}^x \frac{dx}{u_e} \tag{S3.116a}$$

where x_{tr} is the start of flow transition and

$$G = 8.33 \times 10^{-4} \frac{u_e^3}{\nu^2 \text{Re}_x 1.34} \tag{S3.116b}$$

γ is another intermittency factor reflecting the approach from the turbulent boundary layer to the external free stream: that is, the flow near the boundary-layer edge is turbulent for only a fraction of time

$$\gamma = \left[1 + 5.5 \left(\frac{y}{\delta} \right)^6 \right]^{-1} \tag{S3.116c}$$

where δ is the boundary-layer edge. The value y_c is obtained by setting $\varepsilon_i = \varepsilon_0$.

Recommendation
Considering $U(x) = u_e = $ const, program Equation (S3.113) by implementing a single-layer eddy viscosity model, Eq. (S3.114), for $0 \le y \le \delta(x)$. ☐

3.25 Determine the viscous wall-layer thickness and the boundary-layer thickness on the roof of a bus, 6 feet behind the windshield. Assume the boundary layer has a zero thickness at the leading edge of the roof. The bus is traveling at 60 mph in otherwise still air at 68°F. Neglect any pressure-gradient effects.

Solution
The boundary layer undergoes transition at approximately $\text{Re}_{\text{crit}} = 3 \times 10^5 = 88 x_{\text{tr}}/1.6 \times 10^{-4}$. This gives a laminar region of less than 0.5 feet. Let us neglect this laminar portion and assume the boundary layer to be turbulent from the leading edge. This allows us to determine the average wall shear stress from the averaged skin friction expression for moderate (turbulent) Reynolds numbers (cf. Schlichting 1979):

$$\bar{c}_f = \frac{\tau_0}{1/2\rho U_\infty^2} = \frac{0.455}{(\log_{10} \text{Re}_x)^{2.58}} = 0.00292 \qquad \text{for } 0 \le x \le 6 \text{ ft}$$

Hence

$$\frac{\tau_0}{\rho} = 0.00292 \times \frac{88^2}{2} = 11.3 \frac{\text{ft}^2}{\text{s}^2}$$

or

$$u_\tau = \sqrt{\frac{\tau_0}{\rho}} = 3.36 \frac{\text{ft}}{\text{s}}$$

The viscous wall layer has a thickness δ_v, given by

$$\frac{u_\tau \delta_v}{\nu} = 5$$

or

$$\delta_v = \frac{5 \times 1.6 \times 10^{-4}}{3.36} = 2.38 \times 10^{-4} \text{ ft}$$

To determine the turbulent boundary-layer thickness at 6 ft, we find the velocity at the outer edge of the turbulent zone. Using Equation (3.117) in the form of

$$\frac{u_e - u}{u_\tau} = -2.44 \ln \frac{y}{\delta} + 2.5$$

with $y/\delta = 0.15$ to obtain

$$\frac{88 - u}{3.36} = -2.44 \ln 0.15 + 2.5 = 7.1$$

This yields

$$u = 64.2 \text{ ft/s}$$

Then from Eq. (3.115), which is valid in the turbulent zone, we have

$$\frac{64.2}{3.36} = \frac{1}{.41} \ln \frac{3.36y}{1.6 \times 10^{-4}} + 5.0$$

Solving for y, we find the outer edge of the turbulent zone to be located at the distance

$$y = 0.0162 \text{ ft}$$

The edge of the turbulent zone occurs at $y/\delta = 0.15$. Hence,

$$\delta = \frac{y}{0.15} = \frac{0.0162}{0.15} = 0.108 \text{ ft} \qquad \qquad \square$$

3.26 Consider turbulent flat plate flow with zero pressure gradient. Derive correlations between the local skin friction coefficient and the Reynolds number.

Approach
Use the integral method and employ suitable turbulent velocity profiles.

Momentum Integral Relation
For laminar or turbulent flat plate flow with $U = $ const and momentum thickness

$$\delta_2 = \int\limits_0^\infty \frac{u}{U}\left(1 - \frac{u}{U}\right) dy$$

the boundary-layer equation transforms to

$$\tau_w(x) = pU^2 \frac{d\delta_2}{dx} \qquad \qquad (S3.117)$$

or

$$c_f \equiv \frac{2\tau_w}{\rho U^2} = 2\frac{d\delta_2}{dx} \qquad \qquad (S3.118)$$

Log-Law Approach
Assuming that the logarithmic law is valid across the entire turbulent boundary layer, we have

$$\frac{u}{u_\tau} \approx \frac{1}{\kappa} \ln \frac{y u_\tau}{\nu} + B \qquad \qquad (S3.119)$$

where $u_\tau = \sqrt{\tau_w/\rho}$, $\kappa = 0.41$, and $B \approx 5.0$.
   At the boundary-layer edge where $u(y = \delta) = U$, Equation (S3.119) reads

$$\frac{U}{u_\tau} = \frac{1}{\kappa} \ln \frac{\delta u_\tau}{\nu} + B \qquad \qquad (S3.120)$$

Without directly using (S3.118), which should be satisfied by (S3.119), we express Equation (S3.120) in terms of c_f and $\text{Re}_\delta \equiv U\delta/\nu$ using the identities

$$\frac{U}{u_\tau} \equiv \sqrt{\frac{2}{c_f}} \quad \text{and} \quad \frac{\delta u_\tau}{\nu} \equiv \text{Re}_\delta \sqrt{\frac{c_f}{2}} \qquad \qquad (S3.121a,b)$$

Thus, Equation (S3.120) becomes

$$\sqrt{\frac{2}{c_f}} \approx 2.44 \ln\left(\text{Re}_\delta \sqrt{\frac{c_f}{2}}\right) + 5.0 \qquad \qquad (S3.122)$$

Curve fitting of data points for $c_f(\mathrm{Re}_\delta)$ yields the approximation

◆ $$c_f \approx 0.02\,\mathrm{Re}_\delta^{-1/6} \tag{S3.123}$$

Power-Law Approach

Using the 1/7 law originally developed for turbulent pipe flow, we write for a turbulent boundary layer

$$\frac{u}{U} = \left(\frac{y}{\delta}\right)^{1/7} \tag{S3.124}$$

Using (S3.124) and (S3.123) in (S3.118) we obtain with $\delta_2 = 7\delta/27$,

$$0.02\,\mathrm{Re}_\delta^{-1/6} = 2\frac{d}{dx}\left(\frac{7}{27}\delta\right) \tag{S3.125}$$

Now, either $\mathrm{Re}_\delta = U\delta/\nu$ is inserted into (S3.125) or $d\delta/dx$ is expressed as $d(\mathrm{Re}_x)$. In any case, integration of Equation (S3.125) yields with $\delta(x = 0) = 0$

◆ $$\left(\frac{\delta}{x}\right)_{\text{turb}} \approx \frac{0.16}{\mathrm{Re}_x^{1/7}} \tag{S3.126}$$

as compared with $(\delta/x)_{\text{lam}} \approx (4.96/\mathrm{Re}_x^{1/2})$ for Blasius flow.
    Furthermore

◆ $$(c_f)_{\text{turb}} \approx \frac{0.027}{\mathrm{Re}_x^{1/7}} \tag{S3.127}$$

from which

$$(\tau_w)_{\text{turb}} = \frac{0.0135\rho U^2}{\mathrm{Re}_x^{1/7}} \tag{S3.128}$$

Note: These turbulent flat plate results are approximations relying on both log-law and power-law turbulent velocity profiles and on curve fitting.   □

3.3.6   Turbulent Duct Flows

3.27   Calculate the friction velocity and the average velocity for air flow in a pipe ($R = 0.07$ m) when the measured centerline velocity is $U_0 = 5$m/s.

Review

Consider steady, incompressible fully developed turbulent flow in a slanted pipe (radius R, angle α, and length l) with constant (circular) cross-sectional area and wall roughness ε; the log-law is valid throughout. The appropriate integral balance equations are

(continuity) $Q = \bar{v}A = \text{const}$ $\hspace{3cm}$ (S3.129)

where $\bar{v}$ is constant and from the extended Bernoulli equation

(mech. energy balance) $h_f = \Delta z + \dfrac{\Delta p}{\rho g}$ $\hspace{2cm}$ (S3.130)

where

(momentum balance) $\Delta p \pi R^2 + \rho g(\pi R^2)\underbrace{l\sin\alpha}_{\Delta z} - \tau_w(2\pi R)l = 0$ $\hspace{1cm}$ (S3.131)

Combining the last two equations yields the pipe friction loss as

$$\blacklozenge \qquad h_f = \frac{2\tau_w}{\rho g} \frac{l}{R} \qquad\qquad\qquad (S3.132a)$$

In general, $\tau_w = \tau_w(\rho \bar{v}, \mu, R, \varepsilon)$ or from dimensional analysis

$$\frac{8\tau_w}{\rho \bar{v}^2} = f\left(\mathrm{Re}_d, \frac{\varepsilon}{d}\right) \qquad \text{where } d = 2R$$

so that

$$\blacklozenge \qquad h_f = f \frac{l}{d} \frac{\bar{v}^2}{2g} \qquad\qquad\qquad (S3.132b)$$

This is the Darcy–Weisbach equation, where the Darcy friction factor, $f = f(\mathrm{Re}_d, \varepsilon/d)$, is obtained from the Moody diagram. Recall that for *laminar* fully developed flow through smooth pipes, integration of the x-momentum equation yielded

$$\tau = \frac{r}{2} \frac{d}{dx}(p + \rho g z) \qquad\qquad\qquad (S3.133a)$$

or

$$\tau(r = R) = \tau_w = \mu \frac{du}{dr}\bigg|_{r=R} = \frac{2\mu u_{\max}}{R} = \frac{R}{2}\bigg|\frac{d}{dx}(p + \rho g z)\bigg| \qquad (S3.133b)$$

where $u_{\max} = 2u_{av}$ so that

$$f \equiv \frac{8\tau_w}{\rho u_{av}^2} = \frac{64\mu}{\rho u_{av}d} \qquad\qquad\qquad (S3.134a)$$

or

$$f_{\mathrm{lam}} = \frac{64}{\mathrm{Re}_d} \qquad\qquad\qquad (S3.134b)$$

For turbulent flow we employ the log-law in the form

$$\frac{u(r)}{u_\tau} \approx \frac{1}{\kappa} \ln \frac{(R - r)u_\tau}{\nu} + B \qquad\qquad\qquad (S3.135)$$

so that

$$\bar{v} \equiv u_{av} = \frac{Q}{A} = \frac{1}{\pi R^2} \int_0^R u_\tau \left[\frac{1}{\kappa} \ln \frac{(R - r)u_\tau}{\nu} + B\right] 2\pi r\, dr \qquad (S3.136a)$$

or

$$\frac{\bar{v}}{u_\tau} = \left(\frac{1}{\kappa} \ln \frac{Ru_\tau}{\nu} + B - \frac{3}{2\kappa}\right) \qquad\qquad\qquad (S3.136b)$$

Note that $\bar{v}/u_\tau \equiv (\rho \bar{v}^2/\tau_w)^{1/2} = (8/f)^{1/2}$ and that $(Ru_\tau/\nu) = (\bar{v}d/2\nu)(u_\tau/\bar{v}) = 1/2\mathrm{Re}_d$ $(f/8)^{1/2}$. Hence Equation (S3.136b) can be reformulated as an implicit correlation between the Reynolds number and the pipe friction factor for turbulent flow in *smooth* pipes, namely,

$$\blacklozenge \qquad f^{-\frac{1}{2}} = a \log(\mathrm{Re}_d f^{1/2}) - b \qquad\qquad\qquad (S3.137)$$

where $a = 2.0$ and $b = 0.8$ on the basis of measurements.

Equation (S3.137) for smooth pipes has been extended to accommodate also flow in rough pipes for $10 < \mathrm{Re}_d < 10^8$ and $10^{-6} < \varepsilon/d < 10^{-1}$ as

$$\blacklozenge \qquad f^{-1/2} = -2.0 \log((\varepsilon/d)/3.7 + 2.51/(\mathrm{Re}_d f^{1/2})) \qquad\qquad (S3.138)$$

Equation (S3.138) has been plotted and is widely known as the Moody chart for laminar, transitional, and turbulent flow in circular smooth or rough pipes (cf. Fig. 3.20).

Solution
In order to compute the friction velocity u_τ and the average velocity $\bar{v}$, we utilize Equation (S3.135) for $u(r = 0) = U_0$. Thus, we obtain with $\kappa = 0.41$ and $B = 5.0$

$$\frac{U_0}{u_\tau} = \frac{1}{0.41} \ln \frac{Ru_\tau}{v} + 5.0 \tag{S3.139}$$

from which via trial and error ($R = 0.07$ m and $v = 1.51 \times 10^{-5}$ m²/s)

$$u_\tau = 0.228 \text{ m/s}$$

or

$$\tau_w \equiv \rho u_\tau^2 = 0.062 \text{ kg/(ms}^2)$$

The average velocity $u_{av} = \bar{v} = Q/A$ is (cf. Eq. (S3.136b))

$$\bar{v} = 4.17 \text{ m/s}$$

Thus, the Reynolds number is $\text{Re}_d = \bar{v}d/v = 38.700$, which implies $f = 0.022$ for smooth pipes (cf. Eq. (S3.137)).   □

3.28 Consider steady turbulent flow with zero pressure gradient between parallel plates, $2h$ apart, where the upper one moves with a velocity U (cf. sketch).

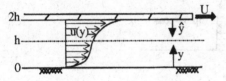

Find the friction factor $f = 2\tau_w h/(\mu U)$ as a function of $\text{Re}_h = 2Uh/v$. This problem has been suggested by White (1974).

Assumptions

- Steep (linear) velocity gradients near the walls (laminar sublayers) and high eddy viscosity around center
- $u(y = h) = U/2$, generating an "S-shaped" profile patched together with two log-profiles

Solution
For the given coordinate systems the turbulent velocity profiles read

$$\frac{U - u}{u_\tau} = \frac{1}{\kappa} \ln \frac{\hat{y}u_\tau}{v} + B \qquad \text{in } 0 \leq \hat{y} \leq h \tag{S3.140a}$$

and

$$\frac{u}{u_\tau} = \frac{1}{\kappa} \ln \frac{yu_\tau}{v} + B \qquad \text{in } 0 \leq y \leq h \tag{S3.140b}$$

Thus,

$$\frac{U}{2u_\tau} = \frac{1}{\kappa} \ln \frac{hu_\tau}{\nu} + B$$

or

$$\frac{U}{u_\tau} = \frac{2}{\kappa} \ln \left(\mathrm{Re}_h \frac{u_\tau}{U} \right) + 2B \tag{S3.141}$$

With $\tau_w = \rho u_\tau^2$, the given friction factor can be expressed as

$$f \equiv \frac{2\tau_w h}{\mu U} = 2\mathrm{Re}_h \left(\frac{u_\tau}{U} \right)^2 \tag{S3.142}$$

or

$$\frac{U}{u_\tau} = \left(\frac{\mathrm{Re}_h}{f} \right)^{1/2} \tag{S3.143}$$

Inserting (S3.143) into (S3.141) yields

$$\left(\frac{\mathrm{Re}_h}{f} \right)^{1/2} = \frac{2}{\kappa} \ln(\mathrm{Re}_h f)^{1/2} + 2B \tag{S3.144}$$

$\square$

3.29 Consider fully developed turbulent smooth pipe flow where the mean velocity distribution follows the log-law (cf. sketch). For a constant (axial) pressure gradient, find the radial location where the eddy viscosity is a maximum.

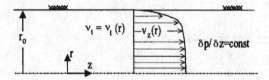

Assumptions
As stated.

Solution
With the given simplifications, the x-momentum equation reduces to

$$0 = -\frac{1}{\rho} \frac{\partial p}{\partial z} + \frac{1}{r} \frac{\partial}{\partial r} \left(r \frac{\tau_{rz}}{\rho} \right) \tag{S3.145}$$

where $\quad \tau_{rz} = \mu_t \dfrac{\partial v_z}{\partial r} \quad$ and $\quad -\dfrac{\partial p}{\partial z} \approx \dfrac{\Delta p}{L} = \mathrm{const}$

The log law can be written as

$$\frac{U_0 - v_z}{u_\tau} = \frac{1}{\kappa} \ln \frac{r_0 - r}{r_0} + B \tag{S3.146}$$

Thus, $v_z(r) = v_z(y)$ where $y = r_0 - r$.

Integrating Equation (S3.145) and using (S3.146) yield

$$Cr = \frac{\tau_{rz}}{\rho} \equiv v_t \frac{\partial v_z}{\partial r} := v_t f(r) = v_t A \frac{r_0}{r_0 - r} \tag{S3.147}$$

from which

♦
$$v_t = g(r) = D\frac{r_0 r - r^2}{r_0}$$

   or

$$\frac{dv_t}{dr} = g'(r_m) = F(r_0 - 2r_m) := 0 \qquad\qquad (S3.149)$$

Equation (S3.148) is satisfied for $r_m = r_0/2$; that is, the eddy viscosity in turbulent pipe flow has in this case a maximum exactly between the wall and the centerline.

Recommendation
Plot a representative eddy viscosity $v_t = g(r)$ for pipe flow.

Problem Assignments

Follow the format for problem solutions: sketch, assumptions, concept, postulates, reduced equations, plus boundary conditions, solution, results with graphs and comments.

3.1 In order to examine the compressibility effect of air (with $a = \sqrt{\gamma R T}$ and $\gamma = c_p/c_v \approx$ 1.4), the equations for the stagnation pressure

$$p_0 = p + 1/2\rho v^2 \qquad \text{for } incompressible \text{ flow}$$

and

$$p_0 = p[1 + (\gamma - 1/2\text{Ma}^2)]\gamma/(\gamma - 1) \qquad \text{for } compressible \text{ flow}$$

are rewritten in terms of $(p_0/p) = \text{fct(Ma)}$. Thus, for incompressible flow

$$\frac{p_0}{p} = 1 + \frac{\gamma}{2}\text{Ma}^2$$

and for compressible flow

$$\frac{p_0}{p} = 1 + \frac{\gamma}{2}\text{Ma}^2\left[1 + \frac{1}{4}\text{Ma}^2 + \frac{(2-\gamma)}{24}\text{Ma}^4 + \cdots\right]$$

(a) Derive the last two equations.
(b) Plot the pressure ratio versus the Mach number and determine the critical Mach number for which both curves (or equations) are still within 5 percent agreement.

3.2 An aircraft flies at Ma = 1.35 with a headwind U = 10 m/s at an altitude of $H = 3$ km. How long after the aircraft passes directly overhead does its sound reach a point on the ground?

3.3 Air enters a combustion chamber at $\text{Ma}_1 = 0.2$, $T_1 = 580$ K, and $p_1 = 1.0$ MPa (abs). With heat addition in the chamber, the exit conditions are $\text{Ma}_2 = 0.4$, $T_2 = 1727$ K, and $p_2 = 862.7$ kPa (abs). Calculate the change in specific entropy across the combustor and plot the process in a $T - s$ diagram.

3.4 It can be shown for isentropic flow that

$$\frac{d\rho}{\rho} = -\frac{\text{Ma}}{1 + [(\gamma - 1)/2]\text{Ma}^2}d\text{Ma}$$

(a) Considering Ma $= O(1)$, reduce this equation.

(b)   Considering Ma $\gg 1$, reduce this equation.

3.5   Consider incompressible, laminar boundary-layer flow with constant U. Assuming a velocity distribution of

$$u = U \left[\frac{2y}{\delta} - \frac{y^2}{\delta^2} \right]$$

determine the wall shearing stress $(\tau_0(x))$ and the boundary-layer thickness $\delta(x)$.

   Hint: A parabolic velocity profile is an approximation to a laminar boundary-layer profile.

3.6   Consider purely radial, steady, ideal fluid flow. Is the flow irrotational or not?
(a)   Determine the functional form of v_r.
(b)   Find the stream function and the equation for the streamlines.

3.7   Examine the nature of the flow given by the velocity potential $\phi = 7x + 2\ln r$.

3.8   Consider an incompressible flow characterized by the velocity field

$$v_r = \frac{A}{r} - B\cos\theta \quad \text{and} \quad v_\theta = \beta\sin\theta$$

Find the velocity potential, the stream function, and the pressure field.

3.9   For a given velocity field for steady, incompressible flow

$$\vec{v} = Ax\hat{i} - Ay\hat{j}$$

with $A = 4\ \text{s}^{-1}$, determine the stream function that will yield this velocity field.

3.10   Consider a porous flat plate of width $w = 1.5$ m and length $l = 2$ m where the constant approach velocity of water at 20°C is $u_\infty = 3$ m/s and the suction velocity is $v_w = -0.2$ mm/s. At $x = l$, the steady laminar boundary layer is measured to be $\delta_l = 1.5$ mm with a velocity distribution $u/u_\infty = 3(y/\delta) - 2(y/\delta)^{1.5}$. Evaluate the mass flow rate across the boundary-layer edge, that is, at $y = \delta_l$ and for $0 \le x \le l$.

3.11   A conical piston (radii r_1 & r_2 and length L) enters with velocity v_p a cylindrical hole (radius r_c and depth d) filled with oil of density ρ and viscosity μ. Obtain an expression for the velocity at which oil escapes from the hole as a function of vertical piston displacement.

3.12   Consider horizontal, axisymmetric slit flow of an ideal fluid that is pumped through two small center holes (from above and below at a combined flow rate of $Q_t = Q/2 + Q/2$ into the narrow gap. The fluid motion is assumed to be only radially dependent.
(a)   Derive an expression for the radial velocity as a function of distance from the holes.
(b)   Obtain an expression for the radial pressure variation.

3.13   Consider a disk of mass m that has been placed flat on a vertical jet where it is levitated to an equilibrium position, h_0. The circular jet has an exit velocity v_0, diameter d_0, and constant density ρ. Obtain a differential equation for the disk height $h(t)$ above the jet exit plane if the disk is released at $H = h(t = 0)$ and plot $h(t)$ for $h_0 \le h(t) \le H$.

3.14   Consider steady incompressible boundary-layer flow with nonzero pressure gradient. Plot the shear stress distribution $\tau(y)$ and the velocity profile $u(y)$ for (i) a favorable pressure gradient and (ii) the case of an adverse pressure gradient before, at, and after the point of flow separation.

3.15 Consider the transient 2-D stagnation flow discussed in Section 3.2.1.B. Derive Equations (P3.4)–(P3.6) and transform the associated boundary conditions.

3.16 Consider steady laminar flow near a rotating disk discussed in Section 3.2.1.B. Write a program or employ suitable software for the two-point BVP, that is, Equations (P3.8), (P3.10), (P3.12), and (P3.14). Check the "starting values" for $f'(0)$ and g.

3.17 Sketch a few streamlines for steady laminar 2-D boundary-layer flow past a horizontal flat plate and show mathematically why they move upward.

3.18 Consider viscous flow between two parallel infinite disks. Initially the fluid and the disks are at rest. At time $t = 0$, one disk is set in motion; its angular velocity is gradually increased to a value Ω and then held constant. Suction is also applied to the rotating disk at $t = 0$ and is gradually increased to a specified level, after which it is held constant.

  (a) Consider this flow after the angular velocity and suction have been held constant for a long time and obtain the set of coupled differential equations from the Navier–Stokes equation.

  (b) Solve the problem-oriented equations using available subroutines.

    Hint: Define the Reynolds number as $\mathrm{Re} = d^2\Omega/\nu$, where d is the disk spacing. Normalize distance, time, velocity, and pressure over density with respect to d, Ω^{-1}, Ωd, and $(\Omega d)^2$, respectively.

3.19 A plane rigid surface wetted with a thin layer of liquid of uniform thickness h_0 is held vertically and the liquid drains off. Show that the layer thickness h at distance x from the upper edge of the plate satisfies the approximate equation

$$\frac{\partial h}{\partial t} + V\frac{h^2}{h_0^2}\frac{\partial h}{\partial x} = 0$$

where $V = \rho g h_0^2/\mu$. Show that at time t after the draining begins

$$h = h_0\left(\frac{x}{Vt}\right)^{1/2} \qquad \text{for } x \leq Vt$$

and

$$h = h_0 \qquad \text{for } x \geq Vt$$

3.20 A circular disk of radius a is parallel to and at distance h from a rigid plane, and the space between them is occupied by fluid. The pressure at the edge of the disk is atmospheric.

  (a) Show that motion of the disk in the direction normal to the plane gives rise to a force on the disk in that direction equal to

$$F = -\frac{3\pi}{2}\frac{\mu a^4}{h^3}\frac{dh}{dt} \qquad \text{provided that } h \ll a \quad \text{and} \quad (\rho h/\mu)/(dh/dt) \ll 1$$

  (b) Show that the constant force F applied to the disk will put it well away from the plane in a time

$$t = \frac{3}{4}\pi\frac{\mu a^4}{h^2 F}$$

3.21 Consider a steady incompressible flow of a viscous fluid between two disks set a distance of $z = l$ apart along the axis. The lower disk, $z = 0$, is stationary while the upper disk is rotating at angular velocity P. Obtain a set of six first-order ordinary differential equations that are required for the solution of this flow problem.

3.22 A long cylinder of radius r_0 is being rotated about its axis with a constant angular speed Ω in an infinite body of viscous fluid of kinematic viscosity ν.

(a) Show that the steady motion induced in the fluid is an irrotational motion with

$$v_\theta = \frac{r_0^2 \Omega}{r}$$

(b) When the rotational motion of the cylinder is suddenly stopped, the fluid motion becomes unsteady. By assuming the radius r_0 of the cylinder as infinitesimally small and the time $t = 0$ at the instant when the rotational motion of the cylinder is stopped, show that the tangential velocity of the unsteady motion of the fluid at any instant t is

$$v_\theta = \frac{r_0^2 \Omega}{r}\left(1 - e^{-r^2/4vt}\right)$$

3.23 Laminar, steady-state flow of an incompressible Newtonian fluid between two coaxial cylinders is created as a result of the axial motion $u_z = U_0$ and rotation $u_\theta = \kappa R \omega_0$ of the inner cylinder.

(a) Employing cylindrical coordinates, develop an expression for $\vec{u}(r)$. Take the radii as κR and R, where κ is less than unity.

(b) Evaluate the volumetric flow rate.

(c) Linearize your answer (a) and obtain an approximate solution for (b).

3.24 Consider the axial flow of a power-law fluid (with $n = 1/3$) in an annulus where the inner cylinder radius is R and the gap width is b, which is much smaller than R. The flow is driven by a constant pressure gradient. Investigate the way in which the volumetric flow rate changes if the inner cylinder is made to rotate at a constant angular velocity ω.

 , Hint: Because of the small gap (thin annulus), curvature can be neglected. Note that the rate-of-strain tensor $\dot{\gamma}$ has more than one nonvanishing component since the fluid flows axially as a result of a pressure gradient and tangentially because of the rotating inner cylinder.

3.25 In order to obtain an estimate for salt water intrusion, solve the simplified problem in the sketch.

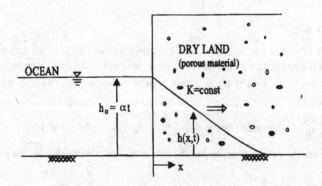

(a) Derive the tailored PDE for the piezometric head which reads

$$\frac{\partial}{\partial x}\left(h\frac{\partial h}{\partial x}\right) = \frac{\partial h}{\partial \tau}$$

with $\tau = Kt/S$, where $K = k\gamma/\mu$ is the hydraulic conductivity and S is the specific storativity.

Hint: Start with the continuity equation and invoke Darcy's law.
(b) State the initial and boundary conditions and employ

$$f(\xi) = \frac{h}{b^2 \tau}$$

where $\xi = x/(b\tau)$, $(b^2 = \alpha S/K)$, to arrive at the final ordinary differential equation.
(c) Postulate a simple trial solution for $f(\xi)$ and compute $h(x, t)$. Comment!

3.26 Consider radial flow generated by a line source at $r = 0$ into a wedge-shaped, semiinfinite channel (cf. sketch). Assume $v_\theta = 0$, $\eta = \theta/\alpha$, $f(\eta) = v_r/v_{max}$, and $Re = v_{max} r \alpha/\nu$.

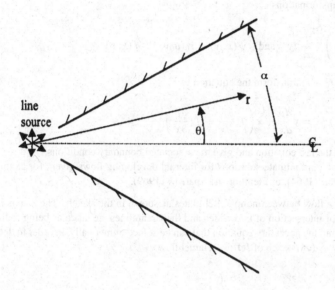

(a) Show that

$$f''' + 2Re\,\alpha f f' + 4\alpha^2 f' = 0$$

subject to

$$f(+1) = f(-1) = 0$$

and $f(0) = 1$ or $f'(0) = 0$ describes the fluid flow.
(b) Solve this equation for creeping flow, that is, $\alpha Re \ll 1$, and sketch a few velocity profiles.
(c) What types of special flows appear when $Re \ll 1$? Sketch profiles for different α's.

3.27 A sphere with a specific gravity ten times that of oil is dropped into a large container of oil. Calculate the terminal (sinking) velocity of the sphere and plot sphere velocity versus height for various release or initial velocities and sphere diameters.

3.28 A porous circular cylinder of radius R submerged in a special fluid rotates at a low angular velocity ω_0 and has a small radial wall-suction velocity of $-v_s = \text{const.}$
(a) Set up this problem for a "special fluid" that is a polymeric liquid that can be described by a power law.
(b) Find the effect of the velocity distribution v_θ on the dimensionless group $Re_w = v_s R/\nu$.

(c) Write a computer program for the solution of (a) and plot v_θ for different power-law viscosity indices n.

3.29 Consider steady laminar developing flow in a circular pipe of radius r_0, that is, the laminar hydrodynamic entry length problem. The governing equation is

$$\rho u \frac{\partial u}{\partial x} + \rho v \frac{\partial u}{\partial r} = -\frac{dp}{dx} + \frac{\mu}{r} \frac{\partial}{\partial r} \left(r \frac{\partial u}{\partial r} \right)$$

subject to $u = u_0 = $ const at $x = 0$ and $u = v = 0$ at $r = r_0$.

(a) Comment on this equation and sketch a few velocity profiles. Using the Falkner–Skan–Mangler transformations

$$d\eta = \left(\frac{u_0}{\nu x} \right)^{1/2} \frac{r}{r_0} dy \quad \text{and} \quad \psi(x, y) = r_0 (u_0 \nu x)^{1/2} f(x, \eta)$$

where $y = r_0 - r$, transform the equation to

$$(bf'')' + \frac{1}{2} ff'' = x \frac{dp}{dx} + x \left(f' \frac{\partial f'}{\partial x} - f'' \frac{\partial f}{\partial x} \right)$$

(b) Define b in the last equation and give the associated boundary conditions.
(c) Comment on approximate solutions for internal developing flows given, for example, by Sparrow et al. (1964), or Fleming and Sparrow (1969).

3.30 Consider the flow between nonparallel plates as shown in the sketch. The x-axis is taken as the line of intersection of two plates and the streamlines are taken as being radial from this axis. Find the necessary equation that can be solved numerically, in order to determine the velocity u_r (cf. sketch of Jeffrey–Hamel flow).

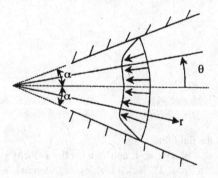

3.31 Suppose that one rigid plane is sliding steadily over another, with constant inclination angle $\theta_0 = 1/2\pi$. Fluid in the region between the planes is set in a two-dimensional motion (cf. sketch below).

(a) Determine the potential velocity field in the neighborhood of the corner. Sketch some of the streamlines $\psi =$ const.
(b) Solve for the velocity distribution and typical boundary-layer parameters; match the outer (potential) flow with the thin shear layer (TSL) flow.

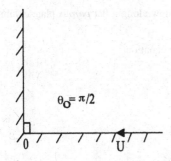

$\theta_0 = \pi/2$

3.32 Provide some representative streamlines for the laminar flow system given in the following sketch.

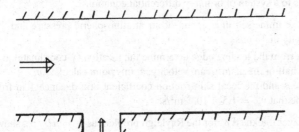

3.33 Consider steady laminar boundary-layer flow past a horizontal flat plate or a solid wedge.

(a) What does the BC $u''(y = 0) = 0$ physically indicate?

(b) Sketch the steps necessary to obtain $\delta(x)$ for a given $u/U\left(\eta = y/\delta\right)$ and $U(x)$.

(c) Justify the (compelling) trial solution for $u(x, y)$ of the form $u = U(x)f'(\eta); \eta = \eta(x, y)$ for the Falkner–Skan problem.

3.34 Assume a flat plate boundary-layer velocity profile of the form

$$\frac{u}{U} = \frac{3}{2}\left(\frac{y}{\delta}\right) - \frac{1}{2}\left(\frac{y}{\delta}\right)^3$$

and compute $(\delta/x\sqrt{Re_x})$, $(\delta_1/x\sqrt{Re_x})$, $(\delta_2/x\sqrt{Re_x})$, and $c_f\sqrt{Re_x}$. Compute the percentage errors when compared with the exact Blasius solutions.

3.35 Use the momentum integral relation and the velocity profile $u/U = \sin[(\pi/2)(y/\delta)]$ to evaluate the boundary-layer parameters δ, δ_1, δ_2, and τ_w for flow over a flat plate.

3.36 Set up the mathematical steps required to find the separation point, x_{sep}, for laminar flow past a flat plate with an assumed velocity distribution

$$U(x) = U_0(1 - x^6)$$

3.37 The outer flow of a specific boundary-layer problem was measured as follows. Estimate the skin friction coefficient at $x/L = 0.10$ and the point of separation x_{sep}/L.

x/L	0.50	0.40	0.30	0.20	0.05
U/U_0	0.50	0.60	0.70	0.80	0.95

3.38 For a flat plate boundary-layer problem, the proposed trial solution is $u(x, y) = f(x)(a\eta + b\eta^3); \eta = y/\delta$. Find the momentum boundary-layer thickness as a function of x for $a = 1.5$.

3.39 Determine the functional form $v(x)$ for laminar flow along a flat *porous* plate exhibiting similarity.

3.40 Given the boundary-layer equation and continuity equation

$$u\frac{\partial u}{\partial x} + v\frac{\partial u}{\partial y} = U\frac{dU}{dx} + v\frac{\partial^2 u}{\partial y^2} \quad \text{and} \quad \frac{\partial u}{\partial x} + \frac{\partial v}{\partial y} = 0$$

where $\quad U = U_0 e^{nx/L}$

show that the transformations

$$u = UF'(\eta) = U_0 e^{nx/L} F'(\eta), \qquad \eta = \frac{y}{L}e^{nx/(2L)}\left(\frac{v}{U_0 L}\right)^{1/2}$$

reduce the equations to a system of ordinary differential equations.

3.41 An ultrathin flat plate immersed in a stream of air at atmospheric pressure and 21°C is moving with a velocity of 15 m/s.

(a) At a distance 20 cm from the leading edge determine the location (y-coordinate) at which the local velocity is half of the mainstream velocity; at this point calculate v.

(b) Find the BL thickness and the local skin-friction coefficient at a distance 1 m from the leading edge; assume that $v_{air} = 1.5 \times 10^{-5}$ m²/s.

3.42 Defining the shear layer thickness δ_s for the Rayleigh problem as $\delta_s = \mu U/\tau_w$, show that

$$\delta_s = \sqrt{\pi}\sqrt{vt}$$

Define also the mean thickness δ_m as $\delta_m = \int_0^\infty (u/U)dy$ and show that

$$\delta_m = \frac{2}{\sqrt{\pi}}\sqrt{vt}$$

Indicate both δ_s and δ_m in a u–y diagram at a fixed time.

3.43 A plate moves in an infinite body of viscous fluid with the velocity $U = At^n$. Employing the similarity (trial) solution

$$u = t^n f(\eta), \qquad \text{where } \eta = y/\sqrt{t}$$

(a) Calculate the velocity distribution $u(y, t)$.

(b) Determine

$$\lim_{t \to 0} u(y, t) \text{ for } y = \begin{cases} \text{zero} \\ \text{nonzero} \end{cases}$$

(c) Evaluate the shear layer thickness $\delta = \mu U/\tau_w$.

(d) Sketch the results and comment.

3.44 For certain transient boundary-layer-type flows, where $u = u(y, t)$ only, similarity solutions exist.

(a) Show that a general class of nonsteady viscous flow problems can be described by

$$\frac{\partial u}{\partial t} + v_0 \frac{\partial u}{\partial y} = \frac{dU}{dt} + v\frac{\partial^2 u}{\partial y^2}$$

(b) Solve the problem of flow in a pipe starting from rest; that is, at $t = 0$ a constant pressure gradient is imposed.

3.45 Using an appropriate code for two-point boundary-value problems, solve the Falkner–Skan equation numerically for a particular value or values of β or $m \neq 0$.

$$f''' + f'' + \beta(1 - f'^2) = 0$$

subject to

$$f(0) = f'(0) = 0 \quad \text{and} \quad f'(\infty) = 1$$

    Note:

$$\beta = \frac{2m}{1+m}, \qquad f = f(\eta); \qquad \eta = y\sqrt{\frac{m+1}{2}\frac{U(x)}{\nu x}} \quad \text{and} \quad U(x) = Kx^m$$

Plot streamlines and a few profiles (u and v) at various x-stations for different exponents m.

3.46 Consider steady turbulent flow with zero pressure gradient between parallel plates where the upper one moves with a velocity U_0.
  (a) Find the centerline velocity $u = u(y = h)$ and the friction factor $f = (2\tau_w h/\mu u_0)$ as a function of $\text{Re}_h = 2U_0 h\nu$, where h is the half-distance between the plates.
  (b) Draw an *accurate* velocity profile $u(y)$ for $\text{Re}_h = 0(10^6)$.

3.47 A tractor and trailer is traveling at 70 mph in otherwise still air at 75°F. Determine the viscous wall-layer thickness and the boundary-layer thickness on the roof of the trailer, 8 ft behind the leading edge. Assume the boundary layer has a zero thickness at the leading edge of the roof. Neglect any pressure gradient effects. What happens if the assumption that the boundary layer has no thickness is withdrawn, and what happens if pressure gradient effects are evaluated?

3.48 Analyze turbulent flow about a rotating cylinder in an infinite fluid. Sketch the system and derive for steady flow an expression for $c_f = 2\tau_w/\rho r_0^2 \omega^2$ as a function of $\text{Re} = \omega r_0^2/\nu$, where $\omega = \text{const}$ and r_0 is the radius.
    Hint: (Cf. White 1974) Neglect transverse curvature effects and assume the 2-D log law to be valid near the cylinder wall with $y = r - r_0$ the wall coordinate. In order to estimate the "penetration depth" δ, approximate δ as the thickness on a flat plate at a point x equal to the circumference of the cylinder, $2\pi r_0$.
    Compare your c_f-value for $\text{Re} = 10^6$ with an empirical formula that yields $c_f(\text{Re} = 10^6) = 0.00288$.

3.49 Analyze the flow through a smooth concentric circular annulus of inner radius R_2 and outer radius R_1, using the logarithmic law, and assuming that the maximum velocity occurs at midradius $(R_1 + R_2)/2$ of the annulus.
  (a) Find the velocity distribution $u(r)$ and use it to find the friction coefficient f.
  (b) Plot a velocity profile in the annular space and indicate why, in reality, it is asymmetric.

3.50 Analyze the turbulent steady flow between parallel plates, using the logarithmic law. Assume that the pressure gradient is zero and the upper plate is moving at speed v. Sketch the profile for $\text{Re} = vh/\nu = 10^6$, where h is the distance between the plates. Compute the dimensionless ratio $(\tau_0 h/\mu v)$ for this condition.

3.51 A thin circular disk with radius R is immersed in a large body of fluid with density ρ and viscosity μ. If a torque T is required to rotate the disk at an angular velocity ω_0 one could define a resistant force F_r as the product of a characteristic area A, a characteristic kinetic energy per unit volume E_{kin}, and a dimensionless friction factor f, namely,

$$\frac{T}{R} = F_r = A E_{kin} f$$

where in our case $A = 2\pi R^2$, $E_{kin} = (\rho/2)R^2\omega_0^2$. Derive a relationship between f and Re ($\mathrm{Re} = R^2\omega_0/\nu$) for *turbulent* boundary-layer flow, using the $1/7$ power-law velocity distribution.

3.52 For a turbulent flow in a round pipe, the shear stress τ_{rz} varies linearly from zero at the pipe axis to a maximum value τ_0 at the pipe wall. Thus we may write

$$\tau_{rz} = \tau_0\left(1 - \frac{y}{r_0}\right)$$

When this expression is equated to Von Karman's eddy shear stress formula, we have

$$\tau_0 = \left(1 - \frac{y}{r_0}\right) = \rho k^2 \frac{(d\bar{u}/dy)^4}{(d^2\bar{u}/dy^2)^2}$$

Starting with this equation, show that Von Karman's universal velocity distribution equation for turbulent pipe flow is

$$\bar{u} = \bar{u}_{max} + \frac{\sqrt{\tau_0/\rho}}{k}\left[\sqrt{\frac{1-y}{r_0}} + \ln\left(1 - \sqrt{\frac{1-y}{r_0}}\right)\right]$$

3.53 Consider turbulent thin film flow on an inclined plate of angle θ and film thickness b.
(a) Show that the shear stress τ_{yx} in the flow at a point that is y above the plane surface can be written as

$$\tau_{yx} = \tau_0\left(1 - \frac{y}{b}\right) = \gamma(b - y)\sin\theta$$

where τ_0 is the shear stress at the plane surface and γ is the specific weight of the liquid.
(b) Equate the preceding expression to Prandtl's mixing length formula to obtain

$$\gamma(b - y)\sin\theta = \rho k^2 y^2\left(\frac{d\bar{u}}{dy}\right)^2$$

and show that the equation for the time-average turbulent velocity distribution for this flow is given by

$$\bar{u} = \bar{u}_{max} + \frac{1}{k}\sqrt{\frac{\gamma b\sin\theta}{\rho}}\left(2\sqrt{1 - \frac{y}{b}}\right) + \ln\frac{\sqrt{b} - \sqrt{b-y}}{\sqrt{b} + \sqrt{b-y}}$$

3.54 Consider fiber stretching. Review Problem 3.14c in Sction 3.3 and solve for $u(x)$, calculate u_L/u_0, and find $R(x)$ when
(a) only viscous forces are important;
(b) viscous and surface tension forces are important.

A Typical Third Homework Set

3.1 Considering the "viscous pump" problem, compute the total moment required to spin the disk (ω_0, r_0).

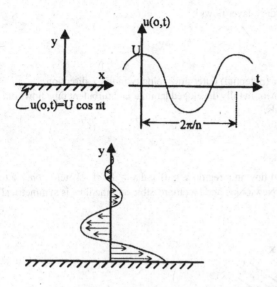

u(o,t)

y

x

u(o,t)=U cos nt

U

t

2π/n

y

3.2 Consider flow near an oscillating plate (i.e., Stokes's second problem) sketched here.

(a) Show that the problem solution is $u(y, t) = Ue^{-ky} \cos(nt - ky)$ if $k = \sqrt{n/2v}$.

(b) Setting $\eta \equiv ky$, plot u/U versus η for a few (nt)-values and comment.

(c) What other problem(s) can be described by the transient 1-D diffusion equation?

3.3 Evaluate the "curvature effect" in cylindrical Couette flow when compared to planar flow (cf. Sample-Problem 3.8 in Sect. 3.3.3).

3.4 Study Sample-Problem Solution 3.11 and plot the velocity vector field or $v_\theta(r)$-profiles for different values of the parameter c.

3.5–10 Solve Sample-Problems 3.10, 3.13, 3.22, 3.26, 3.33, and 3.34 and plot the results in each case.

Fourth Homework Set for Final Exam Preparation

Part A: Proofs, Questions, etc.

1. Prove that normal stresses of incompressible Newtonian fluids are zero at solid surfaces; that is, show that $\tau_{zz}|_{z=0} = 0$ where the z-axis is normal to the wall.

2. Discuss a stress vector versus a stress tensor; sketch and explain!

3. State the mathematical conditions for a fluid flow field in terms of:

    (a) time dependence   (e) type of fluid

    (b) dimensionality   (f) flow regime

    (c) compressibility   (g) system symmetry

    (d) rotationality   (h) directionality

    and explain.

4. **(a)** How are the Navier–Stokes equation and Newton's second law of motion related?

    (b) Is the term $\rho(\vec{v} \cdot \nabla)\vec{v}$ a type of force?

    (c) "Apply" (a) and (b) to Couette flow with a positive, that is, adverse, pressure gradient in order to draw two typical velocity profiles.

5. (a) Show that for boundary-layer flows

$$-\frac{1}{\rho}\frac{\partial p}{\partial x} = \frac{\partial U}{\partial t} + U\frac{\partial U}{\partial x}$$

where $U = U(x, t)$ is the (potential) outer flow velocity in the x-direction.

(b) Show that $\tau(r)$ is linear in fully developed laminar and turbulent pipe flows of incompressible Newtonian fluids.

Part B: Problems

6. Consider "creeping" fluid flow in a region $x \geq 0$ and $y \geq 0$ (cf. sketch) *above* a thin slit or line sink. The fluid is Newtonian and incompressible and the flow is symmetrical about the xz-plane.

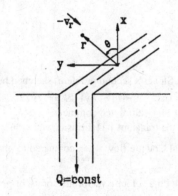

(a) Employing polar coordinates, show that $\vec{v} = [f(\theta)/r, 0, 0]$ and determine the associated boundary conditions, using symmetry, that is, $0 \leq \theta \leq 90°$.

(b) Reduce the r- and θ-components of the momentum equation, cross-differentiate both equations, and show that this leads to

$$\frac{d^3 f}{d\theta^3} + 4\frac{df}{d\theta} = 0$$

(c) Check the solution, $f = K\cos^2\theta$, to the third-order ODE and determine $v_r(Q, W; r, \theta)$.

Note:

$$\sin 2x = 2\sin x \cos x \quad \text{and} \quad \int \cos^2 ax\, dx = \frac{x}{2} + \frac{1}{4a}\sin 2ax$$

7. Consider steady laminar pipe flow of a non-Newtonian fluid (i.e., a plastic) obeying

$$\tau = \tau_0 + \mu_p\dot{\gamma}, \qquad \text{if } \tau > \tau_0$$

however,

if $\tau \leq \tau_0$, $\dot{\gamma} = 0$; that is, the fluid moves as "plug" flow.

(a) Plot $\tau(\dot{\gamma})$ and a representative velocity profile plus $\tau(r)$ for $0 < r < a$.

(b) For a horizontal circular pipe of radius a with a constant pressure gradient, find the axial velocity profile $v_z = v_z(\Delta p/l, \tau_0, \mu_p, a; r)$.

(c) Compute the volumetric flow rate Q and check whether Q collapses to

$$Q_{\text{Hagen-Poiseuille}} = \pi a^4 \frac{\left(\frac{\Delta p}{l}\right)}{(8\mu_p)}$$

when $\tau_0 = 0$.

8. Water $(T = 20°C, \rho = 998 \text{ kg/m}^3, \nu = 1.004 \times 10^{-6} \text{ (m}^2 \text{ /s)})$ flows fully developed through a horizontal pipe of 0.1-m diameter at $Q = 4 \times 10^{-2}$ m³/s and with a pressure drop of $\Delta p/l = 2.59$ kPa/m.

   (a) Determine the ratio $\tau_{\text{turb}}/\tau_{\text{lam}}$ at the midpoint $r_1 = 0.025$ m;

   (b) Determine the approximate thickness of the viscous sublayer, δ_s, provided that the pipe flow is turbulent;

   (c) Calculate the centerline velocity v_c and compare it with v_{av}.

9. Find the location of the maximum eddy viscosity, ν_{turb}, for steady fully developed turbulent flow in a smooth pipe described by

$$\frac{u_0 - \bar{u}(r)}{u_\tau} = \frac{1}{\kappa}\left(\log \frac{R - r}{R} + B\right)$$

where the constant parameter u_0 is the centerline velocity, u_τ is the friction velocity, and R is the pipe radius.

   (a) State the reduced equation of motion.

   (b) Using Boussinesq's EVM in the form $\tau = \rho \nu_t (d\bar{u}/dr)$, develop an expression for $\nu_t = \nu_t(r)$ and draw a profile.

   (c) Determine the radial location where ν_t is a maximum and comment.

10. A suitable time-averaged velocity profile for turbulent boundary layers with constant outer flow may be

$$\frac{\bar{u}}{u_\infty} = \left(\frac{y}{\delta}\right)^{1/7}; \qquad \frac{y}{\delta} = \eta(x, y)$$

Find $\delta(x)$ in order to obtain $u = u(x, y)$.

   Note, τ_w cannot be evaluated with the present form of the 1/7-law. Hence, the Blasius correlation (originally developed for pipe flow)

$$u^* \equiv \frac{\bar{u}}{u_\tau} = 8.74(y^+)^{1/7}; \qquad y^+ = \frac{y u_\tau}{\nu}$$

is evaluated along the boundary-layer edge to obtain $u_\tau \equiv (\tau_w/\rho)^{1/2}$ and then τ_w.

References and Further Reading Material

Abramovich, G. N. 1963. *The Theory of Turbulent Jets.* MIT Press, Cambridge, MA.

Anderson, D. A., J. C. Tannehill, and R. H. Pletcher. 1984. *Computational Fluid Mechanics and Heat Transfer.* McGraw-Hill, New York.

Anderson, J. D. 1990. *Modern Compressible Flow.* McGraw-Hill, New York.

Bird, R. B., R. C. Armstrong, and O. Hassager. 1987. *Dynamics of Polymeric Liquids*, Vol. 1, 2nd ed. Wiley-Interscience, New York.

Blasius, H. 1906. "Grenzschichten in Flüssigkeiten mit kleiner Reibung," *Z. Angew. Math. Phys.*, **56**: 1–37.

Boussinesq, J. 1877. *Mem. Pres. Acad. Sci.*, **23**, 46.

Bradshaw, P., T. Cebeci, and J. H. Whitelaw. 1981. *Engineering Calculation Methods for Turbulent Flows*. Academic Press, New York.

Cebeci, T., and P. Bradshaw. 1977. *Momentum Transfer in Boundary Layers*. Hemisphere/P. McGraw-Hill, Washington, DC.

Cebeci, T., and P. Bradshaw. 1984. *Physical and Computational Aspects of Convective Heat Transfer*. Springer-Verlag, New York.

Cebeci, T., and P. Bradshaw. 1988. *Physical and Computational Aspects of Convective Heat Transfer*. Springer-Verlag, New York.

Cebeci, T., and A. M. O. Smith, 1974. *Analysis of Turbulent Boundary Layers*. Academic Press, New York.

Churchill, S. W. 1988. *Viscous Flows: The Practical Use of Theory*. Butterworth-Heineman, Boston.

Clauser, F. H. 1954. "Turbulent Boundary Layers in Adverse Pressure Gradients," *J. Aeronaut. Sci.*, **21**, 91–108.

Colebrook, C. F. 1938–1939. "Turbulent Flow in Pipes, with Particular Reference to the Transition between the Smooth and Rough Pipe Laws," *J. Inst. Civ. Eng. Lond.*, 11, pp. 133–156.

Cvetkovic, V. D. 1986. "Continuum Approach to High Velocity Flow in a Porous Medium," *Transp. Porous Med.*, **1**(1), 63–98.

Darcy, H. 1856. In chapter 4 of Bear, J. (1979). *Hydraulics of Ground Water*. McGraw-Hill, New York.

Dumas, R., and L. Fulachier, eds. 1983. *Structure of Complex Turbulent Shear Flow*. Springer-Verlag, Berlin.

Durst, F. 1979. *Turbulent Shear Flows I*. Springer-Verlag, New York.

Falkner, V. M., and S. W. Skan. 1931. "Some Approximate Solutions of the Boundary-Layer Equations," *Phil. Mag.*, **12**(7), 865–896.

Fleming, D. P., and E. M. Sparrow. 1969. *Journal of Heat Transfer*, **91**, 345–354.

Fox, R. W., and A. T. McDonald. 1985. *Introduction to Fluid Mechanics*, 3rd ed. John Wiley & Sons, New York.

George, W. K., and R. Arndt. 1989. *Advances in Turbulence*. Hemisphere, New York.

Gerhart, P. H., R. J. Gross, and J. I. Hochstein. 1992. *Fundamentals of Fluid Mechanics*. Addison Wesley, Reading, MA.

Granville, P. S. 1990. "A Near-Wall Eddy Viscosity Formula for Turbulent Boundary Layers," *J. Fluids Eng.*, **112**(2), 240–243.

Hamrock, B. J. 1994. *Fundamentals of Fluid Film Lubrication*. McGraw-Hill, New York.

Hinze, J. O. 1975. *Turbulence*. McGraw-Hill, New York.

Jones, W. P., and Launder, B. E. 1972. *Int. J. Heat Mass Transfer*, **15**, 301.

Kolmogorov, A. N. 1942. *Izv. Akad, Nauk S.S.S.R., Ser. Fiz.*, **6**, 56.

Landahl, M. T., and E. Mollo-Christensen, 1992. *Turbulence and Random Processes in Fluid Mechanics*. Cambridge University Press, New York.

Landau, L. D., and Lifshitz, E. M. 1987. *Fluid Mechanics*. 2nd ed. Pergamon Press, Elmsford, New York.

Laufer, J. 1954. "The Structure of Turbulence in Fully Developed Pipe Flow," *NACA Rep.* 1174.

Lesieur, M. 1990. *Turbulence in Fluids*, 2nd ed. Kluwer Academic, Publishers, Dordrecht, The Netherlands.

Lu, P.-C. 1977. *Introduction to the Mechanics of Viscous Fluids*. McGraw-Hill, New York.

Macosko, C. W. 1994. *Rheology: Principles, Measurements, and Applications*. VCH Publishers, New York.

Merrill, E. W. 1968. "Rheology of Blood," *Physical Rev.*, **49**, 863–888.

Middleman, S. 1977. *Fundamentals of Polymer Processing.* McGraw-Hill, New York.

Moody, L. F. 1944. "Friction Factors for Pipe Flow," *ASME Trans.*, **66**, 671–684.

Nunn, R. H. 1989. *Intermediate Fluid Mechanics.* Hemisphere, New York.

Panton, R. L. 1984. *Incompressible Flow.* Wiley, New York.

Papanastasiou, T. C. 1994. *Applied Fluid Mechanics.* RTR Prentice-Hall, Englewood Cliffs, NJ.

Perry, R. H., D. Green, and J. O. Maloney, eds. 1984. *Perry's Chemical Engineering Handbook.* McGraw-Hill, New York.

Prandtl, L. 1921. "Bemerkungüber die Entstehung der Turbulenz," *Z. Angew. Math Mech.*, 1, 431–436.

Reynolds, W. C. 1976. "Computation of Turbulent Flows." *Annu. Rev. Fluid Mech.*, **8**, 183.

Rodi, W. 1980. *Turbulence Models and Their Application in Hydraulics.* IAHR, Delft, The Netherlands.

Rodi, W. 1984. *Turbulence Models and Their Applications in Hydraulics.* Brookfield, Brookfield, VT.

Schlichting, H. 1979. *Boundary-Layer Theory.* McGraw-Hill, New York.

Schreier, S. 1982. *Compressible Flow.* Wiley, New York.

Spalding, D. G. 1961. "A Single Formula for the Law of the Wall," *J. Appl. Mech.*, **28**, 455–457.

Sparrow, E. M., S. H. Lin, and T. S. Lundgren. 1964. *Physics of Fluids*, **7**, 338–347. Also, *International Journal of Heat and Mass Transfer*, **7**, 583–585.

Tanner, R. I. 1985. *Engineering Rheology.* Oxford University Press, New York.

Tennekes, H., and J. L. Lumley. 1972. *A First Course in Turbulence.* MIT Press, Cambridge, MA.

Thwaites, B. 1949. "Approximate Calculation of the Laminar Boundary Layer," *Aeronaut. Quarterly*, **18**, 245–280.

Van Driest, E. R. 1951. "Turbulent Boundary Layers in Compressible Fluids," J. *Aeronaut. Sci.*, **18**, 145–160.

Van Driest, E. R. 1956. *J. Aeronaut. Sci.*, **23**, 1007.

Warsi, Z. U. A. 1993. *Fluid Dynamics.* CRC Press, Boca Raton, FL.

Whitaker, S. 1986. "Flow in Porous Media. I. A. Theoretical Derivation of Darcy's Law," *Transp. Porous Med.*, **1**(1), 3–26.

White, F. M. 1991. *Viscous Fluid Flow.* McGraw-Hill, New York.

White, F. M. 1986. *Fluid Mechanics*, 2nd ed. McGraw-Hill, New York.

Wylie, C. R. and L. C. Barrett. 1982. *Advanced Engineering Mathematics.* McGraw-Hill, New York.

CHAPTER 4

Nonisothermal Flows

Fluid flows with temperature gradients due to internal heating or heated/cooled walls are examples of thermal flows, which form an integral part of convection heat transfer. Internal heating may result from fluid friction (viscous dissipation), irradiation (thermal radiation), and/or chemical reactions (reactive flows). When the fluid temperature differs from the wall temperature, the thermal wall conditions are commonly expressed as $T_w = $ const (i.e., isothermal wall) or $q_w = $ const (i.e., constant wall heat flux). *Convection heat transfer* is heat conduction, that is, an energy diffusion process, in a moving fluid. Combining heat conduction in solids with thermal convection in a fluid is called a *conjugate heat transfer problem* as it might occur in heat pipes, heat exchangers, fin cooling, tribology, porous media flow, and so forth. *Mixed thermal convection*, that is, simultaneous free and forced convection heat transfer, has to be considered when the buoyancy force is of an order of magnitude comparable to the inertia force. In free and in mixed convection problems a heat source or sink affects the fluid density, typically near a heated or cooled wall, and thus the momentum equation depends via the body force term on the heat transfer equation. Such a two-way coupling may also occur when other fluid properties are temperature-dependent.

Nonisothermal flow problems are typically subdivided into external flows and internal flows where the flow regime is either laminar or turbulent, and the flow is single phase or multiphase. Furthermore, depending upon how the fluid motion is induced, one considers *natural or free convection* when buoyancy is the driving force, and *forced convection* when a mechanical pressure gradient causes fluid flow.

Primary applications of heat transfer research leading directly to "products" include residential/commercial heating or air conditioning, heat exchangers, combustion processes, and noncombustion thermal power; food processing, agricultural drying, phase-change processes, polymer manufacturing; and other thermal processing techniques. Secondary applications are cases in which heat transfer research is needed to overcome thermal obstacles, such as cooling of electronic equipment, engines, and other propulsion systems; space vehicle reentry, nuclear reactors, and environmental heat dissipation for power plants. Figure 4.1 gives an overview of the major aspects and foundations of convection heat transfer (CHT).

A   Thermal Energy Equation

In order to solve convection heat transfer problems, the equation of motion has to be supplemented with the energy equation (cf. Sects. 2.2 and 2.3). In terms of the principle of energy conservation for a closed system (cf. Sect. 1.2.3), the first law of thermodynamics, $dE = \delta Q + \delta W$, can be expressed on a macroscale as

$$E_{\text{total}} = \rho \left(\hat{u} + \frac{v^2}{2} - \vec{g} \cdot \vec{r} \right) = Q + W \tag{4.1}$$

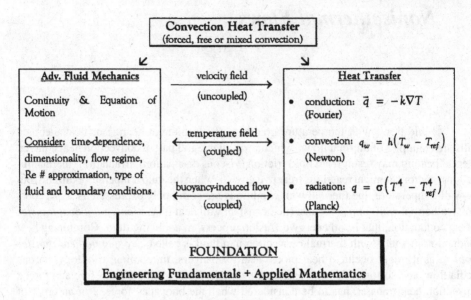

Fig. 4.1. Schematic of CHT fundamentals.

Here, $\hat{u}$ is the internal energy per unit mass, $v^2/2$ is the kinetic energy per unit mass, $\vec{r}$ is the displacement vector of fluid elements, Q is the added heat per unit volume, and W is the work done by the surroundings on the system (e.g., $\delta W = p\,d\forall$, where δ is an "inexact differential" since there is no function of the thermodynamic coordinates representing the "work in a body"). The energy balance (Eq. 4.1) is conveniently written in a time-rate-of-change form following the closed system of identifiable mass, using the material or Stokes derivative D/Dt (cf. Sect. 1.2.1).

$$\frac{DE}{Dt} = \rho\left(\frac{D\hat{u}}{Dt} + \vec{v}\cdot\frac{D\vec{v}}{Dt} - \vec{g}\cdot\vec{v}\right) = \frac{DQ}{Dt} + \frac{DW}{Dt} \tag{4.2}$$

Now we have to find some suitable engineering expressions for the RHS of (4.2) and express the energy balance in terms of Eulerian coordinates, that is, for an open system or control volume, which is typically fixed. An energy balance, considering heat transferred by conduction and work done by fluid flow, yields (cf. Sect. 4.3)

(Rate of heat transferred) $\qquad \dfrac{DQ}{Dt} = -\nabla\cdot\vec{q} \tag{4.3a}$

and

(Rate of work performed) $\qquad \dfrac{DW}{Dt} = -\nabla\cdot\vec{w} \tag{4.3b}$

where $\vec{w} = \vec{v}\cdot\vec{\vec{\pi}}$ with $\vec{\vec{\pi}} = -p\vec{\vec{\delta}} + \vec{\vec{\tau}}$ reflecting the work done, in terms of a flowing stream of a real fluid. While Equation (4.3a) is directly accessible using, for example, Fourier's law $\vec{q} = -k\nabla T$, Equation (4.3b) requires some consideration:

$$\nabla \cdot \vec{w} = \nabla \cdot (\vec{v} \cdot \vec{\pi}) = \vec{v} \cdot (\nabla \cdot \vec{\pi}) + \vec{\pi} \cdot (\nabla \cdot \vec{v}) \tag{4.4}$$

Using the momentum equation

$$\nabla \cdot \vec{\pi} = \rho \left(\frac{D\vec{v}}{Dt} - \vec{g} \right)$$

one term in Eq. (4.4) can be expressed as

$$\vec{v} \cdot (\nabla \cdot \vec{\pi}) = \rho \left(\vec{v} \cdot \frac{D\vec{v}}{Dt} - \vec{g} \cdot \vec{v} \right) \tag{4.5}$$

Clearly, inclusion of the "flowing stream" work (cf. Eq. 4.5) transforms the closed-system energy balance (cf. Eq. 4.2) to an open system, that is, a Eulerian formulation. The second term on the RHS of (4.4) can be split into a pressure term and a viscous dissipation term, via

$$\vec{\pi} \cdot (\nabla \cdot \vec{v}) \equiv \pi_{ij} \frac{\partial u_i}{\partial x_j} = \phi - p(\nabla \cdot \vec{v}) \tag{4.6}$$

where $\phi = \tau_{ij}(\partial u_i / \partial x_j)$ is a viscous dissipation function and the term $p(\nabla \cdot \vec{v})$ can be rewritten using the continuity equation. For compressible flow

$$p(\nabla \cdot \vec{v}) = -\frac{p}{\rho} \frac{D\rho}{Dt} \equiv \rho \frac{D}{Dt} \left(\frac{p}{\rho} \right) - \frac{Dp}{Dt} \tag{4.7}$$

so that, using Equations (4.3) through (4.7), Equation (4.2) can be expressed as

$$\rho \frac{D}{Dt} \left(\hat{u} + \frac{p}{\rho} \right) = \frac{Dp}{Dt} - \nabla \cdot \vec{q} + \phi \tag{4.8}$$

Recall that $\hat{u} + (p/\rho) = \hat{h}$ is the fluid enthalpy, so that with a negligible Dp/Dt term Equation (4.8) is rewritten as

$$\rho c_p \frac{DT}{Dt} = -\nabla \cdot \vec{q} + \phi \tag{4.9}$$

In general, $\vec{q} = \vec{q}_{\text{laminar}} + \vec{q}_{\text{turbulent}}$ where $\vec{q}_{\text{turb}} \approx \rho c_p \overline{v'T'}$ in most CHT problems. Using Fourier's law, Equation (4.9) becomes for constant-fluid-property CHT (cf. Sect. 2.2)

$$\blacklozenge \qquad \rho c_p \left[\frac{\partial T}{\partial t} + (\vec{v} \cdot \nabla)T \right] = k\nabla^2 T + \phi \tag{4.10}$$

For example, in rectangular coordinates with added source terms due to radiation, chemical reaction, and so on, Equation (4.10) reads

$$
\underbrace{\frac{\partial T}{\partial t}}_{\substack{\text{local/transient} \\ \text{temperature} \\ \text{change}}} + \underbrace{u\frac{\partial T}{\partial x} + v\frac{\partial T}{\partial y} + w\frac{\partial T}{\partial z}}_{\substack{\text{net heat convection} \\ \text{due to fluid motion}}} = \underbrace{\alpha\left(\frac{\partial^2 T}{\partial x^2} + \frac{\partial^2 T}{\partial y^2} + \frac{\partial^2 T}{\partial z^2}\right)}_{\substack{\text{net heat conduction} \\ \text{or} \\ \text{thermal diffusion}}} + \underbrace{\phi \pm \sum S_T}_{\substack{\text{heat sinks} \\ \text{or sources}}} \quad (4.11)
$$

Note that for $\alpha = k/(\rho c_p)$, the thermal diffusivity has the same units as D_m, the species diffusivity, or ν, the momentum diffusivity, also known as the kinematic viscosity. Alternative derivations of the thermal energy equation can be found in Sections 2.1, 2.2, and 4.3.

The initial condition specifies the temperature distribution throughout the system:

$$
IC \qquad T(t = 0) = f(x, y, z) \tag{4.12}
$$

The boundary conditions may include isothermal wall or constant wall heat flux, and symmetry, if applicable:

$$
BC \qquad T_w = \text{const} \quad \text{or} \quad T_{\text{wall}} = T_{\text{fluid}}
$$

$$
q_w = \text{const} \quad \text{or} \quad \left(\frac{\partial T}{\partial n}\right)_w = -\frac{1}{k}q_w \tag{4.13a–c}
$$

or

$$
\left(\frac{\partial T}{\partial n}\right)_w = -\frac{h}{k}(T_w - T_{f\infty})
$$

where $T_{f\infty}$ is a reference temperature of the fluid; for example, it is T_{mean} in pipe flows and the external temperature, T_∞ or T_{edge}, in boundary-layer flows. For solids or quiescent fluids, $\vec{v} = \phi = 0$, and Equation (4.10) reduces to the transient heat conduction equation with internal heat sources/sinks. Whereas Equation (4.11) requires the velocity field and a submodel for S_T as input, the equation of motion can be solved independently of the heat transfer equation if the fluid properties are temperature-independent or at least constant w.r.t. a representative temperature.

In general, (two-way) coupling terms that are temperature-dependent may appear in the momentum equation as mentioned earlier. Examples include the following:

- Free convection, in which, on the basis of the Boussinesq assumption, $\rho = \rho(T)$ only in the body force term:

$$
\frac{\vec{F}_{\text{body}}}{\forall} = \underbrace{\rho_0 \beta(T - T_0)\vec{g}}_{\rho(T)} \tag{4.14}
$$

where $\beta = (1/\forall)(\partial \forall/\partial T)_p$ is the volumetric expansion coefficient (cf. Sect. 4.3);

- Temperature dependence of fluid properties, such as the kinematic viscosity; for example,

$$
\nu(T) = \nu_0 + \alpha_1 \Delta T + \alpha_2 \Delta T^2 + \cdots \quad \text{or} \quad k(T) = k_0 + a_1 T + a_2 T^2 + \cdots
$$

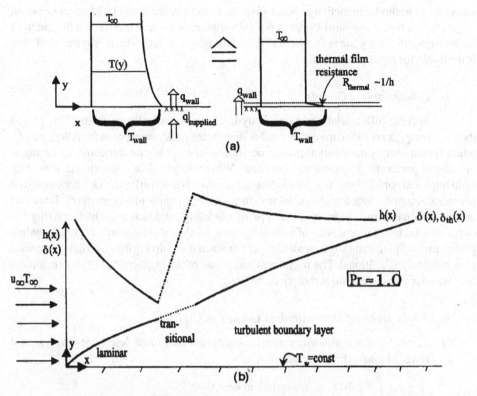

Fig. 4.2. (a) Schematics for Fourier's law and Newton's law of cooling. (b) Boundary-layer thickness and heat transfer coefficient along flat plate.

Heat Transfer Coefficient

The practical goals in convection heat transfer (CHT) are to find the velocity/temperature profiles and to calculate the heat transferred in terms of wall heat flux, heat transfer coefficient, Nusselt number, or Stanton number. Knowing the wall heat flux, heating, cooling, and heat loss problems can be solved. For example, consider steady two-dimensional thermal boundary-layer flow. The objective is to compute $T(x, y) \rightarrow h(x) \rightarrow q_{\text{wall}}$ or $q_{\text{wall}}(\Delta T)$ directly.

At the wall we have (Fig. 4.2a)

Fourier's law   Newton's law of cooling

$$q_w = \underbrace{-k_{\text{fluid}} \frac{\partial T}{\partial y}\bigg|_{y=0}}_{\substack{\text{local conductive} \\ \text{heat flux at the wall}}} = \underbrace{h(T_w - T_\infty)}_{\substack{\text{convective heat flux: local} \\ \text{heat loss due to fluid heating} \\ \text{or wall cooling}}}$$

♦ $\qquad\qquad h = -\dfrac{k(\partial T/\partial y)|_{y=0}}{T_w - T_\infty}$ $\qquad\qquad\qquad\qquad$ (4.15) where k is the fluid conductivity. Naturally, one could eliminate $h(x)$ and concentrate on $q_w(\Delta T)$ directly (cf. Adiutori 1989). A typical variation of the heat transfer coefficient $h(x)$ for an isothermal flat plate is shown in Fig. (4.2b), which indicates that turbulent wall flow is desirable for rapid heat transfer.

B   Dimensionless Groups

Solving differential equations analytically or numerically to obtain $T(x, y)$ and then $h(x)$ or $q_w(x)$ is called the differential or distributed parameter approach. Alternatively, when spatial temperature variations are negligible, at least in one direction, an averaged or lumped parameter approach can be used. When analytical or approximate tools fail, semiempirical correlations, say, for the Nusselt number, $\mathrm{Nu} = \mathrm{Nu}(\mathrm{Pr}, \mathrm{Re}, \mathrm{Gr})$, are employed to evaluate complex heat transfer problems. In any case, dimensionless groups (cf. Table 4.1) are most valuable in determining the type of modeling approach and characterizing the dynamics and transfer processes of a given system. As shown in Section 2.3, dimensionless groups evolve quite naturally in scale analyses or when normalizing the governing equations plus boundary conditions. The usefulness and some of the applications of dimensionless heat transfer parameters are summarized below.

C   Applications of Dimensionless Groups in CHT

(i)   The order of magnitude or critical value of specific groups determines the type and range of convection heat transfer.

$$
\begin{cases}
\mathrm{Re}_{\mathrm{crit}} \begin{cases} 2{,}300 & \text{for turbulent pipe flow} \\ 5 \times 10^5 & \text{for turbulent flat plate BL flow} \end{cases} \\[2ex]
\mathrm{Pr} \begin{cases} \ll 1 & \text{(liquid metals)} \\ \approx 1 & \text{(gases)} \\ \gg 1 & \text{(heavy oils)} \end{cases} \\[3ex]
\mathrm{Gr}/\mathrm{Re}^2 \begin{cases} \ll 1 & \text{in forced convection} \\ \approx 1 & \text{for mixed convection} \\ \gg 1 & \text{in free convection} \end{cases}
\end{cases}
$$

(ii)   A known value of a specific group allows the computation of specific system parameters.

$$
\begin{cases}
\mathrm{Re} \to \delta; & \text{boundary–layer thickness} \\
\mathrm{Nu} \to h; & \text{heat transfer coefficient} \\
\mathrm{Pr}^n \to \delta/\delta_{\mathrm{th}}; & \text{ratio of boundary–layer thicknesses}
\end{cases}
$$

(iii)   A given range of a specific group indicates the validity of a unique modeling approach.

$$
\begin{cases}
\textit{Example}: \text{If } \mathrm{Bi} < 0.1, \\
\text{lumped system analysis} \\
\text{valid for transient heat} \\
\text{conduction problems}
\end{cases}
$$

Table 4.1. *Selected dimensionless groups in heat transfer*

Group	Definition	Interpretation
Biot number (Bi_l)	$\dfrac{hl}{k_s}$	Ratio of the internal thermal resistance of a solid to the boundary-layer thermal resistance
Coefficient of friction (c_f)	$\dfrac{2\tau_w}{\rho u_\infty^2}$	Dimensionless wall shear stress in external flow applications
Friction factor (f)	$\dfrac{\Delta p}{(l/D)\left(\rho u_m^2/2\right)}$	Dimensionless pressure drop due to wall friction for internal flow
Grashof number (Gr_l)	$\dfrac{g\beta(T_w - T_\infty)l^3}{v^2}$	Ratio of buoyancy to viscous forces
Lewis number (Le)	$\dfrac{\alpha}{D_{AB}}$	Ratio of molecular thermal to mass diffusivities
Nusselt number (Nu_l)	$\dfrac{hl}{k_f}$	Ratio of convection heat transfer to conduction in a fluid at wall with length scale l
Peclet number (Pe_l)	$\dfrac{u_\infty l}{\alpha} = \text{Re}_l \text{Pr}$	Dimensionless heat transfer parameter expressing convection vs. thermal diffusion
Prandtl number (Pr)	$\dfrac{c_p \mu}{k} = \dfrac{v}{\alpha}$	Ratio of molecular momentum and thermal diffusivities
Rayleigh number (Ra_l)	$\dfrac{g\beta\Delta T}{\alpha v}l^3 = \text{Gr}_e \text{Pr}$	Ratio of buoyancy to thermal/viscous forces
Reynolds number (Re_l)	$\dfrac{u_\infty l}{v}$	Ratio of inertia to viscous forces
Schmidt number (Sc)	$\dfrac{v}{D_{AB}}$	Ratio of molecular momentum and mass diffusivities
Stanton number (St_l)	$\dfrac{h}{\rho u_\infty c_p} = \dfrac{\text{Nu}_l}{\text{Re}_l \text{Pr}}$	Dimensionless heat transfer coefficient

(iv)   Scale analysis reveals functional dependencies of dimensionless groups. For example, the ratio of momentum and thermal boundary-layer thicknesses depends on the fluid's Prandtl number; heat transfer depends on momentum transfer as expressed in the Reynolds analogy.

$$\begin{cases} \text{Pr}^n = \delta/\delta_{\text{th}} \\ \text{Nu} = \text{Nu (Re, Pr, Gr, or Ra)} \\ c_f = c_f (\text{Re, Nu}); \text{ or} \\ c_f = c_f (\text{St, Pr}) \end{cases}$$

D   Reynolds–Colburn Analogy

The Reynolds analogy and its extended form, the Reynolds–Colburn analogy, relate momentum transfer to heat transfer when certain restrictions apply. Specifically, assuming that compressional work and frictional heat are negligible and that the wall-bounded thermal flows in terms of dimensionless velocity and temperature profiles are similar, the ratio of heat to momentum flux normal to the wall is evaluated as

$$\left.\frac{q_y}{\tau_{yx}}\right|_{y=0} = \left.\frac{-k(\partial T/\partial y)}{\mu(\partial u/\partial y)}\right|_{y=0} \tag{4.16a}$$

In dimensionless form

$$\left.\frac{q_y/(\rho c_p \Delta T)}{\tau_{yx}/(\rho u_\infty)}\right|_{y=0} = \Pr^{-1} \left.\frac{(\partial/\partial y)[(T - T_\infty)/\Delta T]}{\partial/\partial y \left(u/u_\infty\right)}\right|_{y=0} \tag{4.16b}$$

Now, with

$$\mathrm{Nu}_x \equiv \frac{q_w x}{\Delta T k} \quad \text{and} \quad c_f \equiv \frac{2\tau_w}{\rho u_\infty^2} \equiv \frac{f}{4}$$

where $\Delta T = T_w - T_\infty$ and subscript w indicates conditions at $y = 0$, we obtain

$$\frac{2\mathrm{Nu}_x}{c_f \mathrm{Re}_x} = \left.\frac{d[(T - T_\infty)/\Delta T]}{d(u/u_\infty)}\right|_{y=0}$$

Provided that $\Pr \approx 1$ or for turbulent wall flow $\Pr_t \equiv (\varepsilon_m/\varepsilon_h) \approx 1$, the dimensionless profiles may be the same and the RHS is unity. Thus

$$\blacklozenge \qquad \mathrm{Nu}_x = \frac{1}{2}c_f(x)\mathrm{Re}_x \tag{4.17a}$$

Equation (4.17a) represents the Reynolds analogy, which has been empirically extended to the Reynolds–Colburn relation, i.e.,

$$\blacklozenge \qquad \frac{1}{2}c_f(x) = \mathrm{St}_x \Pr^{2/3} \tag{4.17b}$$

for the range $0.6 < \Pr < 60$. Here, the local Stanton number is given as $\mathrm{St}_x \equiv h(x)/(\rho c_p u_\infty)$ $= \mathrm{Nu}_x/(\mathrm{Re}_x \Pr)$. Typically, Equation (4.17b) is used for boundary-layer flows whereas (4.17a) in the form ($\Pr \approx 1$)

$$\blacklozenge \qquad \mathrm{St} = \frac{f}{8} \tag{4.17c}$$

is employed for evaluating thermal pipe flows, where the friction factor $f = f(\mathrm{Re}_d, e/d)$ is obtained from the Moody chart (cf. Fig. 4.13). A few problem solutions given here may help to review some basic aspects of convection heat transfer.

Problem (1): Alternative Derivation of the Heat Transfer Equation
Consider a representative elementary volume (REV): a fluid element $\Delta \forall = \Delta x \, \Delta y \, \Delta z$ with enthalpy convection, storage, and production.
In general,

$$f|_{x+\Delta x}(\Delta y \, \Delta z) = \left[f|_x + \frac{\partial}{\partial x} f|_x \, \Delta x\right](\Delta y \, \Delta z)$$

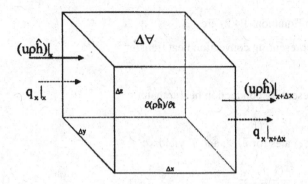

so that

$$f|_x(\Delta y\,\Delta z) - f|_{x+\Delta x}(\Delta y\,\Delta z) = -\frac{\partial f}{\partial x}\Delta\forall$$

Postulate

With $\rho\hat{h} \equiv \rho\hat{u} + p$ the fluid enthalpy per unit volume, the (thermal) energy conservation law can be expressed as

$$\left\{\begin{array}{c}\text{Rate of change}\\ \text{of }(\rho\hat{h})\text{ in REV}\end{array}\right\} - \left\{\begin{array}{c}\text{inflow of}\\ (\rho\hat{h})\text{ into REV}\end{array}\right\} + \left\{\begin{array}{c}\text{outflow of}\\ (\rho\hat{h})\text{ out of REV}\end{array}\right\} = \left\{\begin{array}{c}(\rho\hat{h})\text{ sources}\\ \text{inside REV}\end{array}\right\}$$

(P4.1a)

Solution

Recognizing that the time rate of change in enthalpy is $\partial/\partial t(\rho\hat{h})$, that the net efflux is $\nabla \cdot (\vec{v}\rho\hat{h})$, and that the relevant sources are the net rate of conductive heat transfer, $-\nabla \cdot \vec{q}$, as well as energy dissipation due to viscous stresses, $\tau_{ij}\partial v_i/\partial x_j$, we can write (P4.1a) in the form

$$\frac{\partial}{\partial t}(\rho\hat{h}) + \nabla \cdot (\vec{v}\rho\hat{h}) = -\nabla \cdot \vec{q} + \tau_{ij}\frac{\partial v_i}{\partial x_j}$$

(P4.1b)

Now, with $d\hat{h} = c_p\,dT$, $\vec{q} = -k\nabla T$, and $\tau_{ij}(\partial v_i/\partial x_j) \equiv \mu\Phi$, we obtain, for constant-fluid-property flow, Equation (4.10) in the form

$$\blacklozenge \qquad \frac{\partial T}{\partial t} + (\vec{v} \cdot \nabla)T = \alpha\nabla^2 T + \frac{\mu}{\rho c_p}\Phi$$

(P4.2)

Problem (2): Peclet Number Derivation (cf. Sect. 2.3.1)

Deduce an expression for the Peclet number, Pe $\hat{=}$ (convection h.t./conduction h.t.), from the heat transfer equation in vector form for a system with characteristic velocity U, length l, and temperature T_{ref}.

Postulate

The two relevant terms in Equation (P4.2) are

$$(\vec{v} \cdot \nabla)T \qquad \text{representing convection heat transfer}$$

and

$$\alpha \nabla^2 T \qquad \text{representing conduction heat transfer}$$

Solution

Employing representative scales for $\vec{v}$, ∇, and T yields

$$\text{Pe} = \frac{(\vec{v} \cdot \nabla)T}{\alpha \nabla^2 T} \sim \frac{U(1/l)T_{\text{ref}}}{\alpha(1/l^2)T_{\text{ref}}} \tag{P4.3a}$$

or

♦ $$\text{Pe} = \frac{Ul}{\alpha} = \text{Re}_l \text{Pr} \tag{P4.3b}$$

Problem (3): Application of Reynolds–Colburn Analogy

Calculate the local surface heat flux $q_w(x)$ for air flow past a flat plate where $U = u_\infty = 30.5$ m/s, $T_\infty = 338.6$ K, $T_w = 344.16$ K, $l = 0.1524$ m, and $p = 1$ atm (cf. sketch).

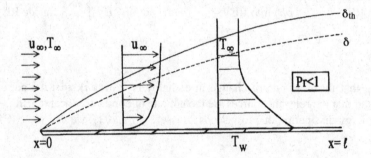

Solution

- The fluid property values are (cf. App. E) $\mu = 2.01 \times 10^{-5}$ kg/m·s, $\rho = 1.039$ kg/m³, $c_p = 1{,}008.2$ J/kg · K, and Pr = 0.6996.
- The Reynolds–Colburn analogy (4.17b) is written in the form

$$\text{St}_x \equiv \frac{q_w(x)}{\rho c_p(T_w - T_\infty)u_\infty} = \frac{1}{2}c_f \text{Pr}^{-2/3} \tag{P4.4a}$$

- With $\text{Re}_l = (\rho u_\infty l/\mu) = 2.4 \times 10^5 < \text{Re}_{\text{crit}} \approx 5 \times 10^5$, we obtain from laminar flat-plate-flow analysis (cf. Sect. 3.2.3)

$$c_f = \frac{0.664}{\sqrt{\text{Re}_x}}; \qquad \text{Re}_x = \frac{u_\infty x}{\nu} \tag{P4.4b,c}$$

- Combining (P4.4a–c) and solving for the wall heat flux yield (cf. graph)

$$q_w(x) = 89.7x^{-1/2} \text{ W/m}^2 \tag{P4.5}$$

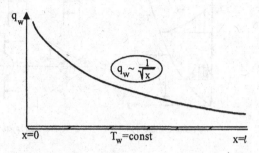

Note to Problem Solution (3)
The wall temperature gradients decrease as x increases because the (thermal) boundary layer thickens; thus; $q_w \sim x^{-1/2}$.

Note to Upcoming Sections 4.1 and 4.2
A thorough review of selected topics discussed in Chapter 1 (i.e., thermodynamic properties and the stress tensor), Chapter 2 (i.e., Sections 2.2 and 2.3) in conjunction with Appendix B, and Chapter 3 (i.e., laminar and turbulent internal/external flow calculations) is highly recommended. Independent solutions of sample problems given in Sections 2.4 and 3.3 should solidify the review material and prepare for this Chapter and Chapter 5.

4.1 Thermal Boundary-Layer Analyses

At sufficiently high Reynolds numbers (as in forced convection) or Grashof numbers (as in natural convection) distinct shear layers develop. Examples include wall-bounded flows past heated/cooled horizontal or inclined plates, thermal entrance flows in conduits, separated thin shear layers, jets, wakes, and mixing layers (cf. Fig. 4.3).

As shown in Chapters 2 and 3, a large class of practical external (and internal) flows are laminar or turbulent high-Reynolds-number flows that can be described by parabolic transport equations. Specifically, for steady two-dimensional ($\kappa = 0$) or axisymmetric ($\kappa = 1$), incompressible thermal flows, the continuity and boundary-layer equations are (cf. Fig. 4.4 and App. B or C)

$$\frac{\partial}{\partial x}(r^\kappa u) + \frac{\partial}{\partial y}(r^\kappa v) = 0 \tag{4.18a}$$

$$u\frac{\partial u}{\partial x} + v\frac{\partial u}{\partial y} = -\frac{1}{\rho}\frac{dp}{dx} + \frac{1}{\rho r^\kappa}\frac{\partial}{\partial y}\left[r^\kappa\left(\mu\frac{\partial u}{\partial y} - \overline{\rho u'v'}\right)\right] + f_x \tag{4.19a}$$

$$u\frac{\partial T}{\partial x} + v\frac{\partial T}{\partial y} = \frac{1}{\rho c_p}\frac{1}{r^\kappa}\frac{\partial}{\partial y}\left[r^\kappa\left(k\frac{\partial T}{\partial y} - \rho c_p\overline{T'v'}\right)\right] \tag{4.20a}$$

where the body force term f_x in (4.19a) may have to be modified for buoyancy-driven flows (cf. Boussinesq assumption outlined in Section 4.1.3). Note that viscous dissipation

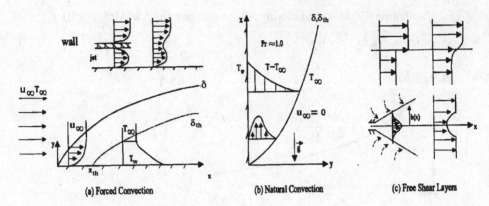

(a) Forced Convection   (b) Natural Convection   (c) Free Shear Layers

Fig. 4.3. Laminar thermal boundary-layer flows.

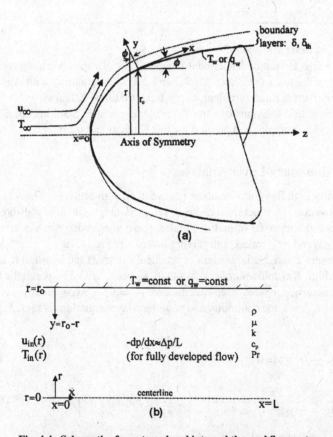

Fig. 4.4. Schematics for external and internal thermal flow systems. (a) coordinate system for external flow past an axisymmetric heated/cooled body where $r(x, y) = r_0(x) + y \cos[\phi(x)]$ and $\phi = \tan^{-1}(dr_0/dx)$ (after Cebeci and Bradshaw 1988); (b) coordinate system for pipe or slit flow with $\phi = 0$ and wall coordinate $y = r_0 - r$.

(cf. Eq. (4.6)) was neglected in Equation (4.20a). For turbulent flow, u, v, p, and T are time-averaged variables and submodels have to be postulated for the apparent stresses, $\overline{\rho u'v'}$, and apparent heat fluxes, $\rho c_p \overline{v'T'}$ (cf. Sects. 3.2.5 and 4.1.2). In thermal boundary-layer flow $(-1/\rho)(dp/dx) \equiv U(dU/dx)$, and for internal flows the axial pressure gradient is directly approximated by $\Delta p/L = $ const for fully developed pipe flow or computed from the condition of mass conservation

$$\dot{m} = \int\!\!\int_A \rho u \, dA = \text{const}$$

4.1.1 Laminar Forced Convection Boundary Layers

The standard problem in external convection heat transfer is steady laminar flow past a heated flat plate (cf. Figs. 4.3a). The governing equations are special cases of Equations (4.18a–4.20a):

(Continuity equation) $\quad \dfrac{\partial u}{\partial x} + \dfrac{\partial v}{\partial y} = 0 \qquad\qquad$ (4.18b)

(Momentum boundary-layer equation) $\quad u\dfrac{\partial u}{\partial x} + v\dfrac{\partial u}{\partial y} = -\dfrac{1}{\rho_0}\dfrac{\partial p}{\partial x} + v\dfrac{\partial^2 u}{\partial y^2} + \rho g_x$ (4.19b)

where $\quad -\dfrac{1}{\rho_0}\dfrac{\partial p}{\partial x} = U\dfrac{dU}{dx} \quad$ and for natural convection $\rho g_x = \rho_0 \beta (T - T_0) g_x$.

(Thermal boundary layer equation) $\quad u\dfrac{\partial T}{\partial x} + v\dfrac{\partial T}{\partial y} = \alpha\dfrac{\partial^2 T}{\partial y^2} + \dfrac{\mu}{\rho c_p}\left(\dfrac{\partial u}{\partial y}\right)^2$ (4.20b)

Typically, the viscous dissipation term in (4.20b) is negligible and for the Blasius problem $U = u_\infty = u_e = $ const. The boundary conditions are at

$$y = 0: \quad u = v = 0 \quad \text{or} \quad v = v_w(x) \quad \text{and}$$

$$T = T_w = \text{const} \quad \text{or} \quad q_w = -k(\partial T/\partial y)\big|_{y=0} = \text{const}$$

and at $\hfill$ (4.21a–g)

$$y = \delta: \quad u = u_\infty \qquad \text{for forced and mixed convection} \qquad (T = T_\infty)$$

$$\text{or} \quad u = 0 \qquad \text{for pure free convection} \qquad (T = T_\infty)$$

Variations to the standard system, that is, "thermal Blasius flow," include variable free-stream velocity, wall suction/injection, variable surface temperature or wall heat flux, variable fluid properties, axisymmetric bodies, arbitrarily shaped surfaces, transient effects, and turbulence effects.

Solution Techniques

(i) Scale analyses (cf. Sects. 2.3.1 and 4.1.3B) can be used to develop (a) expressions for key dimensionless groups (e.g., Re, Pe, or Ra number) and (b) relationships for important system properties, such as

$$\frac{\delta}{l}, \qquad \frac{\delta_{th}}{\delta}, \qquad \text{Nu}, \qquad \text{etc.}$$

(ii) Similarity theory is employed for self-similar flows, typically steady 2-D un-
      coupled TSL flows with $U(x) = Cx^m$ and $T_w(x) - T_\infty = Kx^n$ where m and n
      are constant. It has to be noted that the governing equations are parabolic and
      the transformed equations have no explicit x-dependence. Typically, a stream
      function of the form $\psi(x, y) = (\text{Veloc.} \times \text{Length}) f(\eta)$ is postulated where
      $\eta = y/\delta(x)$ and $f(\eta)$ is the dimensionless stream function. A complete ex-
      ample would be $\psi(x, y) = (u_\infty \nu x)^{1/2} f(\eta)$ where $\eta = y(u_\infty/\nu x)^{1/2}$ and $\theta =$
      $(T - T_\infty)/(T_w - T_\infty)$, which happen to be the Falkner–Skan transformations (cf.
      Sects. 2.3.3 and 3.2.3).

      Notes:
      • Simplify the transformed solution for special case studies: $\text{Pr} \gg 1$ (heavy oils)
         and $\text{Pr} \ll 1$ (liquid metals).
      • In general, use the Runge–Kutta algorithm with suitable starting values to solve
         transformed boundary value problems (cf. App. F.3).

(iii) Nonsimilar flows require the introduction of expanded Falkner–Skan variables, that
       is, $f[\xi(x), \eta]$, the integral method (cf. Sect. 2.3.3) with suitable velocity/tempera-
       ture profiles, or a CFD program. Nonuniformities in the outer flow field $U(x)$ or
       along the surface in terms of $T_w(x)$, $q_w(x)$, $v_w(x)$, as well as internal heat sources,
       flow separation, and so on, cause nonsimilarities. For mild nonsimilar variations in
       $T_w(x)$ the integral method works well when appropriate velocity and temperature
       profiles can be found. In CFD applications, the major hurdle is the generation of
       a suitable (adaptive) mesh or grid in order to obtain accurate results.

Now, the rather simple integral method and the more complex similarity theory are
employed to compute velocity and temperature fields for laminar boundary layers and then
thermal flow quantities such as the wall heat flux and the Nusselt number.

A Integral Solutions

The Von Karman integral method (cf. Sect. 2.3.3) for steady two-dimensional
boundary-layer flow with zero pressure gradient transforms Equations (4.18a) through
(4.20b) to (cf. Sect. 2.3.3B).

$$\blacklozenge \qquad \rho \frac{d}{dx} \int_0^{\delta(x)} (u_\infty - u) u \, dy = \tau_w \qquad\qquad (4.22)$$

and (cf. Problem (4), this chapter) to

$$\blacklozenge \qquad \frac{d}{dx} \int_0^{\delta_{th}(x)} (T_\infty - T) u \, dy = \alpha \frac{\partial T}{\partial y}\bigg|_{y=0} \qquad\qquad (4.23)$$

These integral relations hold for both laminar and turbulent flows; suitable laminar velocity and temperature profiles include

$$\frac{u}{u_\infty} = \frac{3}{2}\left(\frac{y}{\delta}\right) - \frac{1}{2}\left(\frac{y}{\delta}\right)^3 \quad \text{and} \quad \frac{T - T_w}{T_\infty - T_w} = \frac{3}{2}\left(\frac{y}{\delta_{th}}\right) - \frac{1}{2}\left(\frac{y}{\delta_{th}}\right)^3 \tag{4.24a,b}$$

where the standard boundary conditions are already invoked.

Notes

1. Steady laminar thermal flow results in terms of $Nu(x)$ or $St(x)$, do not differ much w.r.t. the type of wall condition employed, that is, $T_w = $ const or $q_w = $ const.
2. The thermal boundary-layer development in terms of $\delta_{th}(x)$ is dependent upon the type of fluid (i.e., Prandtl number). For example, for liquid metals $Pr = \nu/\alpha \ll 1$ and scale analysis (cf. Sect. 2.3.1) shows that $\delta/\delta_{th} \sim Pr^{1/2}$, which implies $\delta \ll \delta_{th}$ and hence within the thermal boundary layer $u(x, y)$ can be replaced by, say, $u_\infty = $ const. On the other hand, for heavy oils $Pr \gg 1$, which leads to $\delta/\delta_{th} \sim Pr^{1/3}$, and it is perhaps sufficient to employ a linear velocity dependence $u(y)$ within the very thin thermal boundary layer.

Applying the integral analysis to the flat plate laminar convection heat transfer problem ($Pr > 1$) depicted in Figure 4.3a, Equation (4.23) with (4.24a,b) reads

$$u_\infty \frac{d}{dx} \int_0^{\delta_{th}} \left[\frac{3}{2}\frac{y}{\delta} - \frac{1}{2}\left(\frac{y}{\delta}\right)^3\right]\left[1 - \frac{3}{2}\frac{y}{\delta_{th}} + \frac{1}{2}\left(\frac{y}{\delta_{th}}\right)^3\right] dy = \frac{3}{2}\frac{\alpha}{\delta_{th}} \tag{4.25a}$$

or with $\Delta \equiv \delta_{th}/\delta$ where $\delta(x) = 4.64x\,Re_x^{-1/2}$ from Section 3.2.3,

$$\frac{d}{dx}\left[\delta\left(\frac{3}{20}\Delta^2 - \frac{3}{280}\Delta^4\right)\right] = \frac{3\alpha}{2\delta\Delta u_\infty} \tag{4.25b}$$

Here, $\Delta^4 \approx 0$ since $Pr \sim \delta/\delta_{th} > 1$, so that

$$x\frac{d\Delta^3}{dx} + \frac{3}{4}\Delta^3 = \frac{39}{56}\frac{\alpha}{\nu} \quad \text{subject to } \Delta(x = x_{th}) = 0 \tag{4.25c}$$

The solution of Eq. (4.25c) is

$$\Delta^3 = \frac{13}{14}Pr^{-1}\left[1 - \left(\frac{x_{th}}{x}\right)^{3/4}\right] \tag{4.25d}$$

or when $x_{th} \to 0$, $\Delta = 0.976\,Pr^{-1/3}$ so that with $\delta(x) = 4.64 \times Re_x^{-1/2}$ from Section 3.2.3, and $h = (3/2)k/\delta_{th}$ obtained with (4.24b), we have an expression for the local Nusselt number of the generic form $Nu = Nu(Re, Pr)$. Specifically,

$$\blacklozenge \qquad Nu_x \equiv \frac{hx}{k} = 0.331Pr^{1/3}Re_x^{1/2} \tag{4.26}$$

Table 4.2.   *Constants for various Prandtl numbers*

Pr	c_1	c_2	c_3
0.7	0.418	0.435	1.87
1.0	0.332	0.475	1.95
5.0	0.117	0.595	2.19
10	0.073	0.685	2.37

Source: Cebeci and Bradshaw (1988).

valid for $\text{Re}_x \le 5 \times 10^5$ with $U = \text{const}$ and $\text{Pr} > 1$. Clearly, Nu $(x) \sim \sqrt{x}$ and $h(x) \sim x^{1/2}$ along the plate surface for laminar constant-property flow (cf. solution of Problem (3), this chapter).

A special integral method for forced convection heat transfer in similar boundary-layer flows with variable outer flow $U(x)$ but isothermal wall temperature, $T_w = \text{const}$, has been proposed by Smith and Spalding (1958) following Thwaites's method (cf. Sect 3.2.3). A "conduction thickness" $\Delta = k/h$ is introduced as

$$\Delta \equiv \frac{T_w - T_\infty}{q_w/k} = -\frac{T_w - T_\infty}{(\partial T/\partial y)_w}$$

Now, similar to Thwaites's correlation (cf. Sect. 3.2.3D),

$$\frac{U}{\nu}\frac{d\Delta^2}{dx} = \text{fct}\left(\frac{\Delta^2}{\nu}, \frac{dU}{dx}, \text{Pr}\right) = A(\text{Pr}) - B(\text{Pr})\frac{\Delta^2}{\nu}\frac{dU}{dx} \qquad (4.27a)$$

Thus, instead of assuming velocity and temperature profiles, an empirical correlation between key parameters is established. Integration of (4.27a) yields

$$\Delta^2 = \frac{\nu A}{U^B}\int_0^x U^{B-1}\,dx + \Delta_i^2\frac{U_i^B}{U^B} \qquad (4.27b)$$

With $\Delta(x = 0) = \Delta_i = 0$ and using the Stanton number as

$$\text{St}_x = \frac{\text{Nu}_x}{\text{Pr}\,\text{Re}_x} = \frac{h(x)}{\rho c_p U} = \frac{k}{\rho c_p U}\Delta^{-1} \qquad (4.27c)$$

we can compute the dimensionless heat transfer coefficient as

$$\blacklozenge \qquad \text{St} = \frac{h(x)}{\rho c_p U} = \frac{c_1 U^{c_2}}{[\int_0^x U^{c_3}\,dx]^{1/2}}\frac{1}{\sqrt{\text{Re}_x}} \qquad (4.27d)$$

where the parameters $c_1 = A^{-1/2}\,\text{Pr}^{-1}$, $c_2 = B/2 - 1$ and $c_3 = B - 1$ are given in Table 4.2 for various Prandtl numbers.

B   Similarity Solutions

The thermal energy integral approach (cf. Equations (4.25) and (4.27)) is a good method for finding approximate solutions to laminar or turbulent (similar or nonsimilar) boundary-layer-type flow problems as long as suitable profiles can be postulated. A restriction to similar flows allows the exact solution of the (laminar) boundary-layer equations. Thus, with the outer flow of the form (cf. Sect. 3.2.3D)

$$U(x) = Cx^m \tag{4.28}$$

we use the Falkner–Skan transformations

$$\psi = (U\nu x)^{1/2} f(\eta) \qquad \text{where } \eta = y \left[\frac{U}{\nu x}\right]^{1/2} \tag{4.29a,b}$$

for Equation (4.19b) to obtain a nonlinear ODE for the velocity field (cf. Fig. 4.5 and Sect. 3.2.3D).

$$\blacklozenge \qquad f''' + \frac{m+1}{2} f f'' + m(1 - f'^2) = 0 \tag{4.30a}$$

$$\text{where} \quad u \equiv \frac{\partial \psi}{\partial y} = U f'(\eta) \text{ and } v \equiv -\frac{\partial \psi}{\partial x} = \frac{1}{2}\left(\frac{\nu U}{x}\right)^{1/2} (\eta f' - f) \tag{4.30b,c}$$

Note: These transformations are based on (i) scale analysis, that is, $\eta \sim y/\delta(x)$ where $\delta(x) \sim x\mathrm{Re}_x^{-1/2}$; (ii) the stream function formulation for incompressible two-dimensional flow; and (iii) the edge condition, which implies that $u \to U = u_e(x)$ when $\eta \to \infty$ and $f' \to 1$. For the Blasius problem, $U(x) = u_\infty = $ const, which implies $m = 0$, and Equation (4.29) reduces to

$$\blacklozenge \qquad 2f''' + ff'' = 0 \tag{4.30d}$$

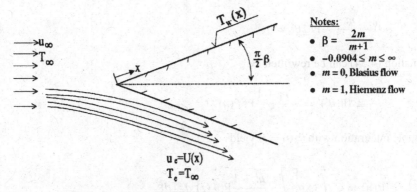

Fig. 4.5. Falkner–Skan flows with heat transfer.

subject to

♦ $f' = f = 0$ at $\eta = 0$ and $f' \to 1$ as $\eta \to \infty$

Similar to (4.28), we assume a power law for the wall temperature distribution

$$T_w(x) - T_\infty = K x^n \tag{4.31}$$

where $n = 0$ for the isothermal wall case and $n = (1 - m)/2$ for the constant wall heat flux case. With the dimensionless temperature $\theta = (T - T_w)/(T_\infty - T_w)$ and the Falkner–Skan transformations, the thermal boundary-layer equation (4.20b) can be rewritten as

$$\theta'' + \frac{m+1}{2}\Pr f\theta' - n\Pr f'\theta = -\Pr \text{Ec } x^{2m-n}(f'')^2 \tag{4.32}$$

where Ec is the Eckert number, $\text{Ec} = u_\infty^2/[c_p(T_w - T_\infty)]$, representing the ratio of flow kinetic energy to boundary-layer enthalpy. The boundary conditions are

$$\theta(\eta = 0) = 0 \quad \text{and} \quad \theta(\eta \to \infty) \to 1$$

Typically, viscous dissipation is negligible, the RHS of (4.32) is zero, and similarity is preserved for cases other than $2m - n = 0$. By definition, the Nusselt number for the isothermal wall case is

$$\text{Nu}_x = \frac{hx}{k} = \frac{x}{T_\infty - T_w}\left(\frac{\partial T}{\partial y}\right)_w = \left(\frac{Ux}{\nu}\right)^{1/2}\theta'(0) \tag{4.33}$$

which requires the computation of the dimensionless temperature gradient at the wall.

Equations (4.30) and (4.32) are two-point boundary-value problems that can be readily solved with appropriate starting values for $f''(0)$ and $\theta'(0)$ employing a Runge–Kutta routine (cf. App. F.3). Thus, such a forward integration scheme starts at $\eta = 0$ and proceeds forward toward a suitable value of $\eta \doteq \infty$, where for flat plate problems with $\Pr = O(1)$: $\Delta\eta = 0.2$ to 0.4, $\eta_\infty = 5$ to 10, $f''(0) = 0.33206$, and $\theta'(0) = -0.29268$.

Applying the similarity analysis to laminar forced convection past a flat plate with uniform surface temperature, that is, $n = 0$ in (4.31), Equation (4.32) reduces to

$$\theta'' + \frac{m+1}{2}\Pr f\theta' = 0 \tag{4.34}$$

Equation (4.34) can be rewritten as

$$\frac{\theta''}{\theta'} \equiv (\ln\theta')' = -\frac{m+1}{2}\Pr f(\eta)$$

Double integration with $\theta(0) = 0$ yields

$$\theta(\eta) = C \int_0^\eta \exp\left[\int_0^\xi -\frac{m+1}{2}\Pr f(z)\, dz\right] d\xi$$

where ξ and z are integration variables. The integration constant can be determined with the edge condition $\theta(\eta_\infty) = 0$ so that

$$\blacklozenge \qquad \theta(\eta) = \frac{\int_0^\eta \exp\left[-(m+1)/2 \, \mathrm{Pr} \int_0^\xi f(z)dz\right]d\xi}{\int_0^{\eta_\infty} \exp[\%] \, d\xi} \qquad (4.35)$$

Clearly, the temperature field (4.35) depends on the dimensionless stream function $f(\eta)$. For thermal Blasius flow, $m = 0$, and the transformed boundary-layer equation (4.30b) can be rewritten as

$$\frac{f'''}{f''} = -\frac{1}{2}f$$

and integrated to

$$-\frac{1}{2}\int_0^\eta f \, d\eta = \ln f'' \qquad (4.36)$$

Using (4.36) in (4.35), we obtain

$$\blacklozenge \qquad \theta = \frac{\int_0^\eta (f'')^{\mathrm{Pr}} d\eta}{\int_0^{\eta_\infty} (f'')^{\mathrm{Pr}} d\eta} \qquad (4.37)$$

For $\mathrm{Pr} = 1.0$ and $m = 0$:

$$\theta(\eta) = \frac{f' - f'(0)}{f'(\eta_\infty) - f'(0)} = f' = \frac{u}{u_\infty} \qquad (4.38)$$

that is, a perfect Reynolds analogy. Now, the local Nusselt number for $T_w = \mathrm{const}$

$$\mathrm{Nu}_x = \frac{x}{T_w - T_\infty}\left(\frac{\partial T}{\partial y}\right)_w = \frac{x}{\sqrt{\nu x/u_\infty}}\theta'(0) = \mathrm{Re}_x^{1/2}\theta'_w \qquad (4.39a)$$

where from Equation (4.38), $\theta'_w \equiv f''(0)$ so that (cf. Eq. (4.26))

$$\blacklozenge \qquad \mathrm{Nu}_x \mathrm{Re}_x^{-1/2} = 0.33206 \qquad (4.39b)$$

For extreme Prandtl numbers the velocity field within the thermal boundary layer can be approximated as

(a) $(u/u_\infty) = f' = 1$ for liquid metals where $\delta \ll \delta_{\mathrm{th}}$ and Equation (4.35) can be simplified to $\theta'(0) = 0.564\sqrt{(m+1)\mathrm{Pr}}$ so that

$$\blacklozenge \qquad \mathrm{Nu}_x = 0.564\sqrt{m+1} \, \mathrm{Pr}^{1/2}\mathrm{Re}_x^{1/2} \qquad (4.39c)$$

and

(b) $(u/u_\infty) = f' = f''_w \eta$ or $f = (1/2) f''_w \eta^2$ for heavy oils where $\delta \gg \delta_{\mathrm{th}}$ and the slope $f''_w = \mathrm{const}$. Now, Equation (4.35) can be evaluated to yield $\theta'(0)$ and as a result

$$\blacklozenge \quad \mathrm{Nu}_x = 1.12 \left(\frac{m+1}{12} f''_w \right)^{1/3} \mathrm{Pr}^{1/3} \mathrm{Re}_x^{1/2} \tag{4.39d}$$

As discussed previously, for $\mathrm{Pr} > 1.0$, the integral method indicates that

$$\mathrm{Nu}_x = \begin{cases} 0.331 \mathrm{Pr}^{1/3} \mathrm{Re}_x^{1/2} & \text{for } m = 0, \text{ i.e., } U = \mathrm{const} \\ 0.570 \mathrm{Pr}^{0.4} \mathrm{Re}_x^{1/2} & \text{for } m = 1, \text{ i.e., } U(x) \text{ linear, or} \\ & \text{for stagnation point flow} \end{cases} \tag{4.39e,f}$$

C Transformations of Nonsimilar Problems

Steady laminar boundary-layer flows disturbed by outer flow variations $U(x) \neq K x^m$ or by wall suction/injection effects ($v_w \neq 0$) or by buoyancy effects ($\mathrm{Ri} = \mathrm{Gr}/\mathrm{Re}^2 = O(1)$) exhibit nonsimilar velocity and temperature profiles. Now, as an alternative to using the integral method, the appropriate boundary-layer equations (cf. Eqs. (4.18b)–(4.20b)), can be transformed, employing expanded Falkner–Skan variables (cf. Eq. (4.29))

$$\eta = \frac{y}{x} \mathrm{Re}_x^{1/2} \tag{4.40}$$

and

$$\psi(x, \eta) = (\nu U x)^{1/2} f(\xi, \eta) \tag{4.41}$$

where, again, $f(\xi, \eta)$ is a dimensionless stream function. Here, $\xi = \xi(x)$ is introduced: (a) to capture the nonsimilarities of boundary layers and (b) to reduce the x-dependence and hence the numerical work for the problem solution. Using the chain rule relations

$$\left(\frac{\partial}{\partial x} \right)_y = \left(\frac{\partial}{\partial x} \right)_\eta + \left(\frac{\partial}{\partial \eta} \right)_x \frac{\partial \eta}{\partial x}$$

and

$$\left(\frac{\partial}{\partial y} \right)_x = \left(\frac{\partial}{\partial \eta} \right)_x \frac{\partial \eta}{\partial y}$$

Equations (4.40) and (4.41) together with the dimensionless temperature

$$\theta = \frac{T - T_\infty}{T_w - T_\infty} \tag{4.42}$$

allow a straightforward transformation of the momentum and thermal boundary-layer equations into stream function form.

Depending upon the type of flow nonsimilarity and thermal boundary condition, different $\xi(x)$ and $\theta(\xi, \eta)$ expressions can be defined, resulting in different (transformed) equations. For example, with

$$\xi(x) = \frac{v_w}{u_\infty} Re_x^{1/2} \tag{4.43}$$

the problem of forced convection over an isothermal flat plate with uniform suction/injection reads

$$\blacklozenge \qquad f''' + \frac{m+1}{2} f f'' + m(1 - f'^2) = \frac{\xi}{2}\left(f' \frac{\partial f'}{\partial \xi} - f'' \frac{\partial f}{\partial \xi} \right) \qquad (4.44)$$

and

$$\blacklozenge \qquad \frac{1}{\mathrm{Pr}} \theta'' + \frac{m+1}{2} f \theta' = \frac{\xi}{2}\left(f' \frac{\partial \theta}{\partial \xi} - \theta' \frac{\partial f}{\partial \xi} \right) \qquad (4.45)$$

subject to

$$\blacklozenge \qquad f'(\xi, 0) = 0, \; f(\xi, 0) + \xi \frac{\partial f(\xi, 0)}{\partial \xi} = -2\xi \quad \text{and} \quad f'(\xi, \infty) = 1$$

$$\blacklozenge \qquad \theta(\xi, 0) = 1 \quad \text{and} \quad \theta(\xi, \infty) = 0$$

Here, m is the dimensionless pressure gradient parameter

$$m = \frac{x}{U} \frac{dU}{dx} \qquad (4.46)$$

Note that similarity is recovered when $v_w = 0$ or $v_w \propto x^{-1/2}$ corresponding to $\xi = 0$ or $\xi = \text{const}$.

In the case of mixed thermal convection along a vertical isothermal plate

$$\xi(x) = \mathrm{Gr}_x \mathrm{Re}_x^{-2} \qquad (4.47)$$

and the coupled transformed equations read for $m = 0$, that is, $U = u_\infty = \text{const}$

$$f''' + \frac{1}{2} f f'' \begin{smallmatrix} (\text{up}) \\ \pm \\ (\text{down}) \end{smallmatrix} \xi\theta = \xi\left(f' \frac{\partial f'}{\partial \xi} - f' \frac{\partial f}{\partial \xi} \right) \qquad (4.48)$$

and

$$\mathrm{Pr}^{-1} \theta'' + \frac{1}{2} f \theta' = \xi\left(f' \frac{\partial \theta}{\partial \xi} - \theta' \frac{\partial f}{\partial \xi} \right)$$

$$f'(\xi, 0) = 0, \qquad f(\xi, 0) + 2\xi \frac{\partial f(\xi, 0)}{\partial \xi} = 0, \qquad f'(\xi, \infty) = 1 \qquad (4.49)$$

and

$$\theta(\xi, 0) = 1, \qquad \theta(\xi, \infty) = 0 \qquad (4.50)$$

These mixed differential equations are best solved with robust finite difference algorithms such as the Keller box method (cf. Appendix F.4). Further discussions on the nonsimilarity method are given by T. S. Chen in Minkowycz et al. (1988).

Problem (4): Derivation of the Thermal Energy Integral Relation
Consider the flat plate thermal boundary-layer equations for steady flow without viscous heating. Derive the heat transfer equation in integral form and then simplify it for laminar constant-fluid-property flow.

Solution

- The continuity and the reduced thermal energy equations

$$\frac{\partial u}{\partial x} + \frac{\partial v}{\partial y} = 0 \quad \text{and} \quad \frac{\partial}{\partial x}(uT) + \frac{\partial}{\partial y}(vT) = -\frac{1}{\rho c_p}\frac{\partial q}{\partial y} \tag{P4.6a,b}$$

are integrated across the thermal boundary layer to

$$\frac{d}{dx}\int_0^\infty u\,dy + v_\infty - v_0 = 0 \quad \text{and} \quad \frac{d}{dx}\int_0^\infty uT\,dy + (vT)_\infty - (vT)_0 = \frac{q_w}{\rho c_p} \tag{P4.7a,b}$$

- Taking $v_0 \equiv v_w = 0$, $T_\infty \equiv T_{\text{edge}}$, and inserting v_∞ of (P4.7a) into Equation (P4.7b) yield

$$\frac{d}{dx}\left[\int_0^\infty \rho c_p u(T - T_e)\,dy\right] = q_w \tag{P4.8}$$

- With Fourier's law for laminar CHT and constant fluid properties, Equation (P4.8) simplifies to Equation (4.23):

$$\blacklozenge \qquad \frac{d}{dx}\left[\int_0^{\delta_{\text{th}}} u(T_e - T)\,dy\right] = \alpha\left(\frac{\partial T}{\partial y}\right)_{y=0} \tag{P4.9}$$

Problem (5): Alternative Formulation of the Thermal Energy Integral Relation
Consider incompressible flow past a heated, isothermal flat plate. Show that for thermal Blasius flow, that is, $dp/dx = 0$, Equation (P4.9) reduces to

$$\frac{d\theta_{\text{th}}}{dx} = \text{St} \quad \text{where} \quad \theta_{\text{th}} \equiv \int_0^\infty \frac{u}{u_\infty}\left(\frac{T - T_\infty}{T_w - T_\infty}\right)dy \quad \text{and} \quad \text{St} = \frac{q_w}{\rho c_p(T_w - T_\infty)u_\infty}$$

- Noting that θ_{th} is a special form of the enthalpy thickness

$$\theta_h \equiv \int_0^\infty \frac{\rho u}{\rho u_\infty}\left(\frac{\hat{h} - \hat{h}_\infty}{\hat{h}_w - \hat{h}_\infty}\right)dy \tag{P4.10}$$

where $\hat{h}_i = c_p T_i$, Equation (P4.8) can be rewritten as

$$\frac{d}{dx}[u_\infty(T_w - T_\infty)\theta_h] = \frac{q_w}{\rho c_p} \tag{P4.11}$$

- With $T_w = \text{const}$, $u_\infty(T_w - T_\infty)$ is constant and Equation (P4.11) takes on the form

$$\frac{d}{dx}\theta_h = \frac{q_w}{\rho c_p u_\infty(T_w - T_\infty)} \equiv St \tag{P4.12a}$$

- Constant fluid properties reduce (P4.12a) to

$$\frac{d\theta_{th}}{dx} = St \tag{P4.12b}$$

♦

Problem (6): Basic Self-Similar Thermal Boundary-Layer Flow

Consider Blasius-type flow of air ($u_\infty = 1$ m/s, $T_\infty = 20°C$, $p_\infty = 1$atm) past a horizontal heated flat plate ($l = 1.5$ m, $T_w = 120°C$) (cf. sketch). Find $Nu(x)$ for $Pr = 0.72, 1.0$.

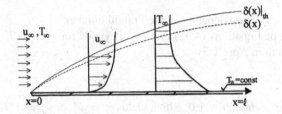

Solution

- Fluid properties at $T_f = 1/2(T_\infty + T_w) := 70°C$ for air (cf. App. E):

$$\rho = 1.009 \text{ kg/m}^3, \qquad \nu = 2.0 \times 10^{-5} \text{ m}^2\text{/s},$$

$$c_p = 1009 \text{ J/kg} \cdot K \quad \text{and} \quad Pr = 0.72$$

- Note that $Re_l = u_\infty l/\nu = 7.5 \times 10^4$ (i.e., laminar flow)
- A special case of the Falkner–Skan flows with $m = 0$ and $n = 0$:

$$\theta'' + \frac{Pr}{2}f\theta' = 0 \qquad \text{(cf. Eq. (4.34))}$$

subject to $\theta(\eta = 0) = 1$ and $\theta(\eta \to \infty) \to 0$ and

$$2f''' + ff'' = 0 \qquad \text{(cf. Eq. (4.30d))}$$

subject to $f' = f = 0$ at $\eta = 0$ and $f' \to 1$ when $\eta \to \infty$.

Manipulation of ODEs

$$\frac{\theta''}{\theta'} \equiv (\ln \theta')' = -\frac{Pr}{2}f(\eta)$$

or

$$\ln(\theta') = -\frac{\Pr}{2} \int_0^\eta f(\eta) \, d\eta$$

in conjunction with

$$\frac{f'''}{f''} = -\frac{1}{2} f \quad \text{or} \quad 2\ln(f'') = -\int_0^\eta f \, d\eta \qquad \text{(cf. Eq. (4.36))}$$

so that

$$\ln(\theta') = \Pr \ln(f'') \quad \text{or} \quad \theta' = (f'')^{\Pr}$$

- Now, the local Nusselt number

$$\mathrm{Nu}_x \equiv -\frac{x}{T_w - T_\infty} \frac{\partial T}{\partial y}\bigg|_{y=0} := \theta'_w \sqrt{\mathrm{Re}_x} \qquad \text{(cf. Eq. (4.33))}$$

depends on $\theta'(0)$, which in turn depends on the Prandtl number.
- A numerical solution of Equations (4.32) and (4.36) yields for $\Pr = 0.72$ (i.e., employing the algorithm in App. F.3)

◆ $\quad \mathrm{Nu}_x = 0.2957 \sqrt{\mathrm{Re}_x}$

However, it is easier to assume $\Pr \approx 1.0$, which yields $\theta' = f''$, or $\theta'_w = f''(0) :=$ 0.332,

$$c_f \equiv \frac{2\tau_w}{\rho u_\infty^2} = \frac{2f''_w}{\sqrt{\mathrm{Re}_x}} = \frac{0.664}{\sqrt{\mathrm{Re}_x}} \qquad \text{(cf. Sect. 3.2.3)}$$

Hence,

◆ $\quad \mathrm{Nu}_x \approx 0.332 \sqrt{\mathrm{Re}_x}$

4.1.2 Turbulent Thermal Boundary Layers

Following Reynolds's idea of decomposing (cf. Sect. 3.2.4) the instantaneous temperature as $T = \bar{T} + T'$, a dominant apparent or turbulent heat flux, q_{turb}, appears in the thermal energy equation. Thus, for steady thermal boundary layers without heat sources (cf. Eqs. (4.9) and (4.20a))

◆ $\qquad \rho c_p \left(\bar{u} \frac{\partial \bar{T}}{\partial x} + \bar{v} \frac{\partial \bar{T}}{\partial y} \right) = -\frac{\partial}{\partial y} q^{\mathrm{total}}$ \hfill (4.51a)

where the total heat flux is given as

◆ $\qquad q^{\mathrm{total}} = q_{\mathrm{lam}} + q_{\mathrm{turb}} = -k \frac{\partial \bar{T}}{\partial y} + \rho c_p \overline{v'T'}$ \hfill (4.51b) neglecting minor heat flux components. Imposing the analog of Boussinesq's eddy viscosity concept in CHT, that is, $q_{\text{turb}} \approx -\varepsilon_h(\partial \bar{T}/\partial y)$, Equation (4.51b) reads

$$\blacklozenge \quad q = -\mu c_p \left(\frac{1}{\text{Pr}} + \frac{\varepsilon_m^+}{\text{Pr}_t} \right) \frac{\partial \bar{T}}{\partial y} \quad (4.52a)$$

where the dimensionless momentum eddy diffusivity $\varepsilon_m^+ = \varepsilon_m/\nu$. The turbulent Prandtl number is defined as the ratio of turbulent momentum to heat transfer eddy diffusivities:

$$\blacklozenge \quad \text{Pr}_t \equiv \frac{\varepsilon_m}{\varepsilon_h} = \frac{c_p \mu_t}{k_t} = \frac{\overline{(u'v')}/(\partial \bar{u}/\partial y)}{\overline{(v'T')}/(\partial \bar{T}/\partial y)} = O(1) \quad (4.52b)$$

Actually, Pr_t varies across the boundary layer from the wall to the edge as $1.9 > \text{Pr}_t \geq 0.5$, where typically an average $\text{Pr}_t = 0.85$ or $\text{Pr}_t = 1.0$ is used in engineering calculations.

Depicting a boundary layer again as a composite consisting of a laminar sublayer, a buffer zone, and a turbulent core, turbulent velocity/temperature profiles in terms of inner variables can be derived and experimentally confirmed (cf. Sect. 3.2.5). For example, for flat plate flow with zero pressure gradient

$$u^+ = y^+ \quad \text{for } 0 < y^+ < 5 \quad \text{where } T^+ = \text{Pr} y^+$$

$$\left.\begin{array}{ll} u^+ = 5.0 \ln y^+ - 3 & \text{for } 5 < y^+ < 30 \\ u^+ = 2.5 \ln y^+ + 5.5 & \text{for } 30 < y^+ < 400 \end{array}\right\} \quad (4.53a\text{–}e)$$

$$\text{where} \quad T^+ = 2.5 \ln y^+ + 12.8 \text{Pr} - 7.3$$

The inner variables are defined as $u^+ = \bar{u}/u_\tau$, $y^+ = u_\tau y/\nu$, and $T^+ = (T_w - \bar{T})\rho c_p u_\tau/q_w$ where $u_\tau = \sqrt{\tau_w/\rho}$. Alternatively, a power law can be used to approximate turbulent temperature profiles (cf. Problem (7)).

In summary, the options for solving turbulent external or internal thermal flow problems include the following:

- Numerical solution of the turbulent boundary-layer equations using appropriate closure models for $\varepsilon_m \equiv \nu_t$ and Pr_t (cf. Sects. 3.2.4 and 3.2.5)
- Use of turbulent velocity and temperature profiles to develop Nusselt number equations, in conjunction with integral relations and/or the Reynolds–Colburn analogy, that is, $c_f/2 = \text{StPr}_t^{2/3}$ (cf. Eq. (4.17b) and Problem (7))
- Use of empirical Nusselt number or Stanton number correlations for turbulent thermal flow inside or around complex system geometries; correlations for turbulent flat plate boundary-layer flows include ($\text{Re} \approx 10^6$ and $\text{Pr} = O(1)$)

$$\blacklozenge \quad \text{StPr}^{0.4} = 0.03 \text{Re}_x^{-0.2} \quad \text{for } q_w = \text{const} \quad (4.54a)$$

and

$$\blacklozenge \quad \text{StPr}^{0.4} \left(\frac{T_w}{T_\infty} \right)^{0.4} = 0.0269 \text{Re}_x^{-0.2} \quad \text{for } T_w = \text{const} \quad (4.54b)$$

In addition to the standard books discussing heat transfer in turbulent fluid flows (e.g., Cebeci and Bradshaw 1988 and Kays and Crawford 1993), Zukauskas and Slanciauskas (1987) address both analytical and experimental techniques for turbulent convection heat transfer on flat plates.

Problem (7): Use of Power-Law Profiles for Turbulent Thermal Boundary Layers

Consider fully turbulent flow with zero pressure gradient past an isothermal flat plate with unheated starting length x_h, that is, for $0 \leq x < x_h$, $T_w = T_\infty$, and for $x \geq x_h$, $T_w > T_\infty$. Employing velocity/temperature power-law profiles, develop an expression for the local dimensionless heat transfer coefficient (cf. sketch).

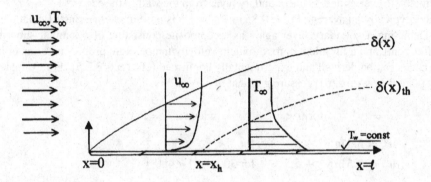

Postulates

- Assume $U(x) \doteq u_\infty = \text{const}$ and postulate power-law profiles of the form

$$\frac{u}{u_\infty} = \left(\frac{y}{\delta}\right)^{1/n} \quad \text{and} \quad \theta^* = \frac{T_w - T}{T_w - T_\infty} = \left(\frac{y}{\delta_{th}}\right)^{1/n} \tag{P4.13a,b}$$

where $n = 7$ for $5 \times 10^5 < \text{Re}_l \leq 10^7$.

- The dimensionless heat transfer coefficient is actually the Stanton number

$$\text{St} \equiv \frac{h}{\rho c_p u_\infty} = \frac{q_w}{\rho c_p u_\infty (T_w - T_\infty)} \tag{P4.14a,b}$$

for which we obtained (cf. Eq. (P4.12))

$$\blacklozenge \quad \text{St} = \frac{d\theta_{th}}{dx}; \qquad \theta_{th} \equiv \int_0^\infty \frac{u}{u_\infty} \theta^* \, dy \tag{P4.15}$$

- A second expression for the Stanton number can be derived from (P4.14b) in conjunction with Equation (4.52b)

$$\text{Pr}_t = \frac{\overline{(u'v')}/(\partial u/\partial y)}{\overline{(v'T')}/(\partial T/\partial y)} = 1.0 \tag{P4.16}$$

where $(\overline{u'v'}) = -\tau/\rho$ so that from (P4.16) with Equations (P4.13a,b)

$$(\overline{v'T'}) = -\frac{\tau}{\rho}\left(\frac{\partial T}{\partial y}\right)\left(\frac{\partial u}{\partial y}\right)^{-1} := \frac{\tau}{\rho}\frac{T_w - T_\infty}{u_\infty}\left(\frac{\delta}{\delta_{th}}\right)^{1/n}$$

Now,

$$q_w \approx \rho c_p(\overline{v'T'}) := \rho c_p \frac{\tau_w}{\rho}\frac{T_w - T_\infty}{u_\infty}\left(\frac{\delta}{\delta_{th}}\right)^{1/n} \tag{P4.17a,b}$$

inserted into (P4.14b) yields

$$St = \frac{c_f}{2}\left(\frac{\delta}{\delta_{th}}\right)^{1/n} \tag{P4.18}$$

where even for $Pr_{fluid} = 1.0$, $\delta_{th} \neq \delta$ because of the delayed plate heating at $x = x_h$.

Solution

- The enthalpy thickness for flow past isothermal surfaces is

$$\theta_{th} = \int_0^\infty \frac{u}{u_\infty}\left(\frac{T - T_\infty}{T_w - T_\infty}\right)dy$$

and the momentum thickness is

$$\delta_2 \equiv \int_0^\infty \frac{u}{u_\infty}\left(\frac{u_\infty - u}{u_\infty}\right)dy$$

where $d\delta_2/dx = c_f/2$ (cf. Sect. 3.2.3).
- Combining Equations (P4.15) and (P4.18) yields

$$\frac{d\theta_{th}}{dx} = \frac{c_f}{2}\left(\frac{\delta}{\delta_{th}}\right)^{1/n} := \frac{d\delta_2}{dx}\left(\frac{\delta}{\delta_{th}}\right)^{1/n}$$

and inserting the power-law profiles into

$$\frac{d\theta_{th}}{d\delta_2} = \left(\frac{\delta}{\delta_{th}}\right)^{1/n} \tag{P4.19}$$

yields an expression for the ratio

$$\frac{\delta_{th}}{\delta} = \left[1 - \left(\frac{x_h}{x}\right)^{(4(n+2)/5(n+1))}\right]^{n/(2+n)} \tag{P4.20}$$

for which the previous result $\delta = \delta_{turb} \sim x^{4/5}$ was employed (cf. Sect. 3.2.5). In Section 3.2.4 it was also shown that for $n = 7$,

$$c_f = 0.059 Re_x^{-1/5} \tag{P4.21}$$

- Inserting (P4.21) and (P4.20) into (P4.18) yields with $n = 7$

◆ $$St_x \equiv \frac{h(x)}{\rho c_p u_\infty} = 0.0295 Re_x^{-0.2} \left[1 - \left(\frac{x_h}{x} \right)^{9/10} \right]^{-1/9} \qquad (P4.22)$$

- It should be noted that Equation (P4.22) collapses for $Pr \equiv 1$ to the Reynolds analogy

$$St_x = \frac{1}{2} cf(x)$$

when $x_h = 0$ and $\delta \approx \delta_{th}$. Thus, the bracket term in (P4.22) represents the "unheated starting-length effect."

Problem (8): Evaluation of Combined Laminar/Turbulent Flat Plate Boundary-Layer Flows

Consider air flow ($u_\infty = 30$ m/s, $T_\infty = 20°C$, and $p = 1$ atm) past a horizontal heated flat plate ($l = 5$ m and $T_w = 120°C$). The Blasius flow is first laminar until at, say, $Re_{x_{trans}} = 2 \times 10^6$ the boundary layer suddenly becomes turbulent with a continuous momentum thickness at the delayed "point" of transition. Find expressions for the laminar/turbulent thermal flow parameters (cf. sketch).

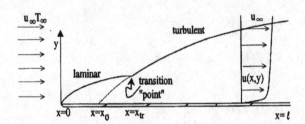

Properties

The air properties at $T_f = 70°C$ are $\rho = 1.009$ kg/m^3, $\nu = 2 \times 10^{-5}$ m^2/s, $c_p = 1009$ J/kg · K, and $Pr = 0.72$.

Postulates

Conceptually, the turbulent boundary layer has an effective origin at $x_0 < x_{tr}$ and at $x = x_{tr}$, $\delta_2|_{lam} = \delta_{2_{trans}} = \delta_2|_{turb}$. Thus, we find from

$$Re_{tr} = \frac{u_\infty x_{tr}}{\nu} \quad \Rightarrow \quad x_{tr} = 1.33 \text{ m}$$

and from

◆ $$\frac{\delta_2|_\ell}{x} = \frac{0.664}{Re_x^{1/2}} \qquad (\text{cf. Sect. 3.2.3D})$$

$$\delta_2|_{trans} = x_{tr} \frac{0.664}{\sqrt{Re_{x_{tr}}}} := 0.626 \times 10^{-3}$$

so that with

♦ $\qquad \dfrac{\delta_2|_t}{x} = \dfrac{0.036}{\mathrm{Re}_x^{1/5}}$ (cf. Sect. 3.2.5)

where here $x \stackrel{\wedge}{=} x_{\mathrm{tr}} - x_0$ and $\delta_2|_{\mathrm{turb}} = \delta_2|_{\mathrm{trans}}$. Hence, we can obtain x_0 from

$$\frac{0.626 \times 10^{-3}}{x_{\mathrm{tr}} - x_0} = 0.036 \left[\frac{30(x_{\mathrm{tr}} - x_0)}{2 \times 10^{-5}}\right]^{-0.2}$$

that is,

$\qquad x_0 = 1.1$ m

Solution

Now, the result of Problem (6) provides the solution for the initial part, for $0 \le x < x_0$:

$\qquad \mathrm{Nu}_x = 0.2957\sqrt{\mathrm{Re}_x}$ (laminar air flow)

Equation (P4.22) yields with $x_h = 0$ the solution for $x_0 \le x \le l$:

$\qquad \mathrm{St}_{\bar{x}} = 0.0295\mathrm{Re}_x^{-0.2}$ (turbulent air flow)

Note that the results can be used to calculate and plot the local laminar/turbulent heat transfer coefficient.

4.1.3 Natural Convection Boundary Layers

When a stagnant fluid reservoir is exposed to a heated or cooled wall or body surface, the temperature differences $\Delta T(x, y) = T_{\mathrm{wall}} - T_{\mathrm{fluid}}$ may change the local fluid density, thus in turn creating a buoyancy force that naturally sets the fluid near the wall into motion. The resulting boundary-layer flow can be laminar or turbulent (cf. Fig. 4.3b). Natural or free convection may also play a significant role in forced convection heat transfer when the Grashof number or the Rayleigh number is large enough, typically when the Richardson number is about unity (cf. page 258 and Table 4.1).

$$\mathrm{Ri}_l \equiv \frac{\mathrm{Gr}_l}{\mathrm{Re}_l^2} = \left(\frac{\Delta \rho}{\rho}\right)\left(\frac{gl}{u_{\mathrm{ref}}^2}\right) = O(1) \tag{4.55}$$

Mathematically, the flow field is coupled with the temperature field (cf. Eq. (4.19b) and Fig. 4.3b) and the Nusselt number also becomes a function of the Grashof number: Nu = Nu(Re, Gr, Pr). Convection heat transfer systems for which natural convection could become important include (cf. Gebhart et al. 1988) the following:

(i) Vertical heated walls where $\delta \approx \delta_{\mathrm{th}}$ for $\mathrm{Pr} \lesssim 1$ and $\delta > \delta_{\mathrm{th}}$ for $\mathrm{Pr} \gg 1$. In the latter case, the higher the Prandtl number, the thicker the layer of unheated fluid driven upward by the heated layer.

(ii) Free convection on a horizontal or inclined flat plate often extended to a mixed thermal convection problem.

(iii)   Heated or cooled standard bodies such as spheres and cylinders of different orientations. Of practical importance are multiple closely spaced cylinders, where interaction effects greatly determine the local heat transfer coefficient.
(iv)   Natural convection in a channel formed by two heated vertical walls creating the chimney effect.
(v)   Free convection in an enclosure largely determined by the thermal boundary condition and the enclosure geometry/orientation.
(vi)   Natural convection in porous materials such as: embedded heated walls, cylinders or spheres, and porous medium enclosures heated from one side.
(vii) Natural convection and mass transfer such as laminar film condensation on a cold vertical wall or moisture movement in granular materials or composites.

Numerous (natural) phenomena such as large-scale circulations in the atmosphere or in the ocean, up-drafts in street canyons or chimneys, heat dissipation of subsurface nuclear waste, city heat islands, natural cooling of electronic equipment, partially heated enclosures, double-pane windows, solar rooms, and natural cooling of casted materials or manufactured goods can be modeled with concepts discussed later.

A   Boussinesq Assumption

For small local density variations it is assumed that $\rho(T) = \rho_\infty = \text{const}$ everywhere in the governing equations except in the body force term, where $\rho(T)$ is approximated by using an equation of state, for example, the ideal gas law. Specifically, combining

$$\rho = \frac{p_\infty/R}{T} \quad \text{and} \quad \rho_\infty = \frac{p_\infty/R}{T_\infty} \quad \text{to} \quad \rho - \rho_\infty = \frac{p_\infty}{R}\left(\frac{1}{T} - \frac{1}{T_\infty}\right)$$

leads to

$$\rho - \rho_\infty = \rho\left(1 - \frac{T}{T_\infty}\right) \tag{4.56a}$$

which can be rewritten for small temperature changes, $T - T_\infty$, as

$$\frac{\rho_\infty - \rho}{\rho_\infty}\left(1 - \frac{\rho_\infty - \rho}{\rho_\infty}\right)^{-1} = \frac{T - T_\infty}{T_\infty}$$

or with $(T - T_\infty) \ll T_\infty$

$$\rho \approx \rho_\infty\left[1 - \frac{1}{T_\infty}(T - T_\infty) + \cdots\right] \tag{4.56b}$$

In general, for all Newtonian fluids

◆ $$\rho \approx \rho_\infty[1 - \beta(T - T_\infty) + \cdots] \tag{4.56c}$$

where the volumetric expansion coefficient is defined as

$$\beta = -\frac{1}{\rho}\left(\frac{\partial \rho}{\partial T}\right)_p \tag{4.56d}$$

For boundary-layer flow with zero pressure gradient (cf. Fig. 4.3b)

$$\frac{\partial p}{\partial x} = \frac{dp}{dx} \approx \frac{dp_\infty}{dx} = -\rho_\infty g$$

Now, combining the fluid static pressure gradient, $-\rho_\infty g$, with the body force term, ρg_x, yields $-(\rho - \rho_\infty)g_x$, or using Equation (4.56c) for $\rho(T)$,

$$-\frac{\partial p}{\partial x} + \rho g_x = \rho_\infty g - \rho g \quad \text{is then replaced by} \quad \rho_\infty g \beta (T - T_\infty),$$

as discussed in Section 4.1 and applied later.

B Scale Analysis (cf. Sect. 2.3.1)

The governing equations for free convection in terms of net heat flux and net force balances are

$$\blacklozenge \qquad \underbrace{u\frac{\partial T}{\partial x} + v\frac{\partial T}{\partial y}}_{\text{convection}} = \underbrace{\alpha\frac{\partial^2 T}{\partial y^2}}_{\text{conduction}} \qquad\qquad (4.57a)$$

and

$$\blacklozenge \qquad \underbrace{u\frac{\partial u}{\partial x} + v\frac{\partial u}{\partial y}}_{\text{inertia}} = \underbrace{v\frac{\partial^2 u}{\partial y^2}}_{\text{friction}} + \underbrace{g\beta(T - T_\infty)}_{\text{buoyancy}} \qquad\qquad (4.57b)$$

With the appropriate scales

$$u \to \tilde{u}, \qquad v \to \tilde{v}, \qquad \partial T \to T - T_\infty = \Delta T, \qquad \partial x \to l, \qquad \partial y \to \delta_{\text{th}} \quad \text{or} \quad \delta$$

we can form several proportionalities depending on the dominance of particular fluxes or forces within the boundary layers. From continuity

$$\frac{\tilde{u}}{l} \sim \frac{\tilde{v}}{\delta} \qquad\qquad (4.58a)$$

so that the thermal energy equation generates

$$\tilde{u}\frac{\Delta T}{l} \sim \alpha\frac{\Delta T}{\delta_{\text{th}}^2} \qquad\qquad (4.58b)$$

and applying the momentum equation near the wall a balance of viscous and buoyancy forces indicates that

$$v\frac{\tilde{u}}{\delta^2} \sim g\beta \, \Delta T \qquad\qquad (4.58c)$$

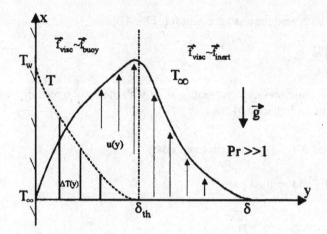

Special Case Studies (cf. Bejan 1984)

(a) For high-Prandtl-number-fluids such as heavy oils, $\delta > \delta_{th}$ and with δ_{th} here the dominant wall layer (cf. sketch), we obtain from Equations (4.58a–c)

$$\tilde{u} \sim \frac{\alpha l}{\delta_{th}^2} \tag{4.59a}$$

and

$$\nu \frac{\alpha l}{\delta_{th}^4} \sim g\beta\Delta T \tag{4.59b}$$

or

◆
$$\frac{\delta_{th}}{l} \sim \mathrm{Ra}_l^{-1/4} \tag{4.60a}$$

where the (local) Rayleigh number is defined as

◆
$$\mathrm{Ra}_x \equiv \mathrm{Gr}_x\mathrm{Pr} = \frac{g\beta\Delta T}{\alpha\nu}x^3 \tag{4.60b}$$

Combining (4.59) and (4.60) yields a proportionality for a reference velocity in natural convection, that is,

$$\tilde{u} = u_{\mathrm{ref}} \sim \frac{\alpha}{l}\mathrm{Ra}_l^{1/2} \quad \text{or} \quad u_{\mathrm{ref}} \sim \sqrt{x}$$

from which we deduce

◆
$$u_{\mathrm{ref}} = \sqrt{g\beta(T_w - T_\infty)l} \tag{4.61}$$

Recalling that the heat transfer coefficient $h = -(k/\Delta T)(\partial T/\partial y)|_w$, it can be scaled as

$$h \sim \frac{k}{\delta_{th}}$$

and we can state with Nu $\equiv hl/k$ that

♦ $\mathrm{Nu}_l \sim \mathrm{Ra}_l^{1/4}$ when Pr $\gg$ 1   (4.62)

Also from Equation (4.60)

♦ $\delta_{th} \sim x^{1/4}$ and $\dfrac{\delta}{\delta_{th}} \sim \mathrm{Pr}^{1/2} > 1$ (4.63a,b)

(b)   For low-Prandtl-number fluids such as gases and liquid metals, $\delta \lesssim \delta_{th}$ and within a very thin shear layer, δ_v, at the wall (cf. sketch):

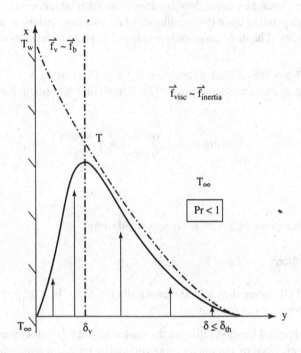

$$\nu \frac{\tilde{u}}{\delta_v^2} \sim g\beta\Delta T$$

while the remaining boundary layer is governed by

$$\tilde{u}\frac{\tilde{u}}{l} \sim g\beta\Delta T$$

from which $\tilde{u} \sim (g\beta\Delta T l)^{1/2}$ so that

$$\nu \frac{(g\beta\Delta T l)^{1/2}}{\delta_v^2} \sim g\beta\Delta T$$

or

$$\frac{\delta_v}{l} \sim \mathrm{Gr}_l^{-1/4} \qquad\qquad\qquad (4.64a)$$

Now, assuming $\delta_{th} - \delta_v \approx \delta_{th}$ we can write

$$\blacklozenge \quad \frac{\delta_{th}}{l} \sim (Ra_l Pr)^{-1/4} \tag{4.64b}$$

and again with $h \sim k/\delta_{th}$ and $Nu_l = hl/k$

$$\blacklozenge \quad Nu_l \sim (Ra_l Pr)^{1/4}; \qquad Pr < 1 \tag{4.65}$$

C Solution Techniques (cf. Sect. 2.3.3)

Again, free convection boundary-layer flow equations can be transformed or directly numerically integrated, depending upon the complexity of the problem and the availability of computational resources. The three commonly employed solution techniques can be described as follows:

(i) For self-similar wall flows, the similarity variable $\eta = y/\delta_{th}(x)$ where $\delta_{th} \sim x^{1/4}$ is obtained from scaling analysis as shown previously. Employing the stream function approach

$$u = \frac{\partial \psi}{\partial y} \quad \text{and} \quad v = -\frac{\partial \psi}{\partial x} \quad \text{where } \psi = \left(\begin{array}{c} \text{veloc.} \\ \text{scale} \end{array} \right) \times \left(\begin{array}{c} \text{length} \\ \text{scale} \end{array} \right) f(\eta, x)$$

or

$$\psi = \nu F(\eta) G(x)$$

The boundary-layer equations (4.57a,b) are transformed using

$$\theta = \frac{T - T_\infty}{T_w - T_\infty} \quad \text{where} \quad T_w - T_\infty = Ax^n$$

The resulting coupled ODEs are then solved numerically (cf. App. F.3) subject to appropriate starting boundary conditions.

(ii) The accuracy for the integral method relies on the choice of suitable velocity and temperature profiles enhanced by parameter proportionalities based on scale analysis. Specifically, Equation (4.57a) is transformed to the thermal energy integral relation

$$\frac{d}{dx} \int_0^\infty u(T_\infty - T)\, dy = \alpha \frac{\partial T}{\partial y}\bigg|_w \tag{4.66}$$

and Equation (4.57b) is transformed to the momentum integral relation

$$\frac{d}{dx} \int_0^\infty u^2\, dy = -\nu \frac{\partial u}{\partial y}\bigg|_w + g\beta \int_0^\infty (T - T_\infty)\, dy \tag{4.67}$$

Possible profiles include

$$T - T_\infty = \Delta T e^{-y/\delta_{th}} \quad \text{and} \quad u = u_{\text{ref}}\, e^{-y/\delta}(1 - e^{-y/\delta_{th}}) \qquad \text{for } Pr \gg 1$$

or

$$T - T_\infty = \Delta T \left(1 - \frac{y}{\delta_{th}} \right)^2 \quad \text{and} \quad u = u_{ref} \left(1 - \frac{y}{\delta} \right)^2 \left(\frac{y}{\delta} \right) \qquad \text{for Pr} \gtrless 1 \; (4.68a\text{--}d)$$

where $\Delta T = T_w - T_\infty$. Depending upon the magnitude of the Prandtl number further simplifying assumptions can be introduced.

(iii) A direct numerical solution of the transport equations is most appropriate when restrictive assumptions are not permissible. Of the numerous finite difference algorithms available, Keller's box method is a good choice for solving boundary-layer-type problems (cf. App. F.4 as well as Cebeci and Bradshaw 1977 and 1988). Applications are discussed in Chapter 5.

Problem (9): Derive the Dimensionless Groups for Free Convection
Considering that the heat transfer coefficient for steady laminar free convection depends on

$$h = h(\rho, \mu, c_p, k, l, g\beta, \Delta T)$$

determine the relevant dimensionless groups by using the Buckingham Pi Theorem.

Solution
The mass–length–time–temperature (M-L-T-θ) dimensions of the variables of interest are

h	ρ	μ	c_p	k	l	$g\beta$	ΔT
$\dfrac{M}{T^3\theta}$	$\dfrac{M}{L^3}$	$\dfrac{M}{LT}$	$\dfrac{L^2}{T^2\theta}$	$\dfrac{ML}{T^3\theta}$	L	$\dfrac{L}{T^2\theta}$	θ

By inspection, we have $n = 8$ dimensional groups and the repeating variable $j\,(j = 5$, i.e., $\rho, \mu, k, l,$ and $\Delta T)$ is the maximum number of variables that do not form a π-group among themselves. Hence, $k = n - j = 3$ nondimensional groups (i.e., Nu, Gr, and Pr) can be expected.

Now, $[\pi_1] = h\rho^a \mu^b k^c l^d \Delta T^e \hat{=} [1]$, from which

$$
\left.
\begin{array}{ll}
\text{mass} & 1 + a + b + c = 0 \\
\text{length} & -3a - b + c + d = 0 \\
\text{time} & -3 - b - 3c = 0 \\
\text{temperature} & -1 - c + e = 0
\end{array}
\right\}
$$

In order to solve five unknowns with four equations, we set $e \equiv 0$.

Hence,

$$c = -1, \qquad b = 0, \qquad a = 0 \quad \text{and} \quad d = 1$$

so that

$$\pi_1 = \frac{hl}{k} \equiv \text{Nu}_l$$

Similarly, the remaining variables c_p and $g\beta$ are nondimensionalized employing the five repeating parameters. The results are with $e \equiv 0$

$$\pi_2 = \frac{\mu c_p}{k} \equiv \mathrm{Pr}$$

and with $e \equiv 1$

$$\pi_3 = \frac{\rho^2 g\beta l^2 \Delta T}{\mu^2} \equiv \mathrm{Gr}_l$$

so that

◆ $\mathrm{Nu}_l = \mathrm{Nu}_l(\mathrm{Gr}_l, \mathrm{Pr})$

Problem (10): Self-Similar Natural Convection along a Vertical Wall
Transform the steady two-dimensional boundary-layer equations for free convection along a vertical heated wall. Assume $\mathrm{Pr} = O(1)$, that is, $\delta \approx \delta_{\mathrm{th}}$.

Solution
The governing equations are

$$\frac{\partial u}{\partial x} + \frac{\partial v}{\partial y} = 0, \qquad u\frac{\partial u}{\partial x} + v\frac{\partial u}{\partial y} = \nu\frac{\partial^2 u}{\partial y^2} + g\beta(T - T_\infty)$$

and

$$u\frac{\partial T}{\partial x} + v\frac{\partial T}{\partial y} = \frac{\nu}{\mathrm{Pr}}\frac{\partial^2 T}{\partial y^2}$$

subject to

$$u = v = 0 \quad \text{and} \quad T = T_w \qquad \text{at} \quad y = 0$$

and

$$u = 0 \quad \text{and} \quad T = T_\infty \qquad \text{at} \quad y = \delta \approx \delta_{\mathrm{th}}$$

We postulate with Eqs. (4.63a and 4.64b) for $\eta = y/\delta_{\mathrm{th}}$, recalling that $\delta_{\mathrm{th}} \sim x^{-1/4}$,

$$\eta = \left[\frac{g\beta(T_w - T_\infty)}{\nu^2 x}\right]^{1/4} y$$

and with Eq. (4.61) for $\psi = [\begin{smallmatrix}\text{veloc.}\\\text{scale}\end{smallmatrix}] \times [\begin{smallmatrix}\text{length}\\\text{scale}\end{smallmatrix}] f(\eta)$

$$\psi = \left[g\beta(T_w - T_\infty)\nu^2 x^3\right]^{1/4} f(\eta)$$

Finally,

$$\theta = \frac{(T - T_\infty)}{(T_w - T_\infty)}$$

Now, employing the stream function approach, we transform with $\Delta T = T_w - T_\infty$

$$u \equiv \left.\frac{\partial \psi}{\partial y}\right|_x = \frac{\partial \psi}{\partial \eta}\frac{\partial \eta}{\partial y} = [g\beta\Delta T x]^{1/2} f'$$

$$v \equiv -\left.\frac{\partial \psi}{\partial x}\right|_y = -\left.\frac{\partial \psi}{\partial x}\right|_\eta - \frac{\partial \psi}{\partial \eta}\frac{\partial \eta}{\partial x} = -(g\beta\Delta T v^2 x^2)^{1/4}\left(\frac{1}{4} + \frac{3f}{x} + f'\frac{\partial \eta}{\partial x}\right)$$

$$\frac{\partial u}{\partial x} = \left.\frac{\partial u}{\partial x}\right|_\eta + \frac{\partial u}{\partial \eta}\frac{\partial \eta}{\partial x} = [g\beta\Delta T x^2]^{1/2}\left(\frac{f'}{2x} + f''\frac{\partial \eta}{\partial x}\right)$$

$$\frac{\partial u}{\partial y} = \frac{\partial u}{\partial \eta}\frac{\partial \eta}{\partial y} = [g\beta\Delta T x]^{\frac{1}{2}} f''\left[\frac{g\beta\Delta T}{v^2 x}\right]^{1/4}$$

$$v\frac{\partial^2 u}{\partial y^2} = [g\beta\Delta T]f'''$$

$$\frac{\partial T}{\partial x} = \left.\frac{\partial T}{\partial x}\right|_\eta + \frac{\partial T}{\partial \eta}\frac{\partial \eta}{\partial x} = \Delta T\left[\frac{\eta}{x}\theta + \theta'\frac{\partial \eta}{\partial x}\right]$$

$$-v\frac{\partial T}{\partial y} = \Delta T[g\beta\Delta T x]^{1/2}\left(\frac{3}{4}\frac{\theta'}{x}f + \theta' f'\frac{\partial \eta}{\partial x}\right)$$

$$\frac{v}{\mathrm{Pr}}\frac{\partial^2 T}{\partial y^2} = \frac{\theta''}{\mathrm{Pr}}\Delta T\left[\frac{g\beta\Delta T}{x}\right]^{1/2}$$

Substitution and rearranging yield (cf. Eq. (4.30a))

◆ $$4f''' + 3ff'' - 2(f')^2 + 4\theta = 0 \tag{P4.23}$$

and

◆ $$\theta'' + \frac{3}{4}\mathrm{Pr}f\theta' = 0 \tag{P4.24}$$

subject to $f(\eta = 0) = f'(\eta = 0) = 0$, $\theta(\eta = 0) = 1$, $f'_\infty(\eta = \eta_\infty) = 0$, and $\theta(\eta = \eta_\infty) = 0$. Of major interest are the local Nusselt number

◆ $$\mathrm{Nu}_x \equiv \frac{q_w}{\Delta T}\frac{x}{k} := -\theta'_w \mathrm{Gr}_x^{1/4} \tag{P4.25}$$

and the skin friction factor

◆ $$c_f(x) \equiv \frac{2\tau_w}{\rho u_{\mathrm{ref}}^2} := \frac{2f''_w}{(\mathrm{Gr}_x)^{1/4}} \tag{P4.26}$$

where $\mathrm{Gr}_x = g\beta\Delta T x^3/v^2$ and $\Delta T = T_w - T_\infty$.

Numerical solution of (P4.23 and P4.24) yields for different Prandtl number fluids (cf. App. F.4)

Pr	$-\theta'_w = \mathrm{Nu}_x \mathrm{Gr}_x^{-1/4}$	$f''_w = \dfrac{c_f}{2}\mathrm{Gr}_x^{1/4}$
0.72	0.357	0.956
1.00	0.401	0.908
10.00	0.826	0.593

Note, for nonuniform surface temperatures $T_w(x)$ and explicit x-dependence of the dimensionless stream function $f(x; \eta)$, Equations (P4.23) and (P4.24) can be generalized to

$$4f''' + 3ff'' - 2(f')^2 + 4\theta - \kappa\left[2(f')^2 - ff''\right] = 4x\left(f'\frac{\partial f'}{\partial x} - f''\frac{\partial f}{\partial x}\right) \qquad \text{(P4.27)}$$

and

$$\frac{\theta''}{\mathrm{Pr}} + \frac{3}{4}f\theta' + \kappa\left[\frac{1}{4}f\theta' - f'\theta\right] = x\left(f'\frac{\partial\theta}{\partial x} - \theta'\frac{\partial f}{\partial x}\right) \qquad \text{(P4.28)}$$

$$\text{where} \qquad \kappa \equiv \left(\frac{x}{T_w(x) - T_\infty}\right)\frac{d}{dx}[T_w(x) - T_\infty]$$

which has to be constant for similar boundary-layer flows.

4.1.4   *Natural Convection in Enclosures*

Heated wall enclosures filled with one or two fluids and possibly some type of porous material are key components in many engineering, food science, and geophysical systems. Buoyancy-induced internal flow and heat transfer are complex phenomena that largely depend on the particular thermal wall condition, the type of enclosed fluid or porous matrix, as well as the enclosure geometry and orientation (cf. Fig. 4.6).

Enclosures heated from the side represent base cases to study air circulation in small rooms, the efficiency of double-pane windows, solar collectors, cooling systems of large

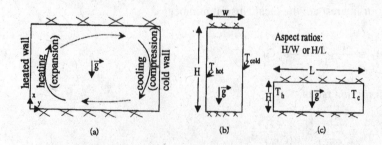

Fig. 4.6. Schematics of two-dimensional enclosures heated from the side: (a) room with one heated vertical wall; (b) vertical slot; (c) shallow enclosure

rotating machinery, railroad containers, and so on. The characteristics of a few special cases are as follows:

(i) Natural convection in tall enclosures or vertical slots (cf. Fig. 4.6b) are determined by the aspect ratio, $A = H/W$; the Rayleigh number, $\mathrm{Ra}_H = g\beta \Delta T H^3/(\alpha\nu)$; and the Prandtl number, $\mathrm{Pr} = \nu/\alpha$. Stable regimes include:

- conduction where the fluid hardly moves, $\mathrm{Ra} \leq O(1)$
- transition to fluid motion, $\mathrm{Ra} = O(10^2)$
- unicellular boundary-layer flow, $\mathrm{Ra} \geq O(10^4)$

For high-Prandtl-number fluids, that is, $\mathrm{Pr} = O(10^2)$, and large aspect ratios, that is, $A = 10\text{--}20$, the unicellular convection breaks down and instabilities of various degrees set in. Specifically,

- at $\mathrm{Ra}_{\mathrm{crit}} \approx (3-6) \times 10^5$, the onset of secondary flows starts and multicellular patterns appear;
- at $3 \times \mathrm{Ra}_{\mathrm{crit}}$, tertiary cells that counterrotate between the secondary cells appear;
- at $10 \times \mathrm{Ra}_{\mathrm{crit}}$, traveling waves disrupt the cellular patterns; and
- at $\mathrm{Ra} > 10^7$, the flow becomes turbulent.

Obviously, modeling the stable regimes is rather straightforward, considering Equations (4.69) to (4.72), which follow; whereas flows at $\mathrm{Ra} > 10^7$ require additional turbulence models for convective heat transfer. The unstable range of $3 \times 10^5 < \mathrm{Ra} \leq 10^7$ is still being experimentally explored.

(ii) Heat transfer in shallow enclosures is dominated by the presence of strong vertical thermal layers at typically high Rayleigh numbers. Assuming quasi-insulated top and bottom walls, horizontal counterflow sets up an elongated core region in which frictional and buoyancy forces balance each other. With the aspect ratio $A = H/L \ll 1$, Equations (4.69) to (4.72) can be significantly reduced for the core region.

(iii) Horizontal parallel plates or narrow enclosures heated from below generate multicellular, counterrotating flow patterns known as *Bénard flow*. The critical Rayleigh number based on the hot/cold wall spacing d is $\mathrm{Ra}_d \approx 1{,}700$ (cf. Tritton 1977).

Using the Boussinesq approximation and neglecting viscous dissipation, the steady laminar two-dimensional thermal flow equations are (cf. Fig. 4.6a)

$$\frac{\partial u}{\partial x} + \frac{\partial v}{\partial y} = 0 \tag{4.69}$$

$$u\frac{\partial u}{\partial x} + v\frac{\partial u}{\partial y} = -\frac{1}{\rho}\frac{\partial p}{\partial x} + \nu\left(\frac{\partial^2 u}{\partial x^2} + \frac{\partial^2 u}{\partial y^2}\right) + g\beta(T - T_c) \tag{4.70}$$

$$u\frac{\partial v}{\partial x} + v\frac{\partial v}{\partial y} = -\frac{1}{\rho}\frac{\partial p}{\partial y} + \nu\left(\frac{\partial^2 v}{\partial x^2} + \frac{\partial^2 v}{\partial y^2}\right) \tag{4.71}$$

$$u\frac{\partial T}{\partial x} + v\frac{\partial T}{\partial y} = \alpha\left(\frac{\partial^2 T}{\partial x^2} + \frac{\partial^2 T}{\partial y^2}\right) \tag{4.72}$$

The associated boundary conditions are at

$$y = 0; \qquad 0 \leq x \leq H: \qquad u = v = 0 \quad \text{and} \quad T = T_h$$
$$y = W; \qquad 0 \leq x \leq H: \qquad u = v = 0 \quad \text{and} \quad T = T_c$$
$$x = 0, H; \qquad 0 \leq y \leq W: \qquad u = v = 0 \quad \text{and} \quad \frac{\partial T}{\partial x} = 0$$

In employing the stream function approach, the pressure gradients are eliminated and equations (4.70) to (4.72) collapse to

$$\nabla^2 \tilde{\psi} = \frac{1}{\text{Pr}} \left[\frac{\partial \tilde{\psi}}{\partial \tilde{y}} \frac{\partial}{\partial \tilde{x}} (\nabla^2 \tilde{\psi}) - \frac{\partial \tilde{\psi}}{\partial \tilde{x}} \frac{\partial}{\partial \tilde{y}} (\nabla^2 \tilde{\psi}) \right] - \text{Ra} \frac{\partial \theta}{\partial \tilde{y}} \tag{4.73}$$

and

$$\nabla^2 \theta = \frac{\partial \tilde{\psi}}{\partial \tilde{y}} \frac{\partial \theta}{\partial \tilde{y}} - \frac{\partial \tilde{\psi}}{\partial \tilde{x}} \frac{\partial \theta}{\partial \tilde{y}} \tag{4.74}$$

where $\quad \tilde{x} = \dfrac{x}{W}, \qquad \tilde{y} = \dfrac{y}{W}, \qquad \tilde{\psi} = \dfrac{\psi}{\alpha} \quad \text{and} \quad \theta = \dfrac{T - T_c}{T_h - T_c}$

The key dimensionless groups are

$$A = \frac{H}{W}, \qquad \text{Pr} = \frac{\nu}{\alpha} \quad \text{and} \quad \text{Ra}_W = g\beta(T_h - T_c)\frac{W^3}{(\alpha\nu)}$$

Experimental and numerical results of the velocity and temperature fields for limiting values of aspect ratio A, Prandtl number Pr, and Rayleigh number Ra are given in Gebhart et al. (1988). A transient three-dimensional case study is discussed in Chapter 5.

4.1.5 Thermal Free Shear Layer Flows

Examples of free shear layers for which Prandtl's boundary-layer assumption may apply include detached thermal shear layers after boundary-layer separation, jet discharge from rectangular or circular ducts into the same but stagnant fluid at different temperature, and mixing layers between two uniform streams at different velocities and temperatures (cf. Fig. 4.7 and Sect. 3.2.5C).

Concentrating on steady laminar/turbulent jets, wakes, or mixing layers, the describing equations are for $(b(x)/x) \sim \nu/u \sim ([\partial^2/\partial x^2]/[\partial^2/\partial y^2]) \ll 1$ (cf. Fig. 4.7a–c)

$$\blacklozenge \qquad \frac{\partial u}{\partial x} + \frac{\partial v}{\partial y} = 0 \tag{4.75}$$

$$\blacklozenge \qquad u\frac{\partial u}{\partial x} + v\frac{\partial u}{\partial y} = \frac{1}{\rho}\frac{\partial \tau}{\partial y} \tag{4.76}$$

$$\blacklozenge \qquad u\frac{\partial T}{\partial x} + v\frac{\partial T}{\partial y} = \frac{1}{\rho c_p}\frac{\partial q}{\partial y} \tag{4.77}$$

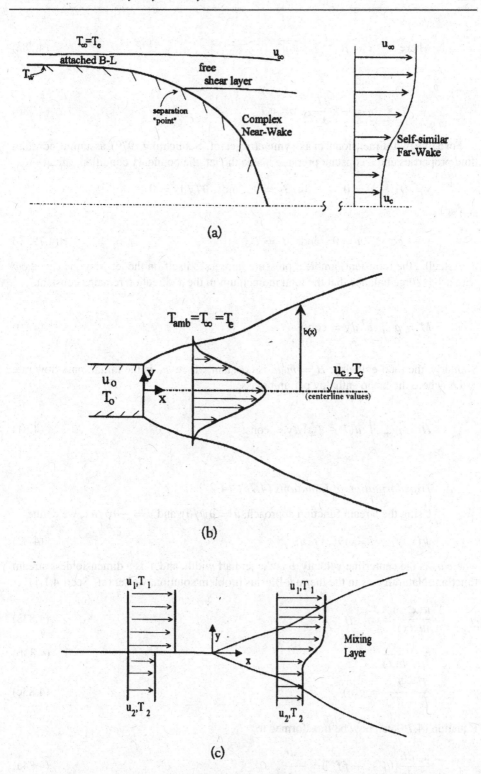

Fig. 4.7. Thermal free shear flows: (a) free shear layer separating outer flow and wake region; (b) symmetric jet; (c) thermal mixing layer.

where $\qquad \tau = \mu \dfrac{\partial u}{\partial y} - \rho \overline{u'v'}$ $\qquad\qquad\qquad\qquad\qquad\qquad$ (4.78a)

and

$$q = -k \frac{\partial T}{\partial y} + \rho c_p \overline{v'T'} \qquad\qquad\qquad (4.78b)$$

For a free two-dimensional or axisymmetric jet (cf. Schlichting 1979), assuming constant fluid properties and a constant pressure inside the jet, the boundary conditions are at

$$y = 0 \qquad v = 0, \qquad \partial u / \partial y = 0 \quad \text{and} \quad \partial T / \partial y = 0$$

and at

$$y \rightarrow \infty \qquad u = 0 \quad \text{and} \quad T = T_\infty \qquad\qquad (4.79a,b)$$

Actually, the (constant) ambient pressure impresses itself on the jet, and one can show with a 1-D force balance that the total momentum in the x-direction remains constant.

$$M = \rho \int_{-\infty}^{\infty} u^2 \, dy = \text{const} \qquad\qquad\qquad (4.80)$$

Similarly, the total enthalpy, $H = \dot{m}_0 \hat{h}$, is constant where $\dot{m}_0$ is the initial mass flow rate and h is here the mean enthalpy per unit mass.

$$H = \rho c_p \int_{-\infty}^{\infty} u(T - T_\infty) \, dy = \text{const} \qquad\qquad\qquad (4.81)$$

A   Transformation of Equations (4.75)–(4.77)

Using the stream function approach, $u = \partial \psi / \partial y$ and $v = -\partial \psi / \partial x$, we define

$$\psi(x, y) = u_c(x) b(x) f(\eta) \qquad\qquad\qquad (4.82)$$

where u_c is the centerline velocity, b is the jet half width, and f is a dimensionless stream function. Naturally, as in the thermal Blasius problem solution, we set (cf. Sect. 4.1.1)

$$\frac{u(x, y)}{u_c(x)} = f'(\eta) \qquad\qquad\qquad (4.83a)$$

$$\eta = \frac{y}{b(x)} \qquad\qquad\qquad (4.83b)$$

$$\frac{T - T_\infty}{T_c - T_\infty} = g(\eta) \qquad\qquad\qquad (4.83c)$$

Equation (4.76) can now be transformed to

$$\frac{u_c}{2} \frac{db}{dx} [(f')^2 + ff''] = -\frac{\tau'}{\rho} \qquad\qquad\qquad (4.84a)$$

subject to

$$f(\eta = 0) = f''(\eta = 0) = 0 \quad \text{and} \quad f'(\eta \to \infty) \to 0 \tag{4.84b}$$

Similarly, the heat transfer equation (4.77) now reads

$$bu_c \frac{dT_c}{dx}(fg)' = -\frac{q'}{\rho c_p} \tag{4.85a}$$

subject to

$$g'(\eta = 0) = 0 \quad \text{and} \quad g(\eta \to \infty) \to 0 \tag{4.85b}$$

Notes

- Now, the constant total momentum and enthalpy take on the form (cf. Sect. 3.2.5C)

$$M = 2\rho u_c^2 b \int_0^\infty (f')^2 d\eta = C_1 \tag{4.86a}$$

and

$$H = 2\rho c_p b(T_c - T_\infty) \int_0^\infty f'g \, d\eta = C_2 \tag{4.86b}$$

which implies that

$$(u_c^2 b) = \text{const} \quad \text{and} \quad [b(T_c - T_\infty)] = \text{const} \tag{4.87a,b}$$

- The stress, τ, and heat flux, q, have to be specified for laminar or turbulent flow.
- For similarity solutions to exist, Equations (4.84) and (4.85) have to be (explicitly) independent of x.

Applying the analysis to steady laminar two-dimensional jets, that is, $\text{Re} \le 30$, we introduce $\tau = \mu \partial u/\partial y$ and $q = -k\partial T/\partial y$ so that (4.84a) and (4.85a) reduce to

$$\blacklozenge \qquad f''' + \frac{u_c}{2\nu}b\frac{db}{dx}\left[(f')^2 + ff''\right] = 0 \tag{4.88}$$

and

$$\blacklozenge \qquad \frac{g''}{\text{Pr}} - \frac{b^2 u_c}{\nu(T_c - T_\infty)}\frac{dT_c}{dx}(fg)' = 0 \tag{4.89}$$

Specific applications to self-similar jets are given later.

B Laminar Self-Similar Jets

Similarity demands that the coefficients on the LHS of (4.88) are constant, that is,

$$\frac{u_c b}{2\nu}\frac{db}{dx} \equiv 1 \tag{4.90a}$$

and for convenience

$$\frac{b^2 u_c}{\nu(T_c - T_\infty)}\frac{dT_c}{dx} \equiv -1 \tag{4.90b}$$

Equations (4.88) and (4.89) now take on the form

$$f''' + (f')^2 + ff'' = 0 \tag{4.91}$$

and

$$g'' + \Pr(fg)' = 0 \tag{4.92}$$

Equation (4.91) can be rewritten as

$$\frac{d}{d\eta}(f'') + \frac{d}{d\eta}(ff') = 0$$

Integration yields

$$f'' + ff' = \text{const}$$

Using the boundary conditions and integrating again yields

$$f(\eta) = \sqrt{2}\, tanh\left(\frac{\eta}{\sqrt{2}}\right)$$

or (cf. Fig. 4.8)

♦ $$\frac{u}{u_c} = f'(\eta) = sech^2\left(\frac{\eta}{\sqrt{2}}\right) \tag{4.93}$$

where $\text{sech}\, \alpha = 2/(e^\alpha + e^{-\alpha}) = [cosh\alpha]^{-1}$. Similarly, the solution to (4.92) is (Fig. 4.8)

♦ $$g(\eta) = \left[sech\left(\frac{\eta}{\sqrt{2}}\right)\right]^{2\Pr} \tag{4.94}$$

Now, (4.86a,b) can be evaluated and used to obtain together with (4.87) and (4.90)

$$b(x) = \left(\frac{12\sqrt{2}\nu^2 x^2}{M/\rho}\right)^{1/3} \sim x^{2/3} \tag{4.95a}$$

$$u_c(x) = \left[\frac{3}{32}\left(\frac{M}{\rho}\right)^2 \frac{1}{\nu x}\right]^{1/3} \sim x^{-1/3} \tag{4.95b}$$

and with $\Pr = 1$

$$T_c(x) - T_\infty = \frac{H}{\rho c_p}(\nu u_c b^2 x)^{-1/3} \sim x^{-1/3} \tag{4.95c}$$

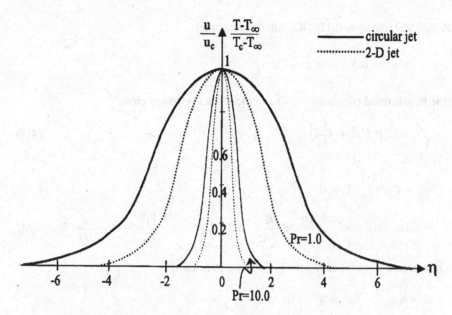

Fig. 4.8. Velocity and temperature profiles of a free jet.

C   Turbulent 2-D Jets

Equations (4.75)–(4.77) also hold for turbulent boundary-layer-type flows. Neglecting molecular transport phenomena (cf. Eqs. 4.78a,b) we postulate

$$\frac{\tau}{\rho} = -\overline{u'v'} = \varepsilon_m \frac{\partial u}{\partial y} = \varepsilon_m \frac{u_c}{b} f'' := u_c^2 F(\eta) \tag{4.96a}$$

and

$$-\frac{q}{\rho c_p} = -\overline{v'T'} = \varepsilon_h \frac{\partial T}{\partial y} = \frac{\varepsilon_m}{\text{Pr}_t} \frac{(T_c - T_\infty)}{b} g' := u_c(T_c - T_\infty)G(\eta) \tag{4.96b}$$

Hence, the governing equations can be transformed to

$$\frac{1}{2}\frac{db}{dx}\left[(f')^2 + ff''\right] + F' = 0 \tag{4.97}$$

and

$$\frac{b}{T_c - T_\infty}\frac{d}{dx}(T_c - T_\infty)(fg)' - G' = 0 \tag{4.98}$$

In order to achieve similarity, the coefficients in (4.97) and (4.98) have to be constant; that implies that

$$b(x) \sim x \quad \text{and} \quad (T_c - T_\infty) \sim x^n$$

Still, ε_m and Pr_t in Equations (4.96a,b) have to be modeled. On the basis of empirical evidence, we take

$$\varepsilon_m = Au_c b, \qquad \text{Pr}_t = \text{const} \quad \text{and} \quad b(x) = Bx$$

with $A \approx 0.037$ and $B = 0.115$; the latter implies that

$$f' \equiv \frac{u}{u_c} = 0.5 \quad \text{at } \eta = 1.0$$

Now, the transformed equations (4.97) and (4.98) can be rewritten as

♦ $\qquad f''' + C_1[(f')^2 + ff''] = 0$ $\hfill (4.99)$

and

♦ $\qquad g'' + C_2 \text{Pr}_t (fg)' = 0$ $\hfill (4.100)$

$\qquad$ where $\qquad C_1 = \frac{u_c b}{2\varepsilon_m} \frac{db}{dx} := \frac{B}{2A}$ and $C_2 \text{Pr}_t = -\frac{\text{Pr}_t}{\varepsilon_m} \frac{u_c b^2}{T_c - T_\infty} \frac{d}{dx}(T_c - T_\infty).$

Triple integration of (4.99) using the boundary conditions at

$$\eta = \eta_{\text{edge}} \qquad f' = f'' = 0$$

and

$$\eta = 0 \qquad f' = 1, \qquad f = 0$$

yields

$$f = \sqrt{\frac{2}{C_1}} tanh\left(\sqrt{\frac{C_1}{2}}\eta\right)$$

With the requirement that $f'(\eta = 1) = 1/2$, the constant $C_1 = 1.5523$, so that

$$f = 1.135 \, tanh(0.881\eta)$$

and

♦ $\qquad \dfrac{u}{u_c} = sech^2(0.881\eta)$ $\hfill (4.101\text{a--e})$

while

$$b(x) = 0.115x, \qquad u_c = 2.40\sqrt{\frac{M/\rho}{x}} \quad \text{and} \quad \dot{m} = 0.625\sqrt{\rho M x}$$

Setting $C_2 \approx C_1$ and employing symmetry, that is, $g(0) = 1$, Equation (4.100) can be integrated to

♦ $\qquad g \equiv \dfrac{T - T_c}{T_c - T_\infty} = [sech(0.881\eta)]^{2\text{Pr}_t}$ $\hfill (4.102)$

Note: Jet growth in laminar flow is $b \sim x^{2/3}$, whereas in turbulent flow $b \sim x$. With all the simplifying assumptions, the profiles are very much the same in both flow regimes.

4.2 Thermal Flows in Conduits

The analysis of internal flow problems with thermal boundary conditions is of importance for a basic understanding and design applications of numerous engineering systems. Primary examples include single or multiple tube flow in heat exchangers, heat pipe systems, commercial/residential duct heating or cooling, and thermal processing of manufactured goods. Secondary examples include systems and transport mechanisms in which excessive heat production is an undesirable side effect as in tribology, densely packed electronic equipment, nuclear reactors, and so forth. Traditionally, thermal duct flow problems are grouped into the following categories: laminar uncoupled, turbulent uncoupled, and coupled duct flows. The latter includes developing flows with fluid property changes, shock-wave interactions, chemically reactive flows, flows with phase change, and possibly conjugate heat transfer problems. The relevant equations, assuming steady two-dimensional or axisymmetric thermal flows, are discussed in Section 4.1 (cf. Eqs. (4.18–4.20)).

The base case is steady laminar hydrodynamically fully developed (HFD) and thermally fully developed (TFD) circular duct flow with either $T_w = $ const or $q_w = $ const. Variations of this standard problem include

- HFD but thermally developing entrance region, known as the Graetz problem;
- the special case of extreme Prandtl number fluids Pr $\to 0$ and Pr $\to \infty$, where Pr $= \nu/\alpha$ and Pr $\sim \delta/\delta_{\text{th}}$ in the entrance region;
- transient flows due to a pulsating pressure gradient;
- hydrodynamically and thermally developing flows;
- variable thermal wall conditions;
- ducts with noncircular cross sections;
- turbulent duct flows;
- flow of non-Newtonian fluids; and
- multiphase flow with heat transfer.

4.2.1 Steady Laminar Pipe Flow

When the hydrodynamic entrance length l_e in a pipe is negligibly small, where $l_e/D \approx 0.06 \, \text{Re}_D$ for laminar flow, the radial velocity $v = 0$, implying $\partial u/\partial x = 0$. In addition, at Peclet numbers $\text{Pe}_l > 100$, where $\text{Pe}_l = \text{Re}_l \, \text{Pr} = u_m l/\alpha$, streamwise convection heat transfer is much greater than thermal diffusion (i.e., axial conduction). As a result, the heat transfer equation in cylindrical coordinates for HFD and TFD flow in a pipe of radius r_0 can be reduced to (cf. Eq. (4.19a) and Fig. 4.4b)

$$\blacklozenge \qquad u\underbrace{\frac{\partial T}{\partial x}}_{\substack{\text{axial} \\ \text{convection}}} = \underbrace{\frac{\alpha}{r}\frac{\partial}{\partial r}\left(r\frac{\partial T}{\partial r}\right)}_{\substack{\text{radial} \\ \text{conduction}}} \qquad\qquad (4.103)$$

where internal heat generation has been omitted. Scaling of Equation (4.103) yields (subscript m indicates mean or mixing-cup values)

$$u_m \frac{\Delta T_m}{\Delta x} \sim \alpha \frac{\Delta T}{r_0^2}$$

From a 1-D enthalpy balance for a cylindrical control volume we obtain (cf. Fig. 4.9a and Eq. (4.113))

$$\frac{\Delta T_m}{\Delta x} \approx \frac{2}{r_0} \frac{q_w}{\rho c_p u_m}$$

so that

$$\frac{2q_w}{r_0 \rho c_p} \sim \alpha \frac{\Delta T}{r_0^2}$$

where $\Delta T = T_w - T_m$ and with $q_w / \Delta T = h$

$$\blacklozenge \qquad h \sim \frac{k}{2r_0} \quad \text{or} \quad \text{Nu}_D \equiv \frac{hD}{k} \sim O(1)$$

This preliminary result implies that for fully developed flow, the circular tube Nusselt number is constant and of the order of 1.

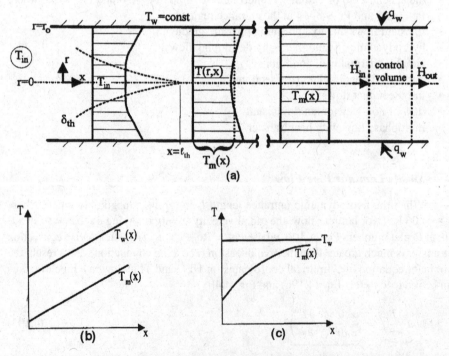

Fig. 4.9. **Uncoupled laminar duct flow: (a) nonisothermal pipe flow; (b) constant heat flux case; (c) isothermal wall case.**

Whereas the condition for hydrodynamically fully developed (HFD) flow is quite tangible (cf. Poiseuille flow in pipes), the criterion for thermally fully developed (TFD) flow requires a generalized temperature profile $\theta(r)$ that is independent of the streamwise coordinate. Such a profile fulfills the condition (Fig. 4.9)

$$\blacklozenge \qquad \frac{\partial}{\partial x}\left(\frac{T_w - T}{T_w - T_m}\right) = \frac{\partial \theta}{\partial x} = 0 \qquad (4.105)$$

Indeed, from

$$\mathrm{Nu}_D = \frac{hD}{k} = \frac{D}{T_w - T_m}\left(\frac{\partial T}{\partial r}\right)_{r=r_0}$$

one can conclude that

$$\mathrm{Nu}_D \sim \theta = \frac{T_w - T}{T_w - T_m} = \frac{f_1(x, r)}{f_2(x)} = f\left(\frac{r}{r_0}\right) = O(1)$$

which implies that

$$\theta \neq \theta(x)$$

Now, from Equation (4.104) $\partial T/\partial x$ can be expressed as (cf. Fig. 4.9b)

$$\blacklozenge \qquad \frac{\partial T}{\partial x} = \frac{dT_w}{dx} - \theta\frac{dT_w}{dx} + \theta\frac{dT_m}{dx} := \begin{cases} \theta\dfrac{dT_m}{dx} & \text{for } T_w = \text{const} \\[2mm] \dfrac{dT_m}{dx} = \text{const} & \text{for } q_w = \text{const} \end{cases} \qquad (4.105a,b)$$

Note the following:

(i)   The mean fluid or mixing-cup temperature is defined as

$$\blacklozenge \qquad T_m = \frac{1}{Au_{\mathrm{av}}}\int_A u(r)T(r, x)\,dA \qquad (4.106)$$

where A is the constant cross-sectional area; the mean or average temperature T_m differs from the arithmetic mean or bulk fluid temperature, often defined as $T_b = (T_{\mathrm{in}} + T_{\mathrm{out}})/2$.

(ii) The mean fluid temperature T_m conveniently replaces the outer flow temperature, T_∞ or T_{edge}, when calculating the wall heat flux, $q_w = h(T_w - T_m)$, in pipes.

(iii) With $\theta = \theta(r)$ only, we can confirm

$$\frac{\partial \theta}{\partial r}\bigg|_{r=r_0} = \text{const} \quad \text{or} \quad \frac{q_w/k}{T_w - T_m} = \text{const}$$

where $q_w = h(T_w - T_m)$, so that

$$\frac{h}{k} = \text{const} \qquad (4.107)$$

Thus, for laminar fully developed pipe flow, h/k is invariant along the tube wall. Furthermore,

$$\mathrm{Nu} = O(1)$$

as shown previously.

(iv) For the constant wall heat flux case, $q_w = $ const which implies that $(T_w - T_m) = $ const since $h = $ const. Hence, $(dT_w/dx) - (dT_m/dx) = 0$ so that Equation (4.105b) yields (Fig. 4.9b)

$$\frac{\partial T}{\partial x} = \frac{dT_w}{dx} = \frac{dT_m}{dx} := \text{const} \tag{4.108}$$

These preliminary results (4.105a,b) are now employed to find the pipe Nusselt numbers for $q_w = $ const and for $T_w = $ const.

A   Constant Wall Heat Flux Case

Substituting Equation (4.108) into (4.103) yields with a Poiseuille profile for $u(r)$

$$\blacklozenge \qquad 2u_{\mathrm{av}}\left(1 - \frac{r^2}{r_0^2}\right)\frac{dT_m}{dx} = \frac{\alpha}{r}\frac{\partial}{\partial r}\left(r\frac{\partial T}{\partial r}\right) \tag{4.109}$$

$$\text{where} \qquad u_{\mathrm{av}} = \left(-\frac{dp}{dx}\right)\frac{r_0^2}{8\mu} \quad \text{with} \quad \left(-\frac{dp}{dx}\right) \approx \frac{\Delta p}{L} = \text{const.}$$

The boundary conditions are

$$k\left(\frac{\partial T}{\partial r}\right)_{r=r_0} = q_w \tag{4.110a}$$

or

$$T(r = r_0) = T_w(x) \tag{4.110b}$$

and

$$\left.\frac{\partial T}{\partial r}\right|_{r=0} = 0 \tag{4.110c}$$

Since dT_m/dx is constant, Equation (4.109) subject to (4.110b,c) can be directly integrated to

$$\blacklozenge \qquad T(r, x) = T_w(x) - \frac{2u_{\mathrm{av}}}{\alpha}\left(\frac{dT_m}{dx}\right)\left(\frac{3}{16}r_0^2 + \frac{r^4}{16r_0^2} - \frac{r^2}{4}\right) \tag{4.111}$$

In order to calculate the heat transfer coefficient $h = q_w/(T_w - T_m)$, we have to evaluate $(T_w - T_m)$ as well as q_w. With Equation (4.111), T_m can be computed as

$$T_m - T_w = \frac{22}{96}\frac{u_{\mathrm{av}}}{\alpha}\left(\frac{dT_m}{dx}\right)r_0^2 \tag{4.112}$$

which leaves us one unknown, dT_m/dx. From an energy balance for a cylindrical control volume in the tube we obtain (Fig. 4.9a)

$$(\pi r_0^2 \rho u_{av})\left(c_p T_m + c_p \frac{dT_m}{dx}\Delta x\right) - (\pi r_0^2 \rho u_{av})c_p T_m - (2\pi r_0 \Delta x)q_w = 0$$

or

$$\blacklozenge \qquad q_w = \frac{1}{2}r_0 \rho u_{av} c_p \frac{dT_m}{dx} \qquad\qquad (4.113)$$

The unknown constant (dT_m/dx) cancels and

$$h = \frac{48}{22}\frac{k}{r_0} \qquad\qquad (4.114)$$

or

$$\blacklozenge \qquad \text{Nu} \equiv \frac{hD}{k} = 4.364 \qquad\qquad (4.115)$$

B   Isothermal Wall Case

Substituting Equation (4.105a) into (4.103) yields

$$\blacklozenge \qquad 2u_{av}\left(1 - \frac{r^2}{r_0^2}\right)\frac{T_w - T}{T_w - T_m}\frac{dT_m}{dx} = \frac{\alpha}{r}\frac{\partial}{\partial r}\left(r\frac{\partial T}{\partial r}\right) \qquad\qquad (4.116)$$

subject to (4.110a,b). Since T_m and dT_m/dx are not known (cf. Eq. (4.106)), an iterative procedure is necessary to solve Equation (4.116). However, a numerical scheme such as Keller's box method (cf. App. F.4) to solve Equation (4.103) directly would be the first choice. Alternatively, Kays and Crawford (1993) describe a successive approximation scheme where previously obtained temperature profiles from the constant heat-flux solution are employed, to arrive at a Nusselt number of

$$\blacklozenge \qquad \text{Nu} = 3.658 \qquad\qquad (4.117)$$

C   Thermally Developing Flow Case

In contrast to the two previous applications, the axial temperature gradient $\partial T/\partial x$ in Equation (4.103) cannot be simplified for this "thermal entrance" or Graetz problem. However, assuming a Poiseuille profile for $u(r)$, that is, HFD flow, the energy equation is linear and the method of separation of variables can be employed. Alternatively, the energy equation can be transformed and numerically solved (cf. App. F.4).

In order to estimate the thermal entrance length, the entrance regions due to a sudden difference between incoming fluid temperature and wall temperature are observed for the limiting cases of steady incompressible flow of low- and high-Prandtl-number fluids without viscous dissipation (cf. Fig. 4.10).

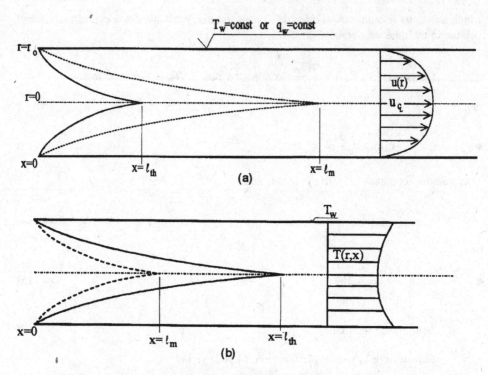

Fig. 4.10. Entrance lengths for low- and high-Prandtl-number fluids: (a) thermal and hydrodynamic entrance lengths for liquid metals; Pr = $\nu/\alpha \ll 1$; (b) thermal and hydrodynamic entrance lengths for heavy oils; Pr = $\nu/\alpha \gg 1$.

By inspection, the ratio of hydrodynamic to thermal entrance length is inversely proportional to the ratio of boundary-layer thicknesses.

$$\frac{l_e}{l_{th}} \sim \frac{\delta_{th}}{\delta} \sim \mathrm{Pr}^{-1} \qquad (4.118a)$$

Replacing $\delta_{th}(x)$ with a flat plate expression (cf. Sect. 4.1.1) for all Prandtl numbers

$$\delta_{th}(x) \sim x\mathrm{Pr}^{-1/2}\mathrm{Re}_x^{-1/2} \qquad (4.118b)$$

and assuming that $\delta_{th}(x = l_{th}) \approx D_h/2$, where D_h is the hydraulic diameter, we obtain

$$D_h \sim 2l_{th}\mathrm{Pr}^{-1/2}\mathrm{Re}_{l_{th}}^{-1/2}$$

or

$$\left(\frac{l_{th}/D_h}{\mathrm{Re}_{l_{th}}\mathrm{Pr}}\right)^{1/2} = \mathrm{const} \qquad (4.119)$$

On the basis of the measurements where the Reynolds number length scale is D_h, we have

$$\blacklozenge \qquad \frac{l_{th}}{D_h} = 0.05\mathrm{Pr}\,\mathrm{Re}_{D_h} \quad \text{and} \quad \frac{l_e}{D_h} = 0.06\mathrm{Re}_{D_h} \qquad (4.120a,b)$$

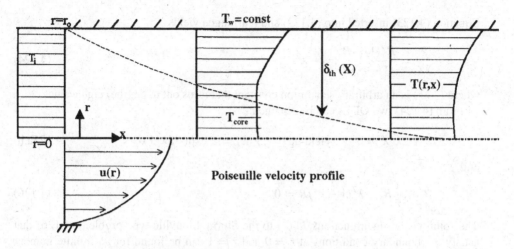

Fig. 4.11. Temperature profiles for Graetz problem.

In case l_{th} is significant when compared with the tube length and the (laminar) flow is fully developed, the Graetz problem of thermally developing Poiseuille flow in isothermal pipes (cf. Fig. 4.11) can be described as (Pe $\gg$ 1)

$$\blacklozenge \qquad u(r)\frac{\partial}{\partial x}T(x,r) = \frac{\alpha}{r}\frac{\partial}{\partial r}\left[r\frac{\partial}{\partial r}T(x,r)\right] \qquad (4.121)$$

where $u(r) = 2\bar{u}[1 - (r/r_o)^2]$ with $\bar{u} = (1/2)u_c = (r_0^2/8\mu)(-dp/dx)$.

At the inlet, a uniform fluid temperature T_i is assumed, encountering a constant wall temperature T_w. Nondimensionalization of (4.121) yields

$$\blacklozenge \qquad \frac{1}{2}(1-\tilde{r}^2)\frac{\partial\theta}{\partial\tilde{x}} = \frac{1}{\tilde{r}}\frac{\partial}{\partial\tilde{r}}\left(\tilde{r}\frac{\partial\theta}{\partial\tilde{r}}\right) \qquad (4.122)$$

subject to

$$\blacklozenge \qquad \theta(\tilde{x}=0) = 1, \qquad \theta(\tilde{r}=1) = 0 \quad \text{and} \quad \left.\frac{\partial\theta}{\partial\tilde{r}}\right|_{\tilde{r}=0} = 0$$

where   (cf. Eq. (4.119))

$$\blacklozenge \qquad \theta = \frac{T-T_w}{T_i-T_w}, \qquad \tilde{r} = \frac{r}{r_0} \quad \text{and} \quad \tilde{x} = \frac{x/D}{\mathrm{Re}_D\mathrm{Pr}}$$

Since this parabolic PDE is linear, the principle of superposition should be applicable and hence a product solution is postulated for the separation of variables method.

$$\theta = R(\tilde{r})\chi(\tilde{x}) \qquad (4.123)$$

Equation (4.123) inserted into (4.122) and rearranged yields

$$\frac{\dot{\chi}}{2\chi} = \frac{R''(1/r)R'}{(1-\tilde{r}^2)R} := -\lambda^2 \tag{4.124}$$

where $(-\lambda^2)$ is an arbitrary separation constant that turns out to be (the) eigenvalues, λ_n. From (4.124) two ODEs can be obtained:

$$\dot{\chi} + 2\lambda^2\chi = 0 \qquad \text{yielding} \qquad \chi(\tilde{x}) = C\exp(-2\lambda^2\tilde{x}) \tag{4.125a,b}$$

and

$$R'' + \frac{1}{\tilde{r}}R' + \lambda^2(1-\tilde{r}^2)R = 0 \tag{4.126}$$

The solutions or eigenfunctions $R_n(\tilde{r})$ to the Sturm–Liouville-type problem (4.126) that satisfy the boundary conditions at $\tilde{r} = 0$ and $\tilde{r} = 1$ can be found for an infinite number of suitable eigenvalues λ_n. Thus the final solution for the temperature distribution can be written as

$$\blacklozenge \qquad \theta = \sum_{n=0}^{\infty} C_n R_n(\tilde{r})\exp(-2\lambda_n^2\tilde{x}) \tag{4.127}$$

Whereas the suitable λ_n's are obtained from the boundary conditions associated with Equation (4.126), the C_n's are constants determined by the "initial" condition at $\tilde{x} = 0$. With the solution (4.127), the mean or bulk temperature, θ_m, and the local Nusselt number Nu_x can be obtained:

$$\theta_m = \frac{T_m - T_w}{T_i - T_w} = 8\sum_{n=0}^{\infty} \frac{G_n}{\lambda_n^2}\exp(-2\lambda_n^2\tilde{x}) \tag{4.128}$$

and

$$\blacklozenge \qquad \text{Nu}_x = \frac{hx}{k} = \frac{\sum_0^\infty G_n\exp(-2\lambda_n^2\tilde{x})}{2\sum_0^\infty \left(\frac{G_n}{\lambda_n^2}\right)\exp(-2\lambda_n^2\tilde{x})} \tag{4.129}$$

where $G_n \equiv -1(C_n/2)R_n'(1)$ and λ_n^2 are listed in Table 4.3.

Table 4.3. *Infinite-series-solution functions*

n^a	G_n	λ_n^2
0	0.749	7.312
1	0.544	44.620
2	0.463	113.800
3	0.414	215.200

Note: [a]For $n > 2$, $\lambda_n = 4n + 8/3$ and $G_n = 1.01276\lambda_n^{-1/3}$.

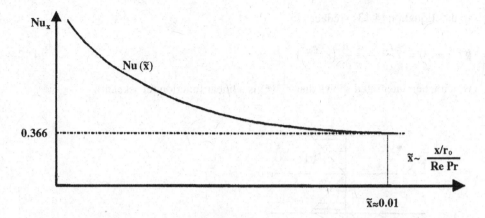

Fig. 4.12. Local Nusselt number in entrance region of circular pipe; solution of the Graetz problem.

Note that for $x \gg 1$, or $\tilde{x} > 0.01$, the asymptotic value of Nu for the isothermal wall case is 3.66 (cf. Fig. 4.12), and for the constant wall heat flux case Nu is 4.364 (cf. Eqs. (4.115) and (4.117)). An extension of the Graetz problem to non-Newtonian fluid flow is discussed in Chapter 5.

4.2.2   Fully Developed Turbulent Pipe Flow

The Reynolds-averaged transport equations for steady incompressible circular duct flow can be written with Boussinesq's eddy viscosity model as

♦
$$\frac{\partial \bar{u}}{\partial x} + \frac{1}{r}\frac{\partial}{\partial r}(r\bar{v}) = 0 \tag{4.130}$$

♦
$$\bar{u}\frac{\partial \bar{u}}{\partial x} + \bar{v}\frac{\partial \bar{u}}{\partial r} = -\frac{1}{\rho}\frac{d\bar{p}}{dx} + \frac{1}{r}\frac{\partial}{\partial r}\underbrace{\left[r(\nu + \varepsilon_m)\frac{\partial \bar{u}}{\partial r}\right]}_{\sim \tau_{\text{total}}} \tag{4.131}$$

and

♦
$$\bar{u}\frac{\partial \bar{T}}{\partial x} + \bar{v}\frac{\partial \bar{T}}{\partial r} = \frac{1}{r}\frac{\partial}{\partial r}\underbrace{\left[r(\alpha + \varepsilon_h)\frac{\partial \bar{T}}{\partial r}\right]}_{\sim q_{\text{total}}} \tag{4.132}$$

A   Velocity Field

For fully developed flow, $\partial \bar{u}/\partial x = 0$, which implies $\bar{v} = 0$, and from a control volume force balance

$$\frac{d\bar{p}}{dx} = 2\frac{\tau_w}{r_0} \tag{4.133}$$

so that Equation (4.131) reduces to

♦
$$0 = 2\frac{\tau_w}{r_0} - \frac{1}{r}\frac{\partial}{\partial r}\left(r\tau_{rx}^{\text{total}}\right)$$
(4.134)

which when integrated shows that $\tau^{\text{tot}}(r)$ is a linear function (cf. sketch).

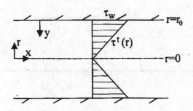

Specifically (cf. Sect. 3.2.4)

$$\tau^{\text{total}} = \rho(\nu + \varepsilon_m)\frac{\partial \bar{u}}{\partial y}$$

where $y = r_0 - r.$

Combining both equations in dimensionless form yields

♦
$$\left(1 - \frac{y^+}{r_0^+}\right) = (1 + \varepsilon_m^+)\frac{du^+}{dy^+}$$
(4.135)

where $u^+ = \dfrac{\bar{u}}{u_\tau},$ $y^+ = \dfrac{u_\tau y}{\nu},$ $\varepsilon_m^+ = \dfrac{\varepsilon_m}{\nu}$ and $u_\tau = \left(\dfrac{\tau_w}{\rho}\right)^{1/2}$

An expression for the momentum eddy diffusivity based on Prandtl's mixing length hypothesis (MHL) together with Van Driest's damping factor $(1 - e^{-y/A})$ can be written as

$$\varepsilon_m = l^2\left(1 - e^{-y/A}\right)^2\frac{d\bar{u}}{dy}$$
(4.136a)

or

♦
$$\varepsilon_m^+ = \left\{l^+\left[1 - \exp\left(-\frac{y^+}{A^+}\right)\right]\right\}^2\frac{du^+}{dy^+}$$
(4.136b)

Inserting (4.136b) and integrating (4.135) yield

♦
$$u^+ = \int \frac{2\left(1 - (y^+/r_0^+)\right)}{1 + \sqrt{1 + 4\left(1 - [y^+/r_0^+]\right)L^{+2}}}dy^+$$
(4.137a)

where $L^+ = l^+\left[1 - \exp\left(-\dfrac{y^+}{A^+}\right)\right],$ $l^+ = \dfrac{lu_\tau}{\nu}$ and $A^+ = 26$

(4.137b–d)

A solution to Equation (4.137) requires a knowledge of the friction velocity $u_\tau = u_\tau[\tau_w(f)]$ where the friction factor $f = f(Re_D, (e/D))$ is obtained from suitable correlations or the Moody chart as discussed later.

B Temperature Field

With $\bar{v} = 0$ in HFD flow, the thermal energy equation (4.132) reduces to

$$\blacklozenge \qquad \bar{u}\frac{\partial \bar{T}}{\partial x} = \frac{\nu}{Pr}\frac{1}{r}\frac{\partial}{\partial y}\left[r\left(1 + \frac{Pr}{Pr_t}\varepsilon_m^+\right)\frac{\partial \bar{T}}{\partial y}\right] \qquad (4.138)$$

Typically, the turbulent Prandtl number $Pr_t \equiv (\varepsilon_m/\varepsilon_h)$ is taken to be constant. For TFD flows, $\partial\theta/\partial x = 0$ where $\theta = (T - T_w)/(T_m - T_w)$ and we can set

$$\frac{\partial \bar{T}}{\partial x} = (1 - \theta)\frac{dT_w}{dx} + \theta\frac{dT_m}{dx} \qquad (4.139)$$

Thus, with

$$\hat{u} = \frac{\bar{u}}{u_m}, \qquad \hat{x} = \frac{2x/r_0}{Re_D Pr}, \qquad \hat{r} = \frac{r}{r_0} \quad \text{and} \quad Re_D = u_m\frac{D}{\nu}$$

we have to solve

$$\blacklozenge \qquad \hat{u}\left[(1 - \theta)\frac{dT_w}{d\hat{x}} + \theta\frac{dT_m}{d\hat{x}}\right] = \frac{1}{\hat{r}}\frac{\partial}{\partial \hat{r}}\left[\hat{r}\left(1 + \frac{Pr}{Pr_t}\varepsilon_m^+\right)\frac{\partial \bar{T}}{\partial \hat{r}}\right]$$

subject to

$$\blacklozenge \qquad \left.\frac{\partial \bar{T}}{\partial \hat{r}}\right|_{\hat{r}=0} = 0 \quad \text{and} \quad \bar{T}(\hat{r} = 1) = T_w \quad \text{or} \quad q_w = k_{\text{turb}}\left.\frac{k\bar{T}}{\partial r}\right|_{r=r_0} \qquad (4.140\text{a--d})$$

The Nusselt number can be computed from

$$\blacklozenge \qquad Nu_D = \frac{q_w}{T_w - T_m}\frac{D}{k} \qquad (4.141)$$

Numerical iterative solutions of (4.140a–d) in conjunction with experimental data for Pr_t and ε_m can be carried out with App. F.4.

C Profiles and Correlations

When relatively simple turbulent-duct-flow-problem solutions are appropriate, the momentum equation and the heat transfer equation could be ignored and semiempirical turbulent velocity and temperature profiles may be directly usable. Next to the log-law profiles, for example,

$$\frac{\rho c_p u_\tau}{q_w}(T_w - T) \equiv T^+ = 2.2\ln y^+ + 13.4Pr^{2/3} - 5.66 \qquad \text{for } 0.7 \le Pr \le 10 \qquad (4.142)$$

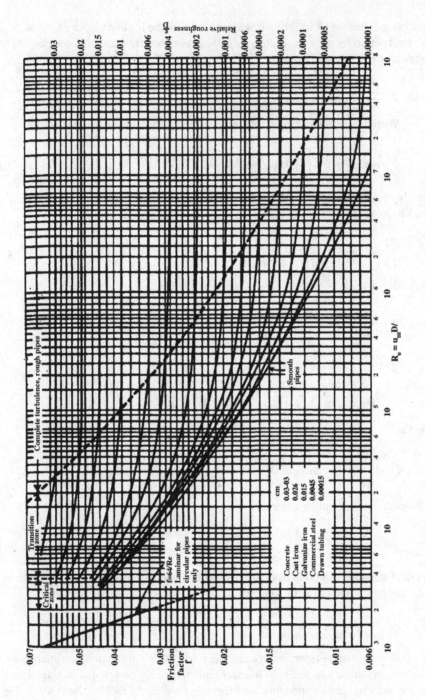

Fig. 4.13. Friction factor for flow inside circular pipes (the Moody Chart after Colebrook and Moody).

semiempirical power-law velocity and temperature profiles are the best choice although the symmetry condition cannot be enforced (cf. Sect. 3.2.5B). Schlichting (1979) developed the following expression for circular pipe flow:

$$\blacklozenge \qquad \frac{u}{U_{\max}} = 8.74 \left(\frac{y u_\tau}{\nu} \right)^{1/7} \qquad\qquad (4.143)$$

where $u_\tau \equiv \sqrt{\tau_w / \rho}$ is the wall friction velocity and $y = r_0 - r$ is the wall coordinate. The friction velocity is related, via the mean velocity, u_m, to the pipe friction factor $f \equiv 8\tau_w / (\rho u_m^2)$. Here, f can be obtained from the Moody chart (Fig. 4.13) or some correlation $f = f(\mathrm{Re}_D, k)$ where $k = e/D$ is the relative wall roughness height.

Alternatively, once f is known for a given problem, the Reynolds–Colburn analogy (cf. Sect. 4.1.1)

$$c_f(x) = 2 \mathrm{St}_x \, \mathrm{Pr}^{2/3}; \qquad \mathrm{St}_x = \frac{\mathrm{Nu}_x}{\mathrm{Re}_x \mathrm{Pr}} \qquad\qquad (4.144 \mathrm{a,b})$$

which relates momentum transfer parameters to the heat transfer coefficient can be employed. With $c_f = f/4 = 2\tau_w / (\rho u_m^2)$,

$$\blacklozenge \qquad \mathrm{St}_{\mathrm{turb}} \mathrm{Pr}^{2/3} = \frac{f}{8} \qquad\qquad (4.144 \mathrm{c})$$

Notes

- The Stanton number is often based on $T_{\mathrm{bulk}} = (T_{\mathrm{in}} + T_{\mathrm{out}})/2$.
- The Prandtl number is typically based on $T_{\mathrm{film}} = (T_w + T_m)/2$.

For example, for smooth tubes (cf. Sects. 3.2.4 and 3.2.5)

$$f = 0.184 \, \mathrm{Re}_D^{-0.2} \qquad\qquad (4.145)$$

and hence

$$\blacklozenge \qquad \mathrm{St}_t \mathrm{Re}_D \mathrm{Pr} \equiv \mathrm{Nu}_D = \frac{hD}{k} = 0.023 \mathrm{Re}_D^{0.8} \mathrm{Pr}^{1/3} \qquad\qquad (4.146)$$

which holds for $0.7 < \mathrm{Pr} < 160$, $\mathrm{Re}_D > 10^4$, and $l/D > 60$. Equation (4.146) compares well with the empirical formula developed by Dittus and Boelter (1930)

$$\blacklozenge \qquad \mathrm{Nu}_D = 0.023 \mathrm{Re}_D^{0.8} \mathrm{Pr}^{\overbrace{0.3}^{T_w < T_f} \ \mathrm{or} \ \overbrace{0.4}^{T_w > T_f}} \qquad\qquad (4.147)$$

Turbulent convection heat transfer in a pipe is discussed further in Chapter 5.

4.3 Sample-Problem Solutions

This end-of-chapter collection of representative CHT problem solutions deals in Section 4.3.1 with external (i.e., boundary-layer-type) flows; Section 4.3.2 focuses on internal (i.e., duct) flow problems. Occasionally, new problem areas are explored and sufficient background information is provided for specific CHT applications not discussed in the text. As usual, the problem solution format follows three basic steps (cf. Fig. 1.1):

1. Problem understanding: system sketch, list of assumptions, and conceptual approach.
2. Model solution: solution method, postulates, transformations, and solution procedure.
3. Modeling results: results, graphs, and/or comments

Variations and extensions of the problem solutions in the form of homework or course project assignments, using computer simulations (cf. App. F) and plotting routines, should further deepen the physical understanding of convection heat transfer (cf. Sect. 5.4).

4.3.1   Thermal Boundary-Layer Flows

4.1   Consider steady laminar flow of a liquid metal past a heated flat plate with $T_w =$ const. Find the Nusselt number distribution $\mathrm{Nu}(x)$ (cf. sketch).

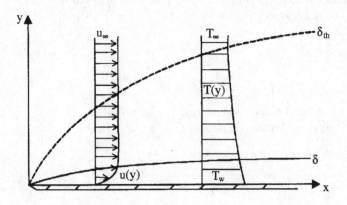

Assumptions

- Thermal Blasius problem (i.e., $u_\infty =$ const)
- Constant fluid properties
- $\mathrm{Pr} = \nu/\alpha \ll 1 \Rightarrow \delta \ll \delta_{\mathrm{th}}$ and hence $u \approx u_\infty =$ const

Approach
Integral method assuming $u \approx u_\infty$ within $0 < y \leq \delta_{\mathrm{th}}$ and a cubic profile for $T(y, \delta_{\mathrm{th}})$.

Solution

(energy integral relation) $\displaystyle \frac{d}{dx} \int_0^{\delta_{\mathrm{th}}} (T_\infty - T) u \, dy = \alpha \frac{\partial T}{\partial y}\bigg|_{y=0}$ (S4.1)

(dimensionless temp. profile) $\displaystyle \theta = \frac{T - T_w}{T_\infty - T_w} = \frac{3}{2}\left(\frac{y}{\delta_{\mathrm{th}}}\right) - \frac{1}{2}\left(\frac{y}{\delta_{\mathrm{th}}}\right)^3$ (S4.2)

Using $u = u_\infty$ and inserting (S4.2) into (S4.1) yield

$$\delta_{th} = \sqrt{\frac{8\alpha x}{u_\infty}} \qquad (S4.3)$$

With (S4.2) we form

$$\left.\frac{\partial \theta}{\partial y}\right|_{y=0} = \frac{3}{2\delta_{th}}$$

so that

$$h \equiv k\frac{(\partial T/\partial y)|_{y=0}}{T_\infty - T_w} = k\left.\frac{\theta}{\partial y}\right|_{y=0} = \frac{3}{2\sqrt{8}}\frac{k}{x}\sqrt{Re_x}\,Pr$$

or

$$\text{Nu}_x = \frac{hx}{k} = 0.530\text{Re}_x^{1/2}\text{Pr}^{1/2} \qquad (S4.4a)$$

Note: Equation (S4.4a) compares quite well with the Falkner–Skan similarity solution discussed in Sect. 4.1.1, which, for zero-pressure-gradient flow ($m = 0$) and very-low-Prandtl-number fluids, is

$$\text{Nu}_x = 0.564\text{Re}_x^{1/2}\text{Pr}^{1/2} \qquad (S4.4b)$$

For low-to-medium-Prandtl-number fluids the Nu(Pr) dependence changes to (cf. first graph)

$$\text{Nu}_x = C\text{Re}_x^{1/2}\text{Pr}^{1/3} \qquad (S4.4c)$$

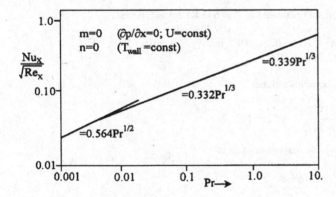

where $C \approx 0.332$ for $Pr = O(1)$ and $C \approx 0.339$ for $Pr = O(10)$. It is convenient to plot $\text{Nu}_x/\sqrt{\text{Re}_x}$ versus Pr on a log–log graph. Recalling that $\text{Nu}_x/\sqrt{\text{Re}_x} = \theta'_{wall}$, that is, the dimensionless temperature gradient at the plate surface, the Nu(Pr) trend is best explained with the second graph. The larger the Prandtl number the thinner the thermal boundary layer and hence the steeper the dimensionless temperature gradient at the wall. In accordance with the Reynolds analogy (cf. Sect. 4.1) the dimensionless temperature and velocity profiles are the same when the Prandtl number is equal to 1.

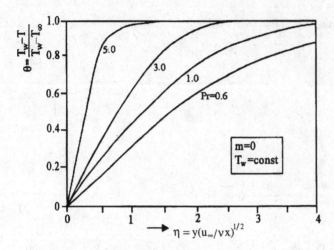

$$\eta = y(u_\infty/\nu x)^{1/2}$$

□

4.2 Molten lead at a temperature of 645 K flows past a flat plate with a temperature of 310 K at a velocity of 15 m/s. The dimensions of the plate are 1 m by 1 m. Determine the Nusselt number from Equation (S4.4a) and then the heat transfer coefficient and total heat transferred in this process.

Assumptions

- Thermal Blasius problem
- Constant fluid properties
- $Pr \ll 1$

Solution

The properties of the metal (evaluated at $T = 645$ K) are as follows:

$\mu = 2.4 \times 10^{-4}$ kg/m · s, $\nu = 0.023 \times 10^{-6}$ m²/s, $\quad k = 16.1$ W/m · °C, $C_p = 0.159$ kJ/kg ·° C, Pr = 0.024

Now, we can find the Reynolds number

$$Re = \frac{u_\infty x}{\nu} = \frac{15(1)}{0.023 \times 10^{-6}} = 652 \times 10^6$$

Using Equation (S4.4a), the Nusselt number is

$$Nu_x = .530 Re^{1/2} Pr^{1/2} = 2,096.8$$

Now, the heat transfer coefficient can be determined from the Nusselt number and thus the total heat transfer can be determined as well.

$$\bar{h} = 2Nu_x \frac{k}{x}\bigg|_{x=L} = 67,520 \frac{W}{m^{2}°C}$$

So,

$$Q = \bar{h}A\,\Delta T = 67,520(1)(645 - 310) = 22.62 \text{ MW}$$

4.3 Consider gravity flow of a cold Newtonian liquid film on an inclined solid wall (cf. sketch). The downflowing liquid film enters at uniform temperature T_0, which is equal to the air temperature T_∞. The wall surface is maintained at a constant temperature T_s. Assume the film thickness, $\delta \ll 1$, to be constant (cf. Bird et al. 1960).

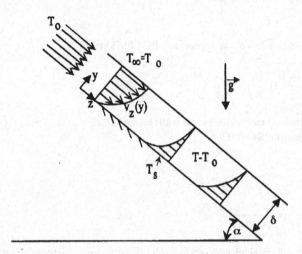

Assumptions

- Steady laminar fully developed flow
- Constant fluid properties
- Thin film approximation $\nabla p \approx 0$
- Constant film thickness δ
- Zero stress at interface
- Negligible viscous dissipation and streamwise thermal diffusion

Postulates
On the basis of the assumptions and boundary conditions (cf. Eq. (S4.7)) we conclude

$$v_z = v_z(y) \quad \text{and} \quad T = T(y, z)$$

Solution
While the continuity equation is satisfied, the Navier–Stokes equations reduce to

(z-momentum) $\qquad 0 = \mu \frac{\partial^2 v_z}{\partial y^2} + \rho g \sin \alpha$ $\qquad$ (S4.5)

and

(energy equation) $\qquad \rho c_p v_z \frac{\partial T}{\partial z} = k \frac{\partial^2 T}{\partial y^2}$ $\qquad$ (S4.6)

The boundary conditions are at

$$
\begin{aligned}
y = 0: & \quad v_z = 0 \quad \text{and} \quad T = T_s \quad \text{for } z > 0 \\
y = \delta: & \quad dv_z/dy \approx 0 \quad \text{and} \quad T = T_\infty = T_0 \\
z = 0: & \quad T = T_0 \quad \text{for } y > 0
\end{aligned}
\tag{S4.7}
$$

Results

Integration of (S4.5) subject to the appropriate BC (S4.7) yields

$$
v_z(y) = \frac{\delta^2 \rho g \sin\alpha}{2\mu} \left[2\left(\frac{y}{\delta}\right) - \left(\frac{y}{\delta}\right)^2 \right]
$$

Since $\delta \ll 1$ and the contact region (at the wall) is quite small, that is, $y < \delta$, $v_z(y)$ can be linearized for Equation (S4.6). Thus

$$
y \frac{\partial T}{\partial z} = K \frac{\partial^2 T}{\partial y^2}
\tag{S4.8a}
$$

where

$$
K = \frac{\mu k}{(\rho^2 c_p g \delta \sin\alpha)}
\tag{S4.8b}
$$

Either a separation of variables with product solution or a coordinate transformation using a combined variable $\eta = \eta(y, z)$ is a suitable analytic solution method. Introducing

$$
\theta = \frac{T - T_0}{T_s - T_0} \quad \text{and} \quad \eta = \frac{y}{(\kappa z)^{1/3}}; \quad \kappa \equiv 3K
\tag{S4.9a–c}
$$

we transform

$$
\frac{\partial T}{\partial z} = \Delta T \theta' \left(\frac{-y}{3\kappa^{1/3} z^{4/3}} \right) = -\Delta T \theta' \frac{\eta}{3z}
$$

and

$$
\frac{\partial^2 T}{\partial y^2} = \Delta T \theta'' (\kappa z)^{-2/3}
$$

so that Eq. (S4.8a) reads

$$
\theta'' + \eta^2 \theta' = 0 \quad \text{or} \quad \frac{\theta''}{\theta'} = -\eta^2
$$

and after double integration

$$
\theta' = C e^{-\eta^3}
$$

and

$$
\theta = \frac{1}{\Gamma(4/3)} \int_\eta^\infty e^{-\eta^3} \, d\eta
\tag{S4.10}
$$

Note: A numerical solution of (S4.10) allows plotting of $T(y, z)$ and calculation of the wall Nusselt number; $\Gamma(4/3)$ is a value of the gamma function. $\square$

4.4 Consider compressible flow with free-stream conditions u_∞, T_∞, and p_∞ brought to rest reversibly at the stagnation point of a submerged body ($u_0 = 0$, $T = T_0$, and $p = p_0$). Evaluate the temperature increase through adiabatic compression.

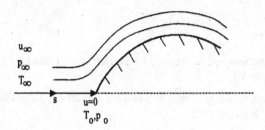

Assumptions

- Constant entropy along a streamline (i.e., no heat conduction and no dissipation)
- 1-D flow along streamline coordinate s

Solution

For compressible flow, we rewrite the energy equation (Sect. 4.1) as

$$\rho \frac{D\hat{u}}{Dt} + p\nabla \cdot \vec{v} = \nabla \cdot \vec{q} + \mu\phi \tag{S4.11}$$

or with $D\hat{u} = c_v DT = c_p DT - D(p/\rho)$, $\nabla \cdot \vec{v} = -1/\rho \, D\rho/Dt$, and $\vec{q} = -k\nabla T$, Equation (S.4.11) reads

$$\rho c_p \frac{DT}{Dt} = \frac{Dp}{Dt} + \nabla \cdot (k\nabla T) + \mu\Phi \tag{S4.12}$$

In our case, $\partial/\partial t = 0$, $\vec{q} = 0$, and $\Phi = 0$ so that for one-dimensional flow along a streamline s

$$\rho c_p u \frac{dT}{ds} = u \frac{dp}{ds} \tag{S4.13}$$

Dividing (S4.13) by (ρu) and integrating along a streamline, we obtain

$$c_p(T - T_\infty) = \int_{p_\infty}^{p} \frac{dp}{\rho} \tag{S4.14}$$

We know from Bernoulli's equation that

$$\frac{u^2}{2} + \int \frac{dp}{\rho} = \text{const} \tag{S4.15}$$

The temperature change can be obtained by combining (S4.15) and (S4.14) to

$$T - T_\infty = \frac{u_\infty^2 - u^2}{2c_p} \tag{S4.16}$$

For a stagnation point $u = 0$ and $T \to T_0$ so that

$$T_0 - T_\infty = \Delta T|_{\text{adiabatic}} = \frac{u_\infty^2}{2c_p}$$
(S4.17a)

or

$$\Delta \hat{h} = c_p \Delta T = \frac{u_\infty^2}{2} = e_{\text{kin}}$$
(S4.17b)

For example, the adiabatic temperature increase for air with $c_p = 1.006$ kJ/(kg $\cdot$ °C) and $u_\infty = 100$ m/s is $\Delta T_{\text{adiab}} \approx 5$°C.   □

4.5 Find an approximate solution for liquid metal flow near the stagnation point of a horizontal, heated cylinder of radius R (cf. sketch).

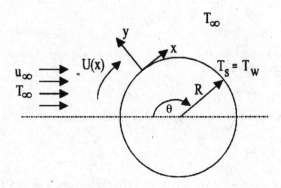

Assumptions

- Steady laminar 2-D stagnation point flow with constant fluid properties and constant surface temperature
- $\text{Pr} \ll 1$ (i.e., $\delta_{\text{th}} \gg \delta$)

Postulate
Flow near the stagnation point of a cylinder is approximated as $U(x) = u = 2u_\infty x/R$ for $x/R \ll 1$ (cf. Sect. 3.3).

Solution
Using curvilinear coordinates x–y, the streamwise velocity is linearly dependent on x, that is, $u(x) = Cx$. Then continuity indicates that $v = -Cy$ so that the thermal boundary-layer equation reads

$$Cx \frac{\partial T}{\partial x} - Cy \frac{\partial T}{\partial y} = \alpha \frac{\partial^2 T}{\partial y^2}$$
(S4.18a)

A combined variable in the form (cf. Hansen et al. 1964)

$$\eta = \frac{yU(x)}{2\sqrt{\alpha \int_0^x U \, dx}}$$
(S4.19)

is introduced; it reduces in this problem to

$$\eta = \frac{yCx}{2\sqrt{\alpha(Cx^2/2)}} = Ky$$

where $\quad K = \frac{1}{2}\sqrt{\frac{2C}{\alpha}}, \quad C = \frac{2u_\infty}{R} \quad$ and $\quad \alpha = \frac{k}{\rho c_p}$

Now we transform $T(x, y) \Rightarrow T(\eta)$:

$$\frac{\partial T}{\partial x} = \frac{\partial T}{\partial \eta}\frac{\partial \eta}{\partial x} = 0$$

$$\frac{\partial T}{\partial y} = \frac{\partial T}{\partial \eta}\frac{\partial \eta}{\partial y} = K\frac{\partial T}{\partial \eta} \quad \text{and} \quad \frac{\partial^2 T}{\partial y^2} = K^2\frac{\partial^2 T}{\partial \eta^2}$$

Substituting the transformed derivatives into (S4.18a) yields

$$\frac{d^2 T}{d\eta^2} + 2\eta\frac{dT}{d\eta} = 0 \tag{S4.18b}$$

subject to $T(\eta = 0) = T_w$ and $T(\eta \to \infty) \to T_\infty$.

Rewriting (S4.18b) with $dT/d\eta \equiv T'$ and employing the trial solution $T' = A\exp(-\eta^2)$ we obtain

$$\frac{dT'}{d\eta} + 2\eta T' = 0 \quad \text{and} \quad \int_{T_w}^{T} T' = A\int_0^\eta e^{-\eta^2}\,d\eta$$

or

$$T - T_w = A\int_0^\eta e^{-\eta^2}\,d\eta$$

Invoking the second boundary condition $T(\eta = \infty) = T_\infty$, we have

$$A = \frac{T_\infty - T_w}{\int_0^\infty \exp(-\eta^2)d\eta} = \frac{T_\infty - T_w}{\sqrt{\pi}/2}$$

Finally

$$\frac{T - T_w}{T_\infty - T_w} = \text{erf}(\eta) \tag{S4.20}$$

In order to compute the Nusselt number $\text{Nu}_x = hx/k$, we evaluate the temperature gradient at the wall

$$\frac{dT}{dy}\bigg|_{y=0} = \frac{2}{\sqrt{\pi}}(T_\infty - T_w)\left[\frac{\partial}{\partial \eta}\int_0^\eta e^{-\eta^2}\,d\eta\right]_{\eta=0}\frac{d\eta}{dy}$$

using Leibnitz's rule

$$\frac{\partial}{\partial \eta} \int_0^\eta e^{-\eta^2} d\eta = \frac{\partial \eta}{\partial \eta} e^{-\eta^2} = e^{-\eta^2}$$

so that

$$\left.\frac{dT}{dy}\right|_{y=0} = \frac{2K}{\sqrt{\pi}}(T_\infty - T_w)$$

Thus

$$Nu = -\frac{(dT/dy)|_{y=0}}{(T_\infty - T_w)}x = \sqrt{\frac{2}{\pi}}\sqrt{\frac{Cx}{\alpha}}x := \sqrt{\frac{2}{\pi}}\sqrt{\frac{\nu}{\alpha}}\sqrt{\frac{2u_\infty}{R\nu}}x \tag{S4.21a}$$

where $Re_x = Ux/\nu := 2u_\infty x^2/(\nu R)$

$$Nu_x = 0.798 Pr^{1/2} Re_x^{1/2} \tag{S4.21b}$$

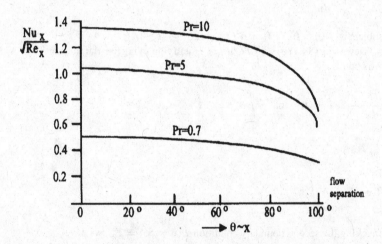

As seen in the third graph, the Nusselt number increases with higher Prandtl numbers. That is, the fluid's thermal diffusivity decreases and so does the thermal boundary layer, which in turn increases the temperature gradient at the surface and hence the Nusselt number. Equation (S4.21) should be compared with Equation (S4.4), which describes heat transfer of very-low-Prandtl-number fluids on a flat plate. □

4.6 Consider laminar heat transfer from a heated vertical wall to air forming a free convection boundary layer of length L (cf. sketch below). Assuming $T_w - T_\infty = Ax^n$, which includes both basic thermal wall conditions, find the mean Nusselt number for case (i), where the heat transfer coefficient is based on the temperature difference between the average surface temperature and the ambient, and case (ii), where $\Delta T - T_w(x = L/2) - T_\infty$. Compare the two $\overline{Nu}$ with those for an isothermal wall at $Pr = 0.1, 0.72, 1.0,$ and 100 (cf. Gebhart et al. 1988).

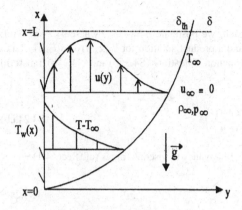

Background and Assumptions

The gravitational field is aligned with the principal direction of motion and hence the hydrostatic pressure field outside the boundary layer can be described by

$$\frac{dp}{dx} = -\rho_\infty g$$

Thus, combining the body force term and the pressure gradient term, the x-momentum equation reads with the Boussinesq approximation for natural convection, that is, $\rho_\infty - \rho = \rho\beta(T - T_\infty)$

$$u\frac{\partial u}{\partial x} + v\frac{\partial u}{\partial y} = g\beta(T - T_\infty) + v\frac{\partial^2 u}{\partial y^2} \tag{S4.22}$$

or in the stream function formulation

$$\frac{\partial \psi}{\partial y}\frac{\partial^2 \psi}{\partial x \partial y} - \frac{\partial \psi}{\partial x}\frac{\partial^2 \psi}{\partial y^2} = g\beta(T_w - T_\infty)\theta + v\frac{\partial^3 \psi}{\partial y^3}$$

The coupled heat transfer equation is

$$u\frac{\partial T}{\partial x} + v\frac{\partial T}{\partial y} = \alpha\frac{\partial^2 T}{\partial y^2}$$

or

$$\frac{\partial \psi}{\partial y}\frac{\partial \theta}{\partial x} + \theta\frac{\partial \psi}{\partial y}\left(\frac{dT_w}{dx}\right) - \frac{\partial \psi}{\partial x}\frac{\partial \theta}{\partial y} = \alpha\frac{\partial^2 \theta}{\partial y^2} \tag{S4.23a,b}$$

where $\quad u = \dfrac{\partial \psi}{\partial y}, \quad v = -\dfrac{\partial \psi}{\partial x}, \quad \theta = \dfrac{T - T_\infty}{T_w - T_\infty} \quad$ and $\quad T_w - T_\infty = Ax^n$

The boundary conditions include

$$u = v = 0 \quad \text{and} \quad T = T_w(x) \qquad \text{at } y = 0$$
$$u = 0 \qquad\quad \text{and} \quad T = T_\infty \qquad\quad \text{at } x = 0$$

and, quite different from the forced convection flow condition at the boundary-layer edge,

$$u \to 0 \quad \text{and} \quad T \to T_\infty \qquad \text{at } y \to \infty$$

Transformations

Employing the stream function approach, we postulate $\eta = y/\delta_{th}$, where $\delta_{th} \sim x^{1/4}$ (cf. Sect. 4.1.3), in the form $\eta = yf(x)$ and a product solution for $\psi(x, \eta) = \nu F(\eta) \cdot G(x)$ where $F(\eta)$ is a dimensionless stream function. Equations (S4.22b) and (S4.23b) transform to ODEs provided that

$$f(x) = \left(\frac{\mathrm{Gr}_x}{4x^4}\right)^{1/4} \quad \text{and} \quad G(x) = 4\left(\frac{\mathrm{Gr}_x}{4}\right)^{1/4} \tag{S4.24a,b}$$

where the Grashof number is the ratio of buoyancy to viscous forces (cf. Sect. 4.1)

$$\mathrm{Gr}_x = \frac{g\beta x^3 (T_w - T_\infty)}{\nu^2} \tag{S4.24c}$$

Thus, similarity has been achieved because x or $G(x)$ does not appear in the transformed equations

$$F''' + \theta + (n+3)FF'' - (2n+2)F'^2 = 0 \tag{S4.25}$$

and

$$\theta'' + \mathrm{Pr}[(n+3)F\theta' - 4nF'\theta] = 0 \tag{S4.26}$$

subject to

$$F(0) = F'(0) = 0, \qquad \theta(0) = 1; \qquad F(\infty) = 0 \quad \text{and} \quad \theta(\infty) = 0$$

Solutions

- For the isothermal wall case, $T_w = \text{const}$, which implies $n \equiv 0$ and with $h(x) = q_w/(T_w - T_\infty)$,

where $\quad q_w = -k\left.\frac{\partial T}{\partial y}\right|_{y=0} = -k(T_w - T_\infty)\left.\frac{\partial \theta}{\partial y}\right|_w = -k\Delta T\left(\frac{d\theta}{d\eta}\frac{\partial \eta}{\partial y}\right)_w$

$$\mathrm{Nu}_x = \frac{hx}{k} = -\theta'(0)xf(x) = -\frac{\theta'_w}{\sqrt{2}}\mathrm{Gr}_x^{1/4} \sim x^{3/4} \tag{S4.27a}$$

or

$$\mathrm{Nu}_x\,\mathrm{Gr}_x^{-1/4} = \frac{\theta'(0)}{\sqrt{2}} \tag{S4.27b}$$

Now, the average heat transfer coefficient is obtained from

$$\bar{h} = \frac{1}{x}\int_0^x h(x)\,dx$$

$$= \frac{k}{x}\int_0^x \frac{-\theta'_w}{x\sqrt{2}}\mathrm{Gr}^{1/4}\,dx$$

$$\bar{h} = \frac{k}{x} \int_0^x \frac{Cx^{3/4}}{x} \, dx$$

$$= \frac{4}{3} h(x)$$

$$\therefore \overline{\mathrm{Nu}} = \frac{4}{3} \mathrm{Nu}_x \tag{S4.28a}$$

- For the constant wall heat flux case

$$q_w = -k(T_w - T_\infty)\theta'(0)f(x) = B(T_w - T_\infty)^{5/4}x^{-1/4} = \text{const}$$

which implies that $(T_w - T_\infty)$ has to vary as $x^{1/5}$ and B is given as

$$B = -k\left(\frac{g\beta}{v^2}\right)^{1/4} \frac{\theta'_w}{\sqrt{2}}$$

Thus, for $q_w = \text{const}$, we require that

$$T_w - T_\infty = C_1 x^{1/5}$$

and the local heat transfer coefficient can be expressed as

$$h(x) = -\frac{k}{\sqrt{2}} \frac{\theta'(0)}{x} \mathrm{Gr}_x^{1/4} = -\frac{k\theta'(0)}{\sqrt{2}} \left(\frac{g\beta}{v^2}C_1\right) \frac{1}{x}\left(x^3 x^{1/5}\right)^{1/4}$$

or

$$h(x) = C_2 x^{-1/5} \qquad \text{where} \qquad C_2 = -k\frac{\theta'(0)}{\sqrt{2}}\left(\frac{g\beta}{v^2}C_1\right)^{1/4}$$

The average heat transfer coefficient is then

$$\bar{h} = \frac{1}{x}\int_0^x h(x)\,dx = \frac{5}{4}C_2 x^{-1/5} = \frac{5}{4}h(x)$$

so that

$$\overline{\mathrm{Nu}} = \frac{5}{4}\mathrm{Nu}_x \tag{S4.28b}$$

Results

Values for $\mathrm{Nu}_x \, \mathrm{Gr}_x^{-1/4}$ at different Prandtl numbers have been compiled for two thermal boundary conditions ($T_w = \text{const}$ and $q_w = \text{const}$) by Kays and Crawford (1993):

	Pr	0.01	0.1	0.72	1.0	10	100	1000
$\mathrm{Nu}_x \, \mathrm{Gr}_x^{-1/4}$	$T_w = \text{const}$	0.057	0.104	0.357	0.401	0.827	1.55	2.80
	$q_w = \text{const}$	0.067	0.189	0.406	0.457	0.931	1.74	3.14

The next graph depicts the influence of the different fluids on $Nu_x Gr_x^{-1/4}$ for the two thermal boundary conditions.

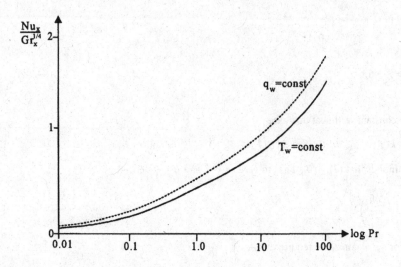

4.7   Consider free convection along an isothermal vertical wall assuming $\delta \approx \delta_{th}$ and the profiles $u = U(1 - y/\delta)^2(y/\delta)$ and $T - T_\infty = (T_w - T_\infty)(1 - y/\delta_{th})$ where the reference velocity $U \sim x^{1/2}$ and $\delta \approx \delta_{th} \sim x^{1/4}$ are obtainable from scale analysis (cf. Sects. 2.3.1 and 4.1.3). Perform an integral analysis to compute $Nu = Nu(Pr, Ra_x)$ where the Rayleigh number $Ra_x = Gr\ Pr = g\beta\Delta T x^3/(\alpha\nu)$ denotes the ratio of buoyancy force to change in momentum flux (cf. sketch).

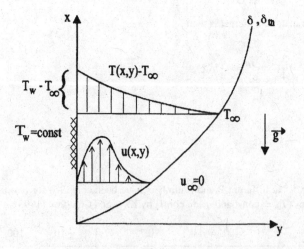

Background Information

The boundary-layer equations (S4.22a and S4.23a) can be integrated across the boundary layer noting that

$$\int\limits_{y=0}^{y=\delta} v\frac{\partial u}{\partial y}\,dy = 0$$

and

$$\int\limits_{y=0}^{y=\delta_{\text{th}}} v\frac{\partial T}{\partial y}\,dy = 0$$

Hence, the momentum integral relation reads

$$\frac{d}{dx}\int\limits_0^\infty u^2\,dy = -v\left(\frac{\partial u}{\partial y}\right)_w + g\beta\int\limits_0^\infty (T - T_\infty)\,dy \tag{S4.29a}$$

and the energy integral relation can be written as

$$\frac{d}{dx}\int\limits_0^\infty u(T_\infty - T)\,dy = \alpha\left(\frac{\partial T}{\partial y}\right)_w \tag{S4.30a}$$

With the given velocity and temperature profiles, (S4.29a) and (S4.30a) transform to a coupled set of ODEs for $\delta \approx \delta_{\text{th}}$ and U.

$$\frac{d}{dx}\left(\frac{U^2\delta}{105}\right) = -v\frac{U}{\delta} + \frac{g\beta}{3}\Delta T\delta \tag{S4.29b}$$

where $\Delta T = T_w - T_\infty$, and

$$\frac{d}{dx}(U\delta) = \alpha\frac{60}{\delta} \tag{S4.30b}$$

With the given postulates $U = C_1 x^{1/2}$ and $\delta = C_2 x^{1/4}$, we can determine the constant coefficients C_1 and C_2 using Equations (S4.29b) and (S4.30b). Thus,

$$C_1 = 5.17v\left(\text{Pr} + \frac{20}{21}\right)^{-1/2}\left(\frac{g\beta\Delta T}{v^2}\right)^{1/4}$$

and

$$C_2 = 3.93\text{Pr}^{-1/2}\left(\text{Pr} + \frac{20}{21}\right)^{1/4}\left(\frac{g\beta\Delta T}{v^2}\right)^{-1/4}$$

Now, the Nusselt number can be evaluated as

$$\text{Nu}_x = \frac{q_w}{\Delta T}\frac{x}{k} = \frac{-k(\partial T/\partial y)_w}{\Delta T}\frac{x}{k} = 2\frac{x}{\delta} = 2\frac{x^{3/4}}{C_2}$$

or

$$\text{Nu}_x = 0.508\left(\frac{0.95}{\text{Pr}} + 1\right)^{-1/4}\text{Ra}_x^{1/4} \qquad (S4.31)$$

Note that $(\text{Nu}_x\text{Ra}_x^{-1/4})$ approaches a constant value with increasing Prandtl number as shown in the graph.

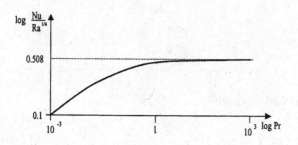

4.8 A square plate 0.4 by 0.4 m, maintained at constant, uniform temperature of 38 K, is suspended vertically in atmospheric air at 315 K. Use Eq. (S4.31) developed in Sample-Problem 4.7 to help determine the total rate of heat loss from the plate.

Assumptions

- $\delta \approx \delta_{\text{th}}$
- Steady flow
- Constant fluid properties

Solutions
The properties of the air are as follows:

$$\nu = 24.36 \times 10^{-6} \text{ m}^2/\text{s} \qquad \alpha = 0.3527 \times 10^{-4} \text{ m}^2/\text{s}, \qquad \text{Pr} = 0.691$$

The Rayleigh number can now be determined

$$\text{Ra}_{x=L} = \frac{g\beta\Delta T L^3}{\alpha\nu} = 146 \times 10^6$$

So

$$\text{Nu}_{x=L} = 0.508\left(\frac{.95}{\text{Pr}} + 1\right)^{-1/4}\text{Ra}_L^{1/4} = 44.98$$

Now, the heat transfer coefficient (h) can be determined from the Nusselt number

$$h = \text{Nu}_L\frac{k}{L} = 3.69\frac{\text{W}}{\text{m}^2{}^\circ\text{C}} \qquad \text{and} \qquad \bar{h} = \frac{4}{3}h = 4.92\frac{\text{W}}{\text{m}^2{}^\circ\text{C}}$$

Thus, the total heat loss is

$$Q = \bar{h}A\Delta T = 4.92[2(0.4)^2](70) = 110.2 \text{ W}$$

4.9   Assuming self-similar profiles, cast Sample-Problem 4.6 in a similarity-solution formulation (this problem was suggested by Bejan 1984).

Derivation

With the similarity variable $\eta = \eta(x, y)$ and the scaling results $\delta \sim x^{-1/4}$ and $u \sim Ra_x^{1/4}(\alpha/x)$ we obtain, using the stream function approach,

$$\eta = Cyx^{-1/4} \quad \text{and} \quad \psi = \alpha Cx^{3/4} f(\eta) \qquad \text{(S4.32 \& S4.33)}$$

where $\quad C = \left[\dfrac{g\beta\Delta T}{(\alpha\nu)} \right]^{1/4}\quad$ is a constant $\quad$ and $\quad f(\eta)$ is dimensionless.

The boundary-layer equation (S4.22b) can be transformed with (S4.32) and (S4.33) to

$$\Pr^{-1}\left(\frac{3}{4} f f'' - \frac{1}{2} f'^2 \right) = f f'' + \theta \qquad \text{(S4.34)}$$

where $\theta(\eta) = (T - T_\infty)/(T_w - T_\infty)$ and $f(0) = f'(0) = 0$ and $f(\infty) = 0$. Similarly, the thermal boundary-layer (S4.23b) equation can be transformed to

$$\frac{3}{4} f\theta' = \theta'' \qquad \text{(S4.35)}$$

where $\quad \theta(0) = 1 \quad$ and $\quad \theta(\infty) = 0.$

Note: Equations (S4.34) and (S4.35) are coupled and have to be solved iteratively (cf. App. F.4) employing the Runge–Kutta algorithm with appropriate "starting" values for f'' and θ'' based on trial and error. $\qquad\qquad\square$

4.10   Consider *fully turbulent flow* past a horizontal flat plate with unheated starting section $0 \leq x \leq x_0$ (cf. sketch). Assuming appropriate turbulent velocity and temperature profiles, derive an expression for the turbulent heat transfer coefficient employing integral relations and the definition of the Stanton number (cf. Reynolds–Colburn analogy in Sect. 4.1).

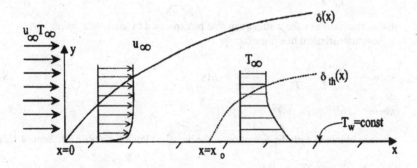

Assumptions

- $U(x) = u_\infty = $ const
- Turbulent Prandtl number $\Pr_t = \varepsilon_m/\varepsilon_h = 1.0$
- Power-law profiles with $n \approx 7$, i.e., valid, require $5 \times 10^5 < Re_x < 10^7$

Postulates

$$\left(\frac{u}{u_\infty}\right) = \left(\frac{y}{\delta}\right)^{1/n} \quad \text{and} \quad \frac{T_w - T}{T_w - T_\infty} = \left(\frac{y}{\delta_{th}}\right)^{1/n} \tag{S4.36a,b}$$

Concept

Express q_w^{turb} in Stanton number definition

$$\text{St} = \frac{h(x)}{\rho c_p u_\infty} = \frac{q_w^{\text{turb}}}{\rho c_p u_\infty (T_w - T_\infty)}$$

and employ integral relations.

Solution

Writing the momentum integral relation as

$$\int_0^y \frac{\partial}{\partial x}\left(\frac{u}{u_\infty}\right)^2 dy - \frac{u}{u_\infty} \int_0^y \frac{\partial}{\partial x}\left(\frac{u}{u_\infty}\right) dy = \frac{1}{\rho u_\infty^2}(\tau - \tau_w) \tag{S4.37a}$$

yields with the velocity power law where $\eta \equiv y/\delta$

$$\frac{\tau}{\tau_w} = 1 + \frac{2}{c_f}\left[\frac{d\delta}{dx}\left(\int_0^\eta \eta^{\frac{2}{n}} d\eta - \eta^{\frac{1}{n}} \int_0^\eta \eta^{\frac{1}{n}} d\eta\right)\right] \tag{S4.37b}$$

Evaluating c_f, $d\delta/dx$, and the integrals with the power law yields

$$\frac{\tau}{\tau_w} = 1 + \frac{(n+1)(n+2)}{n}\left[\frac{n}{n+2}\eta^{(n+2)/n} - \frac{n}{n+1}\eta^{(n+2)/n}\right]$$

or

$$\frac{\tau}{\tau_w} = 1 - \left(\frac{y}{\delta}\right)^{1+2/n} \tag{S4.37c}$$

that is, the turbulent shear stress can also be expressed in terms of a power law.

Now, for turbulent heat transfer

$$\text{St}_x = \frac{q_w^{\text{turb}}(x)}{\rho c_p u_\infty (T_w - T_\infty)} = \frac{\text{Nu}_x}{\text{Re}_x \text{Pr}} \sim h(x) \tag{S4.38}$$

where $\quad q_w^{\text{turb}} \approx \rho c_p \overline{v'T'} \tag{S4.39a}$

which is obtained with the definition of the turbulent Prandtl number (cf. Sect. 4.1.2).

$$\text{Pr}_t \equiv \frac{\varepsilon_m}{\varepsilon_h} = \frac{(\overline{u'v'})/(\partial u/\partial y)}{(\overline{v'T'})/(\partial T/\partial y)}$$

with $\overline{u'v'} = -\tau/\rho$ we obtain for $\text{Pr}_t = 1.0$

$$\overline{v'T'} = -\frac{\tau}{\rho}\left(\frac{\partial T}{\partial y}\right)\left(\frac{\partial u}{\partial y}\right)^{-1} := \frac{\tau}{\rho}\frac{T_w - T_\infty}{u_\infty}\left(\frac{\delta}{\delta_{th}}\right)^{1/n}$$

so that

$$q_w^{\text{turb}} = \rho c_p \frac{\tau_w}{\rho} \frac{T_w - T_\infty}{u_\infty} \left(\frac{\delta}{\delta_{\text{th}}} \right)^{1/n} \tag{S4.39b}$$

and

$$\text{St} = \frac{c_f}{2} \left(\frac{\delta}{\delta_{\text{th}}} \right)^{1/n} \tag{S4.40}$$

Note: $\delta_{\text{th}} \neq \delta$ because of the delayed plate heating at $x = x_0$.

- Find an expression for $\delta/\delta_{\text{th}}$ using the thermal energy equation in integral form as

$$\frac{d}{dx} \left[\int_0^\infty u(T - T_\infty)\, dy \right] = \frac{q_w^{\text{turb}}}{\rho c_p} \tag{S4.41a}$$

or introducing $\theta_{\text{th}} \equiv \int_0^\infty (u/u_\infty)[(T - T_\infty)/(T_w - T_\infty)]\, dy$, for $T_w = \text{const}$ and $u_\infty = \text{const}$

$$\frac{d\theta_{\text{th}}}{dx} = \text{St} \tag{S4.41b}$$

Combining both Stanton number expressions (S4.40) and (S4.41b) yields

$$\frac{d\theta_{\text{th}}}{dx} = \frac{c_f}{2} \left(\frac{\delta}{\delta_{\text{th}}} \right)^{1/n} = \frac{d\delta_2}{dx} \left(\frac{\delta}{\delta_{\text{th}}} \right)^{1/n} \tag{S4.41c}$$

where $\quad \delta_2 \equiv \int_0^\infty \frac{u}{u_\infty} \left(u_\infty - u/u_\infty \right) dy.$

Inserting the power-law profiles in

$$\frac{d\theta_{\text{th}}}{d\theta} = \left(\frac{\delta}{\delta_{\text{th}}} \right)^{1/n} \tag{S4.41d}$$

yields an expression for

$$\frac{\delta_{\text{th}}}{\delta} = \left[1 - \left(\frac{x_0}{x} \right)^{4(n+2)/5(n+1)} \right]^{n/(2+n)}$$

where $\delta \sim x^{4/5}$ was employed. Now, with $c_f = 0.59\, \text{Re}_x^{-0.2}$ and $n = 7$ (cf. p. 279), Equation (S4.40) can be written as

$$\text{St}_x \approx 0.03 \text{Re}_x^{-0.2} \underbrace{\left[1 - \left(\frac{x_0}{x} \right)^{9/10} \right]^{-1/9}}_{\substack{\text{unheated starting} \\ \text{length effect}}} \tag{S4.42a}$$

Note: When $x_0 = 0$, $\delta \approx \delta_{\text{th}}$ for $\text{Pr} \approx 1$ and (S4.42a) reduces to the Reynolds analogy

$$\text{St}_x = \frac{1}{2} c_f(x) \tag{S4.42b}$$

so that

♦ $h(x) = \rho c_p u_\infty \text{St}_x$ (S4.42c)

                                                                                                          □

4.11   Find an expression for the local Nusselt number $\text{Nu} = \text{Nu}(\text{Ra}_x, \text{Pr})$ for turbulent free
convection boundary-layer flow along a vertical flat wall at constant temperature(cf. sketch).

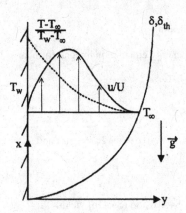

Assumptions

-   Steady fully turbulent boundary layers
-   Negligible inertia terms
-   Analogies to forced convection hold, such as $\delta \approx \delta_{\text{th}}$, $\tau_w = \tau_w(U, \delta)$, and $\text{St} = \text{St}(\text{Pr}, \tau_w)$

Postulates
Following the suggestions by Eckert and Jackson (1951), we postulate

$$\frac{T - T_\infty}{T_w - T_\infty} = \left[1 - \left(\frac{y}{\delta_{\text{th}}}\right)^{1/7}\right] \quad \text{and} \quad u = U\left(1 - \frac{y}{\delta}\right)^4 \left(\frac{y}{\delta}\right)^{1/2}$$

where $U(x)$ is a reference velocity in turbulent natural convection.

Solution
The Reynolds-averaged transport equations can be reduced (cf. Assumptions) to

$$0 = \frac{\partial}{\partial y}\left(\varepsilon_m \frac{\partial u}{\partial y}\right) + g\beta(T - T_\infty)$$ (S4.43)

and

$$u\frac{\partial T}{\partial x} = \frac{\partial}{\partial y}\left(\varepsilon_h \frac{\partial T}{\partial y}\right)$$ (S4.44)

Integrating (S4.43) across the boundary layer and using the postulated temperature profile
yield

$$\frac{\tau_w^{\text{turb}}}{\rho} = \frac{\delta}{8}g\beta(T_w - T_\infty)$$ (S4.45)

Integrating (S4.44) across the boundary layer and using the suggested temperature and velocity profiles yield

$$\frac{q_w^{\text{turb}}}{\rho c_p} = 0.0366(T_w - T_\infty)\frac{d}{dx}(U\delta_{\text{th}}) \tag{S4.46}$$

Notes

- From "forced convection" (Sect. 3.2.5) we know that

$$\tau_w^{\text{turb}} \approx 0.03\rho U^2 \text{Re}_x^{-0.2} \quad \text{and} \quad (\text{Re}_\delta)^{1/4} = \frac{4}{3}\text{Re}_x^{-0.2}$$

so that

$$\tau_w^{\text{turb}} \approx 0.0225\rho U^2\left(\frac{U\delta}{\nu}\right)^{-1/4} \tag{S4.47}$$

Use of Equation (S4.47) implies that the buoyancy-induced reference velocity U is equivalent to a hypothetical outer flow velocity in generating the same turbulent wall shear stress.
- The Reynolds–Colburn analogy

$$\text{StPr}^{2/3} = \frac{c_f}{2} \tag{S4.48a}$$

is rewritten as

$$\frac{q_w^{\text{turb}}}{(T_w - T_\infty)\rho c_p U}\text{Pr}^{2/3} = \frac{\tau_w^{\text{turb}}}{\rho U^2} \tag{S4.48b}$$

Results

The four equations (S4.45) to (S4.48) can be used to compute numerically the four unknowns U, δ, τ_w and q_w, which are functions of x only. Alternatively, the solution procedure can be simplified by introducing the postulates $U = Ax^m$, $\delta = Bx^n$, $\tau_w = Cx^p$, and $q_w = Dx^q$ into these equations, matching the exponents, that is, $m = 1/2$, $n = 7/10$, $p = 7/10$, and $q = 1/5$, and obtaining expressions for the coefficients A through D.

The final result is

$$\text{Nu}_x = \frac{q_w(x)}{T_w - T_\infty}\frac{x}{k} = \frac{x}{\Delta T k}(Dx^{1/5}) \tag{S4.49a}$$

that is,

$$\text{Nu}_x = 0.0391\text{Pr}^{-1/5}\text{Ra}_x^{2/5} \tag{S4.49b}$$

Comments

Equation (S4.49) indicates that

$$\text{Nu}_x^{\text{turb}} \sim x^{6/5} \quad \text{while} \quad \text{Nu}_x^{\text{lam}} \sim x^{3/4}$$

However, the error involved in using Eq. (S4.47) should be analyzed by (a) finding a more accurate expression for τ_{wall} in turbulent natural convection or (b) solving Equations (S4.43,44) directly (cf. App. F.4). □

4.12 Consider high-speed flow ($\text{Ma} = u_\infty/a \to 1$) over a flat plate: Air at $p_\infty = 1/20$ atm and $T_\infty = 275$ K flows with $u_\infty = 700$ m/s past a flat plate of width $W = 1$ m and length $L = 1.2$ m. What is the amount of cooling needed to keep the plate at $T_w = 325$ K maximum?

Background Information

For this high-speed flow problem, temperature increase due to frictional effects has to be included. Specifically, the thermal energy generated near the wall due to viscous dissipation and/or kinetic energy conversion (no-slip) is carried away by thermal diffusion. Under adiabatic steady-state conditions, the wall heats up to the adiabatic wall temperature T_{aw} (cf. sketch). In other words, without cooling, the wall temperature would be

$$T_{aw} = T_\infty + r_c \frac{u_\infty^2}{2c_p} \tag{S4.50}$$

where $r_c = r_c(\text{Pr})$ is a recovery factor and $u_\infty^2/(2c_p)$ is the kinetic energy of which a fraction is envisioned to convert into internal energy. Note,

$$r_c = \text{Pr}^{1/2} \qquad \text{for laminar BL flow} \tag{S4.51a}$$

and

$$r_c = \text{Pr}^{1/3} \qquad \text{for turbulent boundary layers} \tag{S4.51b}$$

The rapidly varying fluid properties are evaluated at a (constant) reference temperature suggested by Eckert (1956)

$$T_{\text{ref}} = T_\infty + \frac{1}{2}(T_w - T_\infty) + 0.22(T_{aw} - T_\infty) \tag{S4.52}$$

The local wall heat flux, $q_w = h\Delta T$, is evaluated with the low-speed heat transfer coefficient $h(x)$ for laminar or turbulent flow and $\Delta T = T_w - T_{aw}$, where T_{aw} is a corrected free stream temperature replacing T_∞. Hence,

$$q_w(x) = h(T_w - T_{aw}) \tag{S4.53}$$

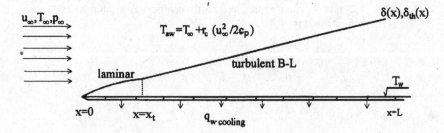

Solution

$$Q = Q_{\text{lam}} + Q_{\text{turb}} = \bar{h}_l A_l (T_w - T_{aw}) + \bar{h}_t A_t (T_w - T_{aw}) \tag{S4.54}$$

where $A_l = x_c W$ and $A_t = (L - x_c)W$. Since T_{aw} is not known, Equation (S4.52) cannot be used yet and we assume $T_{\text{ref}} = T_w = 325$ K $\Rightarrow \text{Pr} = 0.7025$.

- Laminar flow region $0 \le x \le x_t$

  Recovery factor $\qquad r_c = \sqrt{\text{Pr}} = 0.838$

  Adiabatic wall temperature $\qquad T_{aw} = T_\infty + r_c \dfrac{u_\infty^2}{2c_p} := 479 \text{ K}$

  Reference temperature $\qquad (S4.52): T_{ref} = 345 \text{ K} \quad \text{and} \quad p = 1/20 \text{ atm}$

  yields the fluid properties

  $\rho = 0.0508 \text{ kg/m}^3, \qquad \mu = 2.052 \times 10^{-5} \text{ kg/m} \cdot \text{s}, \qquad k = 0.030 \text{ W/m} \cdot ^\circ\text{C}$

  $c_p = 1009 \text{ J/kg} \cdot ^\circ\text{C}, \qquad \text{Pr} = 0.698$

  Point of transition $x_t = \mu \text{Re}_t(\rho u_\infty) := 0.289 \text{ m}$ where we assume
  $\text{Re}_t = 5 \times 10^5$

  As derived in Section 4.1.1, the local Nusselt number for laminar boundary-layer flow with zero pressure gradient past a flat isothermal plate is

  $$\text{Nu}_x = \frac{h(x)x}{k} = 0.332 \text{Pr}^{1/3} \text{Re}_x^{1/2} \tag{S4.55}$$

  and

  $$\bar{h} = \frac{1}{x_t} \int_0^{x_t} h(x)\, dx = 2h(x) \tag{S4.56}$$

  Thus

  $$Q_{\text{lam}} = \bar{h}_l A_l (T_w - T_{aw}) := -1{,}923 \text{ W}$$

- Turbulent flow region $x_t < x \le L$:

  Recovery factor $r_c = \text{Pr}^{1/3} := 0.889$ so that $T_{aw} = 491 \text{ K}$ and the reference temperature $T_{ref} = 347.5 \text{ K}$, which is close to the previous T_{ref}. The maximum Reynolds number is $\text{Re}_L = 2 \times 10^6$, so we employ (cf. Sect. 4.1.2)

  $$\frac{h(x)}{\rho u_\infty c_p} = 0.0296 \text{Re}_x^{-0.2} \text{Pr}^{-2/3} \qquad \text{for } 5 \times 10^5 < \text{Re}_x < 10^7 \tag{S4.57}$$

  Thus,

  $$h(x) = 73.6 x^{-0.2}$$

  so that

  $$\bar{h}_t = \frac{1}{L - x_t} \int_{x_t}^{L} h(x)\, dx = 82.4 \text{ W/m}^2{}^\circ\text{C} \tag{S4.58}$$

  Now we have

  $$Q_t = \bar{h}_t A_t (T_w - T_{aw}) = -12{,}461 \text{W}$$

The final result is

♦ $|Q| = Q_l + Q_t = 14.4 \text{ kW}$ (S4.59)

                                                                                                                        □

4.3.2 Developing and Fully Developed Internal Flows with Heat Transfer

4.13 Consider a parallel-plate lubrication system (slider bearing) with viscous heating and temperature-dependent parameters μ and k (cf. sketch).

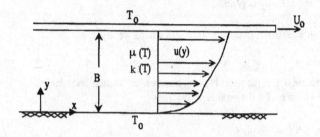

Assumptions

- Steady laminar unidirectional flow without pressure gradient
- $\mu(T)$ and $k(T)$ in form of power series expansions (after Bird et al. 1960)

Concept

Couette-type flow with internal viscous heating and isothermal plates at $T_0 = \text{const.}$

Postulates

For Couette flow (cf. Sect. 3.2.1A), $T(y)$ and $u = u(y)$ only.

Governing Equations

The reduced equations and boundary conditions are

$$\frac{\partial}{\partial y}\left(\mu\frac{\partial u}{\partial y}\right) = 0; \quad u(0) = 0 \quad \text{and} \quad u(B) = U_0$$ (S4.60a)

and

$$\frac{\partial}{\partial y}\left(k\frac{\partial T}{\partial y}\right) + \mu\left(\frac{du}{dy}\right)^2 = 0; \quad T(0) = T(B) = T_0$$ (S4.60b)

Nondimensionalization generates a set of normalized equations and dimensionless groups. Specifically,

$$\eta = \frac{y}{B}, \qquad \phi = \frac{u}{U_0}, \qquad \theta = \frac{T - T_0}{T_0}$$

$$\frac{k}{k_0} = 1 + \alpha_1\theta + \alpha_2\theta^2 + \cdots \alpha_n\theta^n$$

and for convenience (cf. Eq. (S4.62b))

$$\frac{\mu_0}{\mu} = 1 + \beta_1 \theta + \beta_2 \theta^2 + \cdots \beta_n \theta^n$$

where k_0, μ_0, α_i's, and β_i's are known constants.

Inserting (S4.61) into (S4.60) yields

$$\frac{d}{d\eta}\left(\frac{\mu}{\mu_0}\frac{d\phi}{d\eta}\right) = 0 \tag{S4.62a}$$

or

$$\frac{d\phi}{d\eta} = c_1\frac{\mu_0}{\mu} \tag{S4.62b}$$

and

$$\frac{d}{d\eta}\left(\frac{k}{k_0}\frac{d\theta}{d\eta}\right) + \text{Br}\frac{\mu}{\mu_0}\left(\frac{d\phi}{d\eta}\right)^2 = 0 \tag{S4.63}$$

where Br is the Brinkman number,

$$\text{Br} = \text{Pr Ec} = \frac{\mu_0 U_0^2}{k_0 T_0} \hat{=} \frac{\left(\begin{array}{c}\text{momentum transfer}\\\text{due to fluid friction}\end{array}\right)}{\left(\begin{array}{c}\text{heat transfer due}\\\text{to conduction}\end{array}\right)}$$

Typically, Br < 1. The boundary conditions are transformed to

$$\eta = 0 \quad \phi = 1 \quad \text{and} \quad \theta = 0$$
$$\eta = 1 \quad \phi = 0 \quad \text{and} \quad \theta = 0$$

Equation (S4.62b) is inserted into the energy equation (S4.63)

$$\frac{d}{d\eta}\left[(1 + \alpha_1\theta + \alpha_2\theta^2 + \cdots)\frac{d\theta}{d\eta}\right] + \text{Br}c_1^2(1 + \beta_1\theta + \beta_2\theta^2 + \cdots)^2 = 0 \tag{S4.64}$$

This is a nonlinear ODE for $\theta(\eta)$ with unknown integration constant c_1, which cannot be determined via a boundary condition.

Solution

The coupled set of equations (S4.62) and (S4.63) with appropriate submodels for the fluid properties can be solved with a finite difference method such as Keller's box method (cf. App. F.4). Alternatively, Eq. (S4.64) is solved using the Runge–Kutta algorithm with a suitable starting value for $\theta'(\eta = 0)$.

Since $d\phi/d\eta \sim f(\theta)$ and the solution for $\theta(\eta)$, when μ and k are constant, is $\theta(\eta) \sim \text{Br}(\eta - \eta^2)$, we postulate expansions for θ, c_1, and ϕ of the form (cf. Bird et al. 1960)

$$\theta = \text{Br}\theta_1(\eta) + \text{Br}^2\theta_2(\eta) + \cdots$$

$$\phi = \phi_0(\eta) + \text{Br}\phi_1(\eta) + \text{Br}^2\phi_2(\eta) + \cdots \tag{S4.65a–c}$$

$$c_1 = c_{10}(\eta) + \text{Br}c_{11} + \text{Br}^2 c_{12} + \cdots\cdots$$

provided that Br < 1. These power series solutions satisfy the boundary conditions, and the partial solutions c_{1i}, ϕ_i, and θ_i are only functions of η, that is, independent of the Brinkman number.

The trial solutions (S4.65a–c) are inserted into Equation (S4.64) and coefficients of like power of Br are equated to zero. This procedure has two advantages: (i) the solution is broken down into several simpler steps, and (ii) the product form of Equation (S4.64), that is, $d/d\eta[f(\theta)d\theta/dy]$, is resolved into $d^2\theta_i/d\eta^2 + \cdots\cdots$

Br^1 terms

$$\frac{d^2\theta_1}{d\eta^2} + c_{10}^2 = 0 \quad \Rightarrow \quad \theta_1 = \frac{1}{2}c_{10}^2(\eta - \eta^2)$$

Substituting the trial solutions (S4.65a–c) into (S4.62b), we obtain

$$\frac{d}{d\eta}(\phi_0 + Br\phi_1 + Br^2\phi_2 + \cdots) = (c_{10} + Brc_{11} + Br^2c_{12} + \cdots\cdot)$$

$$[1 + \beta_1(Br\theta_1 + Br^2\theta_2 + \cdots) + \beta_2(\ldots)]$$

Br^0 terms

$$\frac{d}{d\eta}\phi_0 = c_{10} \quad \Rightarrow \quad \phi_0 = c_{10}\eta + c_2$$

with the BC $\phi(0) = 0$ and $\phi(1) = 1$, $c_{10} = 1$, and $c_2 = 0$. Thus, $\phi_0 = \eta$ and $\phi_1 = (\eta - \eta^2)/2$.

The Br^2 terms are obtained in similar fashion as

$$\phi_1 = -\frac{\beta_1}{12}(\eta - 3\eta^2 + 2\eta^3)$$

and

$$\theta_2 = -\frac{\alpha_1}{8}(\eta^2 - 2\eta^3 + \eta^4) - \frac{\beta_1}{24}(\eta - 2\eta^2 + 2\eta^3 - \eta^4)$$

Inserting the partial results into the trial solutions yields

$$\phi(\eta) = \underbrace{\eta}_{\phi_0} + \underbrace{-\frac{Br}{12}\beta_1(\eta - 3\eta^2 + 2\eta^3)}_{\sim\phi_1} + \cdots \qquad \text{(S4.66a)}$$

and

$$\theta(\eta) = \underbrace{\frac{Br}{2}(\eta - \eta^2)}_{\sim\theta_1} - \frac{\alpha_1 Br^2}{8}(\eta^2 - 2\eta^3 + \eta^4) - \frac{\beta_1 Br^2}{24}(\eta - 2\eta^2 - 2\eta^3 - \eta^4) \qquad \text{(S4.66b)}$$

□

4.14 Consider (turbulent) flow through a rough pipe ($d = 5$ cm, $L = 9$ cm, $e/d = 0.002$). Water at $T_{in} = 38°C$ and a mass flow rate of $\dot{m} = 6$ kg/s enters a heated pipe with $T_{wall} = 65°C$ (cf. sketch). Find T_{exit} and Q_{total}.

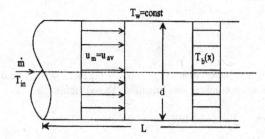

Concept

Use the Reynolds–Colburn analogy, St = St(f, Pr), in conjunction with Moody's chart for the pipe friction factor $f = f(\text{Re}_d, e/d)$ in order to obtain an average heat transfer coefficient $\bar{h}$. Then perform a global enthalpy balance, $\dot{m}c_p \Delta T_{\text{bulk}} = \bar{h}A_s \Delta T_{\text{av}}$, to calculate T_{exit} and Q_{total}.

Assumptions

- Lumped parameter approach using uniform profiles: averaged velocities and temperatures, constant fluid properties, and a global heat balance
- Negligible entrance length: $L_e/d \ll 1$

Solution

The key equation is the Reynolds–Colburn analogy in the form

$$\overline{\text{St}}\,\text{Pr}^{2/3} = f/8, \qquad \overline{\text{St}} \equiv \frac{\bar{h}}{\rho c_p u_m} \tag{S4.67a,b}$$

where the Stanton number is based on the bulk temperature, T_b, and the Prandtl number is based on the film temperature, T_f. Here,

$$T_b = T_{\text{in}} = 38°C \quad \Rightarrow \quad \rho = 993 \text{ kg/m}^3,$$

$$\nu = 0.7 \times 10^{-6} \text{ m}^2/\text{s} \quad \text{and} \quad c_p = 4.17 \text{ kJ/kg°C}$$

and

$$T_f = \frac{1}{2}(T_{\text{in}} + T_w) = 51.5°C \Rightarrow \text{Pr}_f = 3.47$$

Results

With the Reynolds number $\text{Re}_d = 2.24 \times 10^5$ and with $e/d = 0.002$, we obtain from the Moody chart (cf. Fig. 4.13) a friction factor of $f = 0.025$ so that

$$\text{St} \equiv \frac{\bar{h}}{\rho c_p u_{\text{av}}} = \text{Pr}^{-2/3}\frac{f}{8} := 1.363 \times 10^{-3}$$

or

$$\bar{h} = 17{,}388 \text{ W/m}^2 \cdot °\text{C}$$

A global heat balance for the pipe of length L, where heat transferred to the pipe is equal to heat gained by the fluid, reads

$$Q = \underbrace{\bar{h} A_s \, \Delta T_{av}}_{q_w A_s} = \underbrace{\dot{m} c_p \, \Delta T_b}_{\Delta H} \tag{S4.68}$$

Here, $A_s = d\pi L$ is the heat transfer area, $\Delta T_{av} = T_w - (1/2)(T_{in} + T_{out})$, and $\Delta T_b = T_{out} - T_{in}$. Inserting appropriate numbers into Equation (S4.68) yields

♦ $T_{exit} = 55.78°C$ and then $Q = 445.2 \text{ kW}$

Note: Simplification of the two-dimensional problem with variable fluid properties to a one-dimensional constant-property problem required the definition of a number of "representative" or mean temperatures such as T_{film}, ΔT_{bulk}, and $\Delta T_{average}$. It would be of interest to solve this problem numerically and compare the results (e.g., as a computer course project).   □

4.15   Consider steady laminar flow of an extremely viscous fluid through a pipe of radius r_0 and length L. In order to maintain a constant wall temperature T_0 in light of viscous dissipation, a certain amount of heat, Q, for the tube length has to be rejected (cf. sketch). Find the fluid temperature distribution and the total heat transfer rate.

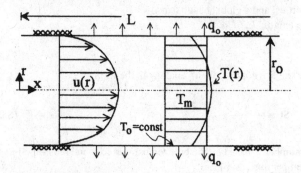

Assumptions

- Fully developed laminar flow
- $T_{mean} = $ const: balance of internal heating and wall cooling (i.e., $\dot{Q}_{gen} \approx \dot{Q}_{cool}$)
- Constant fluid properties.

Postulates

On the basis of the assumptions of Poiseuille flow and $T_m \neq T_m(x)$, we have

$$u = u(r) \quad \text{and} \quad T = T(r) \quad \text{only}$$

Solution

We are assuming Poiseuille flow, that is,

$$u(r) = 2u_{av}\left[1 - \left(\frac{r}{r_0}\right)^2\right] \quad \text{where } u_{av} = \frac{r_0^2}{8\mu}\left(-\frac{d\hat{p}}{dx}\right) \tag{S4.69a,b}$$

and we are reducing the energy equation to the form

$$\frac{k}{r}\frac{d}{dr}\left(r\frac{dT}{dr}\right) + \mu\left(\frac{du}{dr}\right)^2 = 0 \qquad\qquad (S4.70a)$$

subject to $T(r = r_0) = T_0$ and $(dT/dr)|_{r=0} = 0$.
Inserting (S4.69) into (S4.70) and integrating yield

$$T - T_0 = \frac{\mu u_{av}^2}{k}\left[1 - \left(\frac{r}{r_0}\right)^4\right] \qquad\qquad (S4.70b)$$

Now, the wall heat flux is

$$q_0 = -k\frac{dT}{dr}\bigg|_{r=r_0} = 4\frac{\mu u_{av}^2}{r_0} := \text{const}$$

or with $Q_{cooling} = (2\pi r_0 L)q_0$, the heat flow rate can be expressed as

$$Q = 8\pi\mu L u_{av}^2$$

Alternatively, with Eq. (S4.69b),

$$Q_{cool} = \underbrace{\pi r_0^2 u_{av}}_{\dot{m}/\rho}\left(-\frac{d\hat{p}}{dx}\right)L = \dot{Q}_{gen}$$

or

$$Q = \Delta P\frac{\dot{m}}{\rho} \qquad \text{where } \Delta P = \left(-\frac{d\hat{p}}{dx}\right)L$$

Comment
The heat flow rate Q appears to depend on momentum and mass transfer only. Employ the Reynolds analogy to show that Q depends equivalently on heat transfer parameters. □

4.16 Consider heating of a liquid fuel in a tube of finite length (cf. sketch). For steady laminar fully developed flow with constant fluid properties, compute $T_m(x)$ and the maximum fluid temperature to be achieved for the constant wall heat flux case ($D = 0.6$ cm, $L = 1.2$ m, $T_{in} = 10°C$, $T_{out} = 65°C$, $\dot{m} = 1.26 \times 10^{-4}$ kg/s, $Pr = 10$, $\rho = 753$ kg/m^3, $c_p = 2.092$ kJ/kg $\cdot$ K, $k = 0.137$ W/m $\cdot$ K, $\mu = 0.022$ N $\cdot$ s/m^2).

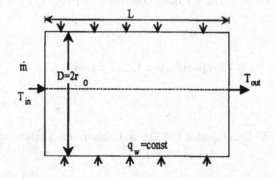

Postulates and Assumptions

In contrast to Sample-Problem 4.15, here we have $T_m = T_m(x)$ where $dT_m/dx = \text{const}$ (cf. Fig. 4.9b). Assume negligible thermal entrance length since $l_{th} = 0.04\,\text{Pe} \approx 1$ cm, whereas $L = 120$ cm. Constant fluid properties and Poiseuille flow prevail.

Governing Equations

$$u\frac{dT_m}{dx} = \frac{\alpha}{r}\frac{\partial}{\partial r}\left(r\frac{\partial T}{\partial r}\right) \tag{S4.71}$$

$$q_w = \underbrace{\dot{m}c_p(T_{\text{out}} - T_{\text{in}})}_{\Delta H} = h[T_w(x) - T_m(x)] \tag{S4.72a}$$

$$h = k\frac{\text{Nu}}{D} \tag{S4.72b}$$

$$T_m = \frac{2}{r_0^2 u_{\text{av}}} \int_0^{r_0} u(r)T(r)r\,dr \tag{S4.73}$$

Solution

With $dT_m/dx = \text{const}$, and employing $u = u(r)$ from Equation (S4.69), as well as the boundary conditions $T(r = r_0) = T_w(x)$ and $\partial T/\partial r = 0$ at $r = 0$, Equation (S4.71) has the solution (cf. Eq. (4.111))

$$T(x,r) = T_w(x) - \frac{2u_{\text{av}}}{\alpha}\left(\frac{dT_m}{dx}\right)\left(\frac{3}{16}r_0^2 + \frac{r^4}{16r_0^2} - \frac{r^2}{4}\right) \tag{S4.74}$$

Hence, with the definition of the mixing cup or mean temperature (cf. Eq. (S4.73))

$$T_m(x) = T_w - \frac{11}{96}\frac{2u_{\text{av}}}{\alpha}\left(\frac{dT_m}{dx}\right)r_0^2 \tag{S4.75}$$

and with (S4.72a,b) the wall temperature is

$$T_w(x) = T_m + \frac{q_w D}{k\text{Nu}} \tag{S4.76}$$

Using the last two equations to eliminate $(T_w - T_m)$, we obtain an ODE for $T_m(x)$

$$\frac{11}{96}\frac{2u_{\text{av}}}{\alpha}r_0^2\frac{dT_m}{dx} = \frac{2q_w r_0}{k\text{Nu}}$$

Now, with $T_m(x = 0) = T_{\text{in}}$ and $\text{Nu} = 4.364$ from Section 4.2.1, integration yields

$$\blacklozenge \quad T_m(x) = T_{\text{in}} + \frac{2\alpha q_w}{kr_0 u_{\text{av}}}x \tag{S4.77a}$$

In this case, $q_w = 641$ W/m^2 from Equation (S4.72a) and with the other values we obtain

$$T_m(x) = 10 + 45.83x \tag{S4.77b}$$

Using $T_m(x)$ in Equation (S4.76) and setting $x = L = 1.2$ m, we obtain

$$T_w(x = L) = T_{max} = 71.4°C$$

Note: Assuming a linear increase of the bulk temperature, Eq. (S4.77b) could have been immediately determined knowing T_{in}, T_{out} and L; in general, $T_{in} < T_{out} < T_w(x = L) = T_{max}$. □

4.17 Consider fully developed turbulent heat transfer in 2-D flow between parallel plates spaced a distance $2B$ apart and subject to a constant wall heat flux (cf. sketch). Derive an expression for the dimensionless temperature $\theta = (T - T_w)/(T_m - T_w)$ and for the average Nusselt number $\bar{N}u = \bar{h}B/k$ in terms of u/u_m, $\eta = 1 - y/B$, and $\alpha/(\alpha + \alpha_t)$, where $\alpha_{turbulent} = \nu_t/Pr_t$. This problem was suggested by Kays and Crawford (1993).

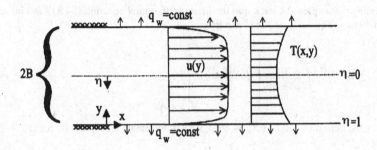

Assumptions

- Steady symmetric flow
- Constant fluid properties
- No viscous dissipation

Governing Equations

$$\rho c_p u \frac{\partial T}{\partial x} = \frac{\partial}{\partial y} q_{toatal} \tag{S4.78}$$

where $\qquad q_{total} = \rho c_p (\alpha + \alpha_t) \dfrac{\partial T}{\partial y} \quad$ and $\quad y = B(1 - \eta)$

Thus

$$u B^2 \frac{\partial T}{\partial x} = \frac{\partial}{\partial \eta}\left[(\alpha + \alpha_t)\frac{\partial T}{\partial \eta}\right]; \qquad 0 \le \eta \le 1 \tag{S4.79a}$$

where $(\partial T/\partial x) = (\partial T_w/\partial x) = (\partial T_m/\partial x) = $ const for $q_w = $ const. The boundary conditions are

$$\left.\frac{\partial T}{\partial \eta}\right|_{\eta=0} = 0 \quad \text{and} \quad \left.\frac{\partial T}{\partial \eta}\right|_{\eta=1} = -\frac{q_w B}{k} \tag{S4.79b,c}$$

Recall that

$$T_m = \frac{1}{u_m A}\int_A uT\,dA \quad \text{and} \quad u_m = \frac{1}{A}\int u\,dA \tag{S4.80a,b}$$

Solution
Integration of (S4.79a) yields

$$\frac{\partial T}{\partial \eta} = \frac{B^2}{\alpha + \alpha_t}\left(\frac{dT_m}{dx}\right)\int_0^\eta u(\eta)\,d\eta \tag{S4.81}$$

which can be evaluated at the wall, that is, $\eta \to 1$ and $\alpha_t \approx 0$, to obtain

$$\frac{dT_m}{dx} = \frac{\alpha q_w}{k\mathrm{Bu}_m} \tag{S4.82}$$

Inserting this expression back into the integrodifferential equation (S4.81) and integrating again where $T(\eta = 1) = T_w(x)$ result in

$$T_w - T = \frac{q_w B}{k}\int_\eta^1 \frac{\alpha}{\alpha + \alpha_t}\left[\int_0^{\eta'}\frac{u(\eta'')}{u_m}d\eta''\right]d\eta' \tag{S4.83}$$

Using the definition for T_m (cf. S4.80a), we can formulate an expression for $(T_w - T_m)$ as

$$T_w - T_m = \int_0^1 \frac{u}{u_m}(T_w - T)\,d\eta \tag{S4.84}$$

into which (S4.83) is inserted so that

$$T_w - T_m = \frac{q_w B}{k}\int_0^1 \frac{u}{u_m}\left\{\int_\eta^1 \frac{\alpha}{\alpha + \alpha_t}\left[\int_0^{\eta'}\frac{u(\eta'')}{u_m}d\eta''\right]d\eta'\right\}d\eta \tag{S4.85}$$

Dividing (S4.83) by (S4.85) yields the dimensionless temperature

$$\theta = \frac{T_w - T}{T_w - T_m} = \frac{\int_\eta^1 \alpha/(\alpha + \alpha_t)\left[\int_0^{\eta'}(u(\eta'')/u_m)d\eta''\right]d\eta'}{\int_0^1 (u/u_m)\left\{\int_\eta^1 \alpha/(\alpha + \alpha_t)\left[\int_0^{\eta'}(u(\eta'')/u_m)d\eta''\right]d\eta'\right\}d\eta} \tag{S4.86}$$

Comments
Expressions for u/u_m and α_t have to be specified in order to determine $\theta(\eta)$. A simple application would be assuming plug flow $u/u_m = 1.0$ and a linear mixing model $\alpha_t = c(1 - \eta)$. Now the integrations in (S4.86) can be completed, and the average Nusselt number $\overline{\mathrm{Nu}} = Bq_w/(k(T_w - T_m))$ can be computed with Equation (S4.85) and compared to experimental values.   □

4.18   Derive a relationship for the Stanton number $\mathrm{St} = \mathrm{St}(f, \mathrm{Pr})$ considering fully developed turbulent pipe flow with $\bar{u} \approx U$ and $q_w = \mathrm{const}$ (cf. Sample Problem 4.17).

Solution

The energy equation (S4.78) takes on the form

$$\rho c_p U \frac{\partial T}{\partial x} = \frac{1}{r} \frac{\partial}{\partial r}(r q_{\text{total}}) \tag{S4.87}$$

where $r = r_0 - y$, $q_{\text{total}} = -\rho c_p(\alpha + \varepsilon_h)\partial T/\partial y$, and $\partial T/\partial x = dT_m/dx = \text{const.}$ Integrating from $r = 0$ to $r = r$ and invoking $q_{\text{app}}(r = 0) = 0$ because of symmetry yield

$$\frac{U}{2}\left(\frac{dT_m}{dx}\right)\frac{r_0 - y}{\alpha + \varepsilon_h} = -\frac{\partial T}{\partial y} \tag{S4.88}$$

A second integration from $y = 0$ to $y = y$ yields in terms of inner variables

$$T^+ \equiv \frac{\rho c_p u_\tau}{q_w}(T_w - T) = \int_0^{y^+} \frac{1 - (y/r_0)}{(\alpha/\nu + \varepsilon_h/\nu)}dy^+ \tag{S4.89}$$

where $y^+ = (u_\tau y/\nu)$, $u_\tau = \sqrt{\tau_w/\rho} = U\sqrt{f/2}$, and $\alpha/\nu = \text{Pr}$.

Submodels

Although we have assumed so far $\bar{u} \approx U$, that is, "turbulent plug flow," we are forced to make postulates on the structure of the turbulent thermal flow and the velocity profile in order to solve (S4.89). Dividing turbulent thermal pipe flow into a "conduction sublayer" where $\varepsilon_h/\nu \ll \text{Pr}^{-1}$ and a "core region" where $\varepsilon_h/\nu \gg \text{Pr}^{-1}$, we rewrite (S4.89) as

$$T^+ = \int_0^{y_{\text{cs}}^+} \frac{1}{\text{Pr}^{-1}} dy^+ + \int_{y_{\text{cs}}^+}^{y^+} \frac{1 - (y/r_0)}{(\alpha/\nu + \varepsilon_h/\nu)} dy^+ \tag{S4.90}$$

where y_{cs}^+ is the dimensionless conduction sublayer and $\varepsilon_h/\nu = (\varepsilon_m/\nu)\text{Pr}_t^{-1}$ where $\varepsilon_m/\nu \sim \kappa y^+$ according to the mixing length model and $\text{Pr}_t \approx \text{const.}$ Thus, (S4.90) can be integrated to

$$T^+ = \text{Pr}\, y_{\text{cs}}^+ + \text{Pr}_t\left(\frac{1}{\kappa}\ln\frac{y^+}{y_{\text{cs}}^+} - \frac{y^+ - y_{\text{cs}}^+}{\kappa r_0^+}\right) \tag{S4.91}$$

where $\kappa = 0.4$, $r_0^+ = (r_0 u_\tau/\nu) = (DU/\nu)\sqrt{f/8} = \text{Re}_D\sqrt{f/8}$ and $\text{Pr}_t \equiv \varepsilon_m/\varepsilon_h \approx 0.9$. At the centerline $y^+ = r_0^+$ and $T^+ = T_{cL}^+$ so that

$$T_{cL}^+ = \text{Pr}\, y_{\text{cs}}^+ + \text{Pr}_t\left(\frac{1}{\kappa}\ln\frac{r_0^+}{y_{\text{cs}}^+} - \frac{r_0^+ - y_{\text{cs}}^+}{\kappa r_0^+}\right) \tag{S4.92}$$

By definition, the dimensionless centerline temperature is

$$T_{cL}^+ \equiv \frac{\rho c_p u_\tau}{q_w}(T_w - T_{cL}) = \frac{T_w - T_{cL}}{T_w - T_m}\frac{\sqrt{f/2}}{\text{St}} \tag{S4.93}$$

where the Stanton number St $= h/(\rho c_p U)$ is regarded as a dimensionless heat transfer coefficient. From a one-dimensional energy balance

$$\pi r_0^2 (T_w - T_m)U = 2\pi \int\limits_0^{r_0} (T_w - T)ur\,dr$$

we can express $(T_w - T_m)$ using, for example, the 1/7 power law for $u/u_{cL} = (1 - r/r_0)^{1/7}$. Thus, with $(T_w - T_m)/(T_w - T_{cL}) = 0.833$, Equation (S4.93) yields

$$\text{St} = \frac{h}{\rho c_p U} = \frac{\sqrt{f/2}}{0.833 T_{cL}^+} \qquad\qquad\qquad\text{(S4.94)}$$

where $f = f(\text{Re}_D, e/D)$ may be obtained from the Moody chart and T_{cL}^+ is given as Equation (S4.92). $\Box$

4.19   Consider steady uniform flow through an annulus with a radius ratio of $r_i/r_0 = 0.6$, constant heat flux at the inner wall, $q_i = $ const, and an insulated outer wall, that is, $q_0 = 0$. Find an expression for the Nusselt number Nu_i of the inner surface (cf. sketch).

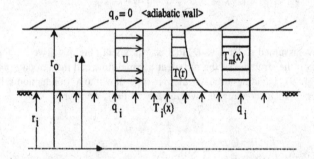

Assumptions

- Steady laminar thermally fully developed slug flow
- No viscous dissipation
- Constant fluid properties
- Negligible axial conduction

Postulates

$(T_i - T)/(T_i - T_m) \equiv \theta = \theta(r)$ only; $\partial\theta/\partial x = 0$, $\partial T/\partial x = dT_m/dx =$ const, and thus $h=$ const.

Governing Equations

$$U\left(\frac{dT_m}{dx}\right) = \frac{\alpha}{r}\frac{\partial}{\partial r}\left(r\frac{\partial T}{\partial r}\right) \qquad\qquad\qquad\text{(S4.95)}$$

$$q = -k\frac{\partial T}{\partial r} = \begin{cases} h_i(T_i - T_m) & \text{for } r = r_i \\ 0 & \text{for } r = r_0 \end{cases} \qquad\qquad\text{(S4.96a,b)}$$

$$T_m = \frac{1}{AU} \int_A UT\,dA = \frac{2}{r_0^2 - r_i^2} \int_{r_i}^{r_0} Tr\,dr \tag{S4.97}$$

$$Nu_i = \frac{h_i D_h}{k} \tag{S4.98a}$$

where the hydraulic diameter is defined as

$$D_h = 2(r_0 - r_i) \tag{S4.98b}$$

Solution

Integrating Equation (S4.95) yields

$$T = \frac{U}{4\alpha} r^2 \left(\frac{dT_m}{dx} \right) + C_1 \ln r + C_2 \tag{S4.99a}$$

and after applying the boundary conditions $T = T_i(x)$ at $r = r_i$ and $\partial T/\partial r = 0$ at $r = r_0$, we obtain

$$T(r,x) = T_i(x) + \frac{U}{2\alpha} \left(\frac{dT_m}{dx} \right) \left[\frac{r^2}{2} - \frac{r_i^2}{2} + r_0^2(\ln r_i - \ln r) \right] \tag{S4.99b}$$

Using (S4.97) we have

$$T_m(x) = T_i + \frac{U}{2\alpha} \left(\frac{dT_m}{dx} \left[\frac{r_0^2 + r_i^2}{4} + \frac{r_0^4}{r_0^2 - r_i^2} \ln \frac{r_i}{r_0} + \frac{r_0^2}{2} - \frac{r_i^2}{2} \right] \right) \tag{S4.100}$$

and Equation (S4.96a) can be expressed as

$$h_i(T_i - T_m) = -k \left(\frac{\partial T}{\partial r} \right)_{r=r_i} \tag{S4.101a}$$

from which with the two previous results for $T(r,x)$ and $T_m(x)$

$$\frac{h_i}{k} = \left(\frac{r_i^2 - r_0^2}{r_i} \right) \left[\frac{3}{4} r_0^2 + \frac{r_i^2}{4} + \frac{r_0^4}{r_0^2 - r_i^2} \ln \frac{r_i}{r_0} \right]^{-1} \tag{S4.101b}$$

Now Equation (S4.98) becomes

$$Nu_i = 2(r_0 - r_i) \frac{h_i}{k} \tag{S4.102a}$$

and for $r_i = 0.6 r_0$

$$Nu_i = 6.176 \tag{S4.102b}$$

Comment

Comparing the slug-flow result with the Poisseuille flow result, $Nu = 5.912$, it is clear that the convenient uniform flow assumption overestimates the Nusselt number by 4.5 percent for this case study. □

Problem Assignments

General

4.1 Study the derivation of Equation (4.10) and comment on the alternative derivation of the heat transfer equation given in Sample-Problem 4.1 and in Sections 2.1, 2.2, and 4.3.

4.2 Would it be more useful to eliminate the heat transfer coefficient $h(x)$ and deal only with $q_w(\Delta T)$?

4.3 Provide additional applications of dimensionless groups in CHT (cf. item (iii)) of Section 4.1C.

4.4 Derive Equation (4.17c).

4.5 Considering Table 4.1 and the solution steps for Sample-Problem 4.2, derive other dimensionless groups employing scale analysis (cf. Sect. 2.3.1).

Forced Convection

4.6 Consider steady laminar flow of water ($T_i = 20°C$, Pr $= 5.0$, Re$_l = 10^5$) past a horizontal flat plate ($l = 1$ m, $T_w = 40°C$). Use Equations (4.26) and (4.27d) to find and plot $h(x)$ and Nu(x).

4.7 Derive Equations (4.30a–d) and (4.32, 33).

4.8 Plot Nu (Pr) for thermal Blasius flow at Re $= 10^4$ (cf. Equations (4.39b–f)).

4.9 Derive Equation (4.51a) and employ (4.52a) to develop a turbulent thermal BL equation with an energy dissipation term.

4.10 Compare Equation (4.54b) and Equation (P4.22) for $x_h = 0$.

4.11 Calculate and plot the local Nusselt numbers and heat transfer coefficients for air flow past a horizontal heated flat plate (cf. Sample-Problem 4.8 solution).

Free Convection

4.12 Reviewing the development of Equation (4.63b), find proportionalities for δ/δ_{th} (Pr) for Pr $\gg 1$ and Pr $\ll 1$ and provide a physical explanation.

4.13 Review the derivation of Equations (P4.23) and (P4.24) plus boundary conditions.

4.14 Derive Equations (P4.27) and (P4.28).

Internal Flows

4.15 Analyze and comment on Equation (4.104) in conjunction with Fig. 4.9b,c.

4.16 Graetz problem: Evaluate Equation (4.127), compute θ_m and Nu$_x$, and plot Nu(x) (cf. Fig. 4.12).

4.17 Turbulent pipe flow profile: Program and integrate Equation (4.137a) and plot u(y) for different Re$_D$ and e/D values.

4.18(a) Plot Equation (4.142) for different Prandtl numbers (i.e., 0.7 and 6.0).

(b) Calculate Nu$_D$ (cf. Eq. (4.141)) employing (4.143) for different Re$_D$ and e/D and compare your results with Eq. (4.147).

Sample-Problem Solutions (cf. Sect. 4.3)

4.19 SPS4.3: Integrate Equation (S4.10), plot $T(y, z)$, and calculate the Nusselt number Nu(z).

4.20(a) SPS4.6: Give a physical explanation why Nu > Nu$_x$ (cf. Eq. (S4.28a,b)).
 (b) SPS4.6: Plot for air and $T_w = const$ Nu(x). Interpret the resulting graph.

4.21 SPS4.7: Derive the coefficients C_1 and C_2 and compare related graphs. Comment!

4.22 SPS4.9: Derive Equations (S4.34) and (S4.35) and the associated boundary conditions. Provide a reasonable initial estimate for $f''(0)$ and $\theta''(0)$.

4.23 Plot Nu(x) for air and water using Equation (S4.42a) and Re$_l = 10^7$.

4.24(a) Derive Equations (S4.49a,b).
 (b) Plot the profiles of SPS4.11 and Nu(x) for Pr $= 0.7$ and Re$_l = 10$.

4.25 Why is $Q_{\text{turb}} \gg Q_{\text{lam}}$ in Sample-Problem 4.12?

4.26 Plot the velocity and temperature profiles of Sample-Problem 4.13 for (a) variable and (b) constant fluid properties. Comment!

4.27 Comment on the solution of Sample-Problem 4.15 and show that Q depends equivalently on heat transfer parameters.

4.28 Integrate Equation (S4.86) assuming two different sets of expressions for u/u_m and α_t. Plot the profiles and calculate $\overline{\text{Nu}}$.

4.29 Plot $T^+(y^+)$ as given with Equation (S4.91) and comment on Equation (S4.94) for typical turbulent pipe flow cases.

4.30 What causes the difference in laminar Nusselt numbers between tubular and annular thermal flows?

Test Questions and Problems

Part A: Insight (Closed Book)

A.1 Deduce a definition of the Peclet number, where Pe = (convection h.t./conduction h.t.) from the thermal energy equation (i.e., heat transfer equation) in vector form. The Pe is a composite of which (two) dimensionless groups?

A.2 Consider the thermal energy equation for steady incompressible boundary-layer flow with negligible viscous dissipation.

(a) Show that its integral form can be written as

$$q_w = \frac{d}{dx}\left[\int_0^\infty \rho u c_p (T - T_e)\, dy\right]$$

(b) Is it valid for both laminar and turbulent flow? yes or no? (Explain.)
(c) Can it be simplified for laminar flow with constant fluid properties?

A.3 Scaling of thermal boundary-layer parameters:

(a) Obtain the heat transfer coefficient h from a wall fluid-layer heat balance and show that

$$h \sim \frac{k}{\delta_{\text{th}}}$$

(b) Recalling that within a TSL

$$\frac{\delta}{l} \sim \mathrm{Re}_l^{-1/2}$$

and that for heavy oils ($\mathrm{Pr} \gg 1$)

$$\frac{\delta_{\mathrm{th}}}{\delta} \sim \mathrm{Pr}^{1/3}$$

find a functional proportionality for

$$\mathrm{Nu}_l = \mathrm{Nu}_l(\mathrm{Re}_l, \mathrm{Pr}); \qquad (\mathrm{Pr} \gg 1)$$

A.4 The graph shows $\mathrm{Nu} = \mathrm{Nu}(\mathrm{Re}_d, \mathrm{Pr})$ for turbulent fully developed flow in smooth and rough pipes with constant wall heat flux (source: Cebeci and Bradshaw 1988).

(a) For a given Pr, explain mathematically/physically $\mathrm{Nu}(\mathrm{Re}_d)$ and

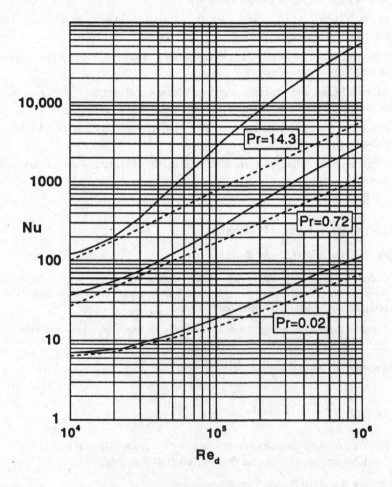

(b) with Re_d fixed, explain $\mathrm{Nu}(\mathrm{Pr})$.

(c) Why is $\mathrm{Nu}_{\mathrm{rough}} > \mathrm{Nu}_{\mathrm{smooth}}$ especially at high Reynolds numbers? Here, the solid lines indicate rough ($r_0/k_s = 100$) and dashed lines smooth surfaces.

A.5 For laminar HFD and TFD flow in a circular pipe of radius r_0

$$u\frac{\partial T}{\partial x} = \frac{\alpha}{r}\frac{\partial}{\partial r}\left(r\frac{\partial T}{\partial r}\right)$$

(a) Assume $(\partial T/\partial x) = (dT_m/dx)$ for $q_{\text{wall}} = $ const and show that $(dT_m/dx) = (2/r_0)(q_w/\rho c_p u_m)$ where subscript m denotes mean values.

(b) Using scale analysis, demonstrate that $h \sim k/d$ and hence that the Nusselt number is constant and $O(1)$.

(c) Is this result also true for turbulent pipe flow?

A.6 The Reynolds analogy

$$\text{Nu}_x = \frac{1}{2}C_{f_x}\text{Re}_x$$

is based on the postulate that the ratio of heat to momentum flux in the direction normal to the surface is unity when evaluated at the wall. Thus, in dimensionless form, evaluate the ratio

$$\left.\frac{q_y/(\rho c_p \Delta T)}{\tau_{yx}/(\rho u_\infty)}\right|_{y=0}$$

and as a result, state:

(a) the underlying assumptions and

(b) the key advantages of the analogy.

A.7 Show that the thermal energy integral relation for steady laminar free-convection boundary layers on a vertical heated wall reduces to

$$\frac{d}{dy}\int_0^\infty v(T_\infty - T)\,dx = \alpha\left(\frac{\partial T}{\partial x}\right)_{x=0}$$

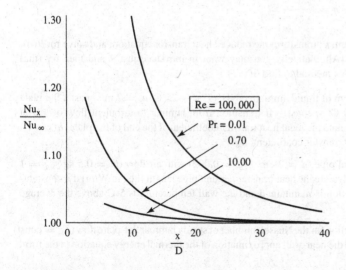

Source: Kays & Crawford (1993)

A.8 For free convection wall flows, sketch

$$\text{Nu Ra}^{-1/4} \quad \text{vs. Pr} \quad \text{where } 10^{-3} \leq \text{Pr} \leq 10^4$$

and comment.

A.9 Discuss Nusselt number-distributions (cf. sketch on page 349) for flow of different Pr fluids in ducts, that is,

(a) classify the conditions and

(b) explain the trends.

A.10 What are the three basic methodologies for obtaining dimensionless groups and what is their use in CHT?

A.11 Show how Pr^n relates to the momentum and thermal boundary-layer thicknesses in laminar forced CHT. Take, for example, $\text{Pr} \gg 1$ and determine n. Give a physical explanation for your results!

Part B: Problems (Open Book; Follow Format: Sketch, Assumptions, Postulates, Solution, Results, Comment)

B.1 Consider steady fully developed gravity flow of a cold Newtonian liquid film of constant thickness, δ, on an inclined (angle α) solid wall of temperature T_s. The liquid enters at a uniform temperature $T_0 = T_\infty \equiv T_{\text{air}}$; all fluid properties stay constant. Select y to be normal to the streamwise coordinate z.

(a) Find the velocity profile.

(b) Considering that $\delta \ll 1$, simplify the velocity field and state the reduced heat transfer equation plus boundary conditions.

(c) Sketch one velocity profile and three temperature profiles.

(d) Introducing

$$\theta = \frac{T - T_0}{T_s - T_0} \quad \text{and} \quad \eta = \frac{y}{(\kappa z)^{1/3}}$$

determine the constant, κ; transform the reduced heat transfer equation; and solve for $\theta(\eta)$, as far as possible. (Alternatively, you may want to introduce just θ and then try (the) separation of variables method to find $\theta(y, z)$.)

B.2 Consider a free stream of liquid ammonia, NH_3 ($T_e = -25°\text{C}$, $u_e = 2$ m/s) past a flat plate ($l = 2$ m, $T_w = 10°\text{C} = \text{const}$). If transition from laminar to turbulent flow occurs at $\text{Re}_{\text{tr}} = 4 \times 10^5$, calculate the local heat transfer coefficient at the end of the plate, $h(x = l)$, and the average heat transfer coefficient, h_m.

B.3 Consider a horizontal pipe ($d = 2$ cm, $e/d = 0.002$) with air flow ($\dot{m} = 0.5$ kg/s, $p = 1$ atm, $T = 10°\text{C}$). Calculate the heat transfer per unit pipe length (i.e., in W/m) if a constant-wall-heat-flux condition is maintained and the wall temperature is $5°\text{C}$ above the average air temperature.

B.4 An approximate solution for the Nusselt number of steady laminar flat-plate flow ($T_w = \text{const}$) can be obtained with the near-wall approximation of the thermal energy equation in the form

$$u \frac{\partial T}{\partial x} \approx \frac{\nu}{\text{Pr}} \frac{\partial^2 T}{\partial y^2} \tag{1}$$

where near the wall

$$u \approx \lambda y, \qquad \lambda = \text{const} \tag{2a,b}$$

Now introducing a combined variable

$$\xi = y \left(\frac{\lambda \text{Pr}}{9\nu x} \right)^{1/3} \tag{3}$$

(a) show that (1) with (2) can be transformed to

$$T'' + 3\xi^2 T' = 0 \tag{4}$$

(b) Determine the BC's for (4). With the solution of (4) as follows:

$$\frac{T - T_w}{T_e - T_w} = \frac{1}{0.893} \int\limits_0^\zeta e^{-\zeta^3} d\zeta \tag{5}$$

(c) show that the local Nusselt number

$$\text{Nu}_x = \frac{x}{0.893} \left(\frac{\lambda \text{Pr}}{9\nu x} \right)^{1/3} \tag{6}$$

(d) Using the Blasius solution to find λ, calculate Nu_x and compare your result with the exact solution

$$\text{Nu}_x \text{Re}_x^{-1/2} = 0.332 \qquad \text{for Pr} = 1.0$$

(e) Is the approximate solution more accurate for low or for high Prandtl numbers?

References and Further Reading Material

Adiutori, E. F. 1989. *The New Heat Transfer*. Ventuno Press, West Chester, OH.

Bejan, A. 1984. *Convection Heat Transfer*. Wiley, New York.

Bird, R. B., W. E. Stewart, and E. N. Lightfoot. 1960. *Transport Phenomena*. Wiley, New York.

Cebeci, T., and P. Bradshaw. 1977. *Momentum Transfer in Boundary Layers*. McGraw-Hill, New York.

Cebeci, T., and P. Bradshaw. 1988. *Physical and Computational Aspects of Convective Heat Transfer*. Springer-Verlag, New York.

Dittus, F. W., and L. M. K. Boelter. 1930. *Univ. Calif. Berkeley Publ. Eng.*, **2**, 443.

Eckert, E. R. G. 1955. J. Aero. Sci., **22**, 585–587; also Eckert, E. R. G. and R. H. Drake. 1959. *Heat and Mass Transfer*. McGraw-Hill, New York.

Gebhart, B., Y. Jaluria, R. L. Mahajan, and B. Sammakia. 1988. *Buoyancy-Induced Flows and Transport*. Hemisphere, New York.

Hansen, A. G. 1964. *Similarity Analyses of Boundary Value Problems in Engineering*. Prentice-Hall, Englewood Cliffs, NJ.

Kays, W. M., and M. E. Crawford. 1993. *Convective Heat and Mass Transfer*. McGraw-Hill, New York.

Minkowycz, W. J., E. M. Sparrow, G. E. Schneider, and R. H. Fletcher. 1988. *Handbook of Numerical Heat Transfer*. Wiley, New York.

Petukhov, B. S., and A. F. Polyakov. 1988. *Heat Transfer in Turbulent Mixed Convection*. Hemisphere, New York.

Schlichting, H. 1979. *Boundary-Layer Theory*. McGraw-Hill, New York.

Smith, A. G., and D. B. Spalding 1958. "Heat Transfer in a Laminar Boundary Layer with Constant Fluid Properties and Constant Wall Temperature," *J. R. Aeronaut. Soc.* **62**, 60–64.

Tritton, D. J. 1977. *Physical Fluid Dynamics*. Van Nostrand Reinhold, New York.

Zukauskas, A., and A. Slanciauskas. 1987. *Heat Transfer in Turbulent Fluid Flows*. Hemisphere Washington, DC.

Selected Case Studies

Chapter 5 attempts to extend the basic knowledge gained and (homework) problem solution skills acquired from studying Chapters 1 through 4 and using Appendices A to F. Section 5.1 deals with mathematical modeling aspects important for the development of computer simulation models, which are becoming more and more acceptable in solving complex fluid mechanics and convection heat transfer problems. The selected case studies begin with turbulent shear-layer flows applied to external curved surfaces (Sect. 5.2.1) and internal walls (Sect. 5.2.2). The basic idea of boundary-layer theory is then extended in Section 5.3 to non-Newtonian fluid flow with exothermal chemical reaction (Sect. 5.3.1) and with wall heat transfer (Sect. 5.3.2). The fifth and sixth case studies are more complicated, focusing on two-phase flows with moving gas–liquid interface due to droplet vaporization (Sect. 5.4.1) and moving liquid–solid interface due to ice formation (Sect. 5.4.2). The last two case studies represent the state of the art in (bio-) fluid dynamics applied to laminar pulsatile flow in branching blood vessels (Sect. 5.5.1) and temperature-driven flow in a local hyperthermia treatment device that is a concentric heated cylinder in a water-filled balloon (Sect. 5.5.2). Both biofluid flow systems are transient three-dimensional and hence require the numerical solution of the complete Navier–Stokes equations.

The open literature provides many additional case studies that encapsulate both basic knowledge in engineering fluid dynamics and computer solution steps for real-world (thermal) flow problems.

5.1 Flow Systems Analysis and Simulation

In Section 1.2.1, we discussed the two frameworks in which fluid flow fields are traditionally described and mentioned that the Eulerian viewpoint is preferred by engineers to derive and apply the conservation laws (cf. Chapter 2). In this section we elaborate and extend the ideas of flow model development (cf. Fig. 1.1) and consider modeling techniques for the simulation of more complex flow systems. The problems or systems under consideration in this text are restricted to

(i) continuum mechanics: velocity, stress, and deformation rates are continuous point functions, ignoring any local molecular effects;

(ii) deterministic processes: random phenomena, chaos, and so on, are time-smoothed and represented by deterministic functions; and, for the most part,

(iii) Eulerian flow descriptions: the velocity field is expressed as $\vec{v} = \vec{v}(x, y, z, t)$ rather than $\vec{v}_{\text{particle}} = \vec{v}_p[t, \vec{r}_p(t)]$, where $\vec{r}_p(t)$ is the particle location vector.

Historical Overview

Unique approaches absorbed in engineering systems analyses were developed in the last three centuries by philosophers and/or scientists, notably Francis Bacon, John

Locke, Immanuel Kant, Isaac Newton, and Karl Popper. Two quite opposing methodologies emerged: the empirical/inductive approach (Bacon 1620) and the theoretical/deductive approach (Popper 1934). Bacon maintained that scientific advancements are achieved by first collecting a wealth of data on some natural phenomena; these observations have to be organized into an orderly array so that certain patterns can be discerned; the patterns, then, are generalized into sound theories, the laws reflecting natural phenomena. Popper, in contrast, proposed that important scientific discoveries are made by formulating a theory/hypothesis first and then using empirical methods to test its validity.

Traditionally, large-scale and/or complex engineering systems were solved by first isolating simplified static or dynamic subsystems (e.g., structural elements, machine parts, or a particular transport phenomenon) and then integrating their partial results into the global system. A modern, more realistic modeling approach should take into account, whenever possible, the coupled transfer processes of the entire system: its complex geometries, transient dynamics, nonlinearities, and internal/external coupling. Examples for such comprehensive, deductive modeling attempts include the simulation of dense droplet sprays in combustion, cooling, or coating; the aerodynamics of entire automobiles or airplanes; fluidized-bed combustion for coal gasification; multilevel, optimal process control of chemical plants or manufacturing centers; and lake ecosystems subjected to natural and anthropogenic stresses. To start with, every theoretical analysis has to be based on *physical insight* usually delineated from experimental or field observations. Preliminary mathematical modeling and computer simulation results should then be verified with measured data sets and/or analytic solutions for accepted base case studies. Hence, engineering systems analysis, modeling, and simulation are a mixture of both methodologies as individually proposed by Bacon and Popper.

Mathematical modeling has matured in recent years to a discipline in its own right as manifested in the number of specialized journals issued and annual conferences held in this field and, most important, the acceptance of computational fluid dynamics (CFD) as a predictive design tool in aerospace, automobile, chemical process, environmental, nuclear, and other industries. Specifically, today's engineering problems or systems are nonlinear, three-dimensional, and transient, requiring interdisciplinary solution approaches.

Although we are predominantly interested in the development and use of appropriate models for the understanding and prediction of complex transport phenomena, a specialized group of researchers in the field of "mathematical modeling and computer simulation" focuses on the development of new modeling methodologies. They may include generalizing theories and error analyses, techniques of parameter estimation, robust numerical methods, automatic adaptive mesh generation, nonlinear optimization techniques, computer algorithms, validation criteria, and presentation of graphical results in color and three dimensions. Thus, they provide scientists and engineers with new tools to represent and study real-world transfer processes. However, as indicated earlier, in many instances, time pressures, limited availability of trained people, low budgets, or insufficient information may restrict a thorough analysis, modeling, and computer simulation of a given system. As a result, complex engineering systems are in reality too often described by very idealized models, utilizing algebraic correlations or linearized steady-state differential equations for problem solutions. However, today's worldwide competition and increasing cost pressures coupled with a high demand for quality products require the use of computer hardware and software that *accurately* predict and prevent product design problems. As a matter of fact, sophisticated computer tools have raised CFD models to such complex levels that they sometimes outstrip our ability to obtain experimental data with which to evaluate them.

Part of a good engineering education requires the advancement of fundamental knowledge to state-of-the-art computer applications. Chapter 5 provides such an extension of the material learned to its application of selected real-world fluid flow problems. To begin with, the important aspect of problem understanding and system conceptualization as a first step in modeling is discussed in Section 5.1.1. Modeling approaches and different stages in model development are outlined in Section 5.1.2. Section 5.1.3 contains a brief discussion of (i) important aspects in the development of system-specific modeling equations from the general field equations and (ii) proven software for computer simulations in fluid mechanics and heat transfer.

5.1.1 Problem Recognition and System Conceptualization

Before starting with the theoretical analysis of any research project (cf. Fig. 5.1), two basic aspects have to be considered:

(1) There has to be sufficient information on the particular project available so that all problem aspects are fully recognized, and the main objectives can be achieved in light of the available resources.

(2) A preliminary assessment has to show that it is cost-effective and manageable to develop an accurate and predictive mathematical model that will have the potential to lead to new discoveries or fundamental advances in the knowledge base and/or technology base.

An amazing amount of waste in terms of money, time, and energy has been generated in the last three decades with the development of computer simulation models that do not address the key problems of a given project and/or do not have any predictive capabilities. In some cases, the important transport mechanisms have never been recognized or the model grew into an inflexible, data-dependent simulator that could only be used by the originators. Well-known large-scale examples can be found in areas such as alternative energy systems, environmental engineering, national defense, and nuclear reactor safety, including radioactive waste disposal, to mention a few. On a smaller scale, the questions of problem recognition and model feasibility/capability prevail because actual fluid dynamics research projects, in contrast to textbook-type problems, are seldom well defined.

As a first step toward system identification (Fig. 5.1), a preliminary literature review together with field observations and/or experimental data sets may help to understand a given project: the issues in question, the important transport phenomena, and the scope of the investigation. Sources and tools for literature search are textbooks, governmental reports, refereed journal articles and conference proceedings, and library resources including machine searches, employing, for example, BITNET, Gopher, and World Wide Web on the INTERNET, and CD-ROM. The resulting review of the relevant background material should contain information on the state of the art in the area of interest, the data base needed for parameter values and model validation, and the potentially successful modeling approach and solution technique to be employed. At this point it is a valuable exercise to describe briefly the specific system to be analyzed (and how it fits into the larger picture, if appropriate), to state the objectives and anticipated results, and to outline the research plan as concisely as possible (cf. first two blocks of Fig. 5.1). Such a short document is similar to a preproposal, which should include a time–activity schedule as well as a preliminary budget.

System Identification

- Problem recognition and understanding of research project
- Statements of specific objectives and expected results
- Advantages and cost-efficiency of developing/using a math model
- Conceptualization and design of modeling framework
- Acquisition of data sets for model parameters and model validation

System Modeling

- Selection of dependent and independent variables
- Selection of model type and modeling approach
- Determination of basic equations plus initial and/or boundary conditions
- List of assumptions and postulates for derivation of modeling equations
- Development of submodels to gain closure

System Solution

- Selection of appropriate solution technique(s) and/or appropriate software package
- Identification of appropriate computer platform (e.g., desk-top, workstation, supercomputer, etc.)
- Validation of computer simulation model (e.g., experimental data sets, flow visualization, exact solutions)
- Display of model results, interpretation and recommendations
- Use of model for design, analysis, process development and decision making

Fig. 5.1. Considerations and steps in model development.

When the research objectives and project significance are known and when the system boundaries, the dependent and independent variables, plus the associated transfer processes are determined, the system identification step has been completed. As indicated in Fig. 5.1, this first step sets the stage for the subsequent phase of project development (cf. Sect. 5.1.2). Although some corrections are possible at a later point, a major redefinition of the system may entail a heavy loss in time and money.

5.1.2   *Types of Models and Modeling Approaches*

Different types of models and related modeling approaches are discussed in this section, together with generic preliminary steps such as literature review and decomposition of complex systems.

A   *Basic Definitions*

Models may be classified as

- verbal (i.e., a concept, hypothesis, or theory);
- physical (e.g., laboratory bench-scale, electric/electronic analog, pilot plant); or

- mathematical (deterministic, empirical, stochastic; continuous, discrete, analytic, numerical, etc.).

Engineers often label their models according to the area of application: lake ecosystem model, urban development model, lubrication model, combustion spray model, and so forth. In general, every equation is a mathematical model if it represents a natural phenomenon. Hence, every researcher is, in a way, a mathematical modeler. This includes experimentalists, who often cast their laboratory/field data sets after some statistical treatment in the form of a theoretical statement expressed in mathematical terms. Some of the more famous "laws" employed by scientists and engineers to describe a system's behavior are actually mathematical models: Hooke's law for elastostatic bodies, Newton's second law of motion of a particle, Fick's law for diffusional processes, Stokes's law of viscosity in fluid mechanics, Fourier's law in heat transfer, Monod's function in microbiology, Henry's law for the solubility of gases, and the ideal gas law, to name a few. Various schematics for engineering system models are given in Figure 5.2. One could define a model as a (mathematical) representation of the real process; the actual operation of the model, for example, the computer program, is the simulation. An alternative definition for system simulation is representation of the system's behavior by moving it from state to state in accordance with well-defined operating rules and subsequent observation of its dynamic performance, using computers.

Design is akin to modeling since it involves the manipulation of elements representing physical systems. Software for computer-aided design (CAD), engineering (CAE), or manufacturing (CAM) tasks can be used to develop a plan (blueprint) from which a device, a piece of machinery, a vehicle, a plant, or a whole city can be built and operated. In contrast, mathematical modeling focuses on the analysis of transfer processes in order to improve existing systems, or modeling assists in designing new systems. An interesting hybrid are expert systems, in which advances in artificial intelligence are combined

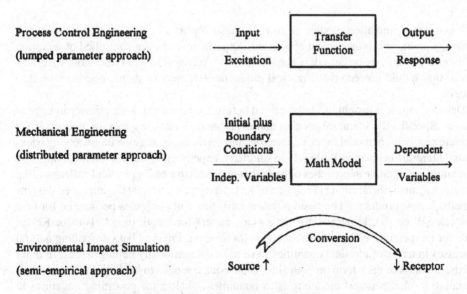

Fig. 5.2. Characteristics of various engineering models.

Table 5.1. *Patterns for review of math modeling publications*

Items to be considered	Steps to be taken
References	Reference reviewed publications on index cards and photocopy three to four key articles or reports; create special data file for program input and model verification. Distinguish between theoretical (analytical/numerical) and experimental papers.
Abstract	Review fundamental features of the project: (i) problem to be solved; (ii) type of mathematical model used (deterministic, stochastic); (iii) approach taken (lumped parameter, distributed parameter); and (iv) how the problem is solved (solution technique).
Governing equations	Examine the basic equations and constitutive equations; initial and boundary conditions; check postulates for principal variables; and check final set of modeling equations.
Assumptions	Make a list of all stated as well as implicitly used simplifications; reflect on their significance; check whether the problem-oriented equations indeed represent the important transport phenomena.
Input data	List data required for model input (geometry, ambient conditions, equation parameters and coefficients, IC, BC, etc.); reflect on possible sources for data acquisition (e.g., analytic solution, laboratory or field measurements).
Solution technique	Review merits of employed transformations and solution techniques (i.e., analytic, asymptotic, approximate. numerical). Check whether mesh resolution is adequate. State computer requirements (e.g., CPU time, RAM) and type of computer used.
Results	Review results critically; check data used. Comment on model validation, output, and display. Do the results gain new physical insight? What are the useful applications?
Limitations	State area(s) of applications with degree of accuracy and reliability.

with computer simulation models in order to make "optimal" decisions and to manage a system, a plant, or a project most efficiently. Experimentation is a controlled observation of either the real system (field data acquisition) or a physical model (laboratory measurements) that should precede mathematical model development or should operate with it in tandem.

Detailed physical insight has to be gained before mathematical descriptions can be postulated. Specifically, literature reviews and experimental data are generally necessary for problem recognition, model input, calibration, and verification. A guide for reviewing background literature is given in Table 5.1. In summary, experimentation, design, modeling, and simulation are iterative approaches or tools that may lead to a better physical understanding of a given system, the accurate prediction of its behavior, and, ultimately, improved designs or perhaps new products. The basic physical units are watts or joules per second for heat flow rate (W or J/s), kilograms for mass (kg), meters for length (m), Celsius or Kelvin scale for temperature ($°C$ or K), and seconds for time (s). Physical laws describing natural processes in terms of physical quantities have to be dimensionally homogeneous; in other words, the units of each term on both sides of an equation have to be the same. A standard operation in mathematical modeling is to nondimensionalize the governing equations in order to generate dimensionless groups on which the solution must depend. In addition,

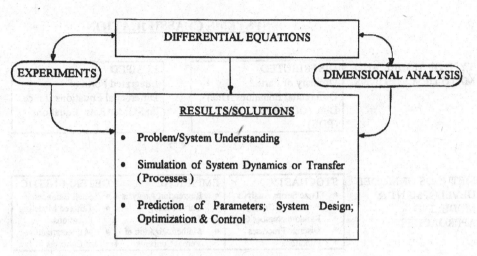

Fig. 5.3. Tools in engineering science research.

dimensional analysis (cf. Buckingham pi theorem) can produce groupings of parameters that determine the system's dynamics and transfer processes. However, dimensional analysis alone will not give the functional form that relates such π terms; but it can reduce the number of process variables significantly, an important capacity in the design of physical or laboratory scale models. Figure 5.3 summarizes the necessary research tools in engineering and lists a few desirable results. The value of dimensional analysis and scale analysis is demonstrated in Section 2.3.1.

A firm notion of equilibrium is important when selecting proper auxiliary equations for a complete system representation. An equation of state relating pressure and density or a parameter equation for the net absorption of a constituent may serve as an example. A system is in a state of equilibrium if no macroscopic changes with respect to time occur. Hence, all energy potentials are balanced and uniform throughout the domain of interest. Thermodynamic equilibrium is the most stringent type, comprising mechanical, thermal, chemical, electrostatic, as well as phase equilibrium. Hence, there is no macroscopic energy, matter, or charge flow within a system in thermodynamic equilibrium, though the molecules are free to produce such flows. For example, equilibrium between the liquid and vapor phases occurs not necessarily when all changes cease, but when these molecular, that is, microscopic, changes just balance each other, so that the macroscopic, that is, the gross or average properties, remain unchanged. As seen in this broader context, equilibrium is actually a dynamic process on a microscopic scale, even though it is treated as a static condition in macroscopic terms. The conservation principles are normally written in the form of partial differential equations expressing the balance of momentum, mass, and energy at any point in space and time. The Navier–Stokes equations, for example, express the condition of equilibrium, namely that for each fluid particle there is equilibrium among body forces, surface forces, and inertia forces. This is a reasonable assumption for fluid flows at normal densities; flows exhibiting shock waves, certain chemical reactions, thermodynamic changes, and/or sorption processes require additional (nonequilibrium) terms for their mathematical description.

SYSTEMS CLASSIFICATION

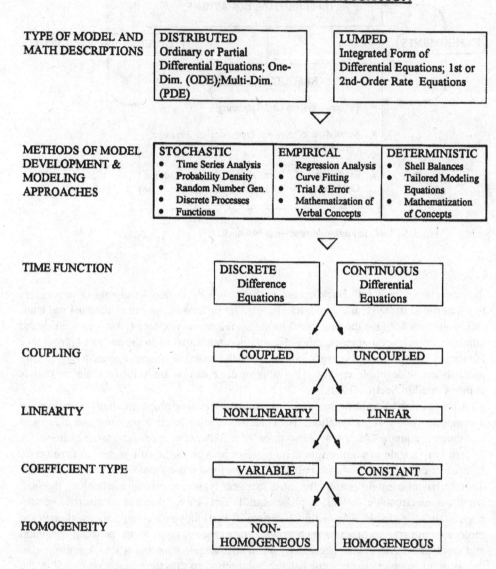

| TYPE OF MODEL AND MATH DESCRIPTIONS | DISTRIBUTED Ordinary or Partial Differential Equations; One-Dim. (ODE);Multi-Dim. (PDE) | | LUMPED Integrated Form of Differential Equations; 1st or 2nd-Order Rate Equations |

| METHODS OF MODEL DEVELOPMENT & MODELING APPROACHES | STOCHASTIC • Time Series Analysis • Probability Density • Random Number Gen. • Discrete Processes • Functions | EMPIRICAL • Regression Analysis • Curve Fitting • Trial & Error • Mathematization of Verbal Concepts | DETERMINISTIC • Shell Balances • Tailored Modeling Equations • Mathematization of Concepts |

| TIME FUNCTION | DISCRETE Difference Equations | CONTINUOUS Differential Equations |

| COUPLING | COUPLED | UNCOUPLED |

| LINEARITY | NONLINEARITY | LINEAR |

| COEFFICIENT TYPE | VARIABLE | CONSTANT |

| HOMOGENEITY | NON-HOMOGENEOUS | HOMOGENEOUS |

Fig. 5.4. Engineering systems classifications for math modeling purposes.

B Modeling Approaches

Model characteristics and modeling approach are directly linked; Figure 5.4 gives an overview of such coupling. The lumped parameter approach is taken when the gross (macroscopic or integral) behavior of a system is of interest. Distributed parameter models are used when the engineering variables at each discrete point in the system have to be evaluated (microscopic or differential approach). For example, in fluid mechanics, if we wish to find the average velocity of a fluid in a duct, we use the Reynolds Transport

Theorem, that is, conservation of mass and momentum in integral form (or use Bernoulli's equation) and thereby ignore spatial variations (lumped parameter results). However, if we wanted to find an expression for the velocity profile, we would use a reduced form of the Navier–Stokes equation (distributed parameter results). In the first case, we are looking at a macroscopic property, such as average velocity, drag force, or mass flux, where only the time dependence may be preserved. In the second case, we are interested in the microscopic properties at each individual point in space. Hence, lumped parameter systems are usually described by algebraic or ordinary differential equations depending on the system's dimensionality. The remaining characteristics of the systems classification in Fig. 5.4 reflect mainly mathematical difficulties, in particular the left branch, which requires special modeling approaches and associated solution techniques.

(i) Coupling: A large or complex system can often be decomposed into simpler sub-systems that are weakly coupled. That is, certain modeling equations can be solved independently. For example, forced convection heat transfer problems can be solved in computing the velocity field independently if the transport properties are temperature-independent. In general, coupled systems of equations have to be solved simultaneously, requiring more computer storage, and might cause (numerical) error magnification. Stability problems might turn up when coupled nonlinear equations representing system components with feedback and time delay have to be solved on digital computers. Packaged simulation languages avoid these problems by feeding the difference equations to the equation solver (Runge–Kutta, Gear, or Simpson) in the correct sequential order.

(ii) Linearity: Linearity of a system is examined by comparing the response of a system to an input. If a system response is directly proportional to an input, for all inputs, then the system is linear. For linear differential equations the principle of superposition can be employed via a product solution:

Total solution = homogeneous solution + particular solution

or

Total solution = steady-state solution + transient solution, etc.

An equation is nonlinear when the exponent of the dependent variable (or its derivative) is not equal to 1 or if dependent variables form products or appear in arguments of any transcendents. Nonlinear algebraic or ordinary differential equations without singularities can be solved with standard software packages (Newton–Raphson, Piccard, or Runge–Kutta, Adams–Bashforth algorithms). Solutions of nonlinear partial differential equations, such as the Navier–Stokes equations, may cause (stability) problems when the nonlinear terms are dominant (cf. Roache 1976).

(iii) Coefficient type: Coefficients of (differential) equations describing a real system are in most cases variable. Elaborate submodels are sometimes necessary to reflect the functional changes of coefficients with independent and principal variables.

(iv) Homogeneity: Representation of systems with "forcing functions" results in homogenous equations.

Although the system classification in Fig. 5.4 groups models into two classes, the modeling approach taken determines the type of model, such as empirical, deterministic, or stochastic. Specifically, empirical models are measured data correlations, or they may serve as "order-of-magnitude" simulators of natural phenomena when only descriptive input/output data sets are available and the system is too complex for deterministic or stochastic modeling approaches. In this sense, phenomenological relationships often surface as empirical submodels for coefficients and parameters in large deterministic and/or stochastic models. The methodologies include statistical/regression analysis, curve fitting techniques, mathematical relationships based on common sense, and a mixture of all of them. Areas of application include correlations for drag coefficients or Nusselt numbers, turbulence models, reaction kinetics, and fluid rheology. A deterministic model is a direct mathematical representation of phenomena that occur in deterministic, continuous, or discrete patterns: the values of variables and parameters are definite fixed numbers. Approaches for the construction of deterministic models are either inductive, starting with material and energy balances, or deductive, using the basic equations as a starting point (cf. App. B). Stochastic models suffice for estimating the behavior of complex engineering systems since transport, conversion, and uptake mechanisms have random features. Turbulent fluid flow systems, perturbations of rocket trajectories, economical forecasting, water resources engineering including flow in geologic media, meteorology, behavior of living organisms, reactor stability and control, and failure of structural elements and machine parts are illustrative examples. Even deceptively simple mechanical systems, such as a fast dripping water faucet, slow fluid motion in homogeneous porous material, or trajectories of balls on a pool table may exhibit strange dynamics including chaos. Stochastic models admit elements of randomness in the mathematical description, so that the model results are probabilistic rather than fixed numbers. In general, stochastic models consist of deterministic parts and randomizing functions that account for unpredictable, that is, irregular, fluctuations of the engineering variables.

In the remainder of this chapter, we focus on deterministic modeling. The main difficulties encountered in the development and validation of deterministic computer simulation models for advanced research projects are

- system decomposition;
- mesh generation and adaptation;
- acquisition of data sets for program input and model validation;
- derivation of the (correct and complete) modeling equations;
- selection of an efficient solution algorithm including a suitable mesh generator; and
- identification of sufficient computing resources.

These topics and others are further discussed in the next section and in subsequent case studies.

5.1.3 Mathematical Representation and System Simulation

A Mathematical Modeling Steps

The core of any mathematical model is the correct or at least adequate representation of all important transport and conversion phenomena occurring in the given real-world system. After selecting a suitable modeling approach (cf. Sect. 5.1.2), this requires the derivation of the system-oriented governing equations plus boundary conditions and necessary submodels for closure (cf. Figs. 1.1 and 5.1). In order to accomplish that, the basic equations (cf. App. B) have to be tailored to problem-oriented equations based on simplifying assumptions and resulting postulates. In addition, associated initial and/or boundary conditions have to be defined. The derivation of modeling equations is a delicate process requiring mathematical skills and sometimes "good engineering feeling":

- Select a coordinate system that simplifies the boundary conditions and problem solution (cf. Chapters 1 and 2).
- Determine significant transport/conversion phenomena and evaluate relevant dimensionless groups (cf. Table 4.1); exclude negligible processes based on physical insight gained from literature reviews, successful solutions to similar problems, and so on.
- Check the significance of certain terms in the governing equations by comparing their value range with observations and measurements; perform a scale analysis if sufficient information is available (cf. Sects. 2.3.1, 2.4, and 4.1); if necessary, do a relative order of magnitude analysis (ROMA) for each term of the basic equations.
- Subdivide the system domain into simpler regions or decouple transfer processes.
- If possible, transform the governing equations plus boundary conditions in order to reduce their mathematical complexity or to make them more tractable for a special solution method.

In case suitable sets of differential equations cannot be directly deduced from the basic equations in terms of fluxes (cf. App. B and C), the control volume approach should be employed as a starting point to derive the system-oriented equations representing energy, material, or force balances in integral form (cf. Sect. 2.1.1).

B Computer Simulation

Most theoretical and industrial problems in fluid mechanics and heat transfer require computer solutions. Once the governing equations, auxiliary relationships, and boundary conditions have been obtained, a suitable numerical method has to be selected in order to program the problem-oriented modeling equations and to simulate the system. All numerical schemes require the subdivision or discretization of the computational domain into a finite difference grid, finite element mesh, or control volume mesh. The gradients (i.e., flux terms) of the governing equations are approximated by finite difference ratios between nodal points or at volume surfaces, as in the finite difference or control volume method, respectively. Alternatively, approximation or shape functions for all dependent variables in association with each finite element are selected, integrated, and assembled, leading again to a coupled system of (nonlinear) algebraic equations. In any case, there are basically three choices for advancing from the mathematical modeling stage to a computer simulation:

(i) Develop your own numerical code based on a proven finite difference, finite element, or control volume method (cf. Roache 1976; Patankar 1980; Bradshaw et al. 1984; Anderson et al. 1995; Cuvelier et al. 1986; Minkowycz et al. 1988; Pepper and Heinrich 1992; Reddy and Gartling 1994). Numerous software routines are available at computer center libraries to aid in executing specific tasks, such as matrix conversion, solution of elemental ODEs, as well as parabolic and elliptic PDEs (cf. App. F; Press et al. 1994; and IMSL software).

(ii) Most universities and R&D companies maintain licenses for multipurpose mathematical software such as Maple, Mathcad, Mathematica, Matlab, Macsyma, and

PDEase, with which most of the chapter problems and some course projects can be solved and the results efficiently plotted.

(iii)   A more powerful, but also cost-intensive option is the utilization of a general purpose software package for fluid flow and heat transfer problems such as FLUENT, PHOENICS, NISA/3-FLUID, CFX, CFD 2000, or FIDAP (cf. O'Connor 1992). Flexible and efficient mesh generators; fast and accurate equation solvers; realistic subroutines for turbulence, two-phase flow, and so on; and vivid postprocessors are the hallmarks of good commercial packages.

With respect to items (i) and (iii), the following aspects should be kept in mind. Computer codes have to account for the problem's unique characteristics, such as complex (3-D) geometries, coupled momentum, heat and mass transfer, moving boundaries, turbulent flows, coupled nonlinear (unknown) interfacial conditions, multiphase flows, chemical reactions, fluid rheology, and/or porous media flow. Local oscillations and "false diffusion" have to be prevented. Efficient and accurate mesh generation is usually the *limiting step* in CFD code applications. Hence, a number of mesh generators, such as ICEM-CFD, I-DEAS, and PATRAN, which may act as pre-processors for most commercial packages, have been developed. If available, the hardware's parallel-processing capabilities should be exploited. In many cases, locally adaptive mesh refinement techniques are more advantageous than implementing elaborate higher-order algorithms.

Generally, transient three-dimensional projects are very CPU-time-intensive when using a finite element method. Finite difference methods, although faster, are too awkward and possibly inaccurate, requiring elaborate transformations to tackle geometrically complex problems. Supercomputers are certainly not the answer for many CFD problems because they become less and less accessible for most researchers because of economic reasons. A practical solution may be the combined use of vectorized numerical codes (e.g. control volume or finite volume algorithms) on superminicomputers (i.e., high-end workstations) with several parallel processors.

C   Model Validation

Ideally, several complete sets of reliable experimental data should be available to compare advanced simulation results with measurements. Data sets for velocity, pressure, and temperature profiles are even more desirable than values for integral or lumped parameters such as skin friction coefficients, flow rates, mass transfer rates as well as average or local dimensionless numbers (e.g., Nu, Sh, C_p). Profiles of the dependent variables for different input conditions allow a very detailed comparison. Field or laboratory data sets have to be scrutinized carefully before they can be used. Information should be available on the experimental set-up, the process conditions, what has been measured and what has been kept constant, and the uncertainty estimates. As indicated earlier, a joint theoretical–experimental venture is the best approach for solving complex problems accurately because it avoids the costly generation of incomplete, that is, often useless or at least superficial data sets (cf. Fig. 5.5). Such an interaction is also helpful in producing plots that display the results most efficiently: insightfully, concisely and usably. If measured data sets are not available, the computer simulation model may be checked against analytical results for a simplified test case.

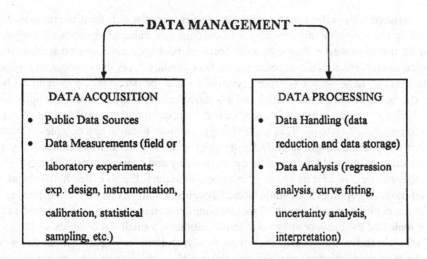

Fig. 5.5. Aspects of data acquisition and processing (after Vemuri 1978).

D Display of Results

Occasionally, computer simulation results are beautifully plotted, employing sophisticated graphics routines, but the display is somewhat void of insightful content. Considerations for the preparation of meaningful graphics include the following:

- Derivation of dimensionless groups for plotting, using scale analysis or nondimensionalized parameters of the governing equations
- Use of the same coordinates for graphs as in benchmark publications to allow for easy comparison
- Plotting of differential parameters, such as profiles, as well as integral parameters in order to display the full range of resolution the simulator is capable of exploring, with emphasis on not the quantity but the quality of figures.

After the computer simulation model has been validated, novel results and the enhanced understanding of the project should be clearly displayed. Parametric sensitivity analyses and unique ways of plotting the research results, using scale analysis, help to achieve these goals. A potentially powerful display of CFD results is animated movies, that is, computer-generated flow visualization employing post processor software such as WAVEFRONT, AVS, SPYGLASS, or FIELDVIEW.

Finally, the graphs or videos have to be interpreted and unique features have to be explained rather than describing the "ups and downs" of single or multiple curves. Conclusive statements summarizing the physical insight and new understanding as well as potential design applications of the research should round out the project work.

Usually as part of "system identification" (cf. Fig. 5.1), the fluid dynamics domain of interest is subdivided into characteristic regions in which dominant transport phenomena can be well represented with simplified equations. External flow of a real fluid past a body is a prime example in which traditionally the computational domain consists of a boundary layer and a potential flow region. Another example is jet discharge into a moving ambient where three subregions are recognized: a zone of flow development and a zone of established flow within the jet plus the ambient from which fluid is entrained by the spreading jet. However, implementation of the complete transport equations in conjunction with the use of powerful fluid dynamics software packages and high-end workstations or supercomputers reduces the dependence on the boundary-layer theory more and more.

In general, large and/or complex systems have to be decomposed or reduced before they can be analyzed in more detail and a solution can be sought. Especially optimization of a dynamic, nonlinear system requires the decomposition of the problem into "simpler" interconnected subsystems. This methodology is similar to building a complex structure or a machine by fabricating or manufacturing single parts that are then assembled. Hence, decomposition might yield reduction in dimensionality and conceptual simplification, which lead to more flexible and more realistic system modeling, the so-called hierarchical, multilevel modeling approach for multifaceted systems. Examples include atmospheric, aerodynamic, ecologic, oceanographic, socioeconomic, manufacturing, or space systems on a large scale and the brain or enzymatic fermentation on a small but complex scale.

Sufficient and reliable measurements are most important for preparing input data sets and for checking the predictive capabilities of the model. Usually, such data sets are incomplete or simply not available and alternative ways have to be found. Occasionally, a few definite values of key input parameters may be replaced by appropriate value ranges. This is done in parametric sensitivity analyses, as the effects of input data variations on the dependent variables are recorded. On the other hand, modeling results can often be compared for special computer simulation case studies with accepted analytic solutions. Because of the desirably strong interactions between theoreticians and experimentalists and the widespread application of microcomputer interfaces with bench-top experimental set-ups or pilot plants, a word on data reduction and analysis is appropriate (cf. Fig. 5.5). For example, in regression analysis, driven data sets are evaluated (usually with a statistical package) for the relative significance of all terms in a postulated equation. The most important terms are retained and their coefficients are evaluated using the least-squares routine. The resulting equation, a lumped parameter, empirical model, possesses the best continuous fit of the discrete data sets. Such an empirical modeling approach is widely employed in parameter estimation and input–output model design for certain subsystems. For more complex multivariable analyses, nonlinear optimization packages such as VMCON (cf. Crane et al. 1980) are recommended. On the basis of physical insight, a specific form of the objective function and a number of constraints are postulated and the optimal coefficients for a "best fit," for example, between predicted and field data, are then evaluated.

5.2   Turbulent Shear-Layer Flows

Turbulence is one of the last unsolved problems in classical physics. Numerous experimental and theoretical research papers have been published in an attempt to enhance the understanding of the mechanisms and structures of turbulence and to postulate turbulence models with which transitional and turbulent flow behaviors can be simulated. Still, even for rather simple geometries the chaotic behavior and coherent structures are still under investigation, and there is no "universal" turbulence model available that will work for any flow. One day, a detailed understanding and accurate mathematical description of turbulence may result from the direct (numerical) solution of the complete transfer equations at time steps smaller than turbulent frequencies and with spatial resolutions small enough to capture the smallest eddy (cf. Zang et al. 1992). In the meantime, one has to select, for a given flow, a suitable turbulence model supported by experimental data (cf. Sect. 3.2.4).

For attached flat-plate two-dimensional boundary-layer flows an extended form of Prandtl's mixing length hypothesis, a zero-equation model, is usually easy to implement and accurate enough (cf. Granville 1989 and 1990). Section 5.2.1 is an example of the acceptable performance of Prandtl's MLH, which has been extended to simulate thick shear layers on curved surfaces.

For turbulent boundary layers at relatively low Reynolds numbers, that is, $Re_l = O(10^6)$, unsteady flows, and separated flow as well as flows over rotating bodies with fluid injection or strong natural convection, more sophisticated turbulence models and data sets are required. A nonequilibrium closure model designed to treat two-dimensional turbulent boundary layers with strong adverse pressure gradients and attendant separation has been proposed by Johnson and King (1985). A review of turbulence models where near-wall and low-Reynolds-number effects are incorporated has been given by Patel et al. (1985). Section 5.2.2 describes extensions to the basic k–ε model for the simulation of such basic internal and external wall flows. Theoretical and experimental investigations of turbulent flows with separation, using patched two-equation and three-equation models, have been reported by Durst and Rastogi (1978), among others. Applications of the extended k–ε model and the algebraic turbulent stress model to swirl flow in particle combustion have been reviewed by Sloan et al. (1986).

Both fluid mechanics projects 5.2.1 and 5.2.2 are discussed within the three-step model development framework (cf. Fig. 5.1) outlined in Section 5.1, with an added component of design-and-future-work considerations.

5.2.1 Thick Boundary-Layer Flow on Curved Surfaces

A System Identification

The accurate and computationally efficient analysis of laminar and turbulent boundary-layer flow developments along strongly curved surfaces is of great importance for superior contour designs of submerged bodies ranging from submarines to gas-turbine blades (cf. Bradshaw 1973; Patel and Lee 1978; Kleinstreuer et al. 1983). However, for many practical applications, Prandtl's "thin-shear-layer" assumption breaks down since the boundary layer becomes thick, that is, $\delta/L_r = O(1)$, where L_r is a reference length such as the length of the body, its longitudinal radius, or its transverse radius of curvature (cf. Van Dyke 1969). External flows past marine craft, airfoils, or turbine blades and internal flows in diffusers or bent conduits may serve as examples. At the (high) Reynolds numbers of interest, nonuniformities of attached boundary-layer flows, such as curvature effects, normal stress and pressure variations, as well as viscous–inviscid flow interactions, additional rates of strain, and the effects of thickening on turbulence intensity, have to be accounted for. Thus, when it is too costly to solve the full elliptic momentum equations, appropriate parabolic higher-order boundary-layer equations have to be solved. It turns out that the second-order boundary-layer equations postulated by Tani (1954) and derived by Van Dyke (1968) are incomplete in the streamwise momentum equation. Thus, a careful reexamination of the derivation process is an integral part of this turbulent flow project. Considering steady incompressible two-dimensional or axisymmetric flow in regions where $v \ll u$ (i.e., the mean flow streamlines remain nearly parallel to the wall), the second-order boundary-layer

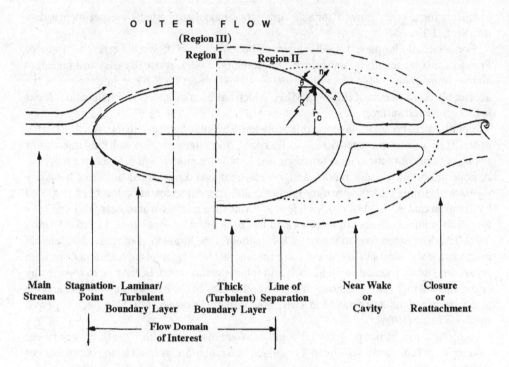

Fig. 5.6. Schematics of submerged axisymmetric body.

equations derived by Kleinstreuer and Eghlima (1985) are discussed for the solution of turbulent flow past 2-D or axisymmetric bodies.

B System Modeling

The starting point for deriving the thick boundary-layer equations (cf. region I in Fig. 5.6) are the steady, incompressible, axisymmetric Navier–Stokes equations. They can be written in an (s, n)-orthogonal coordinate system as (cf. App. B; Goldstein 1943):

$$\frac{\partial}{\partial s}(ur) + \frac{\partial}{\partial n} = 0 \tag{5.1}$$

$$\frac{u}{n}\frac{\partial u}{\partial s} + \frac{\partial u}{\partial n} + \frac{uvK}{h}$$

$$= -\frac{1}{\rho h}\frac{\partial p}{\partial s} + \nu\left\{\frac{1}{h^2}\frac{\partial^2 u}{\partial s^2} + \frac{\partial^2 u}{\partial n^2}\frac{1}{rh}\left[\frac{\partial}{\partial s}\left(\frac{r}{h}\right)\frac{\partial u}{\partial s} + \frac{\partial}{\partial n}(hr)\frac{\partial u}{\partial n}\right]\right\} \tag{5.2}$$

$$\frac{u}{n}\frac{\partial v}{\partial s} + v\frac{\partial v}{\partial n} - \frac{Ku^2}{h}$$

$$= -\frac{1}{\rho}\frac{\partial p}{\partial n} + \nu\left\{\frac{1}{h^2}\frac{\partial^2 v}{\partial s^2} + \frac{\partial^2 v}{\partial n^2}\frac{1}{rh}\left[\frac{\partial}{\partial s}\left(\frac{r}{h}\right)\frac{\partial v}{\partial s} + \frac{\partial}{\partial n}(hr)\frac{\partial v}{\partial n}\right]\right\} \tag{5.3}$$

The geometric parameters are

$$r = r_0 + n\cos\theta, \quad d\theta/ds = -1/R = -K, \quad dr_0/ds = \sin\theta, \quad h = 1 + Kn \tag{5.4}$$

where K is the body curvature, R is the longitudinal, and r_0 is the transverse radius of curvature. We now simplify (5.1) to (5.3) by retaining only first- and second-order terms in the reciprocal of the Reynolds number (cf. Sect. 3.2.3).

In region I where the streamlines stay about parallel to the body surface, the normal velocity component v is significantly smaller than the longitudinal component u. Hence, following Van Dyke (1969), we introduce dimensionless variables that are of the order of unity within the boundary layer as

$$\tilde{u} = u/U, \qquad \tilde{v} = \mathrm{Re}^m \frac{v}{U}, \qquad \tilde{p} = \frac{p}{\rho U^2}, \qquad \tilde{s} = \frac{s}{L},$$

$$\tilde{n} = \mathrm{Re}^m \frac{n}{L}, \qquad \tilde{r} = \frac{r_0}{L}, \qquad \tilde{K} = KL, \qquad \mathrm{Re} = \frac{UL}{v}$$

(5.5)

where L and U are appropriate length and velocity scales, Re is the Reynolds number, and the exponent m is known to be 1/2 for laminar and 4/5 turbulent thin boundary-layer flows over a flat plate with a zero pressure gradient (cf. Sect. 3.2.3D or 3.2.4B).

Substituting (5.5) into (5.1), we obtain the dimensionless form of the continuity equation as

$$\frac{\partial}{\partial \tilde{s}}[(\tilde{r}_0 + \tilde{n}\mathrm{Re}^{-m}\cos\theta)\tilde{u}] + \frac{\partial}{\partial \tilde{n}}[\tilde{r}_0(1 + \tilde{K}\tilde{n}\mathrm{Re}^{-m}) + \tilde{n}\mathrm{Re}^{=m}\cos\theta]\tilde{v}$$

$$+ \mathrm{Re}^{-2m}\frac{\partial}{\partial \tilde{n}}[(\tilde{K}\tilde{n}^2\cos\theta)\tilde{v}] = 0$$

(5.6)

We neglect terms of $O(\mathrm{Re}^{-2m})$ and rewrite (5.6) in terms of the original dimensional variables to obtain the approximate continuity equation

$$\frac{\partial}{\partial s}(r^i u) + \frac{\partial}{\partial n}[v(r^i - r_0^i(1 - h^j))] = 0$$

(5.7)

The exponent i is zero for two-dimensional flow and unity for axisymmetric flow; j is zero for bodies having no longitudinal curvature and unity for bodies with longitudinal curvature. Obviously, we could have simulated (5.1) directly instead of using the approximation equation (5.7); however, that would have been inconsistent with our general derivation.

Our analysis of the momentum equations ((5.2) and (5.3)) proceeds along similar lines. In particular, substituting (5.5) into the streamwise momentum equation (5.2) yields

$$\frac{\tilde{u}}{1 + \tilde{K}\tilde{n}\mathrm{Re}^{-m}}\frac{\partial \tilde{u}}{\partial \tilde{s}} + \tilde{v}\frac{\partial \tilde{u}}{\partial \tilde{n}} + \frac{\tilde{K}\mathrm{Re}^{-m}}{1 + \tilde{K}\tilde{n}\mathrm{Re}^{-m}}$$

$$= \frac{1}{1 + \tilde{K}\tilde{n}\mathrm{Re}^{-m}}\frac{\partial \tilde{p}}{\partial \tilde{s}} + \mathrm{Re}^{-1}\left\{\frac{1}{1 + \tilde{K}\tilde{n}\mathrm{Re}^{-m}}\frac{\partial^2 \tilde{u}}{\partial \tilde{s}^2} + \mathrm{Re}^{2m}\frac{\partial^2 \tilde{u}}{\partial \tilde{n}^2}\right.$$

$$+ \frac{\partial \tilde{u}}{\partial \tilde{s}}\left[\frac{1}{(\tilde{r}_0 + \tilde{n}\mathrm{Re}^{-m}\cos\theta)(1 + \tilde{K}\tilde{n}\mathrm{Re}^{-m})}\frac{\partial \tilde{r}}{\partial \tilde{s}} - \frac{1}{(1 + \tilde{K}\tilde{n}\mathrm{Re}^{-m})^3}\frac{\partial \tilde{h}}{\partial \tilde{n}}\right]$$

$$+ \mathrm{Re}^{2m}\frac{\partial \tilde{u}}{\partial \tilde{n}}\left[\frac{1}{(\tilde{r}_0 + \tilde{n}\mathrm{Re}^{-m}\cos\theta)}\frac{\partial \tilde{r}}{\partial \tilde{n}} + \frac{1}{(1 + \tilde{K}\tilde{n}\mathrm{Re}^{-m})}\frac{\partial \tilde{h}}{\partial \tilde{n}}\right]\right\}$$

(5.8)

where, using (5.4) and (5.5),

$$\frac{\partial \tilde{r}}{\partial \tilde{s}} = \left[\frac{\partial \tilde{r}_0}{\partial \tilde{s}} + \tilde{n}\mathrm{Re}^{-m}\frac{\partial \cos\theta}{\partial \tilde{s}}\right] \qquad \text{and} \qquad \frac{\partial \tilde{h}}{\partial \tilde{n}} = \tilde{K}\mathrm{Re}^{-m}$$

(5.9a,b)

We now expand each individual term in (5.8) in negative powers of the Reynolds number and only retain the two leading terms. Thus, for example, we write the coefficient $(1 + \tilde{K}\tilde{n}\mathrm{Re}^{-m})^{-1}$ as

$$(1 + \tilde{K}\tilde{n}\mathrm{Re}^{-m})^{-1} = 1 - \tilde{K}\tilde{n}\mathrm{Re}^{-m} + O(\mathrm{Re}^{-2m}) \tag{5.10}$$

This derivation technique coincides with the approach outlined by Van Dyke (1969). The expansion of Equation (5.8) yields, after some manipulations,

$$\tilde{u}\frac{\partial \tilde{u}}{\partial \tilde{s}} + v\frac{\partial \tilde{u}}{\partial \tilde{n}} + \mathrm{Re}^{-m}\left(- \tilde{K}\tilde{n}\tilde{u}\frac{\partial \tilde{u}}{\partial \tilde{s}} + \tilde{K}\tilde{u}\tilde{v}\right)$$

$$= -\frac{\partial \tilde{p}}{\partial \tilde{s}}(1 - \tilde{K}\tilde{n}\,\mathrm{Re}^{-m})(\mathrm{Re})^{2m-1}\frac{\partial^2 \tilde{u}}{\partial \tilde{n}^2} + \mathrm{Re}^{m-1}\frac{\partial \tilde{u}}{\partial \tilde{n}}\left(\frac{\cos\theta}{\tilde{r}_0} + \tilde{K}\right)$$

$$+ O\left(\mathrm{Re}^{-1}, \mathrm{Re}^{-m-1}, \mathrm{Re}^{-2m}, \cdots\right) \tag{5.11}$$

Neglecting terms of order Re^{-1} and higher and rewriting (5.11) in terms of the original dimensional variables using (5.5), we find our approximate s-direction momentum equation to be

$$(2 - h^j)u\frac{\partial u}{\partial s} + v\frac{\partial u}{\partial n} + Kuv$$

$$= -\frac{1}{\rho}(2 - h^j)\frac{\partial p}{\partial s} + v\left\{\frac{\partial^2 u}{\partial n^2} + \frac{\partial u}{\partial n}\frac{\partial}{\partial n}\left[\left(\frac{r}{r_0}\right)^i + h^j\right]\right\} \tag{5.12}$$

Finally, substituting (5.5) into the normal momentum equation (5.3)

$$\frac{\mathrm{Re}^{-m}}{(1 + \tilde{K}\tilde{n}\mathrm{Re}^{-m})}\tilde{u}\frac{\partial \tilde{v}}{\partial \tilde{s}} + \mathrm{Re}^{-m}\tilde{v}\frac{\partial \tilde{v}}{\partial \tilde{n}}\frac{\tilde{K}\tilde{u}^2}{(1 + \tilde{K}\tilde{n}\mathrm{Re}^{-m})}$$

$$= -\mathrm{Re}^{m}\frac{\partial \tilde{p}}{\partial \tilde{n}} + \mathrm{Re}^{-1}\frac{\mathrm{Re}^{-m}}{(1 + \tilde{K}\tilde{n}\mathrm{Re}^{-m})}\frac{\partial^2 \tilde{v}}{\partial \tilde{s}^2} + \mathrm{Re}^{m}\frac{\partial^2 \tilde{v}}{\partial \tilde{n}^2}$$

$$+ \mathrm{Re}^{-m}\frac{\partial \tilde{v}}{\partial \tilde{s}}\left[\frac{\partial}{\partial \tilde{s}}\left(\frac{\tilde{s}}{\tilde{h}}\right)\right] + \mathrm{Re}^{m}\frac{\partial \tilde{v}}{\partial \tilde{n}}\frac{\partial}{\partial \tilde{n}}(\tilde{r}\tilde{h}) \tag{5.13}$$

Expanding each term in (5.13) in the Reynolds number while using (5.9) and (5.10) yields, after some manipulations,

$$-\tilde{K}\tilde{u}^2\mathrm{Re}^{-m} + \mathrm{Re}^{-2m}\left(\tilde{u}\frac{\partial \tilde{v}}{\partial \tilde{s}} + \tilde{v}\frac{\partial \tilde{v}}{\partial \tilde{n}} + K^2\tilde{n}\tilde{u}\right)$$

$$= -\frac{\partial \tilde{p}}{\partial \tilde{n}} + \mathrm{Re}^{-1}\frac{\partial^2 \tilde{v}}{\partial \tilde{n}^2} +, \mathrm{Re}^{-1-m}\frac{\partial \tilde{v}}{\partial \tilde{n}}\left(\frac{\cos\theta}{\tilde{r}_0} + \tilde{K}\right)$$

$$+ O(\mathrm{Re}^{-m-1}, \mathrm{Re}^{-2m}, \cdots) \tag{5.14}$$

Neglecting terms of order Re^{-1} and higher and rewriting (5.14) in dimensional variables, we find our approximate n-direction momentum equation to be

$$Ku^2 = \frac{1}{\rho}\frac{\partial p}{\partial n} \tag{5.15}$$

Hence, our boundary-layer equations for regions where $v \ll u$ consist of the continuity equation (5.7) and the momentum equations (5.12) and (5.15). It remains to prescribe boundary conditions for these equations, and these consist of the "no slip" condition on the body surface and the fact that the vorticity vanishes along the (unknown) edge of the boundary layer, that is,

$$\vec{\zeta} \equiv \nabla \times \vec{v} := \frac{1}{h}\left[\frac{\partial v}{\partial s} - \frac{\partial (hu)}{\partial n}\right] = 0 \qquad n \to \infty \tag{5.16}$$

Substituting (5.5) into (5.16), neglecting terms of the order of (Re^{-2m}), and rewriting the result in terms of dimensional variables, we obtain

$$\frac{\partial u}{\partial n} + Ku = 0 \qquad n \to \infty \tag{5.17a}$$

Integration yields

$$u = u_\infty(s)\exp[-nK(s)] \tag{5.17b}$$

where $u_\infty(s)$ is the tangential component of the outer potential flow velocity. Hence the boundary conditions for the system of equations ((5.7), (5.12), and (5.15)) are

$$u = v = 0, \qquad \text{at } n = 0 \quad \text{and} \quad u = u_\infty(s)\exp[-nK] \qquad \text{for } n \geq \delta_0 \tag{5.18}$$

Of the new set of second-order boundary-layer equations only the s-momentum equation (5.12) differs from previously published equations (cf. Tani 1954; or Van Dyke 1969), providing more accurate results when compared with measurements.

C   Turbulent Thick Boundary-Layer Equations

Based on Equations (5.7), (5.12), and (5.15), the continuity and momentum equations for thick shear layers on curved surfaces read (cf. Eghlima and Kleinstreuer 1985)

$$\text{(continuity)} \qquad \frac{\partial}{\partial s}(r^i u) + \frac{\partial}{\partial n}\left[v(r^i - r_0^i(1 - h^j))\right] + O(\mathrm{Re}^{-2m}) = 0 \tag{5.19}$$

$$\text{(s-momentum)} \qquad (2 - h^j)u\frac{\partial u}{\partial s} + v\frac{\partial u}{\partial n} + Kuv$$

$$= -\frac{1}{\rho}(2 - h^j)\frac{\partial p}{\partial s} + v\frac{\partial^2 u}{\partial n^2} + v\frac{\partial u}{\partial n}\left[\left(\frac{r}{r_0}\right)^i + h^j\right]$$

$$+ O(\mathrm{Re}^{-2m}) \tag{5.20}$$

$$\text{(n-momentum)} \qquad Ku^2 = \frac{1}{\rho}\frac{\partial p}{\partial n} + O(\mathrm{Re}^{-2m}) \tag{5.21}$$

where n is the distance normal to the (curved) body surface and s is the distance measured along the wall starting with the stagnation point, for example. The flow indices i and j indicate the cases $i = j = 0$ for flow past a flat plate and $i = 1, j = 0$, for axisymmetric bodies with no longitudinal curvature and $i = 0, j = 1$ for flow without transverse curvature. Finally, $i = j = 1$ when longitudinal and transverse curvature effects exist. The exponent m is equal to 1/2 for laminar and $m = 4/5$ for turbulent thin boundary-layer flow with zero pressure gradient along a flat plate.

Splitting the instantaneous variables into time-smoothed and random components, the turbulent flow version of Equations (5.19) to (5.21) reads

$$\frac{\partial}{\partial s}(r^i \bar{u}) + \frac{\partial}{\partial n}(\vec{v}M) = 0 \tag{5.22}$$

$$(2 - h^j)u\frac{\partial \bar{u}}{\partial s} + v\frac{\partial \bar{u}}{\partial n} + K\bar{u}v$$

$$= -\frac{1}{\rho}(2 - h^j)\frac{\partial \bar{p}}{\partial s} + \frac{\partial^2 u}{\partial n^2} + \frac{\partial \bar{u}}{\partial n}\frac{\partial}{\partial n}\left[\left(\frac{r}{r_0}\right)^i + h^j\right]$$

$$- \frac{1}{M}\left[\left(\frac{M(2 - h^j) + r}{2}\right)\frac{\partial}{\partial s}(\overline{u'^2}) + h^j \sin\theta(\overline{u'^2})\right.$$

$$\left. + \frac{\partial}{\partial n}(\overline{u'v'}M) + KM(\overline{u'v'})\right] \tag{5.23}$$

$$K\bar{u}^2 = \frac{1}{\rho}\frac{\partial p}{\partial n} - K(\overline{u'^2}) \tag{5.24}$$

where $\quad M = r^i - r_0^i(1 - h^j), \qquad K = \left(\frac{1}{R}\right)^{\frac{1}{7}} \quad$ and $\quad \frac{\partial}{\partial s}(\overline{u'^2}) \approx 0$

since $\overline{u'^2}$ is approximately constant for attached boundary layers. This simplification, which actually holds for flat plate flow, makes Eq. (5.23) parabolic. At the wall, the "no slip" condition is usually invoked:

$$n = 0 \rightarrow \bar{u} = \bar{v} = 0 \tag{5.25}$$

In case of suction/injection $\vec{v}(n = 0) = \pm v_{\text{wall}}$. The condition for the tangential velocity at the boundary-layer edge is

$$\bar{u}/\bar{u}_e \rightarrow 1 \qquad \text{for } n \rightarrow \infty \tag{5.26}$$

In order not to mask the fluid mechanics of thick boundary-layer flow developments in region I (cf. Fig. 5.6) only zero-equation models representing turbulence effects are considered. In general, the apparent shear stresses are a function of the bulk flow velocity and velocity gradient as well as the mixing lengths, which in turn are dependent upon the radii of curvature:

$$\overline{u'v'} = \overline{u'v'}[\nu_t(l_m, S); R, r_0; \bar{u}, \nabla\bar{u}]$$

Regarding the thick turbulent boundary layer as a composition of an inner and outer layer (cf. Sect. 3.2.4), we write for the variation of the eddy viscosity in the inner region (Patel and Lee 1978)

$$(v_t)_i = \left(l_m^2\right)_i S^2 \left(\frac{r}{r_0}\right)^i \left|\frac{\partial u}{\partial y}\right| \gamma_{tr}, \quad 0 \leq y \leq y_c \tag{5.27}$$

where γ_c is an intermittency factor that accounts for the transitional region. Streamline curvature effects are incorporated with the function $S(R)$ as postulated by Bradshaw (1969). Bradshaw's correction to the eddy viscosity reads

$$S = \frac{1}{1 + \beta R_i} \quad \text{where } R_i = \frac{2\bar{u}}{R} \left(\frac{\partial \bar{u}}{\partial y}\right)^{-1} \tag{5.28a,b}$$

The coefficient β can range from 2 to 10 or higher. Its significance and best value will be investigated later with a parametric sensitivity analysis. A suitable expression for the inner mixing length for axisymmetric flow is given by Cebeci and Smith (1974) as

$$(l_m)_i = \kappa r_0 \ln\left(\frac{r}{r_0}\right) \left\{1 - \exp\left[-\frac{r_0}{A} \ln\left(\frac{r}{r_0}\right)\right]\right\} \tag{5.29}$$

where $\kappa = 0.4$ and A is a damping length. Sections 3.2.4 and 3.2.4 provide additional details.

D   System Solution

There is a large number of numerical schemes available for solving (parabolic) boundary-layer flow problems. Employing Keller's box method (cf. App. F.4 or Cebeci and Bradshaw 1977), Equations (5.22) to (5.24) are discretized for an arbitrary rectangular grid using centered difference quotients and averages at the midpoints of the net rectangles. The box scheme gives rise to a large block bidiagonal linear algebraic system. The standard technique for solving this system is to use a variant of the block tridiagonal algorithm with row pivoting performed only within blocks. However, this procedure is not guaranteed to be numerically stable, and any attempts to use stable row pivoting algorithms will introduce fill-in outside the blocks. We use a stable numerical technique that was implemented by Flaherty and Mathon (1980) to solve the block bidiagonal system. This approach uses an alternating row and column pivoting strategy to guarantee stability without introducing fill-in. Furthermore, the complexity of the algorithm is comparable to that of the unstable tridiagonal procedure. The main sources of computational errors are caused by round-off error, iteration error, error induced by linearization, and truncation error of the numerical scheme. The first three errors are smaller than the fourth one because the program is in double precision and the iteration and linearization errors are of the order of 10^{-4}. The local accuracy of the numerical algorithm is $O(h^2)$ where h is equal to the grid spacing, because of the centered difference quotients employed. To check the overall accuracy of the numerical solution, we have divided the mesh spacing by two and four and the results have shown that the global error is $O(h^2)$ as well.

The procedure for simulating a general flow system is best illustrated with reference to Fig. 5.7. Starting from the stagnation point, a suitable computer code (e.g., App. F.4, Cebeci et al. 1979; Bradshow et al. 1984) is employed until the boundary layer becomes thick; that is, the condition $\delta/R \ll 1$ is not valid anymore. These intermediate results are then conveyed to the present program, which solves the new parabolic equations of region I. This model remains valid as long as $v \ll u$, which may include the point of separation.

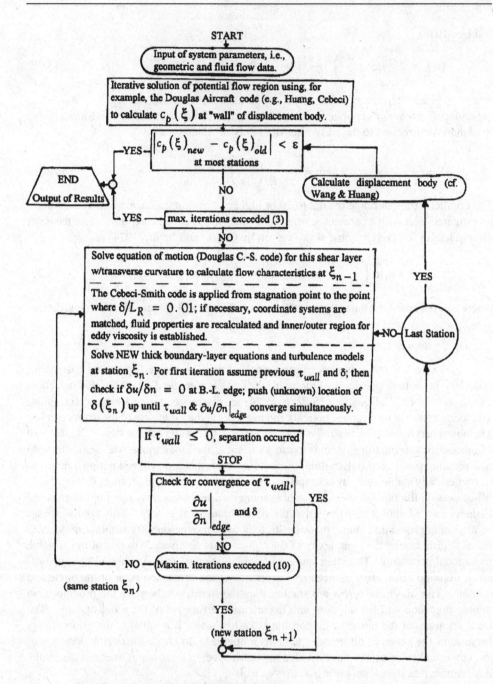

STARTER

Input of system parameters, i.e.,
geometric and fluid flow data.

Iterative solution of potential flow region using, for
example, the Douglas Aircraft code (e.g., Huang, Cebeci)
to calculate $c_p\left(\xi\right)$ at "wall" of displacement body.

—YES— $\left| c_p\left(\xi\right)_{new} - c_p\left(\xi\right)_{old}\right| < \varepsilon$
at most stations

END
Output of Results

Calculate displacement body (cf.
Wang & Huang)

NO

└ YES — max. iterations exceeded (3)

NO

Solve equation of motion (Douglas C.-S. code) for this shear layer
w/transverse curvature to calculate flow characteristics at ξ_{n-1}

The Cebeci-Smith code is applied from stagnation point to the point
where $\delta/L_R = 0.01$; if necessary, coordinate systems are
matched, fluid properties are recalculated and inner/outer region for
eddy viscosity is established.

Solve NEW thick boundary-layer equations and turbulence models
at station ξ_n. For first iteration assume previous τ_{wall} and δ; then
check if $\delta u/\delta n = 0$ at B.-L. edge; push (unknown) location of
$\delta\left(\xi_n\right)$ up until τ_{wall} & $\partial u/\partial n\big|_{edge}$ converge simultaneously.

YES

—NO— Last Station

If $\tau_{wall} \le 0$, separation occurred

STOP

Check for convergence of τ_{wall},
$\dfrac{\partial u}{\partial n}\bigg|_{edge}$ and δ

YES

NO

NO — Maxim. iterations exceeded (10)

(same station ξ_n)

YES

(new station ξ_{n+1})

Fig. 5.7. Computational flowchart.

It has to be noted that boundary-layer thickening might occur for certain body geometries
at quite an early stage as demonstrated by Kleinstreuer and Eghlima (1985) for a laminar
flow case study. Interactions between the growing boundary layer and the potential flow
are computed with the displacement body method using an iterative procedure outlined by
Wang and Huang (1979).

Theoretical and Experimental Results for Turbulent Flow Past
an Axisymmetric Anti-Separation Body (cf. Region I in Fig. 5.6)
Huang et al. (1979) have conducted their measurements at a Reynolds number (Re $=$
UL/ν) of 6.8×10^6 with an axisymmetric convex afterbody shaped in such a way that stern
separation is prevented (cf. Fig. 5.8). Figure 5.9 shows a comparison between measured and
computed displacement plus boundary-layer thicknesses for the afterbody, using Equations
(5.22) to (5.24). It is evident that the boundary-layer thickness in the stern region is
comparable to any local body length scale. Representative predicted and measured velocity
profiles for $u(r)$ and $v(r)$ at a particular station x where v is still much smaller than u are
given in Figures 5.10 and 5.11.

In another study, the effect of a correction (see Eqs. (5.27) and (5.28)) to the inner eddy
viscosity due to boundary-layer thickening on flow parameters has been investigated. It
turns out that the effect of the correction factor β on the computation of velocity profiles
is limited to a region close to the wall and diminishes downstream where $\delta/R \sim O(1)$.
Figure 5.12 compares the computed and "measured" skin friction coefficients along the

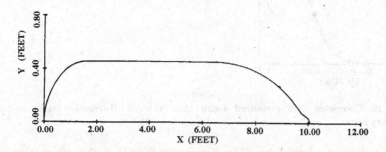

Fig. 5.8. Semi-infinite body with logitudinal curvature.

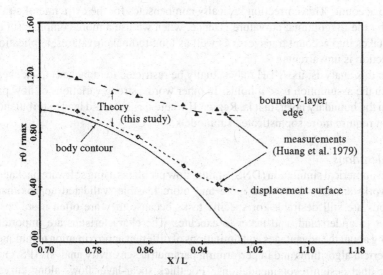

**Fig. 5.9. Schematic of the tail of the body, computed and measured displacement thickness,
and boundary-layer thickness.**

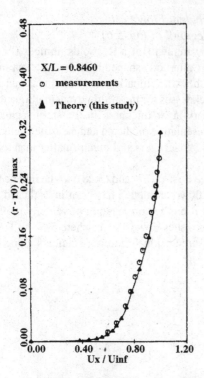

Fig. 5.10. Computed and measured mean axial velocity distribution across stern boundary layer.

body with $\beta = 5$ the best overall value. The discrepancy for $x/L > 0.9$ occurs because Huang et al. (1979) fitted, with a correction factor to the outer eddy viscosity, experimental results with their theoretical computations in which only transverse curvature effects had been taken into account. This correction basically compensates for the extra rate of strain that is caused by the longitudinal curvature. Hence, when we use a more complete set of equations that takes into account transverse as well as longitudinal curvatures, application of such a correction is unwarranted.

As shown in this analysis, the MLH may actually be restricted to (attached) thick shear layers for which the assumption $v \ll u$ holds. In other words, strong variations of flow parameters within the boundary layer, that is, Region II where $v \sim u$, would lead to turbulence phenomena that require more sophisticated submodels.

Design Considerations

Although direct numerical simulation (DNS) of fluid flow problems using software packages on high-end workstations is becoming more and more feasible, validated approximate solution methods are still desirable for specific tasks because they are often more cost-effective, easier to understand, and faster to execute. The characteristics are important assets in learning about the advantages and limitations of different approximation techniques including numerical algorithms and in performing parametric sensitivity analysis (PSA) as a precursor for final design recommendations. For thick shear-layer flows along curved 2-D or axisymmetric surfaces, the present computer simulation model can be directly used

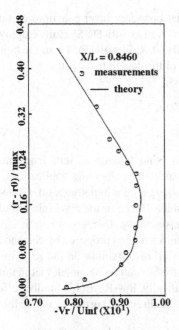

Fig. 5.11. Computed and measured mean radial velocity distribution across stern boundary layer.

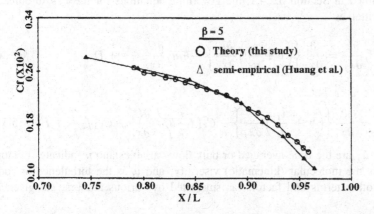

Fig. 5.12. Computed and measured skin friction coefficient along the body.

or modified to analyze and design the following:

- New body contours with possible fluid suction to achieve delayed boundary-layer separation using trial-and-error for a given set of free stream conditions
- Drag reduction by changing a given body contour, flow regime, and/or surface characteristics
- Channel or nozzle entrance shapes to achieve fully developed flow rapidly

As an M.S. computer project, the new higher-order boundary-layer equations could be compared with the ones derived by Van Dyke (1969) as well as with DNS results employing a reliable Navier–Stokes equation solver (e.g., FIDAP, CFX, FLUENT). Then, the higher-order boundary-layer model could be applied to one of the design tasks listed.

5.2.2   Internal and External Turbulent Wall Flows

A   System Identification

As indicated in Sections 3.2.4 and 5.1, there is no "universal" turbulence model that will work for each flow system. However, for specific engineering applications, especially for nonseparating flows, extensions of the mixing length hypothesis (cf. Granville 1989 and 1990), the k–ε model (cf. Nagano and Hishida 1987), or the algebraic turbulent stress model (cf. Rodi 1987) has often performed satisfactorily. Of interest here is an improved two-equation model for predicting near-wall turbulence as proposed by Nagano and Hishida (1987). Correct modeling of the influence wall of proximity on the local turbulence structure is crucial so that accurate wall shear stresses can be obtained; that in turn is important for system design and optimization. Specifically, low-Reynolds-number effects and preferential damping of velocity fluctuations in the direction normal to the wall have to be accommodated.

B   System Modeling

As outlined in Section 3.2.4A, the governing equations for the k–ε turbulence model can be written as

$$\frac{\partial k}{\partial t} + U_j \frac{\partial k}{\partial x_j} = \frac{\partial}{\partial x_j}\left\{\left(v + \frac{v_t}{\sigma_k}\right)\frac{\partial k}{\partial x_j}\right\} - \overline{u_i u_j}\frac{\partial U_i}{\partial x_j} - \varepsilon + D \tag{5.30}$$

and

$$\frac{\partial \varepsilon}{\partial t} + U_j \frac{\partial \varepsilon}{\partial x_j} = \frac{\partial}{\partial x_j}\left\{\left(v + \frac{v_t}{\sigma_\varepsilon}\right)\frac{\partial \varepsilon}{\partial x_j}\right\} - C_{\varepsilon 1} f_1 \frac{\varepsilon}{k}\overline{u_i u_j}\frac{\partial U_i}{\partial x_j} - C_{\varepsilon 2} f_2 \frac{\varepsilon^2}{k} + E \tag{5.31}$$

where the $U_i \equiv \bar{u}_j$ are the time-averaged or bulk flow variables and u indicates random fluctuations; v is the molecular (kinematic) viscosity; and v_t is the turbulence or eddy viscosity. The other terms and factors, as suggested by various authors, are given in Table 5.2.

The apparent (turbulent) stresses are approximated as

$$-\overline{u_i u_j} = v_t\left(\frac{\partial U_i}{\partial x_j} + \frac{\partial U_j}{\partial x_i}\right) - \frac{2}{3}\delta_{ij}k \tag{5.32}$$

The eddy viscosity v_t is related to k and ε through the Kolmogorov–Prandtl relation as

$$v_t = C_\mu f_\mu \sqrt{kL_\varepsilon} = \frac{C_\mu f_\mu k^2}{\varepsilon} \tag{5.33}$$

where $L_\varepsilon = k^2/\varepsilon$ is the eddy length scale (cf. Sect. 3.2.4).

Table 5.2. Constants and functions in k–ε models

Model	ε_w – BC	C_μ	$C_{\varepsilon 1}$	$C_{\varepsilon 2}$	σ_R	σ_ε	f_μ	f_1	f_2	D	E
Basic version	WF	0.09	1.44	1.92	1.0	1.3	1.0	1.0	1.0	0	0
Jones–Launder (1973)	$\varepsilon = 0$	0.9	1.45	2.0	1.0	1.3	$\exp\left\{\dfrac{-2.5}{1+\mathrm{Re}_t/50}\right\}$	1.0	$1-0.3\exp(\mathrm{Re}_t^2)$	$-2\nu\left(\dfrac{\partial\sqrt{k}}{\partial y}\right)^2$	$2\nu\nu_t\left(\dfrac{\partial^2 U}{\partial y^2}\right)^2$
Lam–Bremhorst (1981)	$\varepsilon = \nu\dfrac{\partial^2 k}{\partial y^2}$	0.09	1.44	1.92	1.0	1.3	$\{1-\exp(-0.0165R_k)\}^2\left(1+\dfrac{20.5}{\mathrm{Re}_t}\right)$	$1+(0.05/f_\mu)^2$	$1-\exp(-\mathrm{Re}_t^2)$	0	0
Nagano–Hishida (1987)	$\varepsilon = 0$	0.99	1.45	1.9	1.0	1.3	$\left\{1-\exp\left(-\dfrac{\mathrm{Re}_t}{26.5}\right)\right\}^2$	1.0	$1-0.3\exp(-\mathrm{Re}_t^2)$	$-2\nu\left(\dfrac{\partial\sqrt{k}}{\partial y}\right)^2$	$\nu\nu_t(1-f_\mu)\left(\dfrac{\partial^2 U}{\partial y^2}\right)^2$

The turbulence quantities k, ε, ν_t, and $\overline{u_i u_j}$ can be obtained from Equations (5.30)–(5.33), together with the following continuity and momentum equations for the bulk flow:

$$\partial U_i / \partial x_i = 0 \tag{5.34}$$

and

$$\frac{\partial U_i}{\partial t} + U_j \frac{\partial U_i}{\partial x_j} = -\frac{1}{\rho} \frac{\partial p}{\partial x_i} + \frac{\partial}{\partial x_j}\left(\nu \frac{\partial U_i}{\partial x_j} - \overline{u_i u_j}\right) \tag{5.35}$$

Whereas D and E are some additional terms that may improve predictions in the wall region, the function f_2 increases with higher turbulence Reynolds numbers, $\mathrm{Re}_t = k^2/\nu\varepsilon$, and is part of the term that reflects viscous destruction of vorticity. The function that appears in the eddy viscosity expression, Equation (5.33), has been modified by several researchers to accommodate near-wall effects. In making this function f_μ, dependent upon the distance from the wall, $R_k = \sqrt{k}y/\nu$, and the turbulence Reynolds number Re_t, near-wall turbulence effects are well represented (cf. Lam and Bremhorst 1981). Nagano and Hishida (1987) argued that f_μ should tend to unity as the turbulence Reynolds number and the nondimensional distance from the wall increase and should become much smaller than unity with a decrease of these parameters. Since the turbulence Reynolds number Re_t already appears in Equation (5.33), Nagano and Hishida (1987) suggested

$$f_\mu = [1 - \exp(-\mathrm{Re}_\tau/26.5)]^2 \tag{5.36}$$

which is similar to Van Driest's damping function and emphasizes the importance of the (unknown) wall shear stress on the mechanisms in the near-wall region. The friction Reynolds number $\mathrm{Re}_\tau = u_\tau y/\nu$ with $u_\tau = \sqrt{\tau_w/\rho}$ is equivalent to the inner variable, $\mathrm{Re}_\tau \equiv y^+$. When the bulk flow Reynolds number decreases, the wall sublayer becomes thicker and hence the influence of wall proximity expands.

C Results of Nagano and Hishida (1987)

The improved form of the k–ε model has been examined by Nagano and Hishida (1987) in several test cases. When compared with the log law,

$$U^+ = 2.5 \ln y^+ + 5.5 \tag{5.37}$$

good agreements were achieved for high-Reynolds-number flows in pipes or channels. The same was true for turbulent energy profiles, $k/(u_\tau)^2$, and turbulent shear stress profiles, $-\overline{u'v'}/(u_\tau)^2$. An almost identical match with measured data was achieved for the local skin friction coefficient in flat-plate boundary-layer flow. A fourth application was attached relaminarizing flow. In this case, the skin friction initially increases significantly but then decreases while the boundary-layer flow goes through a retransition from turbulent back to laminar. Again, experimental observations and model predictions were in good agreement.

Project Consideration

As is discussed in Section 5.4.1, partial occlusion of certain branching artery segments due to atherosclerotic plaque formation, called *stenosis*, may cause locally turbulent (blood) flow. For example, Deshpande and Giddens (1980) provided experimental data sets for steady axisymmetric turbulent flow through a tube that was locally 75 percent constricted.

Related papers include Deshpande and Giddens (1983) and Cassonova and Giddens (1978). The major problems of simulating locally turbulent flow in arteries include complex geometries, transients, wall compliance, non-Newtonian fluid effects, transition to turbulence, and relaminarization. As a first step, the flow systems of Deshpande and Giddens (1980) and Cassonova and Giddens (1978) could be simulated by using the Nagona–Hishida model, and then extensions that incorporate physiologically correct features could be developed (cf. Sect. 5.4.1).

5.3 Non-Newtonian Fluid Flows

A few analytic solutions to basic problems concerning non-Newtonian fluid flow are included in Chapters 1 and 3 (cf. Sects. 1.3.1, 3.2.1, and 3.3). It was shown that even for simple "power-law" or generalized Newtonian fluids, the equation of motion in terms of the stress τ, rather than the Navier–Stokes equations, is the starting point. Fluids of interest in engineering that exhibit unusual flow behavior include exotic lubricants, blood in capillaries, syrup and other foodstuff, particle suspensions, pastes and slurries, as well as a wide range of polymeric liquids. In general, these liquids are composed of macromolecules with a molecular weight that is high: $MW > 10^4$. Because Stokes's linear postulate $\tau_{ij} \sim \varepsilon_{ij}$ is invalid for non-Newtonian fluids and because of their structural complexity and diversity, new expressions for the momentum flux, that is, stress tensor $\overset{=}{\tau}$ (cf. Sect. 1.3.1), have to be found. Thus, it is necessary to redefine τ_{ij} in terms of nonlinear expressions for various classes of non-Newtonian fluids in order to simulate/interpret their specific molecular structures, flow phenomena, and material function measurements. Two approaches for the development of such "rheological equations of state" are commonly used: the continuum theory and the molecular theory (cf. Bird et al. 1987).

We only deal with the continuum approach and focus on simple polymeric liquids with a viscosity that is shear-rate-dependent. Flow phenomena such as rod climbing and jet swelling or viscoelastic effects such as fluid recoil, stress relaxation, and stress overshoot are discussed elsewhere (Bird et al. 1987). The steady-state shear flows of interest here can be basically described with an analytical representation of the shear-rate-dependent viscosity, $\eta = \eta(\dot{\gamma})$, which fits specific experimental data sets (cf. Sect. 3.2.1.D).

The two case studies outlined in Sections 5.3.1 and 5.3.2 deal with power-law fluids where rheological and thermal effects change the fluid flow dramatically. In Section 5.3.1, a polymeric solution entering a straight pipe (i.e., a thermal reactor) is undergoing a heat-induced chemical reaction that almost solidifies the slowly flowing material near the wall and hence forces the liquid core at high speeds through the center of the tube. In Section 5.3.2, thermal boundary-layer flow of a power-law fluid past an axisymmetric body with generalized boundary conditions is being considered.

5.3.1 Developing Pipe Flow with Chemical Reaction

A System Identification

Polystyrene has numerous industrial and household applications ranging from absorption to insulation. The concept of a tubular, continuous flow reactor for styrene production in which a pure monomer enters at one end and, after a thermally initiated chemical reaction, leaves as a polymer at the other end is attractive because of its simplicity,

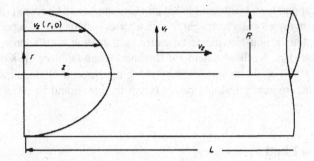

Fig. 5.13. System sketch.

relatively large heat transfer area, and potentially low cost. In reality, however, problems may occur. For example, it can happen that a thick, almost stagnant annular layer of polymer forms along the tube wall and a large portion of unreacted monomer flows rapidly about the centerline through the reactor. In addition, the conversion process becomes unstable at certain flow rates and for large tube diameters. For styrene production the threshold value for thermally induced conversion of a monomer to a polymer is about $T_0 = 100°C$ and thermal reactor instability may occur at $T_c \geq 200°C$. Thus, an analysis of the fluid dynamics, strongly influenced by the coupled heat/mass transfer and variations in system properties, is important to understand the causes of flow restrictions and to assure reactor process stability. Specifically, it was found experimentally (cf. Tien et al. 1985) that a tube of radius up to 2 cm can be very effectively used for continuous flow polymerization. Above this critical radius, thermal runaway, channeling, and very large radial gradients in temperature and concentration may develop.

Most previous theoretical investigations assumed fully developed Poiseuille flow or a uniform velocity profile (i.e., plug flow), in order to concentrate more easily on stability aspects, reactor optimization, or multiple reaction kinetics. In this study, based on the paper by Kleinstreuer and Agarwal (1987), a rather comprehensive analysis is presented to study the effects of variable system properties on the fluid flow field, which in turn affect the process stability and reactor performance. Although this investigation is based on the process dynamics of styrene polymerization, the model is flexible enough to be effectively used for a wide variety of basic studies in chemical/mechanical engineering and physical sciences.

The system consists of a straight circular tube that is maintained at isothermal or adiabatic wall conditions. The monomer solution is elevated to the desired inlet temperature in a preheater before entering the tube continuously with a parabolic velocity profile (cf. Fig. 5.13). The process is that of developing steady laminar flow of a non-Newtonian fluid in a conduit with exothermic reaction where the inlet monomer is (partially) polymerized. The fluid properties vary with temperature and concentration.

B   System Modeling

On the basis of the stated assumptions, the equations of motion in cylindrical coordinates read (cf. App. B and Sect. 3.2.1)

$$\text{(continuity)} \qquad \frac{1}{r}\frac{\partial}{\partial r}(\rho r v_r) + \frac{\partial}{\partial z}(\rho v_z) = 0 \qquad\qquad (5.38)$$

$$(r\text{-momentum}) \qquad \frac{1}{r}\frac{\partial}{\partial r}(\rho r v_r v_r) + \frac{\partial}{\partial z}(\rho v_z v_r)$$

$$= -\frac{\partial p}{\partial r} + \frac{1}{r}\frac{\partial}{\partial r}\left(r\eta\frac{\partial v_r}{\partial r}\right) + \eta\frac{v_r}{r^2} + \frac{\partial}{\partial z}\left(\eta\frac{\partial v_r}{\partial v_z}\right) \qquad (5.39)$$

$$(z\text{-momentum}) \qquad \frac{1}{r}\frac{\partial}{\partial r}(\rho r v_r v_z) + \frac{\partial}{\partial z}(\rho v_z v_z)$$

$$= -\frac{\partial p}{\partial z} + \frac{1}{r}\frac{\partial}{\partial r}\left(r\eta\frac{\partial v_z}{\partial r}\right) + \frac{\partial}{\partial z}\left(\eta\frac{\partial v_z}{\partial z}\right) \qquad (5.40)$$

where for an extended Bingham-type plastic (cf. Sect. 3.2.1)

$$\eta = \begin{cases} \eta_0 & \text{for } 0 \le \dot{\gamma} \le \dot{\gamma}_0 \\[2ex] \eta_0\left[2\left(\left(\frac{\partial v_r}{\partial r}\right)^2 + \left(\frac{v_r}{r}\right)^2 + \left(\frac{\partial v_z}{\partial z}\right)^2\right)\right. \\[2ex] \left. +\left(\frac{\partial v_r}{\partial z} + \frac{\partial v_z}{\partial r}\right)^2\right]^{(n-1)/2} & \text{for } \dot{\gamma} > \dot{\gamma}_0 \end{cases} \qquad (5.41a,b)$$

This patched "power law" is a suitable submodel for the monomer–polymer solution. The low-shear-rate viscosity η_0 is equal to the viscosity correlation for the temperature- and concentration- dependent Newtonian fluid. The flow index has an experimental value of $n = 0.2$, whereas the threshold deformation rate was found to be about $\dot{\gamma}_0 = 5\,\mathrm{s}^{-1}$ for polystyrene.

The boundary conditions include

symmetry	$v_r = 0$ and	$\dfrac{\partial v_z}{\partial r} = 0$	at $r = 0$	for all z
no-slip	$v_r = v_z = 0$	at $r = R$	for all z	

$$\text{\textit{inlet condition}} \qquad v_z = 2\bar{v}\left[1 - \left(\frac{r}{R}\right)^2\right] \qquad (5.42)$$

$$\text{\textit{exit conditions}} \qquad \frac{\partial v_r}{\partial z} = 0 \text{ and } \frac{\partial v_z}{\partial z} = 0 \qquad \text{for all } r$$

The fluid properties as well as the reaction rates are concentration- and/or temperature-dependent as shown later. The semiempirical expressions for these parameters are valid for polystyrene in the temperature range of interest, $100°\mathrm{C} \le T \le 200°\mathrm{C}$. The threshold value for thermally induced conversion of a monomer to a polymer is about $T_0 = 100°\mathrm{C}$ and thermal reactor instability may occur at $T_c \ge 220°\mathrm{C}$.

Density and viscosity are dependent upon temperature T, polymer weight fraction w_p, and average molecular weight MW (cf. Bird et al. 1987):

$$\rho = 1{,}174.00 - 0.918T + (75.3 + 9.313)w_p \qquad (5.43a)$$

$$\eta_0 = \exp\left[-13.04 + \frac{2{,}103}{T}\right.$$

$$\left. +(\mathrm{MW})^{0.18}\left(3.915w_p - 5.437w_p^2 + \left(0.623 + \frac{1{,}387}{T}\right)w_p^3\right)\right] \qquad (5.43b)$$

where MW $= 105R_w$. The local values of the scalar variables T, w_p, R_w, and so forth, are obtained from the solution of appropriate transport equations given later.

The general species transport equation can be written for our axisymmetric system at steady state as

$$\frac{1}{r}\frac{\partial}{\partial r}(\rho r v_r \phi) + \frac{\partial}{\partial z}(\rho v_z \phi) - \frac{1}{r}\frac{\partial}{\partial r}\left(\lambda r \frac{\partial \phi}{\partial r}\right) + \frac{\partial}{\partial z}\left(\lambda \frac{\partial \phi}{\partial z}\right) \pm S \qquad (5.44)$$

where ϕ is the dependent variable, λ is the diffusion coefficient, and S is a sink or source term. Each of the three parameters has to be defined for the auxiliary equations as shown later.

Equation of Thermal Energy

$$\phi = c_p T, \qquad \lambda = \frac{k_{\text{mix}}}{c_p} \quad \text{and} \quad S = \Delta H R_p \qquad (5.45\text{a–c})$$

where the heat of reaction $\Delta H = 670$ kJ/kg and R_p is the polymerization rate discussed in Agarwal and Kleinstreuer (1986). The value for the specific heat was taken as $c_p = 1{,}880$ J/(kg·K), whereas the thermal conductivity of the mixture, k_{mix}, is composed of thermal conductivities for the monomer, k_m, and polymer, k_p, which in turn are both temperature-dependent. Specifically,

$$k_m = 2.72 - 2.8 \times 10^{-3}(T - 150) + 1.6 \times 10^{-5}(T - 150)\,418.4 \times 10^{-4}$$

$$k_p = (2.93 + 5.2 \times 10^{-2}(T - 80))\,418.4 \times 10^{-4} \qquad (5.46\text{a–c})$$

$$k_{\text{mix}} = (1 - X_m)k_m + X_m k_p$$

where X_m is the fraction of monomer conversion. The associated boundary conditions are

$$T(r = R) = T_w \quad \text{or} \quad \tfrac{\partial T}{\partial r}(r = R) = 0; \qquad \tfrac{\partial T}{\partial r}(r = 0) = 0$$

$$\frac{\partial T}{\partial z}(z = L) = 0 \quad \text{and} \quad T(z = 0) - T_i \qquad (5.47)$$

Monomer Mass Transfer Equation

$$\phi = w_m, \qquad \lambda = \rho D_m \quad \text{and} \quad S = R_p \qquad (5.48\text{a–c})$$

where w_m is the monomer mass fraction, $D_m = 2.0 \times 10^{-9}$ m^2/s is the mass diffusivity, and R_p can be interpreted here as a sink term due to monomer consumption. The associated boundary conditions are

$$\frac{\partial w_m}{\partial r} = 0 \quad \text{at } r = 0 \quad \text{and} \quad r = R \quad \text{and} \quad \frac{\partial w_m}{\partial z} = 0 \quad \text{at } z = L \qquad (5.49)$$

Molecular Weight Equation

$$\phi = w_p R_w, \qquad \lambda = \rho D_m \quad \text{and} \quad S = -R_p R_w \qquad (5.50\text{a–c})$$

where w_p is the polymer weight fraction and R_w is the weight average chain length so that $w_p R_w$ is the average weight of the polymer and S is here the source term for the molecular weight due to the newly formed polymer. The associated boundary conditions are

$$\frac{\partial}{\partial r}(w_p R_w) = 0 \quad \text{at } r = 0 \text{ and } r = R \text{ and } \frac{\partial}{\partial z}(w_p R_w) = 0 \quad \text{at } z = L \quad (5.51)$$

In summary, three transport equations for enthalpy, monomer weight fraction, and average molecular weight of the polymer have to be solved with the appropriate submodels representing reaction kinetics and changes in polymer chain length to obtain the local viscosity, density, and conductivity values.

C System Solution

The steady-state two-dimensional momentum, heat, and mass transfer equations subject to the associated boundary conditions have to be solved numerically. The solution technique employed here is the SIMPLER algorithm discussed in Patankar (1980). The calculation domain is divided into a number of nonoverlapping control volumes, each finite volume surrounding one grid point. The differential equations are integrated over each control volume. Piecewise profiles expressing the variation of the dependent variables between grid points are used to evaluate the required integrals. The result is the discretization equation containing the values of ϕ, and so on, for a group of grid points. This algorithm uses an iterative procedure to solve the momentum equation for the staggered control volume mesh. Whereas fluid flow conditions (p, T, w) are evaluated at cell centers, the velocities are computed at cell boundaries. With an initial guess for all unknowns (v, p, T, w) the program calculates the coefficients for each node of the discretized momentum equation. The velocity field, obtained from the momentum equation, will most likely not satisfy the continuity equation, which in turn is then employed to develop a pressure correction equation. This pressure correction equation is solved and the velocity values are updated to satisfy continuity. This velocity field is used to solve the discretized transport equations for the temperature, monomer, and polymer distributions. Once the scalar quantities are determined, the fluid properties (η_0, ρ, and k_{mix}) are updated. The values of all the unknowns are then compared to the previous values to check for convergence. Because of system symmetry, only one-half of the reactor is discretized using a staggered mesh with 50 grid points in the axial and 20 grid points in the radial direction. Mesh independence was checked by increasing the mesh density and observing that (possible) changes in the results did not occur.

The predictive capability of the new model was successfully tested in a comparative study with experimental data for the overall monomer conversion (cf. Agarwal and Kleinstreuer 1986). For the interpretation of the following graphs (cf. Figs. 5.14–5.19), it is helpful to recall that the nonlinear transport and conversion mechanisms are highly coupled and that a typical monomer molecule experiences opposing effects in two distinct regions. Specifically, in the wall region, temperatures are relatively low (cf. thermal conditions) and residence times are relatively high (cf. no-slip condition). The opposite is true for monomer in the core region, that is, around the centerline. Since the threshold temperature for thermal induction is about 100°C and the conversion of monomer to polymer is a chain reaction that takes a finite amount of time, polymer is formed near the wall where the contact time is large. In contrast to the highly viscous, slowly moving polymer region, the unreacted monomer forms a centered core of high velocity. As a result, profiles for the mixture density,

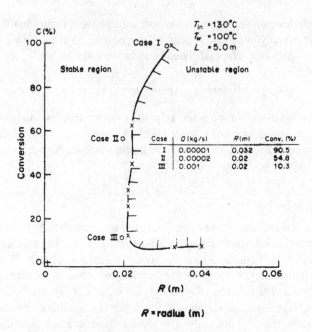

Fig. 5.14. Standard conversion plot with $T_{in} = 130°C$ and $T_w = 100°C$.

viscosity, and thermal conductivity change rapidly in the initial tube section and exhibit steep gradients in radial direction. These variations are strongly influenced by the local monomer fraction of the mixture. As indicated, the velocity distributions follow, similar to the monomer concentration, the two-region patterns and largely determine the temperature profiles, which in turn influence the local fluid properties.

For more detailed analysis of the fluid flow in a tubular reactor characterized by thermal polymerization processes, three representative cases have been selected: reactor operations at high, medium, and low monomer conversions (cf. Fig. 5.14). The conversion-stability plot of Fig. 5.14 was obtained by increasing the tube radius for a given mass flow rate until "thermal runaway" occurs, that is, until the prescribed maximum temperature ($\approx 220°C$ for styrene) has been exceeded somewhere in the system. In running these computer experiments at different (constant) mass flow rates until thermal runaway occurs, different points for the stability limit were obtained. Changes in inlet and/or boundary conditions shift or rotate the stability limit. In case I most of the incoming monomer is converted into polymer and such high conversion rates are only achievable with relatively small tubes and low flow rates. In contrast, cases II and III feature flow rates that are substantially higher and monomer conversion rates that are average (case II) or quite low (case III). The resulting graphs, generated with the operational and boundary conditions summarized in Fig. 5.14, can be divided into two groups. Figures 5.15 and 5.16 show velocity and pressure profiles for the three cases, and Figs. 5.17–5.19 depict temperature, monomer mass, and density profiles for case II.

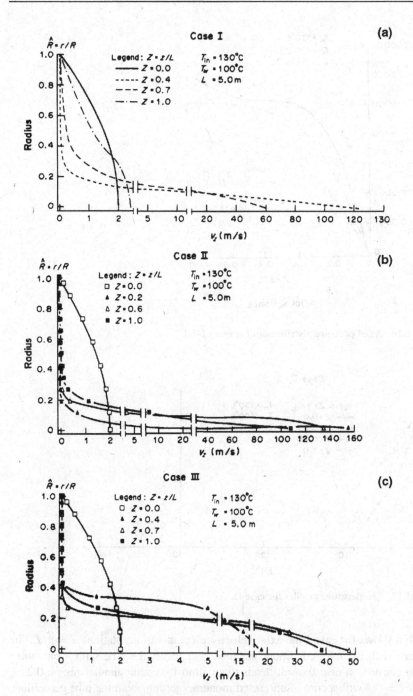

Fig. 5.15. (a) Axial velocity profiles $v_z(r)$ for case I. (b) Axial velocity profiles $v_z(r)$ for case II. (c) Axial velocity profiles $v_z(r)$ for case III.

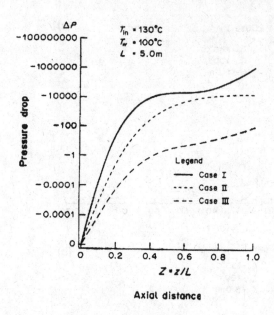

Fig. 5.16. Axial pressure distributions for cases I–III.

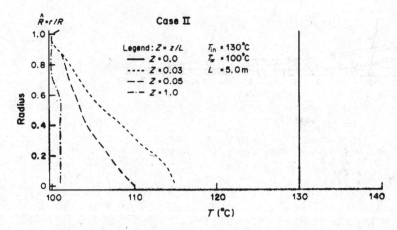

Fig. 5.17. Temperature profiles for case II.

Figure 5.15a shows for case I the axial velocity $v_z(r)$ at various stations $Z = z/L$. In correspondence with the inlet condition, the monomer enters the tube with a parabolic profile. Swift conversion near the wall leads to an almost stagnant annular region $0.2 > r/R \leq 1.0$ at $Z = 0.4$ with a core of unreacted monomer jetting about the tube centerline. Farther downstream, more and more of the remaining monomer is being converted, so that the core velocity reduces and simultaneously shear stresses transfer momentum in radial direction, accelerating the polymer, whose viscosity may be reduced because of higher local temperature (cf. station $Z = 0.7$). About 90 percent of the incoming monomer has been converted at $Z = 1.0$; that is, the tube exit profile is a bit more uniform. Figure 5.15b

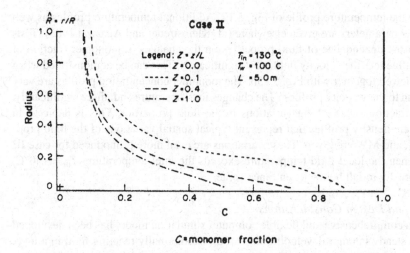

Fig. 5.18. Radial profiles of monomer fraction at various axial stations (case II).

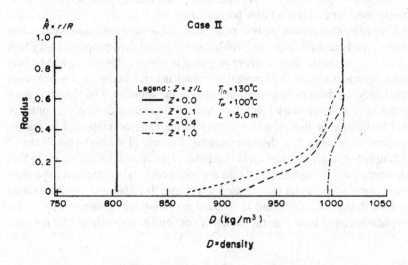

Fig. 5.19. Fluid density profiles at different axial stations (case II).

shows velocity profiles for case II similar to those of the previous case except that because of the moderate conversion the exit profile still exhibits the jetting effect. The velocity scale had to be changed three times in order to give sufficient resolution. The polymerized annular region is smaller for case III, as shown in Figure 5.15c, when compared with case II, where the mass flow rate was much less and hence the contact time was longer. Figure 5.16 depicts the average pressure drop, $\Delta p = p(z) - p(0.0)$, for the tube in all three cases according to their different mass flow rates.

Figure 5.17 depicts, for case II only, the temperature profiles in radial direction near the tube entrance and at the reactor exit. At relatively low flow rates (cf. case II), the bulk temperature drops significantly at the entrance as a result of radial conduction. This behavior is reflected in the temperature profile of Fig. 5.17. Additional temperature profiles as well as heat transfer parameters are given elsewhere (cf. Kleinstreuer and Agarwal 1986). Tests indicated that the dependence of fluid properties on the changes in monomer fraction of the solution influences the velocity field most significantly. This can be demonstrated when Fig. 5.18 is viewed together with Fig. 5.15b. The monomer concentration profiles are very similar in trend to the velocity profiles. The changes in temperature and, more significantly, in monomer fraction cause strong variations in the fluid properties. This is depicted in Fig. 5.19, where density profiles that represent typical spatial variations of the fluid properties η_0, k_{mix}, and MW are given. These variations are even more pronounced for case III conditions, when the local fluid temperature exceeds the inlet temperature, $T_{in} = 130°C$, causing the density to fall below its entrance value.

Conclusions and Design Considerations

In summary, a comprehensive and flexible computer simulation model has been developed to investigate steady laminar developing flow of an exothermally reacting fluid in a tube. Modular testing of the governing equations and a successful comparison with an experimental case study for styrene polymerization (cf. Tien et al. 1985) indicate that the computer code is a predictive tool. Specifically, it was determined that a tube radius up to 2 cm can be very effectively used for continuous flow polymerization.

Furthermore, the present analysis clearly demonstrates that commonly made assumptions of uniform or developed parabolic velocity profiles and constant fluid properties may lead to erroneous results. The momentum transfer equation is strongly coupled with the heat and mass transfer equations via the fluid properties, which depend on the local temperature and, more importantly, on the monomer concentration in the solution. The tubular reactor geometry and the inlet and boundary conditions have to be chosen carefully in order to prevent thermal instability or flow restrictions that in the extreme may lead to reactor clogging or unreacted monomer flow through the tube. Because of the high complexity of this nonlinear coupled system, trial and error computer experiments are necessary to find the appropriate operational conditions to obtain a desired conversion rate and mass flow rate. The present simulation model can be used to achieve this goal and allows detailed studies of the fluid dynamics, which aid in the understanding and interpretation of the basic transport phenomena as well as in the design of continuous flow reactors for polymer production.

5.3.2   *Thermal Boundary-Layer Flows*

The analysis of mixed, that is, free-forced convection heat transfer of a power-law fluid near a spinning axisymmetric body with generalized wall conditions is important for the understanding and design of various engineering systems. Applications include wire or fiber coating, foodstuff processing, and reactor fluidization as well as rotating machinery, transpiration cooling, and projectile behavior. Recent advancements in this field have been briefly reviewed by Kleinstreuer and Wang (1989), among others.

A   *System Identification*

Of interest are the effects of mixed convection coupled with body rotation, wall heating/cooling mode, wall mass transfer, and/or the type of (power-law) fluid on the local

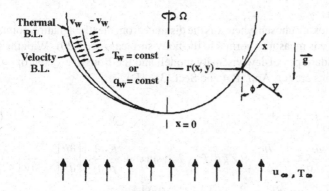

Fig. 5.20. System schematics.

skin friction coefficient and the heat transfer coefficient for the axisymmetric body shown in Fig. 5.20. For example, spin motion enhances convection heat transfer when the centrifugal force pushes the near-surface fluid outward and that fluid is replaced by cooler or warmer fluid depending upon the wall temperature. Momentum and heat transfer rates may also be affected by the buoyancy force, which assists the forced flow for heated surfaces and retards it in the case of cooled surfaces when the fluid is moving upward against the gravitational force. Mass transfer at the porous surfaces, in terms of fluid injection or withdrawal at a prescribed temperature, can alter the local skin friction coefficient and the local Nusselt number significantly. A number of practical applications involve non-Newtonian fluids that can be approximated by the power-law viscosity model. The type of fluid, that is, shear thinning, Newtonian, or shear thickening fluid, has a profound effect on the fluid mechanics and heat transfer parameters.

B   System Modeling

Figure 5.20 depicts axisymmetric boundary-layer flow past a spinning, permeable body placed in a uniform stream moving opposite to the gravitational force and parallel to the axis of body rotation. Considering steady laminar flow of a power-law fluid, the describing equations, boundary conditions, and suitable coordinate transformations are developed for two distinct thermal boundary conditions: constant wall heat flux and isothermal surface (cf. Sects. 4.1.1 and 3.2.1D). In either case, the wall temperature may be higher than the ambient (heating mode; i.e., $Z = 1$) or lower (cooling mode, i.e., $Z = -1$). The fluid properties, which appear in the dimensionless groups, are considered to be constant in each simulation run except for temperature-induced density variations in the body-force term (cf. Boussinesq assumption, Sect. 4.1.3). Wake effects on the attached boundary layer are neglected. Energy dissipation due to fluid friction is not considered, since the maximum Eckert number in this case is $Ec \approx 0.12$, which is below $Ec \leq 0.2$ when viscous heating may come into effect. In addition, axial conduction is negligible when compared with transverse thermal diffusion or axial convection, on the basis of relative order of magnitude analyses (cf. Sects. 2.3.1 and 4.1). The general analysis is then applied to the case of a rotating, permeable sphere.

Governing Equations

Nonrotating coordinates are chosen where x is the distance from the rd stagnation point along a meridian curve and y is measured normal to the body surface (Fig. 5.20). With the stated assumptions, the standard power-law viscosity model, and the Boussinesq approximation, the governing equations are (cf. App. B.4 and Sect. 3.2.1D)

$$\frac{\partial}{\partial x}(ru) + \frac{\partial}{\partial y}(rv) = 0 \tag{5.52}$$

$$u\frac{\partial u}{\partial x} + v\frac{\partial u}{\partial y} - \frac{w^2}{r}\frac{dr}{dx} = u_e\frac{du_e}{dx} + Z\hat{g}\beta|T - T_\infty|\sin\phi + \frac{K}{\rho}\frac{\partial}{\partial y}\left[\left|\frac{\partial u}{\partial y}\right|^{n-1}\frac{\partial u}{\partial y}\right] \tag{5.53}$$

$$u\frac{\partial w}{\partial x} + v\frac{\partial w}{\partial y} + \frac{uw}{r}\frac{dr}{dx} = \frac{K}{\rho}\frac{\partial}{\partial y}\left[\left|\frac{\partial w}{\partial y}\right|^{n-1}\frac{\partial w}{\partial y}\right] \tag{5.54}$$

$$u\frac{\partial T}{\partial x} + v\frac{\partial T}{\partial y} = \alpha\frac{\partial^2 T}{\partial y^2} \tag{5.55}$$

where u_e is the boundary-layer edge velocity, $u_e = U(x)$, $Z = \pm 1$, β is the volumetric expansion coefficient, and $K = m$ is the power-law parameter with units Pa $\cdot$ s^n, where n represents dimensionless power-law index. The associated boundary conditions are

$$\text{at } y \to \infty \quad u = u_e(x), \quad w = 0 \quad \text{and} \quad T = T_\infty \tag{5.56}$$

$$\text{at } y = 0 \quad u = 0, \quad v = \pm v_w, \quad w = r\Omega \quad \text{and} \quad T = T_w \tag{5.57a}$$

or

$$\text{at } y = 0 \quad u = 0, \quad v = \pm v_w, \quad w = r\Omega \quad \text{and} \quad q = -k\frac{\partial T}{\partial y} \tag{5.57b}$$

In order to facilitate the numerical solution, the x-dependence of certain terms in the governing equations is reduced and the boundary conditions are simplified. This is accomplished with coordinate transformations based on a proper choice of transformation parameters derived from scale analysis (cf. Sect. 2.3.1).

Scale Analysis and Transformation

Following the procedure outlined in Section 2.3.1, expressions for the combined variable $\eta(x, y)$, the dimensionless coordinate $\xi(x)$, and the stream function $\psi(\xi, \eta)$ are derived and the governing equations (5.52)–(5.55) are transformed subject to (5.56) and (5.57a). As a first step, appropriate velocity, length, and temperature scales are determined: for example, $u \to u$ or u_∞ or u_e; $x \to L$, $y \to \delta$, or δ_{th}; $T - T_\infty \to \Delta T$.

Continuity equation
$$\frac{\partial(ru)}{\partial x} + \frac{\partial(rv)}{\partial y} = 0$$

$$\frac{ru}{L} \sim \frac{rv}{\delta_{\text{th}}} \Rightarrow \frac{u}{L} \sim \frac{v}{\delta_{\text{th}}}, \quad \text{where } L \text{ is a reference body length}$$

Momentum equation
$$u\frac{\partial u}{\partial x} + v\frac{\partial u}{\partial y} = u_e\frac{du_e}{dx} + Zg\beta(T - T_\infty)\sin\phi$$

$$+ \frac{K}{\rho} \frac{\partial}{\partial y} \left(\left| \frac{\partial u}{\partial y} \right|^{n-1} \frac{\partial u}{\partial u} \right)$$

$$\underbrace{u_\infty \frac{u_\infty}{L}, v \frac{u}{\delta_{th}} \sim u_\infty \frac{u_\infty}{L}}_{\text{inertia}}; \quad \underbrace{u_\infty \frac{u_\infty}{L}}_{\text{pressure}} \quad \underbrace{g\beta \Delta T \sin \phi}_{\text{buoyancy}}; \quad \underbrace{\frac{K}{\rho} \frac{1}{\delta_{th}} \frac{u_\infty^n}{\delta_{th}^n}}_{\text{friction}}$$

The dominant forces in the boundary layer are inertia and viscous forces. Thus, from the momentum balance, we can deduce the proportionality ($\delta_{th} \leq \delta$)

$$u_\infty \frac{u_\infty}{L} \sim \frac{K}{\rho} \frac{1}{\delta_{th}} \frac{u_\infty^n}{\delta_{th}^n}$$

from which

$$\frac{\delta_{th}^{n+1}}{L^{n+1}} \sim \frac{K}{\rho u_\infty^{2-n} L^n} = \frac{1}{\text{Re}}$$

or

$$\delta_{th} \sim L \text{Re}^{-1/(n+1)}$$

where Re is a power-law Reynolds number. Now, with the traditional postulate for the combined variable $\eta \sim y/\delta_{th}$, we form

$$\eta \sim \frac{y}{\delta_{th}} \sim \text{Re}^{1/n+1} \frac{Y}{L}$$

The proportionality is extended into a suitable equation for $\eta(x, y)$ by choosing

$$\eta = \left(\frac{\text{Re}}{\xi} \right)^{1/n+1} \left(\frac{u_e}{u_\infty} \right)^{2-n/n+1} \frac{y}{L} \tag{5.58}$$

where $\xi = x/L$ is the dimensionless streamwise coordinate. The purpose of introducing $(u_e/u_\infty)^{(2-n)/(n+1)}$ is to make the transformation of the viscous term in the momentum equation appear in a very compact form.

The formulation of the stream function has two requirements:

(i)   Satisfying the continuity equation
(ii) Generating the following results:

$$u = u_e F' \quad \text{and at} \quad \eta = 0, \quad v = v_w$$

Thus, for 2-D flow or axisymmetric flow, that is, $\kappa = 0$ or $\kappa = 1$, respectively, we postulate (cf. Sects. 2.3.3.A and 3.2.3.D)

$$\psi = r^\kappa L u_\infty \left(\frac{\xi}{\text{Re}} \right)^{1/n+1} \left(\frac{u_\infty}{u_e} \right)^{1-2n/n+1} F(\xi, n) - \int_0^x x^\kappa v_w \, dx \tag{5.59}$$

with the following relation

$$u = \frac{1}{r^\kappa} \frac{\partial \psi}{\partial y} \quad \text{and} \quad v = -\frac{1}{r^\kappa} \frac{\partial \psi}{\partial y}$$

For the constant wall temperature case of interest here

$$\theta(\xi, \eta) = \frac{T - T_\infty}{T_w - T_\infty} \tag{5.60}$$

where both constant system temperatures have been utilized.

The transverse velocity component is nondimensionalized as

$$g(\xi, \eta) = \frac{w}{r\Omega} \tag{5.61}$$

Equations (5.53) and (5.55) are now transformed for illustration purposes, setting $w = 0$ for convenience. By definition

$$u \equiv \frac{1}{r^\kappa} \frac{\partial \psi}{\partial y} = \frac{1}{r^\kappa} \left[\frac{\partial \psi}{\partial \xi} \frac{\partial \xi}{\partial y} + \frac{\partial \psi}{\partial \eta} \frac{\partial \eta}{\partial y} \right]$$

$$= \frac{1}{r^\kappa} r^\kappa L u_\infty \left(\frac{u_\infty}{u_e} \right)^{1 - 2n/n+1} \left(\frac{\xi}{\mathrm{Re}} \right)^{1/n+1} F' \left(\frac{\mathrm{Re}}{\xi} \right)^{1/n+1} \left(\frac{u_e}{u_\infty} \right)^{2 - n/n+1} \frac{1}{L}$$

$$u = u_e F'$$

$$v \equiv \frac{1}{r^\kappa} \frac{\partial \psi}{\partial x} = -\frac{1}{r^\kappa} \left[\frac{\partial \psi}{\partial \xi} \frac{\partial \xi}{\partial x} + \frac{\partial \psi}{\partial \eta} \frac{\partial \eta}{\partial y} \right]$$

That is,

$$v = -\frac{1}{r^\kappa} \left\{ \frac{1}{L} \kappa r^{\kappa - 1} \frac{dr}{d\xi} L u_\infty \left(\frac{\xi}{\mathrm{Re}} \right)^{1/n+1} \left(\frac{u_\infty}{u_e} \right)^{1 - 2n/n+1} F \right.$$

$$+ \frac{1}{L} r^\kappa L \frac{u_\infty}{(n+1)\xi} \left(\frac{\xi}{\mathrm{Re}} \right)^{1/n+1} \left(\frac{u_\infty}{u_e} \right)^{1 - 2n/n+1} F$$

$$+ \frac{1}{L} r^\kappa L \left(\frac{\xi}{\mathrm{Re}} \right)^{1/n+1} \frac{2n-1}{n+1} \left(\frac{u_\infty}{u_e} \right)^{2 - n/n+1} \frac{du_e}{d\xi} F$$

$$+ \frac{1}{L} r^\kappa L u_\infty \left(\frac{\xi}{\mathrm{Re}} \right)^{1/n+1} \left(\frac{u_\infty}{u_e} \right)^{1 - 2n/n+1} \frac{\partial F}{\partial \xi}$$

$$\left. + r^\kappa L u_\infty \left(\frac{u_\infty}{u_e} \right)^{1 - 2n/n+1} \left(\frac{\xi}{\mathrm{Re}} \right)^{1/n+1} F' \frac{\eta}{L\xi} \left(-\frac{1}{n+1} + \frac{2-n}{n+1} \Lambda \right) \right\} + v_w$$

Finally

$$v = -\frac{u_e}{\xi} \left(\frac{\xi}{\mathrm{Re}} \right)^{1/n+1} \left(\frac{u_\infty}{u_e} \right)^{2 - n/n+1} \left[\gamma(\xi) F + \xi \frac{\partial F}{\partial \xi} \right.$$

$$\left. - u_e \left(\frac{u_\infty}{u_e} \right)^{2 - n/n+1} \left(\frac{\xi}{\mathrm{Re}} \right)^{1/n+1} \frac{\eta}{\xi} F' \left(-\frac{1}{n+1} + \frac{2-n}{n+1} \Lambda \right) \right] + v_w$$

where $\quad \Lambda(\xi) = \dfrac{\xi}{u_e} \dfrac{du_e}{d\xi} \quad$ and $\quad \gamma(\xi) = \dfrac{1}{n+1} + \dfrac{2n-1}{n+1} \Lambda + \kappa \dfrac{\xi}{r} \dfrac{dr}{d\xi}$

Now

$$\frac{\partial u}{\partial y} = \frac{\partial u}{\partial \xi}\frac{\partial \xi}{\partial y} + \frac{\partial u}{\partial \eta}\frac{\partial \eta}{\partial y} = u_e F'' \left(\frac{\text{Re}}{\xi}\right)^{1/n+1} \left(\frac{u_e}{u_\infty}\right)^{2-n/n+1} \frac{1}{L}$$

$$\left|\frac{\partial u}{\partial y}\right|^{n-1}\frac{\partial u}{\partial y} = u_e^n \left(\frac{\text{Re}}{\xi}\right)^{n/n+1} \left(\frac{u_e}{u_\infty}\right)^{(2-n)n/n+1} \frac{1}{L^n}|F''|^{n-1}F''$$

$$\frac{\partial}{\partial y}\left(\left|\frac{\partial u}{\partial y}\right|^{n-1}\frac{\partial u}{\partial y}\right) = u_e^n \left(\frac{\text{Re}}{\xi}\right)\left(\frac{u_e}{u_\infty}\right)^{2-n}\frac{1}{L^{n+1}}\left(|F''|^{n-1}F''\right)'$$

and

$$\frac{\partial u}{\partial x} = \frac{\partial u}{\partial \xi}\frac{\partial \xi}{\partial x} + \frac{\partial u}{\partial \eta}\frac{\partial \eta}{\partial x} = \frac{u_e}{L}\frac{\partial F'}{\partial \xi} + \frac{F'}{L}\frac{du_e}{d\xi} + u_e F''\frac{\eta}{L\xi}\left[-\frac{1}{n+1}+\frac{2-n}{n+1}\Lambda\right]$$

Then

$$u\frac{\partial u}{\partial x} + v\frac{\partial u}{\partial y} = \frac{u_e^2}{\xi L}\left[\xi F'\frac{\partial F'}{\partial \xi} + \Lambda F'^2 - \gamma(\xi)FF'' - \xi F''\frac{\partial F}{\partial \xi} + MPD(\xi)F''\right]$$

where $$MP = \pm\frac{v_w}{u_\infty}\text{Re}^{1/n+1}$$

$$D(\xi) = \xi^{1/n+1}\left(\frac{u_e}{u_\infty}\right)^{1-2n/n+1}$$

$$\frac{L\xi}{u_e^2}u_e\frac{du_e}{dx} = \frac{\xi}{u_e}\frac{du_e}{d\xi} = \Lambda(\xi)$$

$$\frac{L\xi}{u_e^2}\frac{K}{\rho}u_e^n\frac{\text{Re}}{\xi}\left(\frac{u_e}{u_\infty}\right)^{2-n}\frac{1}{L^{n+1}} = 1$$

and

$$Zg\beta(T-T_\infty)\sin\phi\frac{L\xi}{u_e^2} = Z\theta\lambda_T\xi\Omega(\xi)$$

where $$\Omega(\xi) = \frac{\sin\phi}{(u_e/u_\infty)^2} \quad \text{and} \quad \lambda_T = \frac{Gr}{\text{Re}^{2/(2-n)}}$$

Finally, the momentum equation without rotation transforms to

$$\left(|F''|^{n-1}F''\right)' + \gamma(\xi)FF'' + \Lambda(\xi)(1-F'^2) - MPD(\xi)F''$$

$$= -Z\theta\lambda_T\Omega(\xi) + \xi\left[F'\frac{\partial F'}{\partial \xi} - F''\frac{\partial F}{\partial \xi}\right] \tag{5.62}$$

For the transformation of the heat transfer equation we substitute

$$\frac{\partial T}{\partial x} = (T_w - T_\infty)\left[\frac{\partial \theta}{\partial \xi}\frac{\partial \theta}{\partial x} + \frac{\partial \theta}{\partial \eta}\frac{\partial \eta}{\partial x}\right] = (T_w - T_\infty)\left[\frac{1}{L}\frac{\partial \theta}{\partial \xi} + \frac{\theta'}{L}\frac{\eta}{\xi}\left(-\frac{1}{n+1}+\frac{2-n}{n+1}\Lambda\right)\right]$$

$$\frac{\partial T}{\partial y} = (T_w - T_\infty)\left[\frac{\partial \theta}{\partial \xi}\frac{\partial \theta}{\partial y} + \frac{\partial \theta}{\partial \eta}\frac{\partial \eta}{\partial y}\right] = (T_w - T_\infty)\theta'\left(\frac{Re}{\xi}\right)^{1/n+1}\left(\frac{u_e}{u_\infty}\right)^{2-n/n+1}\frac{1}{L}$$

$$\frac{\partial^2 T}{\partial y^2} = (T_w - T_\infty)\theta''\left(\frac{Re}{\xi}\right)^{2/n+1}\left(\frac{u_e}{u_\infty}\right)^{2(2-n)/n+1}\frac{1}{L^2}$$

$$u\frac{\partial T}{\partial x} + v\frac{\partial T}{\partial y} = u_e F'\frac{1}{L}\frac{\partial \theta}{\partial \xi} + u_e F'\theta'\frac{1}{L}\frac{\eta}{\xi}\left[-\frac{1}{n+1} + \frac{2-n}{n+1}\Lambda\right]$$

$$-\theta'\left(\frac{Re}{\xi}\right)^{1/n+1}\left(\frac{u_e}{u_\infty}\right)^{2-n/n+1}\frac{1}{L}\left(\frac{u_e}{\xi}\right)\left(\frac{\xi}{Re}\right)^{1/n+1}\gamma(\xi)F + \xi\frac{\partial F}{\partial \xi}$$

$$-\theta'\left(\frac{Re}{\xi}\right)^{1/n+1}\left(\frac{u_e}{u_\infty}\right)^{2-n/n+1}\frac{1}{L}u_e\left(\frac{u_\infty}{u_e}\right)^{2-n/n+1}$$

$$\times\left(\frac{\xi}{Re}\right)^{1/n+1}\frac{\eta}{\xi}F'\left[-\frac{1}{n+1} + \frac{2-n}{n+1}\Lambda\right]$$

$$= \frac{u_e}{L\xi}\left[\xi F'\frac{\partial \theta}{\partial \xi} - \gamma(\xi)F\theta' - \xi\theta'\frac{\partial F}{\partial \xi}\right]$$

$$\alpha\theta''\left(\frac{Re}{\xi}\right)^{2/n+1}\left(\frac{u_e}{u_\infty}\right)^{2(2-n)/n+1}\frac{1}{L^2}\frac{L\xi}{u_e}$$

$$= \frac{\alpha}{Lu_\infty}Re^{2/n+1}\theta''\left(\frac{u_e}{u_\infty}\right)^{2(2-n)/n+1}\left(\frac{u_\infty}{u_e}\right)\xi^{n-1/n+1} = \frac{E(\xi)}{Pr}\theta''$$

where $\quad E(\xi) = \left(\frac{u_e}{u_\infty}\right)^{2(2-n)/n+1}\xi^{n-1/n+1}\quad$ and $\quad Pr = \frac{u_\infty L}{\alpha}Re^{-2/n+1}$

Thus, the energy equation transforms to

$$\frac{E(\xi)}{Pr}\theta'' + \gamma(\xi)F\theta' - MPD(\xi)\theta' = \xi\left[F'\frac{\partial \theta}{\partial \xi} - \theta'\frac{\partial F}{\partial \xi}\right] \tag{5.63}$$

The corresponding transformed boundary conditions are

(a) at $\eta = 0 \qquad F'(\xi, 0) = 0$
   Note: $u = u_e F'$ and $u = 0$ at $y = 0$ $(\eta = 0)$.
(b) at $\eta = 0 \qquad F(\xi, 0) = 0$
   Note: From the condition $v = v_w$ at $y = 0 (\eta = 0)$ and with

$$v = \frac{u_e}{\xi}\left(\frac{\xi}{Re}\right)^{1/n+1}\left(\frac{u_\infty}{u_e}\right)^{2-n/n+1}\left[\gamma(\xi)F + \xi\frac{\partial F}{\partial \xi}\right]$$

$$- u_e\left(\frac{u_\infty}{u_e}\right)^{2-n/n+1}\left(\frac{\xi}{Re}\right)^{1/n+1}\frac{\eta}{\xi}F'\left(-\frac{1}{n+1} + \frac{2-n}{n+1}\Lambda\right) + v_w$$

we know the following relation:

$$\gamma(\xi)F + \xi\frac{\partial F}{\partial \xi}\bigg|_{\eta=0} = 0$$

Furthermore,

at $\quad \xi = 0 \qquad \gamma(\xi = 0) F(\xi = 0, 0) = 0$

then $\quad F(\xi = 0, \eta = 0) = 0$

For $\quad \xi > 0$

$$\left. \frac{\partial F}{\partial \xi} \right|_{\eta=0} = \frac{\gamma(\xi)}{\xi} F(\xi, \eta = 0)$$

Integrating the preceding equation with respect to ξ we obtain

$$F(\xi, 0) - F(\xi \to 0, 0) = - \int_{\xi \to 0}^{\xi} \frac{\gamma(\xi)}{\xi} F(\xi, \eta = 0) \, d\xi$$

$$F(\xi, 0) = - \int_{\xi \to 0}^{\xi} \frac{\gamma(\xi)}{\xi} F(\xi, 0) \, d\xi$$

Thus, the best choice for $F(\xi, 0)$ satisfying the preceding equation is

$$F(\xi, 0) = 0 \qquad \text{for any } \xi$$

(c) at $\quad \eta = 0 \qquad \theta(\xi, 0) = 1$
Note: $\theta = (T - T_\infty)/T_w - T_\infty)$ and $T = T_w$ at $y = 0(\eta = 0)$.

(d) at $\quad \eta \to \infty \qquad F'(\xi, \infty) = 1$
Note: $u = u_e F'$ and $u = u_e(x)$ at $y \to \infty(\eta \to \infty)$.

(e) at $\quad \eta \to \infty, \qquad \theta(\xi, \infty) = 0$
Note: $\theta = (T - T_\infty)/T_w - T_\infty)$ and $T = T_\infty$ at $y \to \infty$.

The system parameters, such as the skin friction coefficient and the Nusselt number, can be transformed as follows. By definition

$$c_f = \frac{\tau_w}{1/2 \rho u_\infty^2}$$

where $\quad \tau_w = K \left(\left| \frac{\partial u}{\partial y} \right|^{-1} \left. \frac{\partial u}{\partial y} \right|_{y=0} \right) = u_e^n K \left(\frac{\text{Re}}{\xi} \right)^{n/n+1} \left(\frac{u_e}{u_\infty} \right)^{n(2-n)/n+1}$

$$\frac{1}{L^n} |F''(\xi, 0)|^{n-1} F''(\xi, 0)$$

But $F''(\xi, 0) > 0$. Then

$$\tau_w = u_e^n K \left(\frac{\text{Re}}{\xi} \right)^{n/n+1} \left(\frac{u_e}{u_\infty} \right)^{n(2-n)/n+1} \frac{1}{L^n} [F''(\xi, 0)]^n$$

Therefore

$$c_f = 2 \frac{1}{\text{Re}} \left(\frac{\text{Re}}{\xi} \right)^{n/n+1} \left(\frac{u_e}{u_\infty} \right)^{n(2-n)/n+1} [F''(\xi, 0)]^n \left(\frac{u_e}{u_\infty} \right)^n$$

or the skin friction group (SFG) for power-law fluids is

$$\text{SFG} \equiv \frac{1}{2} c_f \text{Re}^{1/n+1} = \xi^{-n/n+1} \left(\frac{u_e}{u_\infty} \right)^{3n/n+1} [F''(\xi, 0)]^n \tag{5.64}$$

The Nusselt number is defined as

$$\text{Nu}_L = \frac{hL}{k}$$

$$q_w = -k \frac{\partial T}{\partial y} \bigg|_{y=0} = h(T_w - T_\infty)$$

$$-k \frac{\partial T}{\partial y} \bigg|_{y=0} = -k\theta'(\xi, \eta = 0) \left(\frac{\text{Re}}{\xi} \right)^{1/n+1} \left(\frac{u_e}{u_\infty} \right)^{2-n/n+1} \frac{1}{L}$$

Thus

$$\text{Nu}_L = -\theta'(\xi, \eta = 0) \left(\frac{\text{Re}}{\xi} \right)^{1/n+1} \left(\frac{u_e}{u_\infty} \right)^{2-n/n+1}$$

or the heat transfer group for the isothermal wall case (HTG_T) is

$$\text{HTG}_T \equiv \text{Nu}_L \text{Re}^{-1/n+1} = -\theta'(\xi, 0)\xi^{-1/n+1} \left(\frac{u_e}{u_\infty} \right)^{2-n/n+1} \tag{5.65}$$

In summary, the original PDEs (5.52) to (5.55) are transformed to a set of mixed differential equations of the form

$$(|F''|^{n-1} F'')' + \gamma F F'' + \Lambda(1 - F'^2) + C_g^2 + Z\lambda_T S\theta - \text{MP}DF''$$

$$= \xi \left[F' \frac{\partial g}{\partial \xi} - F'' \frac{\partial F}{\partial \xi} \right] \tag{5.66}$$

$$G \left(|g'|^{n-1} g' \right)' + \gamma g' F - Hg F' - \text{MP}Dg' = \xi \left[F' \frac{\partial g}{\partial \xi} - g' \frac{\partial F}{\partial \xi} \right] \tag{5.67}$$

$$\frac{E}{\text{Pr}} \theta'' + \gamma F \theta' - \text{MP}D\theta' = \xi \left[F' \frac{\partial \theta}{\partial \xi} - \theta' \frac{\partial F}{\partial \xi} \right] \tag{5.68}$$

The prime denotes differentiation with respect to η. The associated boundary conditions are

$$F'(\xi, 0) = 0, \qquad F(\xi, 0) = 0; \qquad g(\xi, 0) = 1 \quad \text{and} \quad \theta(\xi, 0) = 1$$
$$F'(\xi, \infty) = 1, \qquad g(\xi, \infty) = 0 \quad \text{and} \quad \theta(\xi, \infty) = 0 \tag{5.69a–g}$$

The coefficients in Equations (5.66)–(5.68) are defined as

$$C(\xi) = \frac{\xi}{r} \frac{dr}{d\xi} \text{BP} \left(\frac{r/L}{u_e/u_\infty} \right)^2, \qquad \text{where } \text{BP} = \left(\frac{L\Omega}{u_\infty} \right)^2$$

$$D(\xi) = \xi^{n/n+1} \left(\frac{u_e}{u_\infty} \right)^{1-2n/n+1}$$

$$E(\xi) = \left(\frac{u_e}{u_\infty}\right)^{3(1-n)/n+1} \xi^{n-1/n+1}$$

$$G(\xi) = \left(BP^{1/2}\frac{r/L}{u_e/u_\infty}\right)$$

$$H(\xi) = 2\frac{\xi}{r}\frac{dr}{d\xi}$$

$$N(\xi) = \frac{1}{n+1} + \frac{n-2}{n+1}\Lambda(\xi)$$

$$Re = \rho u_\infty^{2-n}\frac{L^n}{K}$$

$$S(\xi) = \frac{\xi \sin\phi}{(u_e/u_\infty)^2}$$

C Rotating Porous Sphere Application

For a sphere, $R(x) = R\sin(x/R)$ where $x/R = \phi = \xi$ and the sphere's radius, R, corresponds to the characteristic length, L. Typically, two options are available for the outer flow distribution:

$$\frac{u_e}{u_\infty} = \begin{cases} 3/2 \sin\xi & \text{from potential flow theory} \\ 1.5\xi - 0.4371\xi^3 + 0.1481\xi^5 - 0.0423\xi^7 & \text{empirical correlation} \end{cases} \quad (5.70\text{a,b})$$

With $r(x)$ and L identified, and choosing Equation (5.70a) for $u_e(x)/u_\infty$ in order to compare our predictions with accepted results from special case studies, the ζ-dependent coefficients in Equations (5.66)–(5.68) become

$$\Lambda(\xi) = \xi\cot\xi$$

$$\gamma(\xi) = \frac{1}{n+1} + \frac{3n}{n+1}\xi\cot\xi$$

$$C(\xi) = \frac{4}{9}\xi BP\cot\xi$$

$$D(\xi) = \xi^{n/n+1}\left(\frac{3}{2}\sin\xi\right)^{1-2/n+1}$$

$$E(\xi) = \xi^{n/n+1}\left(\frac{3}{2}\sin\xi\right)^{3(1-n)/n+1}$$

$$G(\xi) = \left(\frac{2}{3}BP^{1/2}\right)^{n-1}$$

$$H(\xi) = 2\xi\cot\xi$$

$$S(\xi) = \frac{4}{9}\frac{\xi}{\sin\xi}$$

$$N(\xi) = \frac{1}{n+1}\frac{n-2}{n+1}\xi \cot \xi$$

and

$$S(\xi) = \xi^{n+2/n+1}\frac{\sin \xi}{(3/2 \sin \xi)^{n+4/n+1}}$$

D System Solution

The system of transformed equations with the appropriate coefficients for-the rotating sphere was solved with the Keller box method (cf. App. F.4), a two-point finite difference method outlined in more detail by Cebeci and Bradshaw (1977). It is very convenient and efficient to use an implicit finite difference scheme with orthogonal body contour coordinates for the accurate and stable solution of thin-shear-layer problems. The two-dimensional grid is nonuniform in order to accommodate both the steep velocity and temperature gradients at the wall and the vicinity of the singular point, $\xi = 0$, which are regions of high mesh density. Numerical error testing has been accomplished by straightforward repeat calculations with finer meshes to check grid independence of the results and by local mesh refinement with smooth transitions to coarser regions.

Since measured data sets or analytical/numerical solutions are not available to check the accuracy of the present system, our computer simulation results are compared with those of Rajasekaran and Palekar (1985) and of Lien et al. (1986a) for Newtonian fluids ($n = 1$). Figures 5.21a and 5.21b demonstrate the excellent agreement between predicted results and their particular case studies.

Considering an impermeable sphere, the effects of rotation and fluid characteristics on the local SFG and HTG are depicted in Figs. 5.22 and 5.23. The rotating sphere of constant temperature is heated ($Z = 1$) and the buoyancy force is present ($\lambda_T = 2.0$). As can be concluded from the graphs (Fig. 5.22), body spin and shear-thickening fluids ($n > 1.0$) accelerate flow separation, generating steeper c_f-gradients and higher maxima when compared to stationary spheres and shear-thinning fluids, respectively. Figure 5.23 shows the unique HTG_T distribution for non-Newtonian fluids. As discussed earlier, an increase in angular velocity enhances heat transfer. The contrasting behavior of the two types of power-law fluids near the stagnation point can be explained as follows. For small ξ values, $u_e(\xi)/U_\infty$ is approximated and Equation (5.65) can be written as

$$HTG_T \sim \theta'(\xi, 0)\xi^{1-n/n+1}$$

which implies that for $\xi \to 0$

$$HTG\begin{cases} \to 0 & \text{for} < 1.0 \\ \to \infty & \text{for } n > 1.0 \end{cases}$$

In comparing the effects of the two distinct thermal boundary conditions, computer experiments indicate that for mixed convection past a rotating impermeable heated sphere, larger HTG values are achieved in the constant wall heat flux condition than in the isothermal wall case. In contrast, the constant surface temperature condition generates higher SFG values than in the constant wall heat flux case.

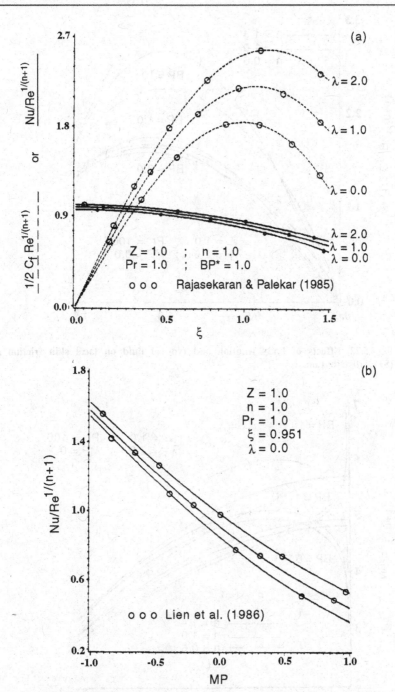

Fig. 5.21. (a) Data comparison for Newtonian fluid flow past a rotating isothermal sphere. (b) Data comparison for Newtonian fluid flow past a rotating sphere with constant wall heat flux.

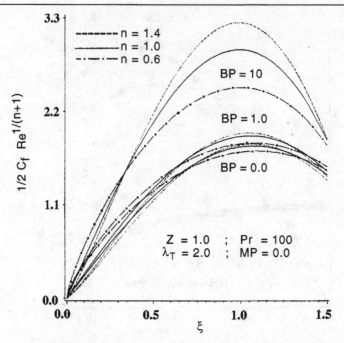

Fig. 5.22. Effects of body rotation and type of fluid on local skin friction group (SFG) distribution.

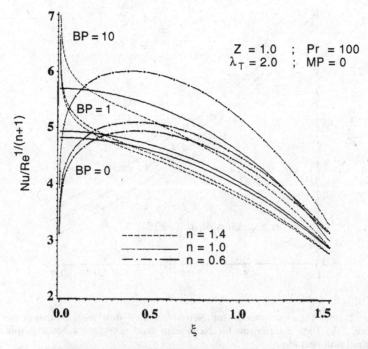

Fig. 5.23. Effects of body rotation and type of fluid on local heat transfer group for the isothermal wall case (HTG$_T$) distribution.

Concentrating on injection and suction of a shear-thinning fluid at constant wall temperature, the effects of wall mass transfer, buoyancy, rotation, and generalized Prandtl number are as follows. Injection increases and withdrawal reduces SFG (ξ) slightly. However, for pure forced convection, the trend is just the opposite. Thus, the buoyancy force, especially in aiding flow, may change the injection/suction effect on SFG.

Rotation enhances heat transfer, but so does steady withdrawal of fluid at wall temperature from the boundary layer. Clearly, fluid suction generates steeper temperature gradients near the sphere's surface and thus higher Nusselt numbers since Nu $\sim \theta'(\xi, 0)$. The fluid temperature near the surface of a sphere with blowing is higher than that with suction at larger angular positions, which produces steeper wall velocity gradients and hence higher c_f values. It can be shown that the buoyancy effect on the skin friction is stronger than that of directional mass transfer. A high Prandtl number for pseudoplastics ($n < 1.0$) in the presence of fluid withdrawal has a dramatic effect on the heat transfer group for reasons discussed previously.

To illustrate how the buoyancy force and the centrifugal force affect the velocity field inside the boundary layer, representative profiles at $\xi = 1.0$ are shown in Fig. 5.24. It is evident that for aiding flow ($Z = 1$), the velocity gradient at the wall increases as the buoyancy force or the centrifugal force increases. This may lead to a velocity overshoot beyond its local free stream value (cf. curve 6). For opposing flow ($Z = -1$), a relatively strong buoyancy force ($\lambda = 5.0$) directed downward retards the forced flow, which is assisted by a centrifical force (BP $= 1.0$) as shown in Figure 5.25. It can be deduced from Figure 5.25 that flow separation may occur for $\xi \geq 1.374$.

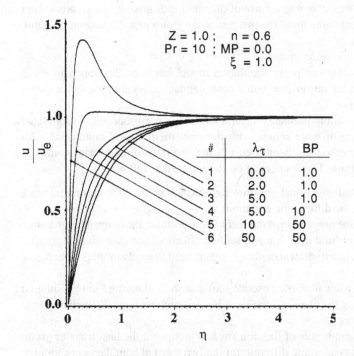

Fig. 5.24. Boundary-layer velocity profiles for rotating heated sphere.

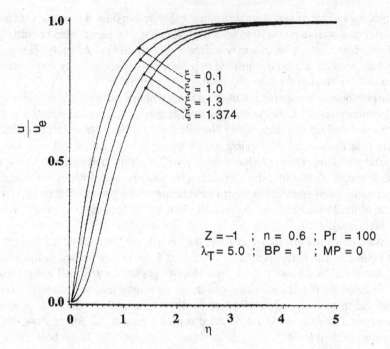

Fig. 5.25. Boundary-layer velocity profiles for rotating cooled sphere.

The constant wall heat flux case and the use of dilatant fluids give similar trends except for the significantly different behavior of the two power-law fluids near the stagnation point.

Conclusions and Design Considerations

A comprehensive and flexible computer simulation model has been developed to study mixed thermal convection of power-law fluids past standard axisymmetric bodies with generalized wall conditions.

The steady laminar nonsimilar boundary-layer analysis has been successfully validated and applied to a rotating permeable sphere with different thermal wall conditions. An implicit finite difference method has been used to solve the transformed set of nonlinear coupled differential equations. The results can be summarized as follows:

- The powerful coordinate transformation significantly reduces the numerical work required to solve the differential equations.
- Depending upon the magnitude of the surface temperature, the buoyancy force may reverse the roles of fluid injection and suction effects on the skin friction group.
- Body rotation, fluid withdrawal, and high generalized Prandtl numbers all enhance heat transfer.
- The type of power-law fluid (i.e., pseudoplastic or shear-thinning fluid vs. dilatant or shear-thickening fluid) affects the local Nusselt number most distinctively near the stagnation point.
- The magnitudes and trends of the skin friction group and the heat transfer group for rotating spheres are quite different for the two thermal boundary conditions.
- The portable computer code for this model is an accurate, flexible, and easy-to-

use tool for studying basic mixed convection phenomena in power-law fluids. Furthermore, the computer simulation model can be used for predicting fluid flow and heat transfer parameters applicable to chemical and mechanical systems design.

5.4   Convection Heat Transfer around Bodies with Moving Boundaries

The moving boundary problems of interest here are actually phase-change problems. In Section 5.4.1, a spherical (fuel) droplet evaporates in a hot gas stream, and in Section 5.4.2, a fixed cooled cylinder in cross-flow solidifies the surrounding fluid, forming nonconcentric ice layers. In both cases, the interfacial transfer processes, expressed in terms of mass, momentum, and energy conservation at the interface, are most important in the accurate description of the system dynamics. Specifically, in case of the droplet that is assumed to stay spherical at all times, the (unknown) vaporization flux $(\rho v)_i$ couples all transport equations and boundary conditions in the vaporizing droplet problem. In the case of nonuniform ice formation around a cylinder, the location of the solid–liquid interface, $R = R(\theta, t)$, is the (unknown) coupling variable. In both cases, the transient interface radius has to be found iteratively, starting with a guessed value for $R(t_i)$ or $R(\theta, t_i)$. The projects are extensions of the basic problems of *thermal flow past submerged bodies*. Thus, the fundamental problem solutions for ideal flows past a sphere (cf. Sect. 3.2.1 and Sect. 3.3) and past a horizontal cylinder (cf. Sect. 4.3) are starting points. The thermal boundary-layer analysis of Section 5.3.2 should be of particular interest. Additional sphere/cylinder solutions are summarized in White (1991) and Cebeci and Bradshaw (1988).

5.4.1   Fuel Droplet Vaporization in a Hot Gas Stream

A   System Identification

An accurate and flexible approximate analysis of vaporizing droplets is most desirable for the basic understanding and improvement of a variety of dispersed flow systems with phase change. Specifically, reliable two-phase flow predictions of single and multiple interacting (fuel) droplets in a hot, relatively high-Reynolds-number gas stream are useful in the development of more complex computer simulations, in laboratory investigations, and in design applications of particular spray processes. Laminar forced convection heat transfer for a single vaporizing droplet, possibly with internal droplet circulation, variable gas-side properties, and transient heat conduction in the liquid phase, has been a cornerstone in the assessment of spray systems for decades (cf. Kuo 1986 or Clift et al. 1978). The present boundary-layer analysis utilizes new transformations that yielded a very efficient, robust, and accurate numerical code for the simultaneous solution of momentum, heat, and mass transfer of a vaporizing droplet. Thus, the present study by Wang and Kleinstreuer (1989) can be readily extended to multiple interacting vaporizing droplets and closely spaced droplet streamers (cf. Kleinstreuer and Wang, 1990; Chiang and Kleinstreuer 1991; Chiang and Kleinstreuer 1992; Kleinstreuer et al. 1993).

The approximate analysis in this study is based on an implicit finite difference solution (cf. App. F.4) of the uniquely transformed boundary-layer equations for momentum, heat, and mass transfer in the gas phase and the transient heat conduction equation inside the

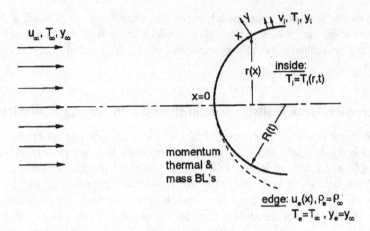

Fig. 5.26. System schematics and coordinates.

liquid droplet. The outer flow is represented by an empirical correlation rather than the simple potential flow solution used by others. The simultaneous solution of the coupled transport equations allows a local update of gas properties, of which the density changes are the most important. Specific assumptions made are justified and referenced at the end of the chapter.

We consider quasi-steady laminar compressible flow of constant free-stream velocity past a liquid vaporizing droplet. The system schematics and orthogonal coordinates measured from the forward stagnation point are shown in Fig. 5.26. The quasi-steady gas phase assumption is based on the fact that the characteristic time for changes in the gas phase ($t_{res} \approx 2\mu s$) is much smaller than the droplet lifetime ($t_{life} \approx 50 \mu$ s). The gas-side boundary layer is assumed to be laminar because a typical droplet size (i.e., $d_p = O(10~\mu m)$), is much smaller than a typical large eddy, which may be of the Kolmogorov scale (i.e., $l_K \approx 200~\mu m$). Radiation effects and mass diffusion due to temperature and pressure gradients have been neglected. Internal droplet circulation, if not a priori present, is insignificant. Clift et al. (1978) argued that omnipresent fuel droplet contaminants damp out liquid-phase motion, and Prakash and Krishan (1984) showed that internal circulation does not affect the droplet lifetime measurably. The effect of near-wake heat transfer, which may contribute up to 7 percent for solitary particles, has not been considered here (cf. Ramachandran et al. 1989).

B   System Modeling

For a computationally effective solution, the gas-phase boundary-layer equations have been reduced by using new transformation parameters based on scale analysis (cf. Sects. 2.3.1 and 4.1.3). The unsteady liquid-phase equation is coupled to the boundary-layer equation via mass and energy conservation conditions at the interface of the shrinking spherical droplet.

Neglecting curvature effects and assuming axisymmetric flow, the governing equations in a surface-attached inertial coordinate frame are (cf. Fig. 5.26)

$$\text{(continuity)} \qquad \frac{\partial(r\rho u)}{\partial x} + \frac{\partial(r\rho v)}{\partial y} = 0 \qquad\qquad (5.71)$$

$$\text{(momentum)} \quad \rho\left(u\frac{\partial u}{\partial x} + v\frac{\partial u}{\partial y}\right) = \rho_e u_e \frac{du_e}{dx} + \frac{\partial}{\partial y}\left(\mu\frac{\partial u}{\partial y}\right) \tag{5.72}$$

$$\text{(energy)} \quad \rho\left(u\frac{\partial h}{\partial x} + v\frac{\partial h}{\partial y}\right) = \frac{\partial}{\partial y}\left(k\frac{\partial T}{\partial y}\right) \tag{5.73}$$

$$\text{(species)} \quad \rho\left(u\frac{\partial Y}{\partial x} + v\frac{\partial Y}{\partial y}\right) = \frac{\partial}{\partial y}\left(\rho D\frac{\partial Y}{\partial y}\right) \tag{5.74}$$

where the subscript e denotes boundary-layer edge and the subscript i indicates interfacial conditions, $h = c_p T$ is the enthalpy of the gaseous mixture, Y is the vapor mass fraction, and D is the vapor diffusivity.

The associated boundary conditions are

$$\text{at } y = 0 \quad u = 0, \quad v = v_i, \quad T = T_i, \quad Y = Y_i \tag{5.75a}$$

and

$$\text{at } y \to \infty \quad u = u_e(x), \quad T = T_\infty, \quad Y = Y_\infty = 0 \tag{5.75b}$$

where the "blowing" velocity v_i includes the shrinking droplet effect, dR/dt. The distance from the axis of symmetry to the droplet surface is

$$r(x) = R\sin\theta = R\sin\left(\frac{x}{R}\right) \tag{5.76}$$

The edge velocity distribution has been experimentally determined (White 1991) as

$$\frac{u_e}{u_\infty} = 1.5\left(\frac{x}{R}\right) - 0.4371\left(\frac{x}{R}\right)^3 + 0.1481\left(\frac{x}{R}\right)^5 - 0.0423\left(\frac{x}{R}\right)^7 \tag{5.77}$$

With the stream function approach

$$\rho r u = \frac{\partial \psi}{\partial y} \quad \text{and} \quad \rho r v = -\frac{\partial \psi}{\partial x} \tag{5.78a,b}$$

and the transformations

$$\xi = \frac{x}{R}(t) \tag{5.79}$$

$$\eta = \left(\frac{\text{Re}}{\xi}\right)^{1/2}\left(\frac{u_e}{u_\infty}\right)^{1/2}\left(\frac{\rho}{\rho_\infty}\right)\frac{y}{R(t)} \tag{5.80}$$

$$\psi = r R \rho_\infty u_\infty \left(\frac{\xi}{\text{Re}}\right)^{1/2}\left(\frac{u_e}{u_\infty}\right)^\infty F(\xi, \eta) - \int_0^x (\rho r v)_i \, dx \tag{5.81}$$

$$\theta = \frac{T - T_\infty}{T_i - T_\infty} \tag{5.82}$$

$$S = \frac{Y}{Y_i} \tag{5.83}$$

Equation (5.71) is automatically satisfied and Equations (5.72)–(5.74) become

$$F''' + \gamma(\xi)FF'' + \Lambda(\xi)\left(\frac{\rho_\infty}{\rho} - F'^2\right) - MPD(\xi)F'' = \xi\left[F'\frac{\partial F'}{\partial \xi} - F''\frac{\partial F}{\partial \xi}\right] \tag{5.84}$$

$$\theta'' + \gamma(\xi)F\theta' - MPD(\xi)\theta' = \xi\left[F'\frac{\partial \theta}{\partial \xi} - \theta'\frac{\partial F}{\partial \xi}\right] \tag{5.85}$$

and

$$S'' + \gamma(\xi)FS' - MPD(\xi)S' = \xi\left[F'\frac{\partial S}{\partial \xi} - S'\frac{\partial F}{\partial \xi}\right] \tag{5.86}$$

A two-way coupling between these nonsimilar boundary-layer equations is most noticeably given via the mass transfer parameter, MP, which is defined as

$$MP = \frac{\rho_i v_i}{\rho_\infty u_\infty}Re^{1/2} \tag{5.87a}$$

where $\quad Re = \dfrac{u_\infty R}{v_\infty} \tag{5.87b}$

and $v_i = v_i[Y_i(T_i), R(t)]$ as shown later. The coefficients in Equations (5.84)–(5.86) are defined as

$$\gamma(\xi) = \frac{1}{2} + \frac{1}{2}\Lambda(\xi) + \frac{\xi}{r}\frac{dr}{d\xi} \tag{5.88a}$$

$$\Lambda(\xi) = \frac{\xi}{u_e}\frac{du_e}{d\xi} \tag{5.88b}$$

and

$$D(\xi) = \xi^{1/2}\left(\frac{u_\infty}{u_e}\right)^{1/2} \tag{5.88c}$$

The transformed boundary conditions are

$$F'(\xi, 0) = 0, \quad F(\xi, 0) = 0, \quad \theta(\xi, 0) = 1, \quad S(\xi, 0) = 1 \tag{5.89a}$$

and

$$F'(\xi, \infty) = 1, \quad \theta(\xi, \infty) = 0, \quad S(\xi, \infty) = 0 \tag{5.89b}$$

Since the pressure across the boundary layer is constant and assuming that the average molecular weight of the gas mixture (i.e., ideal gas) is constant, the density ratio inside the boundary layer can be related to the local temperature ratio, via

$$\frac{\rho_\infty}{\rho} = \frac{T}{T_\infty} \tag{5.90}$$

The mass fraction at the interface is a function of the drop surface temperature following the Clausius–Clapeyron equation.

$$Y_i = \exp\left[\frac{1}{\lambda}\left(\frac{1}{T_b} - \frac{1}{T_i}\right)\right] \tag{5.91}$$

where L is the latent heat of vaporization, λ is the gas constant, T_b is the boiling point temperature, and T_i is the interfacial temperature.

Considering transient one-dimensional heat conduction in a constant-property liquid, the conservation equation in spherical coordinates reads

$$\frac{\partial T_l}{\partial t} = \frac{\alpha_l}{r^2} \frac{\partial}{\partial r}\left(r^2 \frac{\partial T_l}{\partial r}\right) \tag{5.92}$$

where α_l is the thermal diffusivity of the droplet, which initially is at the temperature T_0. Hence, the initial condition is

$$T_l(t=0) = T_0 \qquad \text{for all } r \tag{5.93a}$$

and the boundary conditions for $t > 0$ are

$$\text{at } r = 0 \qquad \frac{\partial T_l}{\partial r} = 0 \tag{5.93b}$$

and

$$\text{at } r = R(t) \qquad \frac{\partial T_l}{\partial r} = \hat{C} \sim q_i \tag{5.93c}$$

Here $\hat{C}$ is proportional to the interfacial heat flux and is therefore coupled to the gas-phase solution.

With the transformed coordinates

$$\xi = \frac{r}{R(t)}, \qquad \tau = \frac{\alpha_l t}{R_0^2} \quad \text{and} \quad \phi = \frac{T_l - T_0}{T_\infty - T_0} \tag{5.94--5.96}$$

Equation (5.92) and conditions (5.93a–c) read

$$\left(\frac{R(t)}{R_0}\right)^2 \frac{\partial \phi}{\partial \tau} = \frac{\partial^2 \phi}{\partial \zeta^2} + \frac{2}{\zeta} \frac{\partial \phi}{\partial \zeta} \tag{5.97}$$

$$\text{where} \qquad \phi(\tau = 0) = 0 \qquad \text{for all } \xi \tag{5.98a}$$

and for $\tau > 0$

$$\left.\frac{\partial \phi}{\partial \zeta}\right|_{\zeta=0} = 0 \quad \text{and} \quad \left.\frac{\partial \phi}{\partial \zeta}\right|_{\zeta=1} = C \tag{5.98b,c}$$

At the interface, convective and diffusive (fuel) mass fluxes as well as heat fluxes have to be conserved. Thus,

$$(\rho v)_i (Y_i - 1) = \left(\rho D \frac{\partial Y}{\partial y}\right)_i \tag{5.99}$$

and

$$(\rho v)_i A L + \left(k_l A \frac{\partial T_l}{\partial r}\right)_i = \left(k_g A \frac{\partial T}{\partial y}\right)_i \tag{5.100}$$

Since $(\partial T/\partial y)_i$ and $(\rho v)_i$ are functions of the streamwise position, Equations (5.99) and (5.100) are combined and written as

$$\left(k_l \frac{\partial T}{\partial y}\right)_i = \frac{1}{A_t}\left[\int_0^{\theta_s}\left(k_g \frac{\partial T}{\partial y}\right)_i 2\pi R \sin\theta\, d\theta - \int_0^{\theta_s}(\rho v)_i L 2\pi R^2 \sin\theta\, d\theta\right] \qquad (5.101)$$

where θ_s is the flow separation angle and the heat transfer area A_t is

$$A_t = \int_0^{\theta_s} 2\pi R_-^2 \sin\theta\, d\theta \qquad (5.102)$$

The interface mass flux, which is proportional to the blowing velocity, reads in dimensionless form

$$(\rho v)_i = \frac{\rho_\infty D_\infty}{R} \frac{Y_i}{Y_i - 1} S'(\xi, 0)\left(\frac{\mathrm{Re}}{\xi}\right)^{1/2}\left(\frac{u_e}{u_\infty}\right)^{1/2} \qquad (5.103)$$

Now, the coefficient $MPD(\xi)$ in Equations (5.84)–(5.86) can be written as

$$MPD(\xi) = \frac{Y_i}{Y_i - 1} \frac{S'(\xi, 0)}{\mathrm{Sc}} \qquad (5.104)$$

where $\mathrm{Sc} = \nu_\infty/D_\infty$ is the Schmidt number.

The constant in Equation (5.98c) can be expressed as

$$C = \left.\frac{\partial\phi}{\partial\zeta}\right|_{\zeta=1} = \frac{T_i - T_\infty}{T_\infty - T_0}\frac{k_g}{k_l}\theta'(\xi, 0)\left(\frac{\mathrm{Re}}{\xi}\right)^{1/2}\left(\frac{u_e}{u_\infty}\right)^{1/2}\frac{\rho_i}{\rho_\infty}$$

$$-\frac{L\rho_\infty D_\infty}{k_l(T_\infty - T_0)}\frac{Y_i}{Y_i - 1}S'(\xi, 0)\left(\frac{\mathrm{Re}}{\xi}\right)^{1/2}\left(\frac{u_e}{u_\infty}\right)^{1/2} \qquad (5.105)$$

The overall vaporization rate at any instant is calculated by integrating the vaporization flux $(\rho v)_i$ over the droplet surface. Thus, the time rate of change of the droplet radius can be expressed as

$$\frac{d}{dr}R(\tau) = \frac{1}{2\rho_l}\int_0^{\theta_s}(\rho v)_i \sin\theta\, d\theta \qquad (5.106)$$

With the definition of the local Nusselt number

$$\mathrm{Nu} = \frac{hR(t)}{k} \qquad (5.107)$$

we obtain

$$\text{Nu}(\xi, \tau) = -\left[\frac{\text{Re}(\tau)}{\xi}\right]^{1/2} \left(\frac{u_e}{u_\infty}\right)^{1/2} \frac{\rho_i}{\rho_\infty} \theta'(\xi, \tau, \eta = 0) \tag{5.108}$$

with which

$$\overline{\text{Nu}}(\tau) = \frac{1}{A_t} \int_0^{\theta_s} \text{Nu}(\xi, \tau) 2\pi R^2 \sin\theta \, d\theta \tag{5.109}$$

As mentioned earlier, the error induced by neglecting near-wake effects is not significant for single droplets.

C System Solution

The system of gas-phase boundary-layer equations is solved by using Keller's box method (cf. App. F.4 and Cebeci and Bradshaw 1974), which has been extended for this study to solve momentum, heat, and mass transfer simultaneously. On the other hand, the Crank–Nicolson algorithm has been employed to solve transient heating in the liquid phase. Figure 5.27 summarizes the computational steps executed. Specifically, Equations (5.84)–(5.86) with associated boundary conditions (5.89a) and (5.89b) are solved at any given time with known surface temperature. Thus, the liquid-side, dimensionless temperature gradient can be evaluated from Equation (5.105), where Equation (5.101) is solved using Simpson's rule. Solving Equation (5.97) in conjunction with (5.98a–c) yields the interior droplet temperature distribution with time. Then the new droplet surface temperature is being used in the solution of the gas-phase equations until, iteratively, convergence has been achieved, that is, $|[(T_i)_{\text{new}} - (T_i)_{\text{old}}]/(T_i)_{\text{new}}| \leq 10^{-4}$. The new interface temperature and the new droplet radius are then used for the next time step, and so on.

A nonuniform mesh in the η and ξ directions has been designed where the highest grid density is near the stagnation point ($\xi = 0$) along the (unknown) boundary-layer edge ($\eta = \eta_\infty$) and at the interface ($\eta = 0$). The parabolic code predicts the point of flow separation very accurately as demonstrated in Wang and Kleinstreuer (1988). In choosing $R(t)$ as the characteristic length in both phases, mesh renewal for the transient process is not required. Thus, except for the changing boundary-layer thickness, the computational domain remains the same during the droplet lifetime (cf. Equation (5.79) and (5.80)). The location of the boundary-layer edge, η_∞, is strongly dependent upon the magnitude of the blowing velocity, v_i. Under worst-case conditions (i.e., maximum boundary-layer thickness and smallest droplet radius considered), $\delta/R \approx 0.009$, which is within Prandtl's boundary-layer assumption. Numerical error testing has been accomplished by straightforward repeat calculations with finer meshes to test grid independence of the results and by local mesh refinements with smooth transitions to coarser regions.

Computer simulation studies were carried out for three different fuels (cf. Table 5.3). The initial droplet temperature in all cases is $T_0 = 300$ K and the gas-phase characteristics include Pr = 1, Sc = 1, and $T_\infty = 1,000$ K.

Results and Discussion
The predictive capabilities of the present computer simulation model have been successfully verified with special case studies of laminar mixed convection heat transfer in a sphere

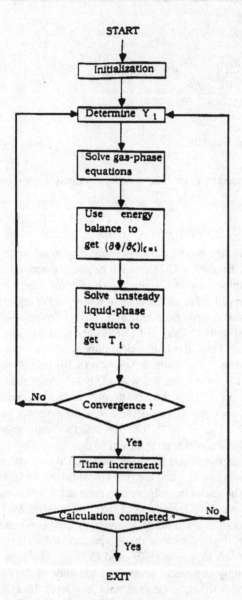

Fig. 5.27. Computational flowchart.

with blowing (cf. Kleinstreuer and Wang 1988). Figure 5.28 shows a comparison for fuel droplet shrinking between the present model and Spalding's "$d^{3/2}$ law." Spalding's correlation

$$\frac{\dot{m}}{4\pi \mu_l R} = 0.265 B^{3/5} \, \mathrm{Re}_d^{1/2} \tag{5.110a}$$

Table 5.3. *Property values*

Property[a]	Fuel		
	n-Hexane	*n*-Decane	*n*-Hexadecane
Molecular weight (g/g · mole)	86.178	142.286	226.338
Boiling point at 10 atm (K)	493	565	693
Density (g/cm³)	0.659	0.73	0.773
Heat of vaporization (cal/g)	80.02	65.98	54.05
Thermal diffusivity (cm²/s)	919×10^{-4}	7.2×10^{-4}	5.8×10^{-4}
Diffusion coefficient (cm²/s)	0.093	0.1155	0.133

[a]Selective properties evaluated at $T_{\text{ref}} = 0.5(T_0 + T_b)$.

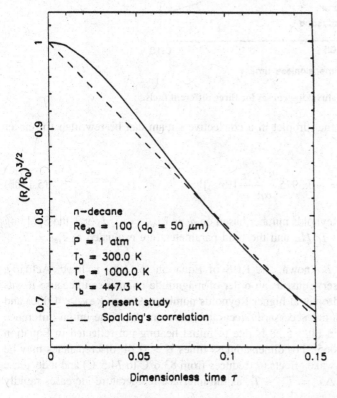

n−decane
$Re_{d0} = 100$ ($d_0 = 50 \ \mu m$)
$P = 1$ atm
$T_0 = 300.0$ K
$T_\infty = 1000.0$ K
$T_b = 447.3$ K
—— present study
— — Spalding's correlation

Dimensionless time τ

Fig. 5.28. Comparison of present results for $R(\tau)$ with experimental correlation.

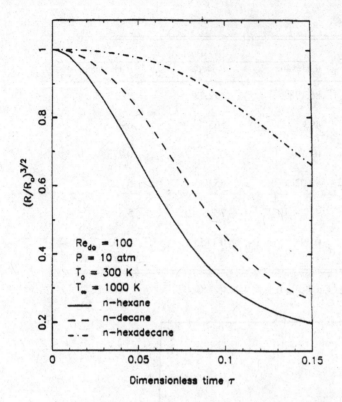

Fig. 5.29. Droplet shrinking curves for three different fuels.

which holds for a liquid fuel droplet in a convective stream can be rewritten for easier plotting as

$$\frac{d}{d\tau}\left[\left(\frac{R}{R_0}\right)^{3/2}\right] = -0.3975\frac{B^{3/5}}{\alpha_l}\frac{\mu_l}{\rho_l}[\text{Re}_{d0}]^{1.2} \tag{5.110b}$$

where the initial droplet Reynolds number based on d_0 is $\text{Re}_{d0} = u_\infty d_0/\nu_\infty$, the Spalding number is $B = c_p(T_\infty - T_i)/L$, and the fluid parameters are evaluated at $T_{\text{ref}} = T_i + 1/3(T_\infty - T_i)$.

With Re_{d0} given and B known, the RHS of Equation (5.110b) is fixed. Actually, Spalding's formula may serve just as an order-of-magnitude comparison because it was developed for a burning droplet at higher Reynolds numbers ($900 \geq \text{Re}_{d0} \geq 4,000$) and with fluid property values not necessarily accurately given at T_{ref}. The initial difference between the two graphs in Fig. 5.28 is due to initial heating not reflected in Equation (5.106b) for $0 \geq \tau \geq 0.065$. For dimensionless times $\tau > 0.14$, discrepancies may be caused by enhanced near-wake effects (θ_s reduces from 83.6°C to 74.5°C) and a decrease in thermal driving force, $\Delta T = T_\infty - T_i$, as the interface temperature increases rapidly with time.

The time-dependent changes in droplet radius for different fuels are given in Fig. 5.29, where n-hexane is the light fuel and n-hexadecane is the heavy fuel. Figure 5.30 shows the inter-facial velocity, which is at times large enough to change the (momentum) boundary-

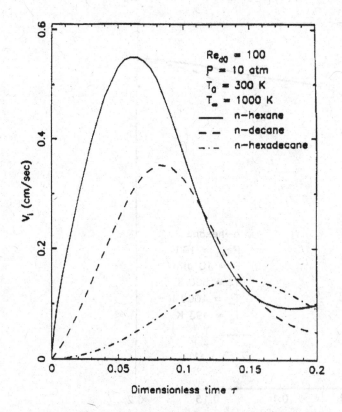

Fig. 5.30. Interfacial or blowing velocities for different fuels.

layer characteristics (e.g., the local drag coefficient, the point of flow separation, and the boundary-layer thickness) significantly. Consequently, the mass transfer parameter $MP(v_i)$ strongly influences heat transfer and gas-side species concentration of the fuel droplet. With an n-hexane fuel droplet as an example, Fig. 5.31 depicts the droplet center temperature and surface temperature with time. As expected, the interface temperature increases very quickly, requiring very small time steps in the calculation procedure for $0 \geq \tau \geq 0.03$. In contrast, the droplet center is unaffected during this initial period until thermal diffusion causes elevation of T_{center} close to $T_{interface}$ for $\tau > 0.15$.

Figure 5.32, depicting the transient, spatially averaged Nusselt number for an n-hexane droplet exposed to different free stream Reynolds numbers, confirms this explanation. However, for $\tau > 0.05$, the effects of the initial droplet Reynolds number, Re_{d0}, on the vapor boundary layer reverse the trend for $\overline{Nu}(\tau)$, causing lower average Nusselt numbers for higher free-stream Reynolds numbers. Volatile droplets (e.g., n-hexane) reduce in size with larger Reynolds numbers more rapidly (cf. Fig. 5.29) because of increasing mass fluxes at the interface (cf. Fig. 5.28), which in turn generate milder temperature gradients at the gas–liquid surface. This phenomenon is reflected in Equation (5.30), where for $\tau \leq 0.05$ the first term on the RHS decreases and the second term increases with higher approach velocities and as a result, $C \sim Nu$. Figure 5.33 shows the transient, spatially averaged Nusselt number for three different fuels. As expected, the "heavier" fuels (e.g., n-hexadecane) exhibit higher Nusselt numbers throughout the droplet lifetime because of steeper interfacial temperature gradients.

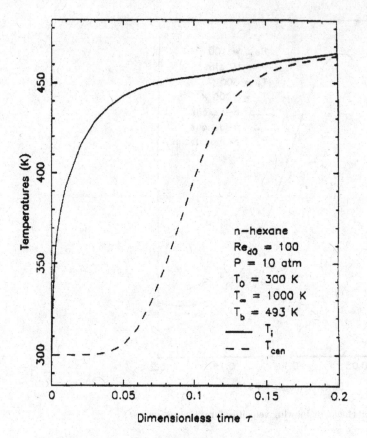

Fig. 5.31. Transient droplet temperatures for *n*-hexane.

Conclusions and Design Considerations

A realistic, computationally efficient analysis of vaporizing droplets in a hot gas stream, or volatile droplets moving through stagnant air, is important for the basic understanding of the process dynamics and for improved design applications to numerous dispersed flow systems including combustion sprays. By concentrating in this section on a single fuel droplet, the gas-side momentum, heat, and mass boundary-layer equations have been transformed and numerically solved together with the unsteady heat conduction equation for the liquid side. Assuming spherically shrinking droplets, the transport equations are coupled via the transient interfacial mass flux and the variable gas density. The new approximate solution to this two-phase convection heat transfer problem is more flexible, accurate, and efficient than previous studies. It may serve as a submodel in broader spray combustion simulations.

The strongly coupled nonlinear transport phenomena, depicted in terms of droplet radius, temperature, blowing velocity, and average Nusselt number, are highly transient during the droplet lifetime for all (three) fuels and initial Reynolds numbers considered. From comparisons with measured data sets, it appears that, indeed, internal droplet circulation and near-wake effects can be neglected for solitary droplets.

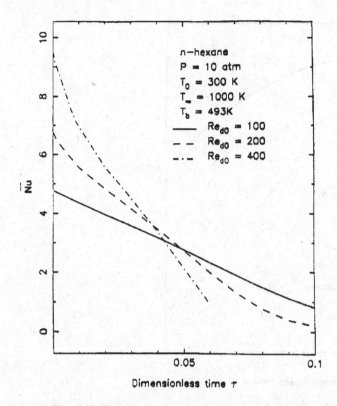

Fig. 5.32. Transient average Nusselt number (n-hexane) for different initial Reynolds numbers.

5.4.2 Ice Formation on a Cylinder in Cross-Flow

A   System Identification

Liquid–solid phase change problems with moving boundaries have numerous applications. Examples include ice cover formation, metal casting, crystal growing, frost layer formation, food stuff freezing, heat shield ablation, ice melting, latent heat storage, and soil freezing or thawing. One common characteristic of these Stefan-type problems is that phase change and an associated source or sink of latent heat occur at the (unknown) moving interface. Furthermore, conductive heat fluxes governed by the Stefan number (Ste $= c_{solid}(T_{fluid} - T_{wall})/L$: i.e., ratio of sensible heat to latent heat) and the property value ratios of the phase change material are important in all of these processes. However, subcooling of the solid, possible density abnormalities of the liquid, and natural convection may be dominant in the melting process, whereas outward solidification in a liquid stream past a cooled wall may be better determined by mechanisms of forced convection heat transfer and/or by the characteristics of the superheated fluid. For example, Lunardini (1980) analyzed freezing and thawing of concentric regions around a cylinder assuming transient thermal diffusion only. On the basis of a rather simple numerical solution, employing the heat balance integral method, phase change locations and surface heat transfer rates are discussed for different system parameters. Solomon et al. (1984) replaced the transient conduction heat transfer equation for the Stefan problem with the telegrapher's equation.

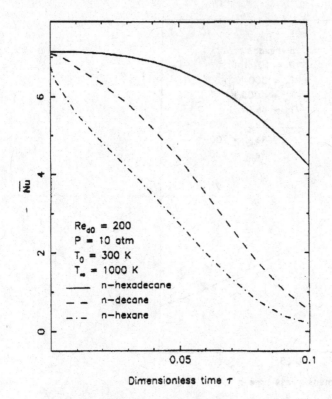

Fig. 5.33. Transient Nusselt numbers averaged over attached boundary-layer surface of different fuel droplets.

They argued that Fourier's law should be more realistically replaced by $\vec{q}(\vec{x}, t + \tau) - k\nabla T$, where τ is called the response or relaxation time of the phase change material. Additional references are given later.

Concentrating on the freezing process of water at low Stefan numbers around a cylinder in cross-flow, the complete transport equations for laminar two-dimensional flow are transformed and solved numerically using an ADI finite difference scheme (cf. Anderson et al. 1984). Specifically, it is of interest to analyze forced convection heat transfer along the moving interface and to compute the growth rate of solidifying material for different system parameters. Because of the lack of experimental data sets for the complete system, our predictive results from special case studies are compared with measurements of single-phase as well as two-phase conduction-dominated and convection-controlled heat transfer processes (cf. Chiang and Kleinstreuer 1989).

B System Modeling

Consider a cylinder of diameter D in uniform cross-flow with an approach velocity U of a phase-change fluid of temperature T_0. The cylinder's wall temperature is suddenly reduced below the fusion, that is, freezing, temperature of the liquid: $T_w < T_f = 0°C$. Thus a solid layer is formed around the cylinder, changing the immediate velocity field. While the cylinder surface remains at the constant temperature T_w, the local velocity field and temperature distributions in the solid and liquid are changing with time t, coupled via the unknown interface location $R = R(\theta, t)$, which stays at $T = T_f$. Figure 5.34 depicts the problem of interest and the coordinate system used. Assuming two-dimensional laminar flow of a Newtonian fluid with constant properties, the differences in volume associated with phase change are neglected as well as cylinder end effects, natural convection, and vortex shedding.

Employing the stream-function/vorticity approach, the transient Navier–Stokes equations for two-dimensional flow in polar coordinates are written as (cf. App. C or D)

$$\frac{\partial \tilde{\omega}}{\partial t} + \frac{1}{r}\left[\frac{\partial}{\partial \theta}\left(\tilde{\omega}\frac{\partial \tilde{\psi}}{\partial r}\right) - \frac{\partial}{\partial r}\left(\frac{\tilde{\psi}}{\partial \theta}\right)\right] = \nu \nabla^2 \tilde{\omega} \tag{5.111}$$

and

$$\tilde{\nabla}^2 \tilde{\psi} = -\tilde{\omega} \tag{5.112}$$

where $\tilde{\omega}$ is the (dimensional) vorticity, $\tilde{\psi}$ is the stream function, and ν is the kinematic viscosity. The associated definitions are

$$\frac{\partial \tilde{\psi}}{\partial \theta} = -r\tilde{v}, \qquad \frac{\partial \tilde{\psi}}{\partial r} = \tilde{u}, \qquad \tilde{\nabla}^2 = \frac{\partial^2}{\partial r^2} + \frac{1}{r}\frac{\partial}{\partial r} + \frac{1}{r^2}\frac{\partial^2}{\partial \theta^2}$$

and

$$\tilde{\omega} = \frac{1}{r}\left[\frac{\partial}{\partial r}(r\tilde{v}) - \frac{\partial \tilde{u}}{\partial \theta}\right] \tag{5.113a–d}$$

The associated boundary conditions are for $t \leq 0$

(symmetry) $\quad \tilde{\omega} = \tilde{\psi} = 0 \quad$ at $\theta = 0$ and $\pi = 0 \tag{5.114a}$

(no-slip) $\quad \tilde{\psi} = \dfrac{\partial \tilde{\psi}}{\partial r} = 0 \quad$ for $r = R(\theta, t) \tag{5.114b,c}$

(potential flow) $\quad \tilde{\psi} = -rU\sin\theta \quad$ and $\quad \tilde{\omega} = 0 \quad$ as $r \to \infty \tag{5.114d,e}$

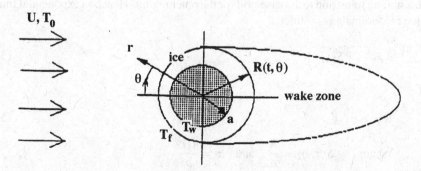

Fig. 5.34. Schematics for outward solidification around a subcooled cylinder.

The temperature field in the liquid phase is described by the transient convection–conduction equation

$$\frac{\partial \tilde{T}}{\partial t} + \frac{1}{r}\left[\frac{\partial}{\partial \theta}\left(\tilde{T}\frac{\partial \tilde{\psi}}{\partial r}\right) - \frac{\partial}{\partial r}\left(\tilde{T}\frac{\partial \tilde{\psi}}{\partial \theta}\right)\right] = \alpha_l \tilde{\nabla}^2 \tilde{T} \tag{5.115}$$

where $\tilde{T}$ is the fluid temperature and α_l is the thermal diffusivity of the liquid.

$$\text{At } t < t_0, \quad \tilde{T} = \tilde{T}_0 \quad \text{and} \quad \text{at } t \geq t_0 \quad \tilde{T} = \tilde{T}_f \quad \text{at } r = R(\theta, t) \tag{5.116a,b}$$

$$\text{As } r \to \infty, \quad \tilde{T} = \tilde{T}_0 \quad \text{and} \quad \frac{\partial \tilde{T}}{\partial \theta} = 0 \quad \text{at } \theta = 0 \quad \text{and} \quad \pi \tag{5.116c,d}$$

For the solid region, the transient heat conduction equation is appropriate, namely,

$$\frac{\partial \tilde{T}_s}{\partial t} = \alpha_s \tilde{\nabla}^2 \tilde{T}_s \tag{5.117}$$

subject to

$$\tilde{T}_S = \tilde{T}_f \quad \text{at } r = R(\theta, t) \quad \text{and} \quad \tilde{T}_S = \tilde{T}_w \quad \text{for } r = a \tag{5.118a,b}$$

Symmetry requires that

$$\frac{\partial \tilde{T}}{\partial \theta} = 0 \quad \text{at } \theta = 0 \quad \text{and} \quad \pi \tag{5.118c}$$

A heat balance at the solid–liquid interface $R(\theta, t)$ yields

$$\rho_s L \frac{dR}{dt} = \left[1 + \left(\frac{1}{R}\frac{\partial R}{\partial \theta}\right)^2\right]\left(k_s \frac{d\tilde{T}_s}{dr} - k_l \frac{d\tilde{T}}{dr}\right) \tag{5.119}$$

where subscript s indicates solid, L is the latent heat, and k is the conductivity.

In order to facilitate the numerical work, the irregular solidification front in the physical domain is transformed to a rectangular shape in the computational domain using body-fitted coordinates. Considering first the liquid region, a fine grid is required near the interface with a smooth transition to a coarse grid for the outer region. Hence an exponential function for the r-coordinate is postulated

$$\bar{r} \cong \frac{r}{a} = S(\tau, \eta)e^{\pi\xi} \tag{5.120}$$

and

$$\theta = \pi\eta \tag{5.121}$$

$$\text{where} \quad S(\tau, \eta) = \frac{R}{a} \quad \text{and} \quad \tau = \frac{Ut}{a} \tag{5.122a,b}$$

are the dimensionless interface location and the convective time, respectively.

Nondimensionalization of Equations (5.111), (5.112), and (5.115) yields

$$g(S, \xi)\left(\frac{\partial \omega}{\partial \tau} - \frac{1}{\pi S}\frac{\partial S}{\partial \tau}\frac{\partial \omega}{\partial \xi}\right) + \left(\frac{\partial \psi}{\partial \xi}\frac{\partial \omega}{\partial \eta} - \frac{\partial \psi}{\partial \eta}\frac{\partial \omega}{\partial \xi}\right) = \frac{2}{\mathrm{Re}}\nabla^2 \omega \tag{5.123}$$

$$-\nabla^2\psi = \omega \tag{5.124}$$

and

$$g(S, \xi)\left(\frac{\partial T}{\partial \tau} - \frac{1}{\pi S}\frac{\partial S}{\partial \tau}\frac{\partial T}{\partial \xi}\right) + \left(\frac{\partial \psi}{\partial \xi}\frac{\partial T}{\partial \eta} - \frac{\partial \psi}{\partial \eta}\frac{\partial T}{\partial \xi}\right) = \frac{2}{\mathrm{Re}\,\mathrm{Pr}}\nabla^2 T \tag{5.125}$$

where $\quad g(S, \xi) = \pi^2 S^2 e^{2\pi\xi}, \qquad \omega = \frac{\tilde{\omega}a}{U}, \qquad \psi = \frac{\tilde{\psi}}{(Ua)} \tag{5.126a–c}$

$$T = (\tilde{T} - \tilde{T}_w)/(\tilde{T}_f - \tilde{T}_w), \qquad \mathrm{Re} = \frac{UD}{\nu}, \qquad \mathrm{Pr} = \nu/\alpha \tag{5.126d}$$

and

$$\nabla^2 = \left(1 + \left(\frac{1}{\pi S}\frac{\partial S}{\partial \eta}\right)^2\right)\frac{\partial^2}{\partial \xi^2} - \frac{2}{\pi S}\left(\frac{\partial S}{\partial \eta}\right)\frac{\partial^2}{\partial \xi \partial \eta}$$

$$+ \frac{\partial^2}{\partial \eta^2} + \frac{1}{\pi S}\left(\frac{1}{S}\left(\frac{\partial S}{\partial \eta}\right)^2 - \frac{\partial^2 S}{\partial \eta^2}\right)\frac{\partial}{\partial \xi} \tag{5.126e}$$

The boundary conditions (5.114) and (5.116) become

$$\psi = \frac{\partial \psi}{\partial \xi} = 0 \quad \text{and} \quad -\nabla^2\psi|_{r=R} = \omega \qquad \text{for } r = R$$
$$\omega = 0 \quad \text{and} \quad \psi = 2\cosh(\pi\xi)\sin(\pi\eta) \qquad \text{at infinity} \tag{5.127}$$
$$\tau \geq \tau_0 \qquad \psi = \omega = 0 \qquad \text{on } \eta = 0, 1$$

and

$$\tau < \tau_0 \qquad T = T_0$$
$$\tau \geq \tau_0 \qquad T = 1 \qquad \text{for } \xi = 0$$
$$T = T_0 \qquad \text{at infinity} \tag{5.128}$$
$$\frac{\partial T}{\partial \eta} = 0 \qquad \text{on } \eta = 0, 1$$

For the solid region, the transformations reported by Ho and Chen (1986) have been used

$$\zeta = \frac{\bar{r} - 1}{S(\mathrm{Fo}, \eta) - 1} \qquad \text{and} \qquad \theta = \pi\eta \tag{5.129a,b}$$

While $S(\mathrm{Fo}, \eta)$ stays the same, a new (diffusive) time for the solid region, the Fourier number, is introduced

$$\mathrm{Fo} \equiv \frac{\alpha t}{a^2} := \frac{2}{\mathrm{Re}\,\mathrm{Pr}}\left(\frac{\alpha_s}{\alpha_l}\right)\tau \tag{5.129c}$$

Equation (5.117) now reads

$$\frac{\partial T}{d\text{Fo}} - \frac{\xi}{S-1} \frac{\partial S}{\partial \text{Fo}} \frac{\partial T}{\partial \xi} = \nabla^2 T \tag{5.130}$$

where

$$\nabla^2 = \left(\left(\frac{1}{S-1} \right)^2 + \frac{1}{\pi^2 \bar{r}^2} \left(\frac{\zeta}{S-1} \right)^2 \left(\frac{\partial S}{\partial \eta} \right)^2 \right) \frac{\partial^2}{\partial \zeta^2} - \frac{2}{\pi^2 \bar{r}^2} \frac{\zeta}{S-1} \frac{\partial S}{\partial \zeta} \frac{\partial^2}{\partial \zeta \partial \eta}$$

$$+ \frac{2}{\pi^2 \bar{r}^2} \frac{\partial^2}{\partial \eta^2} + \frac{1}{\bar{r}(S-1)} \left(1 + \frac{\zeta}{\pi^2 \bar{r}} \left(\frac{2}{S-1} \left(\frac{\partial S}{\partial \eta} \right)^2 - \frac{\partial^2 S}{\partial \eta^2} \right) \right) \frac{\partial}{\partial \zeta} \tag{5.131}$$

The nondimensional energy balance for the liquid–solid interface becomes

$$\frac{\partial S}{\partial \text{Fo}} = \text{Ste} \left(1 + \left(\frac{1}{\pi S} \frac{\partial S}{\partial \eta} \right)^2 \right) \left(\frac{1}{S-1} \frac{\partial T}{\partial \zeta} - \frac{k_l}{k_s} \frac{1}{\pi S} \frac{\partial T}{\partial \xi} \right) \tag{5.132a}$$

where the Stefan number is given as

$$\text{Ste} = \frac{c_S(T_f - T_w)}{L} \tag{5.132b}$$

Defining an effective Nusselt number as

$$\text{Nu}_e = \frac{h \bar{S} D}{k} \tag{5.133a}$$

which reflects heat transfer at the moving boundary location, the local heat transfer coefficient can now be computed from

$$\text{Nu}_e = \frac{2}{\pi} \frac{\partial T/\partial \xi}{\lambda} \tag{5.133b}$$

C   System Solution

Using the previously described coordinate transformations, the solid region is mapped onto a 1 × 1 square domain that is discretized by using a 51 × 21 grid where $\Delta\eta = 1/50$ and $\Delta\zeta = 1/20$. The liquid region, bounded by a radius of $\xi = 1$, which corresponds to about 23 times the cylinder radius, also forms a square domain with $\Delta\eta = 1/50$ and $\Delta\xi = 1/40$ or $\Delta\xi = 1/30$, depending upon the Reynolds number. An alternative-direction implicit finite difference code has been developed to solve Equations (5.123) to (5.125), (5.130), and (5.132) subject to the boundary conditions (5.127) and (5.128). An upwind difference scheme is applied for the convective terms in Equation (5.125). In general, central differencing is used for the spatial derivatives, whereas time derivatives are approximated by a forward difference operator. As an initial condition, a thin ice layer on the cylinder surface is assumed to get the numerical computation started. The thickness of this initial layer depends on the Stefan number and the superheat of the fluid. Prusa and Yao (1984) pointed out that the effect of the incorrect initial interface decays very rapidly. It was found that setting $S = 1.05$ initially resulted in stable and accurate numerical solutions.

Table 5.4. *Data comparison for single-phase thermal cross-flow past a cylinder*[a]

	Separation angle θ_{sep}		Drag coefficient C_D		Nusselt numbers (Pr = 1.0) $Nu_{s.p}$ Nu_{av}			
Re_d	Model	Grove et al. (1964)	Model	Dennis et al. (1970)	Model	Okada et al. (1978)	Model	Eckert (1972)
10	30°	32°	2.92	2.85	—	—	—	—
20	—	—	—	—	12.5	12.9	7.1	6.5
40	54.6°	54.5°	1.52	1.52	—	—	—	—
50	—	—	—	—	16.4	17.1	10.7	10.3
70	62.5°	62.5°	1.22	1.21	—	—	—	—
100	67.7°	67.7°	1.08	1.06	22.2	24.2	14.5	14.5

[a] θ_{sep} is measured counterclockwise.

The location of the moving boundary is calculated from the temperature distributions of the liquid and solid regions at the end of each time step. Using small time steps ($\Delta\tau < 0.05$), the discontinuity caused by the moving solidification front is negligible for low Stefan number problems; hence intermediate iterations are not necessary. Results from repeat calculations with finer grids indicated that mesh independence had been achieved.

Results and Discussion

The predictive capabilities of the new model have been tested by comparing numerical results with experimental data sets from related case studies. Then the validated computer simulation model has been used to analyze flow patterns, isotherms, and ice layer formations for different system parameters.

In Table 5.4, characteristic parameters for single-phase thermal flow past a horizontal cylinder have been summarized. The comparison between predicted and measured data is very good except for higher Reynolds numbers (Re > 100), when vortex shedding may influence θ_{sep}, C_D, and Nu values.

Conduction-dominated ice formation around circular cylinders (Re = 0) has been analyzed by Lunardini (1980). Figure 5.35 shows a comparison of the transient interface location for freezing water between Lunardini's results and our model predictions.

Ice layer growth as well as the temporal development of streamlines and isotherms are shown in a sequence of graphs (Fig. 5.36a–c) for typical values of the key parameters. The results confirm the assumption of concentric ice layer growth in the beginning stage. While the free stream velocity stays constant, the local Reynolds number increases as the solid–liquid interface moves outward with time, $1 \leq Fo \leq 7$. Thus, the recirculation zone expands, in turn affecting the overall heat transfer process. The isotherms are quite evenly spaced along the front part of the cylinder, $0° \leq \theta \leq 50°$, and near the rear stagnation point, $160° < \theta \leq 180°$. As a result, the local heat transfer coefficient and, in turn, the ice layer thickness should be rather uniform within these two angular sections (cf. Figs. 5.37b and 5.38).

Conduction heat transfer in the solid region plays an important role during the freezing process. Because of the very steep temperature gradients across the initially thin ice layer, large amounts of heat are withdrawn from the adjacent liquid. Figure 5.37a shows the

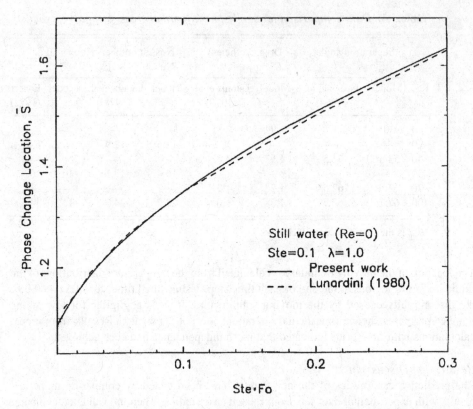

Fig. 5.35. Data comparison of transient solid–liquid interface location for conduction-controlled process.

transient interfacial heat flux ratio for different Reynolds numbers, where

$$\frac{q_s}{q_l} = 1 + \frac{k_s/k_l}{\lambda \mathrm{Nu}_e \mathrm{Ste}} \frac{\partial \bar{S}^2}{\partial \mathrm{Fo}} \tag{5.134}$$

In the early stages, the concentric ice buildup is almost entirely controlled by the thermal diffusion and latent heat balance. After some time (i.e., Ste · Fo > 0.1), when the thermal resistance of the growing ice layer, $R_{\mathrm{th}} - \ln(R/a)$, increases, the influence of variations in convection heat transfer becomes significant and the ice layer thickness varies with location and Reynolds number (Fig. 5.37b). Such a dependence of $S = S(\theta, \mathrm{Re})$ is even more dramatic when the fluid is superheated or the cylinder is subcooled, that is, when λ increases. Figure 5.37c indicates the increase of "eccentricity" in ice formation around a cylinder with dimensionless time. Here, S_{max} is the maximum dimensionless interface location (cf. Fig. 5.37b) and $S_{\mathrm{av}} = S$ is the average dimensionless radius for the cylinder plus ice layer configuration.

The temporal changes of the local Nusselt number (cf. Equation (5.133b)) for different free-stream Reynolds numbers are shown in Fig. 5.38. The graphs $\mathrm{Nu}_e(\theta)$ reflect the

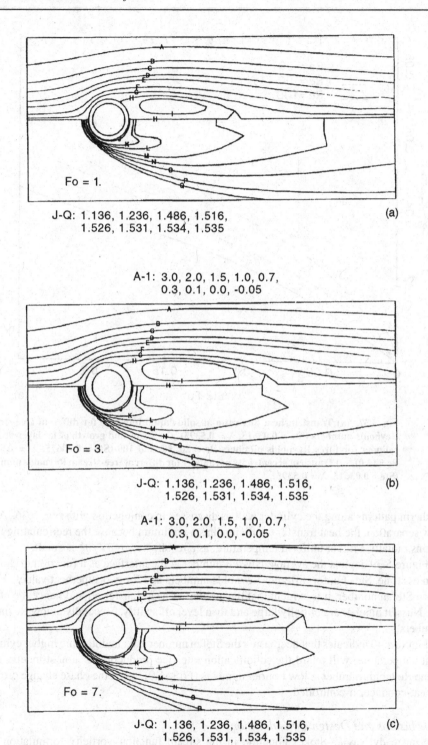

Fo = 1.

J-Q: 1.136, 1.236, 1.486, 1.516,
1.526, 1.531, 1.534, 1.535

(a)

A-1: 3.0, 2.0, 1.5, 1.0, 0.7,
0.3, 0.1, 0.0, -0.05

Fo = 3.

J-Q: 1.136, 1.236, 1.486, 1.516,
1.526, 1.531, 1.534, 1.535

(b)

A-1: 3.0, 2.0, 1.5, 1.0, 0.7,
0.3, 0.1, 0.0, -0.05

Fo = 7.

J-Q: 1.136, 1.236, 1.486, 1.516,
1.526, 1.531, 1.534, 1.535

(c)

Fig. 5.36. Streamlines, isotherms, and ice formation around a: cylinder at three time levels Fo = 1, 3, and 7 (Re = 50, Pr = 13, Ste = 0.03616, λ = 0.5395).

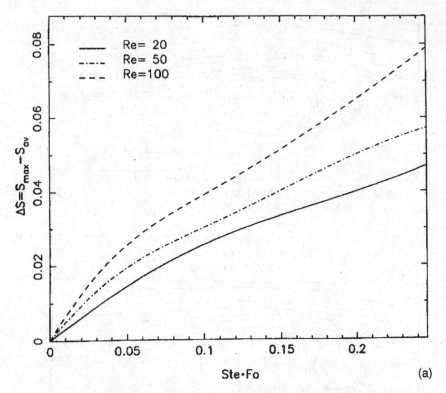

Fig. 5.37. (a) Transient heat flux ratio at solid–liquid interface for different free-stream Reynolds numbers (Ste = 0.03615, λ = 0.5395). (b) Shape and growth of ice layer around cylinder for three Reynolds numbers Re = 20, 50, and 100 (Ste = 0.03615, λ = 0.5395, Pr = 13.0). (c) Transient eccentricity parameter for different free-stream Reynolds numbers (Ste = 0.03615, λ = 0.5395).

isotherm patterns along the cylinder wall as discussed in conjunction with Fig. 5.36. After flow separation, the heat transfer coefficient has a minimum because the recirculating fluid attains radially a rather uniform temperature; as a result, $Nu_e \sim \partial T/\partial\xi$ is small.

Figure 5.39 depicts the average Nusselt number at the interface as a function of dimensionless time, Ste · Fo, for different Reynolds numbers and for two superheat values. For a given Stefan number, it is evident that the amount of heat transferred and hence the effective Nusselt number are initially large and then level off swiftly, especially at low Reynolds numbers.

Figure 5.40 indicates that decreasing the Stefan number (i.e., maintaining higher cylinder wall temperature) will retard the solidification rate. Ice production is almost independent of the Reynolds number at low Fourier numbers (Fo ≤ 1.0), when the phase change process is heat-conduction-controlled.

Conclusions and Design Considerations

The unsteady Navier–Stokes equation in the stream function–vorticity formulation and the thermal energy equation have been solved for laminar flow past a subcooled cylinder with outward solidification. Suitable coordinate transformations allowed mapping of the

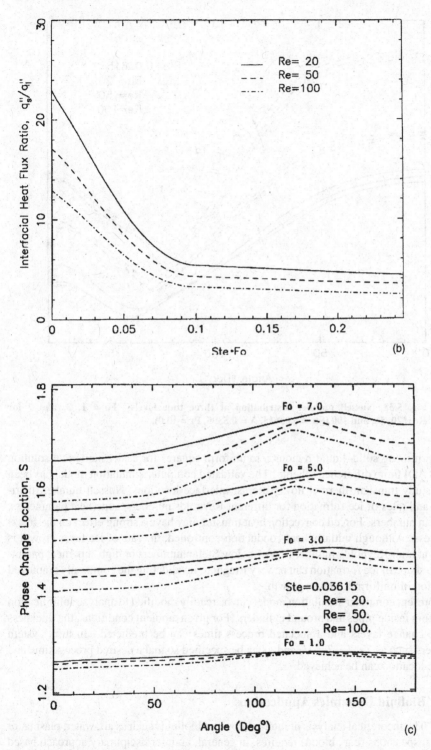

Fig. 5.37. (*Cont.*)

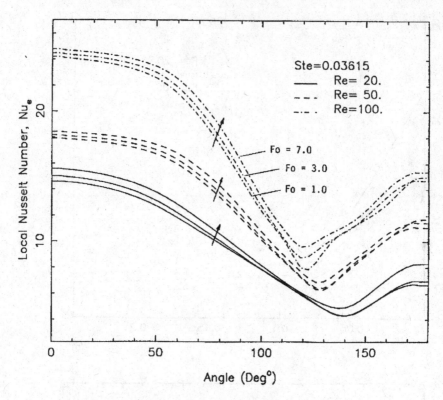

Fig. 5.38. Nusselt number distribution at three time levels: Fo = 1, 3, and 7 for Re = 20, 50, and 100 (Ste = 0.03615, λ = 0.5395, Pr = 13.0).

irregular, moving solid–liquid regions into uniform squares for easier numerical solution using an ADI finite difference scheme. The validated computer simulation model has been used to study transient thermal flow patterns, interface locations, Nusselt number distributions, and rates of ice formation for different Reynolds numbers, superheat parameters, and Stefan numbers. Forced convection heat transfer may have a strong effect on the freezing process. Although initially heat-conduction-controlled, the rate of ice layer growth is retarded at high-Reynolds-number flows. At low Stefan numbers or high superheat parameters, the shape of ice formation can be very irregular, invalidating the commonly employed assumption of uniform ice layer growth.

The present computer simulation model can be readily modified to analyze solidification and melting inside pipes and around cylinders. For given ambient conditions, the thickness of phase-change layers and associated process times can be predicted. In turn, system parameters such as T_0, T_f, T_w, U, and a can be specified so that a desired process time and interface location can be achieved.

5.5 Biofluid Dynamics Applications

The theoretical analysis of internal flows of biofluids such as air, water, plasma, or particle suspensions (e.g., blood) requires, in general, an interdisciplinary approach based on experimental/computational fluid mechanics, biomechanics, and heat and mass transfer.

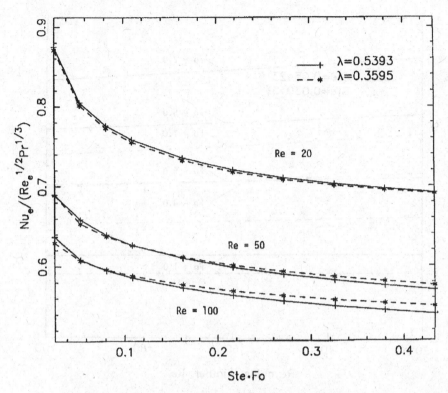

Fig. 5.39. Average Nusselt number profiles as a function of Stefan number and Fourier number for different free-stream Reynolds numbers and superheat parameters.

Specifically, the understanding of and remedy to pathologic interactions of fluid flow and cell biologic characteristic in blood vessels (cf. Sect. 5.5.1) and the removal via heat treatment of undesirable tissue layers in body cavities (cf. Sect. 5.5.2) rely on the collaboration of engineers, scientists, and physicians. Thus, research and development work in the broad area of biofluid dynamics are among the most challenging, rewarding, and future-oriented activities for engineers interested in fluid mechanics, heat/mass transfer, and composite mechanics.

5.5.1 Critical Flow Parameters in Graft–Artery Junctions

Local occlusions developing over decades in large- and medium-size arteries are directly or indirectly the chief cause of death in the Western world. It is widely accepted now that local "disturbed" flow patterns, characterized by high or low oscillatory wall shear stresses, recirculation zones, long particle residence times, and/or significant wall shear stress gradients, play major roles in the onset and development of atherosclerosis (i.e., intimal thickening and plaque formation), myointimal hyperplasia (i.e., local tissue overgrowth after bypass surgery), and thrombosis (i.e., blood clot formation). Several aspects of links between biofluid dynamics and atherosclerosis have been documented by McIntire (1991), Nerem (1992), Giddens et al. (1993), and Davies and Tripathi (1993), among others. Interactions between flow-induced mechanical factors and hyperplasia are

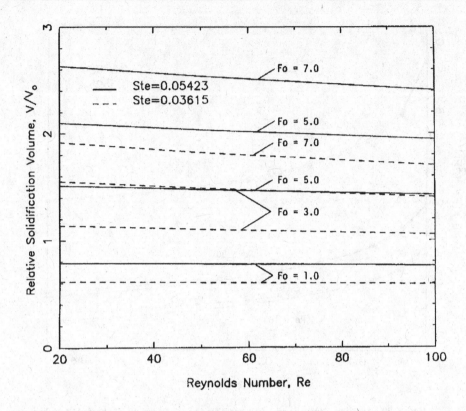

Fig. 5.40. Relative solidification volume as a function of Reynolds number for different Stefan numbers and Fourier numbers.

discussed by Dobin et al. (1989) and Bassiouny et al. (1992). For exact hemodynamics simulations one has to consider that the flow is transient and three-dimensional, blood exhibits non-Newtonian (i.e., shear-thinning) characteristics, and arterial walls are nonlinear viscoelastic composites. In addition, transition, turbulence, and relaminarization effects have to be considered in regions of severe blood vessel constrictions. When arteries become locally weak and severely occluded, it is often necessary to bypass the diseased artery segment using vein graft or synthetic graft material. Unfortunately, hyperplasia plus atherosclerotic plaque formation, labeled restenosis, may cause early graft failure within a few months or years after a bypass operation. Nonuniform hemodynamics, especially at the distal/receptor end of the bypass junction, play a major role in the development of restenosis.

A starting point for studying arterial hemodynamics would be an extension to Newtonian and non-Newtonian fluid flow in straight pipes (cf. Sects. 3.2 and 3.3 as well as He and Ku 1994) and a review of pulsatile flow in curved tubes (cf. Lin and Tarbell 1980; Talbot and Gong 1983; Banerjee et al. 1992). The input waveform characteristics and the degree of wall curvature strongly influence the interaction between the axial and secondary fluid motions. Flow nonuniformities are further amplified in bifurcating flows, as discussed later.

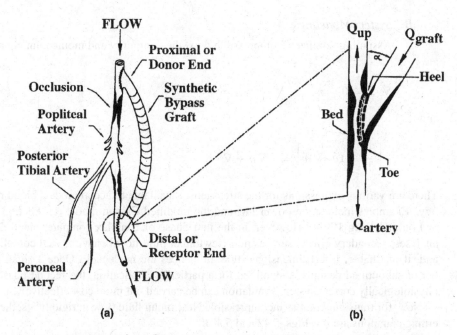

Fig. 5.41. Occluded femoral artery section and bypass graft: (a) actual flow conditions, (b) receptor end of graft–artery bypass.

A System Identification

Figure 5.41a,b shows the schematics of an occluded femoral artery and a blow-up of the distal or receptor graft–artery junction with four regions most susceptible to restenosis. It is postulated that restenosis, that is, pathologic tissue overgrowth and new plaque formation after bypass surgery, can be prevented or at least mitigated with optimal blood vessel geometries, which in turn largely reduce nonuniform hemodynamics.

The geometric factors to be considered include the bifurcation angle α, the wall curvatures k_i, the graft-to-artery area ratio κ, as well as the flow rates Q_i and their directions (cf. Fig. 5.41b). The single best indicator of nonuniform hemodynamics, that is, disturbed flow patterns, is the wall shear stress gradient (WSSG) (cf. Lei et al. 1995; Kleinstreuer et al. 1994, 1996). Within certain limits, different wall shear stress levels can be accommodated by the endothelial cells via cell deformation and alignment with the blood flow. However, locally significant and sustained changes in wall shear stress may aggravate the dysfunctional endothelium, allowing enhanced mass transfer of lipoproteins with oxidized cholesterol, monocytes, and so forth, into the arterial wall. This process is widely regarded as the onset of atherosclerosis, because once inside the arterial tissue, monocytes become scavenger cells (i.e., macrophages) that consume the altered lipoproteins at an accelerated rate and turn into foam cells. Foam cell clusters beneath the endothelium bulge out the inner arterial wall, produce harmful chemicals, and cause proliferation of underlying smooth muscle cells and fatty streaks. Similarly, after injury of the inner arterial wall surface during bypass operations, the rates of myointimal hyperplasia due to scar tissue formation, smooth muscle cell proliferation, and renewed atherosclerotic plaque formation greatly increase because of disturbed flow parameters, best represented by the WSSG.

B System Modeling

Assuming laminar incompressible flow, the continuity and momentum equations are

$$\nabla \cdot \vec{v} = 0 \tag{5.135}$$

and

$$\rho \left[\frac{\partial \vec{v}}{\partial t} + (\vec{v} \cdot \nabla)\vec{v} \right] = -\nabla p + \nabla \cdot \vec{\vec{\tau}} \tag{5.136}$$

There are various expressions for the stress tensor available to accommodate blood rheology. Examples include the power law and the modified Casson model (cf. Sect. 3.2.1D or Lou and Yang 1992). However, in the first phase of complex computer-aided design analyses, secondary effects such as non-Newtonian fluid flow behavior, wall compliance, and, if justifiable, three-dimensional flow influences are neglected. Once a small number of suboptimal designs is identified for a particular application, the second phase of a physiologically correct system simulation can be carried out more cost effectively.

Now, for transient laminar incompressible Newtonian fluid flow in rigid tubes, the governing equations are (cf. Figs. 5.42 and 5.43):

$$\nabla \cdot \vec{v} = 0 \tag{5.137}$$

and

$$\frac{\partial \vec{v}}{\partial t} + (\vec{v} \cdot \nabla)\vec{v} = -\frac{1}{\rho}\nabla p + \nu\nabla^2\vec{v} \tag{5.138}$$

The auxiliary conditions include

$$(\text{inlet}) \qquad -\vec{v} \cdot \hat{n} = u_i(t) \tag{5.139a}$$

$$\vec{v} \cdot \hat{s} = 0 \tag{5.139b}$$

$$(\text{outlet}) \qquad \frac{\partial \vec{v}}{\partial n} = \vec{0} \tag{5.140}$$

$$(\text{wall boundary}) \qquad \vec{v} = \vec{0} \tag{5.141}$$

The inlet condition (5.139a) is a transient parabolic velocity profile changing with $\text{Re}(t)$ as measured here for the femoral artery under active conditions (cf. Fig. 5.42). The inlet Reynolds number is defined as

$$\text{Re} = \rho u_m(t)\frac{d_0}{\mu} \sim Q_{\text{graft}} \tag{5.142}$$

where $Q_{\text{graft}}:Q_{\text{artery}}:Q_{\text{upstream}} = 100:80:20$ for each configuration.

The average blood properties ρ and μ are obtained from Merrill (1968). The influx $u_m(t)$ is the area average of $u_i(t)$ in Eq. (5.139a). The inlet and outlet sections of the branching blood vessel have to be sufficiently long to match the appropriate auxiliary conditions.

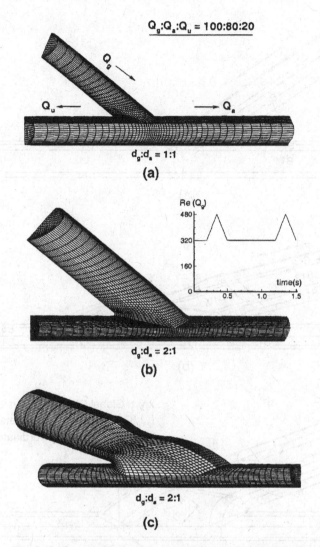

Fig. 5.42. Input pulse and control volume meshes of different receptor ends of femoral graft–artery junctions: (a) base case I ($d_{\mathrm{graft}} : d_{\mathrm{artery}} = 1.1$, $\alpha = 30°$, $Q_{\mathrm{graft}} : Q_{\mathrm{artery}} : Q_{\mathrm{up}} = 100 : 80 : 20$); (b) base Case II (large-diameter graft $d_{\mathrm{graft}} : d_{\mathrm{artery}} = 2 : 1$); (c) Taylor patch connector ($d_g : d_a = 2 : 1$, $\alpha \approx 10°$).

The local absolute value of the wall shear stress gradient is given as

$$|\mathrm{WSSG}| = \left[\left(\frac{\partial \tau_w}{\partial s} \right)^2 + \left(\frac{\partial \tau_w}{r \partial \theta} \right)^2 \right]^{1/2} \tag{5.143}$$

$$\text{where} \quad \tau_w = \begin{cases} \mu \left(\dfrac{\partial v_r}{\partial s} + \dfrac{\partial v_s}{\partial r} \right) \\[2mm] \mu \left(\dfrac{1}{r} \dfrac{\partial v_r}{\partial \theta} + \dfrac{\partial v_\theta}{\partial r} - \dfrac{v_\theta}{r} \right) \end{cases} \tag{5.144a,b}$$

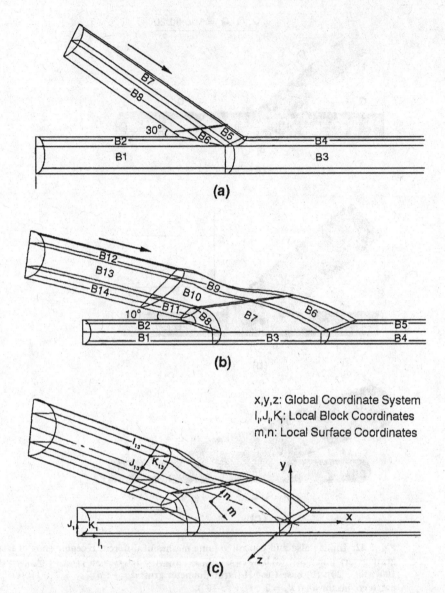

Fig. 5.43. Multiblock structures of base case I junction and Taylor patch connector, as well as schematics of different coordinates used.

and r, θ, and s are the local (cylindrical) coordinates based on the corresponding surface point. Nondimensionalization and time averaging of |WSSG| over the input cycle of duration T yield

$$\overline{|WSSG|}_{nd} = \frac{1}{T} \int_0^T \left[\frac{|WSSG|}{M} \right] dt \qquad (5.145)$$

where M is a scale factor, for example, $M = \tau_0/d_0$; τ_0 is the average inlet wall shear stress; and d_0 is the inlet tube diameter.

C   System Solution

The numerical calculations have been carried out with the finite volume-based algorithm CFDS-FLOW3D (cf. AEA 1991). The computational domain is subdivided into multiple control volumes or finite volumes. Proper mesh generation is one of the most challenging tasks in computational fluid dynamics analyses. In this study, a body-fitted coordinate system has been used, generating multiblock grid structures. Each block has a local coordinate system assigned. A multiblock grid is generated by merging all the blocks. For the basic 3-D graft–artery junction, six blocks were used; of them two were created to minimize mesh element distortion and to allow better mesh size control in the vicinity of the corners (cf. Fig. 5.43). Repeat calculations with finer meshes revealed the mesh independence of the results. As a result of the use of the multiblock grid structure, the total elements are greatly reduced in comparison with those of single-block grid generators. The convection coefficients are obtained by using the Rhie–Chow interpolation formula (Rhie and Chow 1983). It allows the use of standard primitive-variable algorithms, such as SIMPLER (cf. Patankar 1980) (however, without a staggered grid), while avoiding possible oscillations in pressure and velocity. The resulting algebraic equations are of the form

$$A_p\phi_p - \sum_{nb} A_{nb}\phi_{nb} = S^u \tag{5.146a}$$

Here subscript nb denotes each of the six neighboring nodes: U (up), D (down), N (north), S (south), E (east), and W (west); A_p and A_{nb} are coefficients. The diagonal coefficient of the matrix, A_p, is given as

$$A_p = \sum_{nb} A_{nb} - S_p + C_U - C_D + C_N - C_S + C_E - C_W + \frac{\rho V}{\Delta t} \tag{5.146b}$$

where C_U, and so on, are the convection coefficients at corresponding interfaces denoted by the subscripts; the last term stands for the transient effect carried from the previous time step, in which V is the volume of the control volume. In the steady-state case, the sum of the convection coefficients in this expression describes the conservation of mass and thus goes to zero in a converged solution.

The continuity equation (5.137) is replaced by the pressure-correction equation derived by using the SIMPLE or SIMPLEC algorithm (cf. Patankar 1980). The resulting algebraic equation has the form

$$b_p p'_p = \sum_{nn} b_{nn} p'_{nn} + S' - m_p \tag{5.147}$$

where p' is the corrected pressure, b_{nn}, and $b_p = \sum_{nn} b_{nn}$ are coefficients, m_p is the residual mass source, and S' is the additional source term due to nonorthogonality.

The discretized transport equations (5.146a,b) are solved iteratively. There are two levels of iteration: an inner iteration to solve for the spatial coupling (nonlinearity) for each variable and an outer iteration to solve for the coupling between variables. Thus each variable is taken in sequence, regarding all other variables as fixed. The coefficients of the discretized equations are always reformed, using the most recently calculated values of the variables before each inner iteration. There are several solvers available for the solution of the linearized transport equations. For the present problem, the Block Stone method (cf. Stone 1968) has been used. The velocity–pressure coupling is complicated and the SIMPLEC algorithm has been chosen because it has been proved to be less sensitive to the selection of underrelaxation factors and has required less underrelaxation. The convergence of the iterations is controlled by the mass source residual alone. For all calculations, varying time steps have been employed, depending upon the sign and magnitude of the gradient of the input pulse. Specifically, more time steps have been used for the deceleration phase than for the acceleration phase. To eliminate the start-up effect of transient flow, the computation is repeated over at least four periods.

The computational work has been carried out on a DECstation 5000 machine. A typical run for transient 3-D flow in a graft–artery junction constructed with eight blocks and 8,953 elements required a total CPU time of about 10 hours with the convergence criterium of mass residual less than 1.0×10^{-6} kg/s and maximum outer iterations of 200.

Results and Discussion

The flow simulation model has been validated with measured transient wall shear stress distributions given by Ojha (1993) at three sites of an occluded graft–artery receptor junction. Figure 5.44a shows the midplane of the 3-D end-to-side anastomosis with input pulse and the $\tau_w(t)$ variations. The cyclic branch inflow produces a moving stagnation point on the bed of the host artery with a major outflow, $\tau_w > 0$, and a minor reverse flow, $\tau_w < 0$, plus stagnant fluid, $\tau_w \approx 0$, in the occluded end part. A corresponding midplane velocity vector plot at a decelerating time level, $t = 0.5$ s, is given in Fig. 5.44b.

Figures 5.45 and 5.47a,b show midplane velocity fields at time level $t = 0.5$ s as well as 3-D surface contour plots of two time-averaged predictors for three end-to-side anastomoses that differ in graft-to-artery diameter ratios and junction curvatures. The scale factor is $M = 30$ N/M^3 (cf. Eq. (5.145)). The geometry of the two base case junctions differs in subtle ways. In case (a), the circular graft turns gradually into an ellipse at the junction as a result of the slanted cut to form $\alpha = 30°$ and the smaller axis, which is less than the diameter of the artery. Furthermore, the inner edge of the graft–artery junction is rather sharp. Thus, this design mimics a commonly used end-to-side anastomosis. In case (b), the larger circular graft transforms more swiftly to an "oval" where the maximum transverse extent is equal to the diameter of the artery. Case (c) is modeled after Taylor et al. (1991), and known as the Taylor patch, where a synthetic graft with a vein patch forms with $\alpha \approx 10°$ the graft–artery connector.

Figure 5.45 depicts midplane velocity vector plots at $t = 0.5$ s or Re $= 320$ for the two graft–artery junctions. The upstream flow rate, $Q_u = 0.2Q_g$, due to a lower pressure field behind the artery occlusion, contributes to a complicated velocity field and hence a highly varying surface stress field (cf. Fig. 5.45a). The large graft (cf. Fig. 5.45b) generates a slower incoming flow and, for the most part, a smoother flow field. The Taylor patch connector produces even more uniform velocity profiles (cf. Fig. 5.45c).

Figure 5.46a,b depicts constant $|\tau_w/\tau_0|$ contours and associated regions of wall shear stress maxima H and minima L. These extrema are near the clinically observed regions of myointimal hyperplasia, that is, toe, heel, and bed of a graft–artery junction, as well as along the suture line. With a larger graft or branch (cf. case (b)), the wall shear stress field is much smoother, with the exception of a new stress peak at the toe of the junction. The

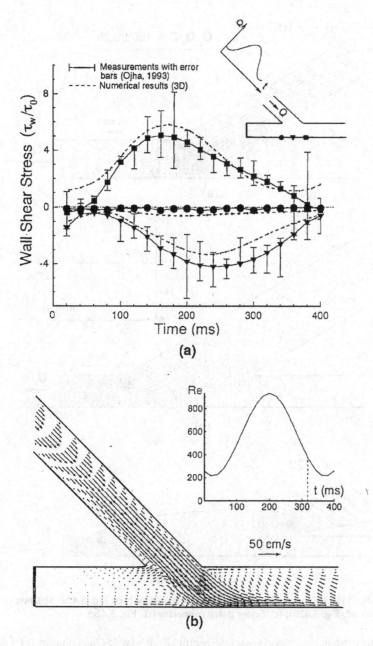

Fig. 5.44. Comparison between wall shear distributions and predicted results for occluded graft–artery bypass receptor (cf. Ojha 1993).

stagnation point on the artery bed in design (a), which actually moves around with time, is much less pronounced in case (b), and the stress peak at the heel of the junction has almost vanished. The stress maximum at the toe of the junction in case (b) is measurably reduced with the Taylor patch design (cf. Fig. 5.46c).

Although the wall shear stress is a good indicator of nonuniform arterial hemodynamics, its gradient is a more accurate and general predictor of sites susceptible to abnormal

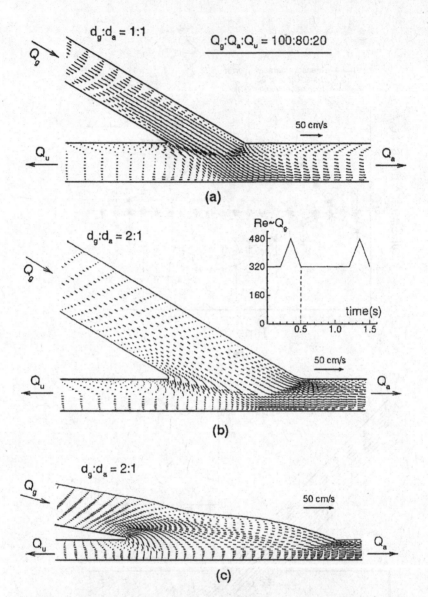

Fig. 5.45. Velocity vector plots at Re = 300: (a) base case I (cf. Fig. 5.42a); (b) base case II junction (cf. Fig. 5.42b); (c) Taylor patch connector (cf. Fig. 5.42c).

biological events in branching blood vessels. Specifically, the $|\overline{\text{WSSG}}|_{\text{nd}}$ contours (cf. Fig. 5.47) help to determine critical regions for case (a), especially at the relatively sharp heel junction. Designs (b) and (c) are clearly improvements, although further geometric alterations are warranted (cf. Lei 1995).

Conclusions and Design Considerations
Over the last two decades it has been well documented that nonuniform hemodynamics trigger abnormal biological events in large- and medium-size branching blood vessels. Of

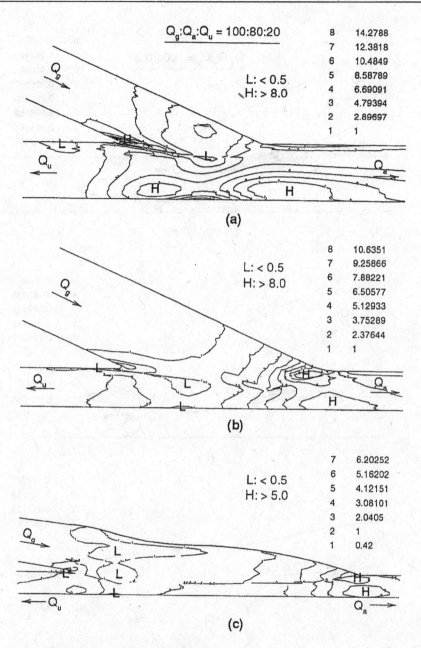

			8	14.2788
			7	12.3818
			6	10.4849
			5	8.58789
			4	6.69091
			3	4.79394
			2	2.89697
			1	1

$Q_g:Q_a:Q_u = 100:80:20$

L: < 0.5
H: > 8.0

(a)

		8	10.6351
		7	9.25866
		6	7.88221
		5	6.50577
		4	5.12933
		3	3.75289
		2	2.37644
		1	1

L: < 0.5
H: > 8.0

(b)

		7	6.20252
		6	5.16202
		5	4.12151
		4	3.08101
		3	2.0405
		2	1
		1	0.42

L: < 0.5
H: > 5.0

(c)

Fig. 5.46. Surface contours of time-averaged wall shear stress: (a) base case I; (b) base case II junction; (c) Taylor patch connector.

interest is the identification of a single best predictor linking disturbed flow patterns with arterial diseases, that is, atherosclerotic plaque formation in bifurcating arteries and/or myo-intimal hyperplasia plus atheroma in graft–artery junctions. Such a predictive parameter or dimensionless group would be beneficial in (i) focusing future experimental research on the mechanisms and causes of the disease process and (ii) developing a computational

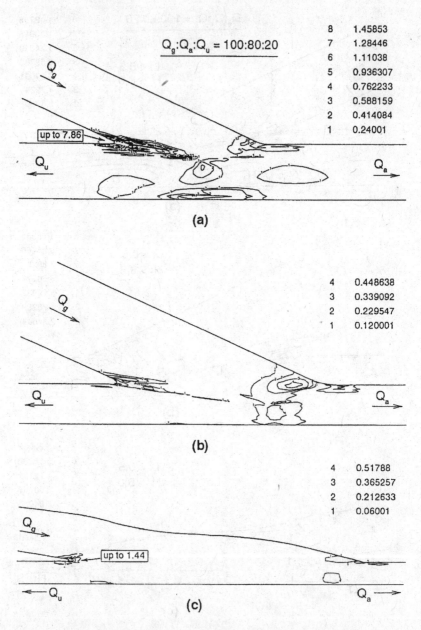

8	1.45853
7	1.28446
6	1.11038
5	0.936307
4	0.762233
3	0.588159
2	0.414084
1	0.24001

$Q_g : Q_a : Q_u = 100:80:20$

up to 7.86

(a)

4	0.448638
3	0.339092
2	0.229547
1	0.120001

(b)

4	0.51788
3	0.365257
2	0.212633
1	0.06001

up to 1.44

(c)

Fig. 5.47. Surface contours of time-averaged wall shear stress gradient: (a) base case I; (b) base case II junction; (c) Taylor patch connector.

procedure for the optimal geometric design of branching blood vessels, significantly reducing restenosis after vascular surgery.

It is postulated that significant and sustained values of the wall shear stress gradient, that is, locally WSSG $\neq$ 0, contribute greatly to "injury" of the endothelium (cf. Ross 1986). Thus, the WSSG plays a key role in high endothelial cell (e-cell) turnover (cf. Weinbaum and Chien 1993) and/or individual bond rupture of the e-cells (cf. Satcher and Dewey 1991; and DePaola et al. 1992), both processes resulting in lesion-prone sites of enhanced wall permeability to macromolecules and cells, such as lipids and monocytes. In addition to the locally increased lipid/cell wall flux and subsequent foam cell formation, sustained WSSGs may trigger deoxyribonucleic acid (DNA) changes that lead to excessive release of growth factors and subsequent smooth muscle cell proliferation, platelet aggregation, and thrombus formation. Similar events triggered by nonuniform hemodynamic forces may lead to myointimal hyperplasia, which occurs at the toe and heel of the junction of graft–artery bypass configurations and on the bed of the host artery across from the junction. Hyperplasia followed by atherosclerosis, additional cell deposition aggregation, intraplaque hemorrhage, and fibrosis all produce rapid progression of restenosis after bypass surgery.

In this study, considering geometrically different femoral receptor anastomoses, the transient three-dimensional velocity fields and associated wall shear stresses and wall shear stress gradients are computed using a validated finite volume program. With respect to the base case junction (i.e., $d_{graft}:d_{artery} = 1.1$, $\alpha = 30°$, relatively sharp corners), significant flow field improvements have been achieved, leading almost to an elimination of WSSG values, via the following geometric changes:

- Larger flow area ratios
- Smaller but variable bifurcation angles
- Smooth junction curvatures wherever possible

Clearly, the type of input waveform, junction geometry, and flow ratio Q_{graft}: Q_{artery}: $Q_{upstream}$ are the major physical factors influencing the hemodynamics and hence the response to endothelial cell dysfunction. Future work may focus on further geometric improvements and more realistic flow simulations incorporating the correct blood rheology, wall compliance, and additional input waveforms (cf. Lei 1995).

5.5.2 Free Convection in a Local Hyperthermia Treatment Device

Ablation of undesirable layers of tissue with a single-use local hyperthermia treatment (LHT) device is an effective, economical, and safe alternative to surgical procedures or laser applications. The device consists of a cylindrical electric-resistance heater, surrounded by a heat shield, and a thermistor, all forming the axis of a water-filled latex balloon (cf. Fig. 5.48). The device is inserted into the body cavity and brought in contact with the diseased area. An aqueous 5 percent dextrose solution is injected and the heater is turned on. The inflated water balloon takes on the shape of the individual body cavity while the heater settings control the heat shield surface temperature $T_h(x, t)$. As a result of free convection, heat is transferred to the undesirable tissue layers where cells are killed off in regions for which $T_c \geq 45°C$. Thus, reaching sufficiently high tissue temperatures and maintaining a quasi-uniform heat flux, say, $q_w \geq 30$ kW/m^2, for a certain period are the key goals. A specific application of the LHT device is discussed in Kleinstreuer and Lei (1994) and Lei and Kleinstreuer (1994).

A   System Identification

The present problem is that of natural convection heat transfer in a *horizontal*, nonuniform finite annulus with concentric heat source (cf. Fig. 5.48). Less complicated

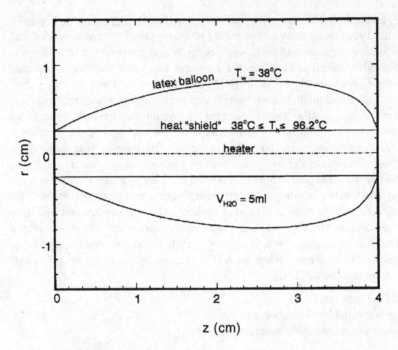

Fig. 5.48. Schematics and dimensions of streamline-shaped latex balloon with concentric heater and heat shield.

configurations are discussed in Sections 4.1 and 4.3 and reviewed by Gebhart et al. (1988). The LHT device in vertical position generates axisymmetric flow fields and temperature distributions. In general, boiling is avoided, that is, $T_{h,\max} = 96.2°C$, and the balloon surface temperature is assumed to be equal to the body temperature, $T_w = 38°C = $ const. The balloon shape has, within limits, only secondary effects on the free convection heat transfer.

Table 5.5. *System parameters for water at 38°C*

Balloon length (axial)	$L = 4$ cm
Maximum radius (streamline shape)	$r_{\max} = 8.2$ mm
Balloon volume	$\forall \geq 5$ ml
Initial or reference temperature	$T_0 = 38°C$
Prandtl number	Pr $= 4.608$
Aspect ratio, $L/r_{\max}$ (vertical orientation)	$A \approx 7$
Volumetric expansion coefficient	$\beta = 1.8 \times 10^{-4}$ K^{-1}
Grashof numbers (horizontal)	Gr$_r = 3.874 \times 10^4$
(vertical)	Gr$_L = 2.73 \times 10^5$
Rayleigh number	Ra$_L = $ Gr$_L$Pr $= 1.26 \times 10^6$
Reference velocity (steady state)	$u_{\text{ref}} = 1.123 \times 10^{-2}$ m/s
Reynolds number, $u_{\text{ref}} r_{\max}/\nu$	Re$_r = 92$

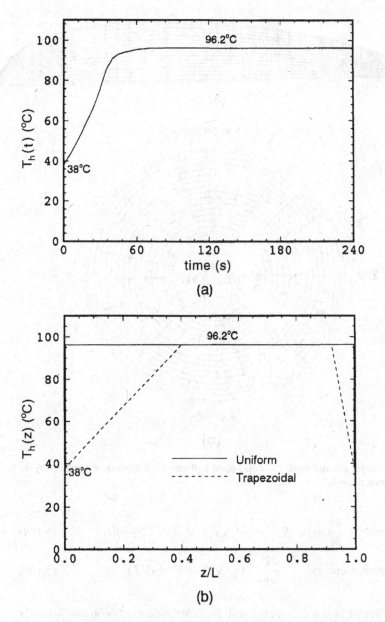

Fig. 5.49. Temporal and spatial heat shield temperature distributions.

B   System Modeling

Considering transient three-dimensional incompressible flow with the Boussinesq approximation for the body force term and iterative updating of the temperature-dependent viscosity and thermal diffusion coefficients, the governing equations are

$$(\text{continuity}) \qquad \nabla \cdot \vec{v} = 0 \tag{5.148}$$

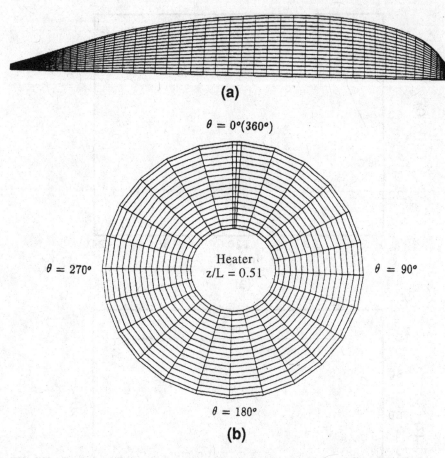

(a)

$\theta = 0°(360°)$

$\theta = 270°$ Heater
                          z/L = 0.51 $\theta = 90°$

$\theta = 180°$

(b)

Fig. 5.50. Control volume mesh for water-filled balloon: (a) midplane of upper half; (b) cross section at z/L = 0.5.

$$(momentum)\qquad \frac{\partial \vec{v}}{\partial t} + (\vec{v} \cdot \nabla)\vec{v} = -\frac{1}{\rho}\nabla p + \nabla \cdot (\nu \nabla \vec{v}) + \vec{g}\beta \Delta T \qquad (5.149)$$

$$where\quad (thermal\ energy)\qquad \frac{\partial T}{\partial t} + (\vec{v} \cdot \nabla)T = \nabla \cdot (\alpha \nabla T) \qquad (5.150)$$

where the velocity vector is $\vec{v} = (v_r, v_\theta, v_z)$ and the volumetric expansion coefficient is $\beta = -1/\rho(\partial \rho/\partial T)_p$. The momentum equation (5.149), for which $\Delta T = T - T_0$, depends on the thermal energy equation (5.150), which in turn requires the velocity field for solution. Typical system parameter values for water at $T_0 = 38°C$ are listed in Table 5.5.

Initially, the preheated water is at $T_0 = 38°C$ and the concentric heat source temperature is $T_h(t = 0) = 38°C$. Then the cylindrical wall temperature increases with time to $T_{h,\max}(t \geq 80s) = 96.2°C$ with either a uniform or a trapezoidal axial distribution (cf. Fig. 5.49). These surface temperature characteristics are measured design values from actual electrical resistance heaters. The isothermal balloon surface is kept at body temperature, $T_w = 38°C$.

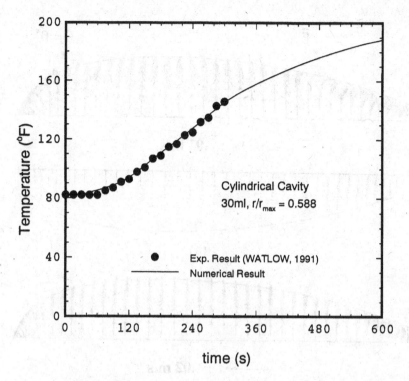

Fig. 5.51. Experimental data comparison for transient point temperature in cylindrical cavity.

C System Solution

The transport equations (5.149) and (5.150) subject to the mass conservation law (5.148) and the initial/boundary conditions outlined can be rewritten in compact form for the control volume solution method as follows (cf. Patankar 1980):

$$\frac{\partial}{\partial t}(\rho\phi) + \text{div}(\rho\vec{v}\phi) = \text{div}(\Gamma_\phi\nabla\phi) + S_\phi \tag{5.151}$$

where the four terms represent the local time rate of change of ϕ, net convection, net diffusion, and all sinks and sources affecting ϕ. The dependent variable ϕ is equivalent to the velocity components v_r, v_θ, and v_z as well as the temperature T. The generalized diffusion coefficient, Γ_ϕ, could be the viscosity μ when ϕ represents a velocity component and k/c_p when ϕ is the temperature. Whatever cannot be accommodated with the first three terms can always be expressed as (a part of) the source term. For example, the pressure gradient and the buoyancy force term in the momentum equations, or the heat fluxes across boundaries are represented as source terms.

Employing a nonlinear operator $L(\cdot)$, Eq. (5.151) can be rewritten in normal form as

$$L(\phi) = 0 \tag{5.152a}$$

or with an unknown approximate solution $\bar{\phi}$

$$L(\bar{\phi}) = R \tag{5.152b}$$

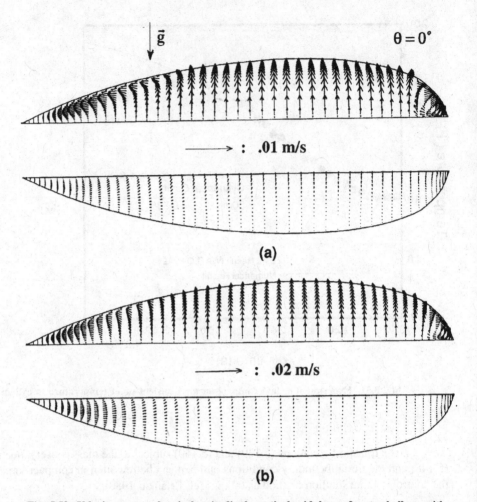

Fig. 5.52. Velocity vector plots in longitudinal, vertical midplane of water balloon with trapezoidal heat shield temperature: (a) transient state at $t = 30$; (b) steady state at $t = 240$ s.

where R is the residual of $L(\bar{\phi})$. Now, the weighted residual is forced to be zero over the entire computational domain. Specifically, the flow field is divided into nonoverlapping control volumes V_i ($i - 1, 2, \ldots, N$), and the weighting function is set to be unity over one subdomain at a time and zero elsewhere.

$$\int_{V_i} R_i \, dV = 0; \qquad i = 1, 2, \ldots, N \tag{5.153}$$

As a result, N algebraic weighted-residual equations are obtained to calculate the value of $\bar{\phi}$ for each control volume. Specifically,

$$a_p\phi_p = a_E\phi_E + a_W\phi_W + a_N\phi_N + a_S\phi_S + a_H\phi_H + a_L\phi_L + b \tag{5.154}$$

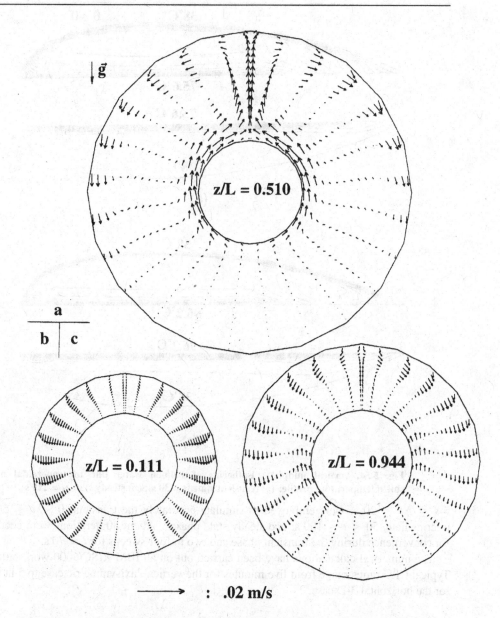

Fig. 5.53. Steady-state flow patterns in three azimuthal planes of horizontal water balloon with trapezoidal heat shield temperature.

where ϕ_p, ϕ_E, and so on, are the unknown values of the dependent variable ϕ at the nodal point p and its neighboring nodal points in the directions E, W, N, S, H, and L for 3-D flows; a_p, a_E, and so forth, are coefficients that depend on the type of finite difference approximation employed to evaluate gradients at the control surfaces.

The variable-density control volume meshes for one-half of a nonuniform finite annulus and its cross-sectional area at $z/L = 0.5$ are shown in Figs. 5.50a,b. A total of 19,552 elements were necessary. Trial-and-error runs with further mesh refinements did not change the results, indicating that mesh independence had been achieved. Variable time steps,

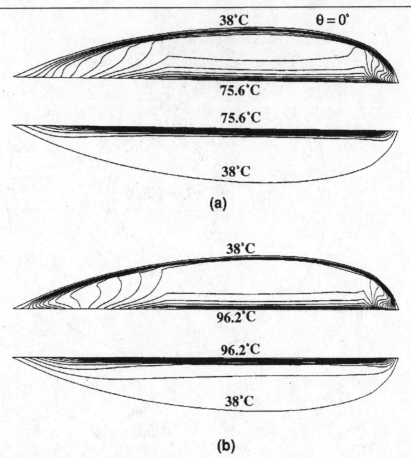

Fig. 5.54. Vertical midplane isotherms of balloon water due to trapezoidal heat shield temperature profile: (a) transient state ($t = 30$ s); (b) steady state ($t = 240$ s).

5 s $\leq \Delta t \leq 20$ s, were employed to simulate accurately the transient, $0 < t \leq 50$ s intermediate, $50 \leq t \leq 100$ s; and steady-state phases. About 30 iterations were needed for convergence during the transient phase and two to four sweeps for $t > 70$ s.

The numerical calculations have been carried out on an IBM RISC/6000 workstation. Typical CPU times range from five minutes for the vertical, axisymmetric case to 5 hours for the horizontal 3-D case.

Results and Discussion

Figure 5.51 shows a comparison between a computed point temperature with time and measurements (cf. Watlow 1991). The experiment consisted of a small water-filled beaker with a concentric cylindrical heater element. A thermocouple was used to measure $T(t)$ while the device was in a constant-temperature bath of $T_w = 38°C$. The agreement between the predicted results and the experimental observations is excellent.

The transient thermal boundary condition, trapezoidal $T_h(t, x)$ along the concentric heater surface, generates three-dimensional velocity and temperature fields as depicted in Figs. 5.52–5.56. Initially, the fluid rises mainly in the upper half from the heated cylinder surface as a result of buoyancy effects (cf. Figs. 5.52 and 5.53). Then the confining balloon

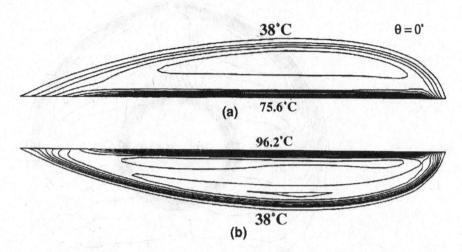

Fig. 5.55. Horizontal midplane isotherms due to trapezoidal heat shield temperature profile at t = 30 s and t = 240 s.

surface forces the fluid to turn simultaneously in longitudinal and downwards in azimuthal directions. A strong recirculation zone develops in the upper half of the front part of the device (i.e., $z/L \geq 0.9$), which is characterized by a rapid change in wall curvature. The lower front end shows hardly any fluid motion. In the tail region (i.e., $0 \leq z/l \leq 0.2$), some of the rising fluid is pushed from the balloon center toward the end, where it turns downward and flows in the lower end part back toward the center.

Isotherms in the vertical plane are shown in the time-dependent phase at t = 30 s (cf. Fig. 5.54a) and at steady state (cf. Fig. 5.54b). At t = 30 s, the heat shield temperature with a trapezoidal axial distribution (cf. Fig. 5.49) has reached T_h = 75.6°C; heat is convected upward in the main part of the upper balloon half ($u_{ref}(t) \approx 0.8 \times 10^{-2}$ m/s) and convected sideward toward the head and tail portions of the device. The resulting temperature field in the upper half is entirely free-convection-dominated. Although all rising fluid is moving down along the balloon wall, the heat convection effect is hardly noticeable in the lower half of the balloon. There, the heat transfer is due to conduction near the heat shield, and at the bottom most of the water stays at T_w = 38°C (Fig. 5.54a). Steady state is reached after $t \approx 80$ s when the maximum heater surface or heat shield temperature $T_{h,max}$ = 96.2°C. Buoyancy-induced flow in the upper half has increased somewhat with $u_{ref} = 1.123 \times 10^{-2}$ m/s. This influences the lower half somewhat; however, a large portion remains at T_w = 38°C (cf. Fig. 5.54b). In the horizontal plane, symmetry allows grouping of the isotherms at two different time levels into one graph. Thus, Figure 5.55 depicts narrowing temperature contours and hence a rapid increase in wall heat flux with time.

Steady-state isotherms at three different cross sections, $z/L = 0.11$ (tail end), 0.510 (center), and 0.944 (head), are shown in Figs. 5.56a–c. The contours at $z/L = 0.5$ are very similar to the isotherms in infinite cylindrical annuli, whereas significant differences appear in the front part of the balloon and especially in the tail end because of 3-D effects.

The previous graphs can be summarized in terms of (steady-state) temperature profiles in axial direction along the balloon surface at $r/r_{max} = 0.9$ (cf. Fig. 5.57) for different circumferential locations: $\theta = 0°$ (top), 90° (side), and 180° (bottom). As expected, the near top temperature is highest, with a rather uniform distribution. Almost the entire upper half of

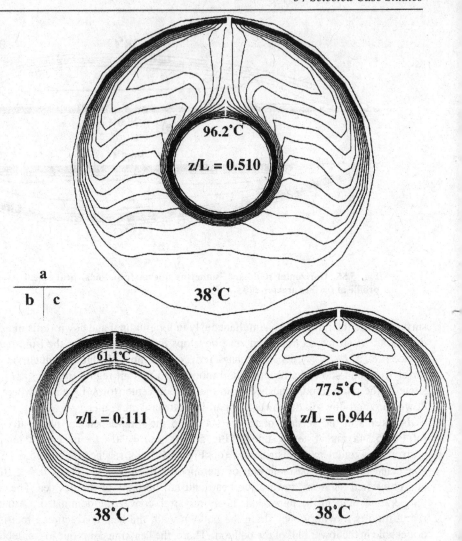

Fig. 5.56. Steady-state isotherms in three cross sections due to trapezoidal heat shield temperature profile: (a) $z/L = 0.11$ (tail end); (b) $z/L = 0.510$ (center); (c) $z/L = 0.944$ (head).

the device exceeds the critical temperature, $T_c = 45°C$, for necrosis. The resulting balloon surface heat fluxes with time, averaged for the upper and lower halves, are given in Fig. 5.58. The maximum $\bar{q}_w$ for the upper part of the device is about eight times larger than for the lower portion.

Conclusions and Design Considerations

The problem of transient three-dimensional free convection in an arbitrary enclosure has been solved numerically using the control-volume method. The validated computer simulation model has been applied to the analysis and design of a water-filled latex balloon that can be used in the local hyperthermia treatment (LHT) of undesirable tissue. Of interest are the liquid-side temperature profiles near the balloon surface and the wall heat flux distributions

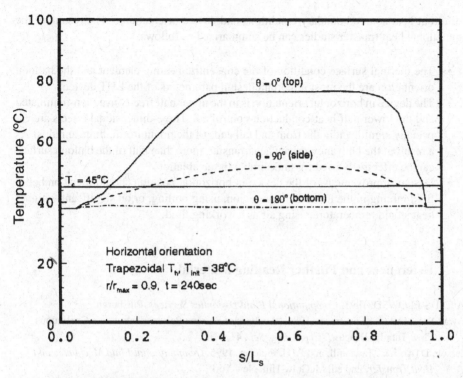

Fig. 5.57. Steady-state temperature profiles at three circumferential locations in near-wall axial direction.

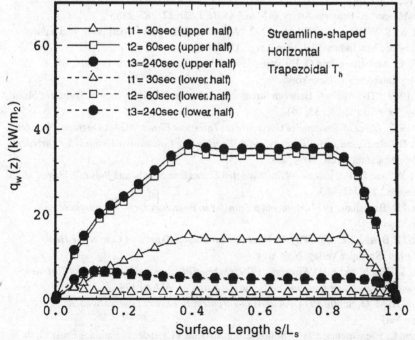

Fig. 5.58. Area-averaged balloon surface heat distribution for upper half and lower half at different time levels due to trapezoidal heat shield temperature profile.

for different heat source boundary conditions and device orientations. The results of the computational heat transfer studies can be summarized as follows:

- The thermal surface condition of the concentric heating element and the balloon orientation are the most important design parameters of the LHT device.
- The device in horizontal orientation is in the upper half free-convection-dominated and the lower half heat-conduction-controlled. Three-dimensional effects are especially significant in the front and tail ends of the nonuniform, finite annulus. As a result of the buoyancy-driven heat transfer, more than half of the balloon surface can be effectively used for undesirable tissue ablation.
- Design improvements of the device in horizontal position should focus on better fluid mixing in the lower balloon region, using boiling, or on significantly higher heat shield temperature, using air as a working fluid.

References and Further Reading Material

AEA CFDS-FLOW3D. 1991. *Computational Fluid Dynamics Services*, Pittsburgh.

Agarwal, S. S., and C. Kleinstreuer. 1986. "Analysis of Styrene Polymerization in a Continuous Flow Tubular Reactor," *Chem. Eng. Sci.*, **41**, 310.

Anderson, D. A., J. C. Tannehill, and R. H. Pletcher. 1995. *Computational Fluid Mechanics and Heat Transfer*, 2nd ed. McGraw-Hill, New York.

Bacon, F. (1620), cited in P. Rossi. 1973. "Baconianism." In: *Dict. History of Ideas*, Vol. 1, pp. 172–197. P. P. Wiener (ed). C. Scribner and Sons, New York.

Banerjee, R. K., Y. I. Cho, and L. H. Back. 1992. "Numerical Analysis of 3-D Arterial Flows in Double-curved Femoral Artery of Man," *ASME BED*, **22**, 285–288.

Bassiouny, H. S., S. White, and S. Glagov. 1992. "Anastomotic Intimal Hyperplasia: Mechanical Injury or Flow Induced," *J. Vasc. Surg.*, **15**, 708–717.

Bird, R. B., R. C. Armstrong, and O. Hassager. 1987. *Dynamics of Polymeric Liquids*, Vol. 1. Wiley-Interscience, New York.

Bradshaw, P. 1969. "The Analogy Between Streamline Curvature and Buoyancy in Turbulent Shear Flow," *J. Fluid Mech.*, **33**, 1612.

Bradshaw, P. 1973. *Effect of Streamwise Curvature on Turbulent Flows*. AGAR Dograph, No. 169.

Bradshaw, P., T. Cebeci, and J. H. Whitelaw. 1984. *Engineering Calculation Methods for Turbulent Flow*, Academic Press, New York.

Cassonova, R. A., and D. P. Giddens. 1978. "Modeled Stenosis in Steady and Pulsatile Flows," *J. Biomech.*, **11**, 441–453.

Cebeci, T., and D. Bradshaw, 1977. *Momentum Transfer in Boundary Layer*. Hemisphere, Washington, DC.

Cebeci, T., and P. Bradshaw. 1988. *Physical and Computational Aspects of Convective Heat Transfer*. Springer Verlag, New York.

Cebeci, T., R. S. Hirsch, and J. H. Whitelaw. 1979. *On the Calculation of Laminar and Turbulent Boundary-Layers on Longitudinally Curved Surfaces*. AIAAJ 174.

Cebeci, T., and A. M. O. Smith. 1974. *Analysis of Turbulent Boundary Layers*. Academic Press, New York.

Chiang, H., and C. Kleinstreuer. 1989. "Solidification Around a Cylinder in Laminar Cross Flow," *J. Heat and Fluid Flow*, **10**(4), 322–327.

Chiang, H., and C. Kleinstreuer. 1991. "Convection Heat Transfer of Collinear Interacting Droplets with Surface Mass Transfer," *Int. J. Heat Fluid Flow*, **12**(3), 223–239.

Chiang, H., and C. Kleinstreuer. 1992. "Computational Analysis of Interacting Vaporizing Fuel Droplets on a One-Dimensional Trajectory," *Comb. Sci. Technlol.* **86**, 289–309.

Clift, R., J. R. Grace, and M. E. Weber. 1978. *Bubbles, Drops and Particles*. Academic Press, New York.

Crane, R. L., K. E. Hillstrom, and M. Minkoff. 1980. *Solution of General Nonlinear Programming Problems with Subroutine VMCON*. Argonne National Laboratories Report ANL-80-64, NTIS, Springfield, VA.

Cuvelier, C., A. Segal, and A. A.Van Steenhoven. 1986. *Finite Element Methods and Navier-Stokes Equations*. D. Reidel, Dordrecht, The Netherlands.

Davies, P. F., and S. C. Tripathi. 1993. "Mechanical Stress Mechanisms and the Cell," *Circul. Res.*, **72**, 239–245.

Dennis, S. C. R., and Gau-Zu Chang. 1970. "Numerical Solutionals for Steady Flow Past a Cylinder at Reynolds Number up to 100," *J. Fluid Mech.*, **42**, 471–489.

DePaola, N., M. A. Gimbrone, P. F. Davies, and C. F. Dewey. 1992. "Vascular endothelium response to fluid shear stress gradients." *Arterioscler. Thromb.*, **12**(11), 1254–1257.

Deshpande, M. D., and D. P. Giddens. 1980. "Turbulence Measurements in a Constricted Tube," *J. Fluid Mech.*, **97**(I), 65–89.

Deshpande, M. D., and D. P. Giddens. 1983. "Computation of Turbulent Flow Through Constrictions," *J. Eng. Mech.*, **109**(2), 466–478.

Dobrin, P. B., F. N. Littoony, and E. D. Endean. 1989. "Mechanical Factors Predisposing to Intimal Hyperplasia and Medial Thickening in Autogeneous Vein Grafts," *Surgery*, **105**, 393–400.

Durst, F., and A. K. Rastogi, 1977. "Theoretical and Experimental Investigations of Turbulent Flows with Separation." In: *Turbulent Shear Flows*, Vol. I, Durst et al., Springer Verlag, New York.

Eckert, E. R. G., and R. M. Drake. 1972. *Analysis of Heat and Mass Transfer*. McGraw-Hill, New York.

Eghlima, A., and C. Kleinstreuer. 1985. "Numerical Analysis of Attached Turbulent Boundary Layers near the Tail of Axisymmetric Bodies," *AIAA J.* **23**(2): 177–184.

Flaherty, J. E., and W. Mathon. 1980. "Collocation with Polynomial and Tension Spheres for Singular-Perturbed Boundary Value Problems," *SIAMJ. Sci. Stat. Comp.*, I, 260.

Gebhart, B., Y. Jaluria, R. L. Mahajan, and B. Sammmakia. 1988. *Buoyancy Induced Flows and Transport*. Hemisphere, New York.

Giddens, D. P., C. K. Zarnis, and S. Glagov. 1993. "The Role of Fluid Mechanics in the Localization and Detection of Atherosclerosis," *J. Biomech. Engl.*, **115**, 558–594.

Goldstein, S., ed. 1943. *Modern Developments in Fluid Dynamics*, Vol. I. Clarendon Press, Oxford.

Granville, P. S. 1989. "A Modified VanDriest Formula for the Mixing Length of Turbulent Boundary Layers in Pressure Gradients," *ASME J. Fluid Eng.*, **111**, 94–97.

Granville, P. S. 1990. "A Near-Wall Eddy Viscosity Formula for Turbulent Boundary-Layers in Pressure Gradients Suitable for Momentum, Heat or Mass Transfer," *ASME J. Fluid Eng.*, **112**, 240–243.

Grove, A. S., F. H. Shair, and A. Acrivos. 1964. "An Experimental Investigation of the Steady Separated Flow Past a Circular Cylinder," *J. Fluid Mech.*, **19**, 60–80.

He, X., and D. N. Ku. 1994. "Unsteady Entrance Flow Development in a Straight Tube," *ASME J. Biomech. Eng.*, **116**, 355–360.

Ho, C. J., and S. Chen, 1986. "Numerical Simulation of Melting of Ice Around a Horizontal Cylinder," *Int. J. Heat Mass Transfer*, **29**, 1359–1369.

Huang, T. T., et al. 1979. *Stern Boundary-Layer Flow on Axisymmetric Bodies*. 12th Symposium on Naval Hydrodynamics, Washington DC, June 5–9, 1978, p. 127. Available from National Academy of Sciences, Washington, DC.

Johnson, D. A., and L. S. King. 1985. "A Mathematically Simple Turbulence Closure Model for Attached and Separated Turbulent Boundary-Layers," *AIAAJ*, **23**(11), 1684.

Jones, W. P., and B. E. Launder. 1983. "The Calculation of Low-Reynolds Number Phenomena with a Two-Equation Model of Turbulence," *Int. J. Heat Mass Transfer*, **16**, 1119.

Kleinstreuer, C., and S. Agarwal. 1987. "Fluid Dynamics of a Tubular Polymerizer Largely Determined by Coupled Heat and Mass Transfer Processes," *Int. J. Eng. Sci.*, **25**(5), 597–607.

Kleinstreuer, C., J. K. Comer, and H. Chiang. 1993. "Fluid Dynamics and Heat Transfer with Phase Change of Multiple Spherical Droplets in a Laminar Axisymmetric Gas Stream," *Int. J. Heat Fluid Flow*, **14**(3), 292–300.

Kleinstreuer, C., and A. Eghlima. 1985. "Analysis and Simulation of New Approximation Equations for Boundary-Layer Flow on Curved Surfaces," *Math. Comp. in Simul.*, **27**, 307.

Kleinstreuer, C., A. Eghlima, and J. E. Flaherty. 1983. *Computer Simulation of Thick Incompressible Boundary Layers*. Proceedings of the 3rd International Conference on Numerical Methods in Laminar and Turbulent Flow, University of Washigton, Seattle.

Kleinstreuer, C., and M. Lei. 1994. "Transient Laminar Three-Dimensional Fluid Flow Fields in Slender Enclosures with Coaxial Heated Cylinder," *Int. J. Eng. Sci.*, **32**(10), 1635–1646.

Kleinstreuer, C., M. Lei, and J. P. Archie, Jr. 1996. "Flow Input Waveform Effects on the Temporal and Spatial Wall Shear Stress Gradients in a Femoral Graft-Artery Connector," *ASME J. Biomech. Engr.*, **118**(3).

Kleinstreuer, C., M. Lei, D. R. Wells, and G. A. Truskey. 1994. "Computational Flow Analysis and Prediction of Atherogenic Sites in Branching Arteries." In: *Biomedical Engineering—Recent Developments*, J. Vossoughi (ed.), Proc. 13th Southern Biomed. Eng. Conf., April 16–17, 1994, Washington, DC, pp. 995–998.

Kleinstreuer, C., and T.-Y. Wang. 1988. "Heat Transfer Between Rotating Spheres and Flowing Power-Law Fluids with Suction and Injection," *Int. J. Heat Fluid Flow*, **9**, 328–333.

Kleinstreuer, C., and T.-Y. Wang. 1989. "Mixed Convection Heat and Surface Mass Transfer between Power-Law Fluids and Rotating Permeable Bodies," *Chem. Eng. Sci.*, **44**(12), 1987–1994.

Kleinstreuer, C., and T.-Y. Wang. 1990. "Approximate Analysis of Dynamically Interacting Vaporizing Droplets," *Int. J. Multiphase Flow*, **16**(2), 295–304.

Kuo, K. K. 1986. *Principle of Combustion*. Wiley-Interscience, New York.

Lam, C. K. G., and K. Bremhorst. 1981. "A Modified Form of the k–ε Model for Predicting Wall Turbulence," *ASME, J. Fluids Eng.*, **103**, 456.

Lei, M. 1995. "Computational Flow Analyses and Optimal Designs of Branching Blood Vessels," Ph.D. thesis, MAE Dept., North Corolina State University, Raleigh.

Lei, M., and C. Kleinstreuer. 1994. "Natural Convection Heat Transfer in a Nonuniform Finite Annulus with Concentric Heat Source," *Int. J. Heat Fluid Flow*, **15**(6): 456–461.

Lei, M., C. Kleinstreuer, and G. A. Truskey. 1995. "A Numerical Investigation of Pulsatile Flow and the Prediction of Atherogenic Sites in Branching Arteries," *ASME J. Biomech. Engr.*, **117**: 350–357.

Lien, F. S., C. K. Chen, and J. W. Cleaver, 1986a. "Forced Convection over Rotating Bodies with Blowing and Suction," *AIAA*, **24**, 854–856.

Lien, F. S., C. K. Chen, and J. W. Cleaver. 1986b. "Mixed and Free Convection over a Rotating Sphere with Blowing and Suction," *ASME J. Heat Transfer*, **108**, 398–404.

Lin, J. Y., and J. M. Tarbell. 1980. "An Experimental and Numerrical Study of Periodic Flow in a Curved Tube," *J. Fluid Mech.*, **100**(3), 397–416.

Lou, Z., and W.-J. Yang. 1992. "Biofluid Dynamics at Arterial Bifurcations," *Crit. Rev. Biomed. Eng.*, **19**(6), 455–493.

Lunardini, V. J. 1980. "Phase Change Around a Circular Cylinder," *ASME Winter Annual Mtg.*, Nov. 1980, Chicago, IL, Paper 80-WA/HT-5, ASME, New York.

McIntire, L. V. 1991. "Bioengineering and Vascular Biology," *Bioeng. Sci. News, BMES Bull.*, **15**(4), 51–53.

Merrill, E. W. 1968. "Rheology of Blood," *Physiol. Rev.*, **49**, 863–888.

Minkowycz, W. J., E. M. Sparrow, G. E. Schneider, and R. H. Pletcher. 1988. *Handbook of Numerical Heat Transfer*. Wiley Interscience, New York.

Nagano, Y., and M. Hishida. 1987. "Improved Form of the k-E Model for Wall Turbulent Shear Flows," *ASME J. Fluids Eng.*, **109**, 156.

Nerem, R. M. 1992. "Vascular Fluid Mechanics and the Arterial Wall, and Atherosclerosis," *J. Biomech. Eng.*, **114**, 274–282.

O'Connor, L. 1992. "Computational Fluid Dynamics Codes," *ASME Mech. Eng.*, May 44–50.

Ojha, M. 1993. "Spatial and Temporal Variations of Wall Shear Stress Within an End-to-Side Arterial Anastomosis Model," *J. Biomechanics*, **26**(12), 1377–1388.

Okada, M., K. Katayama, and K.Terasaki. 1978. "Freezing Around a Cooled Pipe in Crossflow," *Bull. JSME*, **21**, 160–172.

Patankar, S. V. 1980. *Numerical Heat Transfer and Fluid Flow*. Hemisphere, Washington, DC.

Patel, V. C., and Y. T. Lee. 1978. *Thick Axisymmetric Boundary Layers and Wakes: Experiment and Theory*. Paper 4, International Symposium on Ship Viscous Resistance, Goteberg, Sweden.

Patel, V. C., W. Rodi, and G. Scheuere. 1985. "Turbulent Models for Near-Wall and Low Reynolds Number Flows: A Review," *AIAA J.*, **23**(9), 1308–1319.

Pepper, D. W., and J. C. Heinrich. 1992. *The Finite Element Method*. Hemisphere, Washington, DC.

Popper, K. (1934), cited in H. Feigle 1973. "Positivism in the Twentieth Century." In: *Dict. History of Ideas*, Vol. III, pp. 545–552. P. P. Wiener (ed.). C. Scribner and Sons, New York.

Prakash, S., and G. Krishan. 1984. *Convective Droplet Vaporization with Transient Non-Convective Liquid-Phase Heating*. 20th Symposium on Combustion, 1735–1742.

Press, W. H., B. P. Flannery, S. A. Teukolsky, and W. T. Vetterling. 1994. *Numerical Recipes Example Book–FORTRAN Diskette*. Cambridge University Press, New York.

Prusa, J., and L. S. Yao. 1984. "Melting Around a Horizontal Heated Cylinder," *Int. J. Heat Mass Transfer*, **106**, 467–472.

Rajasekaran, R., and M. G. Palekar. 1985. "Mixed Convention about a Rotating Sphere," *Int. J. Heat Mass Transfer*, **28**, 959–968.

Ramachandran, R. S., C. Kleinstreuer, and T.-Y. Wang. 1989. "Forced Convection Heat Transfer of Interacting Spheres," *Nummer. Heat Transfer*, **15**, 471–487, Part A.

Reddy, J. N., and D. K. Gartling. 1994. *The Finite Element Method in Heat Transfer and Fluid Dynamics*. CRC Press, Boca Raton, FL.

Rhie, C. M., and W. L. Chow. 1983. "Numerical Study of the Turbulent Flow Past an Airfoil with Trailing Edge Separation," *AIAA J1.* **21**, 1527–1532.

Roache, P. J. 1976. *Computational Fluid Dynamics*. Hermosa, Albuquerque, NM.

Rodi, W. 1987. "A New Algebraic Relation for Calculating the Reynolds Stresses," *ZAMM*, **56**, T219–T221.

Rodi, W. 1980. *Turbulence Models and their Application in Hydraulics: A State-of-the-Art Review.* IAHR, Delft, The Netherlands.

Ross, R. 1986. "The Pathogenesis of Atherosclerosis: An Update," *N. Eng. J. Med.*, **314**(8), 488–500.

Satcher, R. L. Jr., and C. F. Dewey Jr. 1991. "The Distribution of Fluid Forces on Arterial Endothelial Cells," 1991 *Adv. Bioeng. ASME*, **20**, 595–598.

Schlichting, H. 1979. *Boundary-Layer Theory.* McGraw-Hill, New York.

Sloan, D. G., P. U. Smith, and L. D. Smooth. 1986. "Modeling of Swirl in Turbulent Flow Systems," *Prog. Energy Combust. Sci.*, **12**, 163.

Solomon, A. D., V. Alexiades, D. C. Wilson, and J. Drake. 1984. "The Formulation of a Hyperbolic Stefan Problem," Tedin. Report ORNL-6065, Oak Ridge Natl. Lab., Oak Ridge, TN.

Stone, H. L. 1968. "Iterative Solution of Implicit Approximations of Multidimensional Partial Equations," *SIAM J. Numer. Anal.* **5**, 530–558.

Talbot, L., and K. O. Gong. 1983. "Pulsatile Entrance Flow in a Curved Pipe," *J. Fluid Mech.*, **127**, 1–25.

Tani,1. 1954. *J. Japan Sci. Mech. Eng.*, **57**, 596.

Taylor, R. S., A. Loh, R. J. McFarland, M. Cox, and J. F. Chester. 1991. "Improved Techniques for PTFE Bypass Grafting: Long-Term Results Using Anastomatic Vein Patches," *Br. J. Surg.*, **79**, 348–354.

Tien, N. K., E. Flaschel, and A. Renken. 1985. "Bulk Polymerization of Styrene in a Static Mixer," *Chem. Eng. Comm.*, **36**, 251.

Van Dyke, M. 1969. *Higher-Order Boundary-Layer Theory.* Annu. Rev. Fluid Mech., Vol. 1, 265–292.

Vemuri, V. 1978. *Modeling of Complex Systems.* Academic Press, New York.

Vemuri, V., and W. J. Karplus. 1984. *Digital Computer Treatment of Partial Differential Equations.* Prentice-Hall, Englewood Cliffs, NJ.

Wang, H. T., and T. T. Huang. 1979. *Calculation of Potential Flow-Boundary-Layer Interaction on Axisymmetric Bodies.* Report by the Ship Performance Dept., David W. Taylor Naval Ship R and D Center, Bethesda, MD.

Wang, T.-Y., and C. Kleinstreuer. 1988. "Local Skin Friction and Heat Transfer in Combined Free-Forced Convection from a Cylinder or Spheres to a Power-Law Fluid," *Int. J. Heat Fluid Flow*, **9**, 182–187.

Wang, T.-Y., and C. Kleinstreuer. 1989. "Mixed Convection over Rotating Bodies with Blowing and Suction," *Int. J. Heat Mass Transfer*, **32**:1309–1316.

Watlow (anon.) 1991. Experimental Data Sets for Cartridge Heater and Beaker Temperature Study. Consulting Report for Ed. WECK, Research Triangle Park, NC.

Weinbaum, S., and S. Chien. 1993. "Lipid Transport Aspects of Atherogensis," *J. Biomech. Eng.*, **115**, 602–610.

White, F. M. 1974. *Viscous Fluid Flow.* McGraw-Hill, New York.

White, F. M. 1991. *Viscous Fluid Flow.* McGraw-Hill, New York.

Zang, T. A., R. B. Dahlburg, and J. P. Dahlburg. 1992. "Direct and Large-Eddy Simulations of 3-D Compressible Navier–Stokes Turbulence," *Physics Fluids A*, **4**(1).

Differential Operators and Cartesian Tensor Applications

Cylindrical System

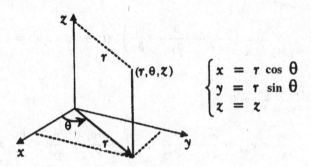

$$\begin{cases} x = r \cos \theta \\ y = r \sin \theta \\ z = z \end{cases}$$

Spherical System

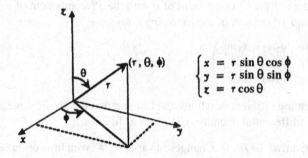

$$\begin{cases} x = r \sin \theta \cos \phi \\ y = r \sin \theta \sin \phi \\ z = r \cos \theta \end{cases}$$

A.1   Introduction

Operators are command symbols implying a well-defined mathematical manipulation of variables. Differential operators such as the material (or substantial) derivative, D/Dt; the total time derivative, d/dt; the del operator, ∇; the Laplacian operator, ∇^2; and the linear operator, $\mathcal{L}$, are well-known examples.

Tensors are used as shorthand notation for the description of and operation with physical properties. Tensors of rank n have 3^n components. For example, scalars are zero-order, vectors are first-order, and dyads (i.e., dyadic or tensor products of two vectors) are second-order tensors. Coordinate transformations, fluid flow properties, or transport phenomena can be depicted in a generalized, compact form employing tensors and tensor operations. Their use prevents errors and saves time. Physical laws describing transfer processes are inherently coordinate-invariant; so are tensors, in general. However, the (scalar) components of a tensor are defined with reference to a particular coordinate system.

Governing equations in engineering systems analysis are written in either expanded form
for a particular coordinate system, in vector notation (also known as the symbolic notation),
or in tensor notation, that is, index notation, where usually Einstein's summation convention
for repeated indices is employed. It is important to be familiar with both notations in order
to be able to follow advanced textbooks and research articles and to use them as time- and
effort-saving codes. For example, the equation div $\vec{v} = 0$ (i.e., divergence-free vector field)
or $\nabla \cdot \vec{v} = 0$, known as the *continuity equation for incompressible flow*, can be written in
various forms. Using here rectangular coordinates, with unit vectors $\hat{i}$, $\hat{j}$ and $\hat{k}$, one could
write the continuity equation

$$\textit{in vector form} \quad \nabla \cdot \vec{v} = \left(\hat{i}\frac{\partial}{\partial x} + \hat{j}\frac{\partial}{\partial y} + \hat{k}\frac{\partial}{\partial z}\right) \cdot (\hat{i}u + \hat{j}v + \hat{k}w) = 0$$

that is,

$$\nabla \cdot \vec{v} = \frac{\partial u}{\partial x} + \frac{\partial v}{\partial y} + \frac{\partial w}{\partial z} = 0 \tag{A.1a}$$

$$\textit{or in tensor notation} \quad \frac{\partial v_j}{\partial x_j} = v_{j,j} = \frac{\partial v_1}{\partial x_1} + \frac{\partial v_2}{\partial x_2} + \frac{\partial v_3}{\partial x_3} = \frac{\partial v_x}{\partial x} + \frac{\partial v_y}{\partial y} + \frac{\partial v_z}{\partial z} = 0$$

$$\tag{A.1b}$$

To illustrate the index notation, we consider a matrix formed by "multiplying" two
vectors: the dyadic product of the ith component of $\vec{v}$ with the jth component of $\vec{w}$. It can
be found in the ith row and j^{th} column of the following array:

$$(\vec{v}\vec{w}) = \begin{pmatrix} v_1 w_1 & v_1 w_2 & v_1 w_3 \\ v_2 w_1 & v_2 w_2 & v_2 w_3 \\ v_3 w_1 & v_3 w_2 & v_3 w_3 \end{pmatrix} \tag{A.2}$$

Another example, illustrating different notations and illuminating the physical meaning of
various time derivatives (differential operators), is given as follows:

- Partial time derivative: $\partial \#/\partial t \triangleq$ Changes in variable "#" with time observed from
  a fixed position in space.
- Substantial or material time derivate: $D\#/Dt \triangleq$ Changes of variable "#" with time
  following the fluid/material motion; by definition

$$\frac{D\#}{Dt} \equiv \frac{\partial \#}{\partial t} + (\vec{v} \cdot \nabla)\# \tag{A.3}$$

Hence the Lagrangian time rate of change is expressed in Eulerian derivatives. For
example, if c is the concentration $[M/L^3]$, then the material time derivative is

$$\frac{Dc}{Dt} \equiv \frac{\partial c}{\partial t} + (\vec{v} \cdot \nabla)c$$

In rectangular coordinates

$$\frac{Dc}{Dt} = \frac{\partial c}{\partial t} + u\frac{\partial c}{\partial x} + v\frac{\partial c}{\partial y} + w\frac{\partial c}{\partial z}$$

whereas in tensor notation

$$\frac{Dc}{Dt} = \frac{\partial c}{\partial t} + v_k \frac{\partial c}{\partial x_k}; \qquad k = 1, 2, 3$$

where $\frac{\partial c}{\partial t} \triangleq$ local time derivative (i.e., accumulation of species c) and $v_k \frac{\partial c}{\partial x_k} \triangleq$ convective derivatives (i.e., mass transfer by convection)

- Total time derivative: $d\#/dt \triangleq$ Changes of # with respect to time observed from a point moving differently than the flow field. For example:

$$\frac{dc}{dt} = \frac{\partial c}{\partial t} + \frac{dx}{dt}\frac{\partial c}{\partial x} + \frac{dy}{dt}\frac{\partial c}{\partial y} + \frac{dz}{dt}\frac{\partial c}{\partial z}$$

where $dx/dt, dy/dt$, and dz/dt are the velocity components of the moving observer.

A.2 Conversions and Operations with Unit Vectors

Of the broad field of tensor calculus, the concepts of gradient (grad), divergence (div), and rotation (curl) of a tensor are very important. Their definitions and applications require basic skills in manipulating products of unit vectors and conversions from rectangular to curvilinear coordinates.

1. *Definition of the ∇-operator and conversions to curvilinear coordinates:* The vector differential operator del is defined in rectangular coordinates as

$$\nabla \equiv \vec{\delta}_1 \frac{\partial}{\partial x_1} + \vec{\delta}_2 \frac{\partial}{\partial x_2} + \vec{\delta}_3 \frac{\partial}{\partial x_3} \qquad (A.4)$$

where $\vec{\delta}_i$ are the unit vectors. From trigonometry we recall that, for example, for cylindrical coordinates

$$\vec{\delta}_r = \vec{\delta}_x \cos\theta + \vec{\delta}_y \sin\theta; \qquad \vec{\delta}_\theta = -\vec{\delta}_x \sin\theta + \vec{\delta}_y \cos\theta; \qquad \vec{\delta}_z = \vec{\delta}_z;$$
$$\vec{\delta}_x = \vec{\delta}_r \cos\theta - \vec{\delta}_\theta \sin\theta; \qquad \vec{\delta}_y = \vec{\delta}_r \sin\theta + \vec{\delta}_\theta \cos\theta \qquad (A.5a\text{--}e)$$

Taking the spatial derivative of the unit vectors, we obtain

$$\frac{\partial}{\partial r}\vec{\delta}_r = \frac{\partial}{\partial r}\vec{\delta}_\theta = \frac{\partial}{\partial r}\vec{\delta}_z = \vec{O} \quad \text{and} \quad \frac{\partial}{\partial z} \cdots = \vec{O},$$

but

$$\frac{\partial}{\partial \theta}\vec{\delta}_r = \vec{\delta}_\theta \quad \text{and} \quad \frac{\partial}{\partial \theta}\vec{\delta}_\theta = -\vec{\delta}_r \qquad (A.6a,b)$$

For example, inserting (A.5d,e) in (A.4) yields the del operator in cylindrical coordinates

$$\nabla = \vec{\delta}_r \frac{\partial}{\partial r} + \vec{\delta}_\theta \frac{1}{r}\frac{\partial}{\partial \theta} + \vec{\delta}_z \frac{\partial}{\partial z} \qquad (A.7)$$

Similarly, in spherical coordinates

♦
$$\nabla = \vec{\delta}_r \frac{\partial}{\partial r} + \vec{\delta}_\theta \frac{1}{r}\frac{\partial}{\partial \theta} + \vec{\delta}_\phi \frac{1}{r\sin\theta}\frac{\partial}{\partial \phi} \tag{A.8}$$

Note: By introducing "metric-scale factors"

$$h_i = \left|\frac{\partial \vec{r}}{\partial x_i}\right|; \qquad \vec{r} = \vec{\delta}_1 x_1 + \vec{\delta}_x x_2 + \vec{\delta}_3 x_3$$

operators and tensor manipulations can be defined for any orthogonal curvilinear coordinates (cf. Bird et al. 1987; Currie 1993).

2. *Manipulations with unit vectors:* There are several basic types of product operations, four of which are illustrated here with the unit vectors $\vec{\delta}_i$.

(a) *dot product:* $\qquad \vec{\delta}_i \cdot \vec{\delta}_j = \delta_{ij} := \begin{cases} 1 & \text{if } i = j \\ 0 & \text{if } i \neq j \end{cases}$ (Kronecker delta).

For example

$$\nabla \cdot \vec{v} = \left(\hat{i}\frac{\partial}{\partial x} + \hat{j}\frac{\partial}{\partial y} + \hat{k}\frac{\partial}{\partial z}\right) \cdot \left(\hat{i}u + \hat{j}v + \hat{k}w\right)$$

♦
$$= \frac{\partial u}{\partial x} + \frac{\partial v}{\partial y} + \frac{\partial w}{\partial z} \tag{A.9}$$

(b) *cross-product:* $\qquad \vec{\delta}_i \times \vec{\delta}_j = \sum_{k=1}^{3} \varepsilon_{ijk}\vec{\delta}_k$

where $\quad \delta_{ij}$ is the Kronecker delta and ε_{ijk} is the permutation symbol.

$$\varepsilon_{ijk} = \begin{cases} +1 & \text{if } ijk = 123, 231, \text{ or } 312 \\ -1 & \text{if } ijk = 321, 132, \text{ or } 213 \\ 0 & \text{if any two indices are alike} \end{cases}$$

For example

$$\nabla \times \vec{v} = \left(\hat{i}\frac{\partial}{\partial x} + \hat{j}\frac{\partial}{\partial y} + \hat{k}\frac{\partial}{\partial z}\right) \times \left(\hat{i}u + \hat{j}v + \hat{k}w\right)$$

With

$$\begin{array}{lll} \hat{i} \times \hat{j} = \hat{k} & \hat{j} \times \hat{i} = -\hat{k} & \hat{i} \times \hat{i} = 0 \\ \hat{j} \times \hat{k} = \hat{i} & \hat{k} \times \hat{j} = -\hat{i} & \hat{j} \times \hat{j} = 0 \\ \hat{k} \times \hat{i} = \hat{j} & \hat{i} \times \hat{k} = -\hat{j} & \hat{k} \times \hat{k} = 0 \end{array}$$

we obtain

$$\nabla \times \vec{v} = \hat{i}\left(\frac{\partial w}{\partial y} - \frac{\partial v}{\partial z}\right) + \hat{j}\left(\frac{\partial u}{\partial z} - \frac{\partial w}{\partial x}\right) + \hat{k}\left(\frac{\partial v}{\partial x} - \frac{\partial u}{\partial y}\right) \tag{A.10}$$

or

$$\nabla \times \vec{v} = \operatorname{curl} \vec{v} = \begin{vmatrix} \hat{i} & \hat{j} & \hat{k} \\ \dfrac{\partial}{\partial x} & \dfrac{\partial}{\partial y} & \dfrac{\partial}{\partial z} \\ u & v & w \end{vmatrix} \equiv \vec{\zeta} \quad \text{(vorticity vector)}$$

(c) *dyadic product:* $\vec{\delta}_i \vec{\delta}_j$ indicates the *location* (i.e., $\vec{\delta}_i$ is the unit normal to the particular surface) and *direction* (i.e., $\vec{\delta}_j$ gives the direction) of a tensor of rank 2 (i.e., with nine components).

For example

$$\nabla \vec{v} = \left(\hat{i} \frac{\partial}{\partial x} + \hat{j} \frac{\partial}{\partial y} + \hat{k} \frac{\partial}{\partial z} \right) \left(\hat{i} u + \hat{j} v + \hat{k} w \right)$$

$$= \hat{i}\hat{i} \frac{\partial u}{\partial x} + \hat{i}\hat{j} \frac{\partial v}{\partial x} + \hat{i}\hat{k} \frac{\partial w}{\partial x} + \hat{j}\hat{i} \frac{\partial u}{\partial y} + \hat{j}\hat{j} \frac{\partial v}{\partial y} + \hat{j}\hat{k} \frac{\partial w}{\partial y}$$

$$+ \hat{k}\hat{i} \frac{\partial u}{\partial z} + \hat{k}\hat{j} \frac{\partial v}{\partial z} + \hat{k}\hat{k} \frac{\partial w}{\partial z}$$

Hence

$$\nabla \vec{v} = \operatorname{grad} \vec{v} = \begin{vmatrix} \dfrac{\partial u}{\partial x} & \dfrac{\partial v}{\partial x} & \dfrac{\partial w}{\partial x} \\ \dfrac{\partial u}{\partial y} & \dfrac{\partial v}{\partial y} & \dfrac{\partial w}{\partial y} \\ \dfrac{\partial u}{\partial z} & \dfrac{\partial v}{\partial z} & \dfrac{\partial w}{\partial z} \end{vmatrix} \tag{A.11}$$

(d) *double dot product of two tensors:* $\vec{\delta}_i \vec{\delta}_j : \vec{\delta}_k \vec{\delta}_l$ generating a scalar; for example, energy dissipation in a moving viscous fluid.

$$\vec{\delta}_i \vec{\delta}_j : \vec{\delta}_k \vec{\delta}_l = \left(\vec{\delta}_i \cdot \vec{\delta}_l \right) \left(\vec{\delta}_i \cdot \vec{\delta}_k \right) = \delta_{il} \delta_{jk}$$

For example:

$$\vec{\vec{\tau}} : \nabla \vec{v} \qquad = \nabla \cdot \left(\vec{\vec{\tau}} \cdot \vec{v} \right) - \vec{v} \cdot \left(\nabla \cdot \vec{\vec{\tau}} \right)$$

$$\sum_i \sum_j \tau_{ij} \frac{\partial}{\partial x_i} v_j = \sum_i \sum_j \frac{\partial}{\partial x_i} (\tau_{ij} v_j) - \sum_j \sum_i v_j \frac{\partial}{\partial x_i} \tau_{ij}$$

$$\sum_i \sum_j \tau_{ij} \frac{\partial}{\partial x_i} v_j = \sum_i \sum_j \tau_{ij} \frac{\partial}{\partial x_i} v_j \qquad \text{where } \tau_{ij} = \tau_{ji}$$

Note: It is very important to practice these tensor manipulations, including conversions from rectangular to curvilinear coordinates, in order to understand the following definitions and to work with the transport equations derived in Chapter 2. See Section A.2.2 and Section 2.4 for Sample-Problem solutions.

A.2.1 Definitions

1. The gradient of a tensor, R, of rank r yields a tensor of rank$(r + 1)$. For example, the gradient of a scalar, ∇s, such as the pressure, temperature, or concentration gradient, is a vector (cf. (A.12)); the dyad of a vector field, $\nabla \vec{v}$ (note that $\nabla \vec{v} \neq \vec{v}\nabla$), produces a tensor of rank 2, having nine components (cf. (A.11)). Specifically

   - gradient of a scalar field:

$$\text{grad } s = \nabla s = \sum_i \vec{\delta}_i \frac{\partial s}{\partial x_i} = \vec{\delta}_1 \frac{\partial s}{\partial x_1} + \vec{\delta}_2 \frac{\partial s}{\partial x_2} + \vec{\delta}_3 \frac{\partial s}{\partial x_3} \qquad \text{(vector)} \qquad \text{(A.12)}$$

   Note: $\nabla(\dot{r} + s) = \nabla r + \nabla s$ but $\nabla s \neq s\nabla$ and $(\nabla r)s \neq \nabla(rs)$.

   - Laplacian of a scalar field (divergence of ∇s):

$$\nabla \cdot \nabla s = \nabla^2 s = \sum_i \frac{\partial^2 s}{\partial x_i^2} \qquad \text{(scalar)} \qquad\qquad \text{(A.13)}$$

   Note: ∇^2 is the Laplacian operator.

   - Dyadic product of two vectors:

$$\nabla \vec{v} = \text{grad } \vec{v} = \sum_i \sum_j \vec{\delta}_i \vec{\delta}_j \frac{\partial}{\partial x_i} v_j \qquad \text{(2nd-order tensor)} \qquad \text{(A.14)}$$

   Thus the components of $\nabla \vec{v}$ are $\partial v_j / \partial x_i$. Its transpose is $\partial v_i / \partial x_j$. In general, let $\vec{\vec{a}}$ be a second-order tensor with components a_{ij}. Then the transpose of $\vec{\vec{a}}$ is denoted by $\vec{\vec{a}}^T$ or $(\vec{\vec{a}})^{tr}$ and is defined by

$$\left[a^T \right]_{ij} = a_{ji}$$

   If $\vec{\vec{a}} = -\vec{\vec{a}}^T$ or $a_{ij} \neq a_{ji}$, then $\vec{\vec{a}}$ is said to be *antisymmetric*, whereas $\vec{\vec{a}} = \vec{\vec{a}}^T$ or $a_{ij} = a_{ji}$ means that $\vec{\vec{a}}$ is *symmetric*.

2. The divergence of a tensor, R, of rank r results in a tensor of rank $(r - 1)$. For example, the divergence of a vector field, $\nabla \cdot \vec{v}$, generates a scalar and $\nabla \cdot \vec{\vec{\tau}}$ results in a vector. Specifically,

   - Divergence of a vector field:

$$\text{div } \vec{v} \equiv \nabla \cdot \vec{v} = \sum_i \sum_j (\vec{\delta}_i \cdot \vec{\delta}_j) \frac{\partial}{\partial x_i} v_j = \sum_i \sum_j \delta_{ij} \frac{\partial v_j}{\partial x_i} = \sum_i \frac{\partial v_i}{\partial x_i} \quad \text{(A.15)}$$

   Note: $\nabla \cdot \vec{v} \neq \vec{v} \cdot \nabla$; $\nabla \cdot s\vec{v} \neq \nabla s \cdot \vec{v}$.

   - Divergence of a tensor field:

$$\text{div } \vec{\vec{\tau}} = \nabla \cdot \vec{\vec{\tau}} = \sum_k \delta_k \left\{ \sum_i \frac{\partial}{\partial x_i} \tau_{ik} \right\} \qquad\qquad\qquad \text{(A.16)}$$

Note: The kth component of $[\nabla \cdot \vec{\vec{\tau}}]$ is $\sum_i (\partial \tau_{ik}/\partial x_i)$.

- Laplacian of a vector field:

$$\text{div (grad } \vec{v}) = [\nabla \cdot \nabla \vec{v}] = \sum_k \vec{\delta}_k \left(\sum_i \frac{\partial^2}{\partial x_i^2} v_k \right) \qquad (A.17)$$

3. If R is a tensor of rank r, tensor rotation or the curl operation will produce an antisymmetric tensor of rank $(r + 1)$. Note, a second-rank tensor can be split into a symmetric and an antisymmetric part: $\vec{\vec{R}} = 1/2(\vec{\vec{R}} + \vec{\vec{R}}^T) + 1/2(\vec{\vec{R}} - \vec{\vec{R}}^T)$; for example, the strain-rate tensor, $\vec{\vec{\varepsilon}} \equiv \nabla \vec{v}$, can be split into two parts, one reflecting deformation due to strain and the other due to rotation, namely,

$$\vec{\vec{\varepsilon}} \equiv \frac{\partial v_i}{\partial x_j} = \frac{1}{2}\left(\frac{\partial v_i}{\partial x_j} + \frac{\partial v_j}{\partial x_i} \right) + \frac{1}{2}\left(\frac{\partial v_i}{\partial x_j} - \frac{\partial v_j}{\partial x_i} \right) \qquad (A.18)$$

As indicated earlier, tensors of rank 2 are frequently encountered in thermofluid transfer processes. Examples include the stress, deformation, and diffusion tensors as well as the dyadic product, $\nabla \vec{v}$. As a *vector* can be defined as a quantity that associates a scalar magnitude of a vector component with each coordinate direction (the unit vectors), a tensor (rank 2) associates a vector with each coordinate direction. For example,

$$\vec{\vec{\tau}} = \vec{\delta}_1 \vec{\tau}_1 + \vec{\delta}_2 \vec{\tau}_2 + \vec{\delta}_3 \vec{\tau}_3 = \sum_{i=1}^{3} \vec{\delta}_i \vec{\tau}_i = \sum_{i=1}^{3} \sum_{j=1}^{3} \vec{\delta}_i \vec{\delta}_j \tau_{ij} \qquad (A.19)$$

where $\vec{\tau}_i$ are vectors,

         $\vec{\delta}_i$ are unit vectors,

         $\vec{\delta}_i \vec{\tau}_i$ are dyadic products, and

         τ_{ij} are the (nine) components of the (stress) tensor $\vec{\vec{\tau}}$.

Another part of the traditional definition of a cartesian tensor is the *transformation rule*. Consider two sets of cartesian axes (x_1, x_2, x_3 and x_1', x_2', x_3'), rotated with respect to one another. Either can be used to describe any point in space

$$x_i = \sum_j x_j' \cos(x_j', x_i) \quad \text{and} \quad x_j' = \sum_i x_i \cos(x_i, x_j')$$

or, with the direction cos $l_{ij} = \cos(x_i, x_j')$, we obtain

$$x_j' = x_i l_{ij} \quad \text{and} \quad x_i = x_j' l_{ij}$$

where $$\sum_{i=1}^{3} \sum_{j=1}^{3} \cos(x_i, x_j') \cdot \cos(x_k, x_j') = l_{ij} \cdot l_{kj} = \delta_{ik} = \begin{cases} 1 & \text{if } i = k \\ 0 & \text{if } i \neq k \end{cases}$$

A.2.2   Summary

Operations with dyads (i.e., tensors of rank 2) can be summarized for our purposes as follows:

(i) Scalar (or double dot) product, $(\vec{\vec{A}} : \vec{\vec{B}})$, results in a scalar; examples are the dissipation term of the heat equation. Thus

$$(\vec{\vec{A}} : \vec{\vec{B}}) = \left(\underbrace{\left\{ \sum_i \sum_j \vec{\delta}_i \vec{\delta}_j A_{ij} \right\}}_{9 \text{ terms}} : \underbrace{\left\{ \sum_k \sum_l \vec{\delta}_k \vec{\delta}_l B_{kl} \right\}}_{9 \text{ terms}} \right)$$

$$= \sum_i \sum_j \sum_k \sum_l (\vec{\delta}_i \vec{\delta}_j : \vec{\delta}_k \vec{\delta}_l) A_{ij} B_{kl}$$

Recall from vector analysis:

$$\vec{\delta}_n \cdot \vec{\delta}_m = \delta_{mn} = \begin{cases} 1 & \text{if } n = m \\ 0 & \text{if } n \neq m \end{cases}$$

and from the operation for unit vectors

$$\vec{\delta}_i \vec{\delta}_j : \vec{\delta}_k \vec{\delta}_l = (\vec{\delta}_j \cdot \vec{\delta}_k)(\vec{\delta}_i \cdot \vec{\delta}_l) = \delta_{jk} \delta_{il}$$

Recall from vector analysis (scalar product of two vectors)

$$\sum_n \sum_m (\vec{\delta}_n \cdot \vec{\delta}_m) v_n w_m = \sum_n \sum_m \delta_{nm} v_n w_m = \sum_n v_n w_m$$

that is, only when $n = m$ are values (terms) collected; similarly when $k = j$ and $l = i$.

$$\therefore \ (\vec{\vec{A}} : \vec{\vec{B}}) = \sum_i \sum_j A_{ij} B_{ji} \tag{A.20}$$

(ii) Vector (dot) product $[\vec{v} \cdot \vec{\vec{T}}]$ resulting in a vector: similar to the divergence of a stress field $[\nabla \cdot \vec{\vec{\pi}}]$, the diffusion of momentum in viscous fluid flow.

$$[\vec{v} \cdot \vec{\vec{T}}] = \left[\left\{ \sum_i \vec{\delta}_i v_i \right\} \cdot \left\{ \sum_k \sum_l \vec{\delta}_k \vec{\delta}_l T_{kl} \right\} \right] = \sum_i \sum_k \sum_l [\vec{\delta}_i \cdot \vec{\delta}_k \vec{\delta}_l] v_i T_{kl}$$

$$= \sum_i \sum_k \sum_l \delta_{ik} \vec{\delta}_l v_i T_{kl} = \sum_l \vec{\delta}_l \left\{ \sum_k T_{kl} v_k \right\} \tag{A.21}$$

(iii) Magnitude of a tensor

$$|\vec{\vec{\tau}}| = \tau = \sqrt{\frac{1}{2}(\vec{\vec{\tau}} \cdot \vec{\vec{\tau}}^T)} = \sqrt{\frac{1}{2} \sum_i \sum_j \tau_{ij}^2}$$

(iv) Invariants of a tensor are scalars independent of the coordinate system

$$I_1 = \text{trace of } \vec{\vec{T}} = \sum_i T_{ii}$$

$$I_2 = \frac{1}{2}[(\operatorname{tr} T)^2 - \operatorname{tr} T^2] = \frac{1}{2}(T_{ii}T_{jj} - T_{ij}T_{ji})$$

$$I_3 = \det(\vec{\vec{T}}) = \text{determinant of matrix of } T_{ij}$$

A useful theorem (Cayley–Hamilton) for the determination of the principal values (eigenvalues) of the symmetrical second-rank tensor T is the characteristic equation:

$$T^3 - I_1 T^2 - I_2 T - I_3 = 0$$

Its solution gives the three characteristic values of the tensor T.

(v) Given the components T_{ij} of a tensor $\vec{\vec{T}}$ in an x_i ($i = 1, 2, 3$) coordinate system, we can always find three principal directions (or axes) x_i' in which the tensor $\vec{\vec{T}}'$ with components T_{ij}' will have a diagonal form, that is, $T_{ij}' = 0$ for $i \neq j$. For the 2-D case, these relationships are:

$$T_{11}' = -\frac{1}{2}(T_{11} - T_{22}) \sin 2(x_1 x_1') + T_{12} \cos 2(x_1, x_1')$$

$$T_{22}' = \frac{1}{2}(T_{11} + T_{22}) - \frac{1}{2}(T_{11} - T_{22}) \cos 2(x_2 x_2') - T_{12} \sin 2(x_2, x_2')$$

$$T_{12}' = T_{21}' = 0; \qquad \tan 2(x_i, x_i') = 2\frac{T_{12}}{(T_{11} - T_{22})}$$

Thus, selecting the three principal directions as the coordinate axes x, y, and z of the flow domain reduces the number of unknown tensor components from six $(T_{ij} = T_{ji})$ to three.

Additional Relations

$$\nabla rs = r\nabla s + s\nabla r$$

$$(\nabla \cdot s\vec{v}) = (\nabla s \cdot \vec{v}) + s(\nabla \cdot \vec{v})$$

$$(\nabla \cdot [\vec{v} \times \vec{w}]) = (\vec{w} \cdot [\nabla \times \vec{v}]) - (\vec{v} \cdot [\nabla \times \vec{w}])$$

$$[\nabla \times s\vec{v}] = [\nabla s \times \vec{v}] + s(\nabla \times \vec{v})$$

$$[\nabla \cdot \nabla \vec{v}] = \nabla(\nabla \cdot \vec{v}) - [\nabla \times [\nabla \times \vec{v}]]$$

$$[\vec{v} \cdot \nabla \vec{v}] = \frac{1}{2}\nabla(\vec{v} \cdot \vec{v}) - [\vec{v} \times [\nabla \times \vec{v}]]$$

$$[\nabla \cdot \vec{v}\vec{w}] = [\vec{v} \cdot \nabla \vec{w}] + \vec{w}(\nabla \cdot \vec{v})$$

$$\left[\nabla \cdot s\vec{\vec{\delta}}\right] = \nabla s$$

$$\left[\nabla \cdot s\vec{\vec{\tau}}\right] = \left[\nabla s \cdot \vec{\vec{\tau}}\right] + s\left[\nabla \cdot \vec{\vec{\tau}}\right]$$

$$\nabla(\vec{v} \cdot \vec{w}) = [(\nabla \vec{v}) \cdot \vec{w}] + [(\nabla \vec{w}) \cdot \vec{v}]$$

$$(s\vec{\vec{\delta}} : \nabla \vec{v}) = s(\nabla \cdot \vec{v})$$

Sample Problems

To illustrate a few tensor manipulations, the following sample problems are solved. Given the components of a symmetric tensor $\vec{\vec{\tau}}$:

$$\tau_{xx} = 3 \qquad \tau_{xy} = 2, \qquad \tau_{xz} = -1,$$
$$\tau_{yy} = 2, \qquad \tau_{yz} = 1,$$
$$\tau_{zz} = 0$$

and the components of a vector $\vec{v}$: $v_x = 5$, $v_y = 3$, $v_z = 7$; evaluate:

(a) $[\vec{\vec{\tau}} \cdot \vec{v}]$

(b) $[\vec{v} \cdot \vec{\vec{\tau}}]$

(c) $[\vec{\vec{\tau}} : \vec{\vec{\tau}}]$

(d) $\vec{v}\vec{v}$

(e) $\{\vec{\vec{\tau}} \cdot \vec{\vec{\delta}}\}$

where $\vec{\vec{\delta}}$ is the unit tensor:

$$\delta_{ij} = \begin{pmatrix} 1 & & \phi \\ & 1 & \\ \phi & & 1 \end{pmatrix}$$

Solution

(a) $\displaystyle [\vec{\vec{\tau}} \cdot \vec{v}] = \sum_i \vec{\delta}_i \left\{ \sum_j \tau_{ij} v_j \right\} = (\vec{\delta}_1 \vec{\delta}_2 \vec{\delta}_3) \begin{pmatrix} 3 & 2 & -1 \\ 2 & 2 & 1 \\ -1 & 1 & 0 \end{pmatrix} \begin{pmatrix} 5 \\ 3 \\ 7 \end{pmatrix}$

$$= 14\vec{\delta}_1 + 23\vec{\delta}_2 - 2\vec{\delta}_3$$

where $\vec{\delta}_i \triangleq$ unit vector in i-direction.

(b) $[\vec{v} \cdot \vec{\vec{\tau}}] = [\vec{\vec{\tau}} \cdot \vec{v}]$ since $\vec{\vec{\tau}}$ is symmetric

(c) $[\vec{\vec{\tau}} : \vec{\vec{\tau}}] \sum_i \sum_j \tau_{ij} \tau_{ji}$

Using symmetry

$$i = x \qquad \tau_{xx}^2 + \tau_{xy}^2 + \tau_{xz}^2$$

$$i = y \qquad \tau_{yx}^2 + \tau_{yy}^2 + \tau_{yz}^2$$

$$i = z \qquad \tau_{zx}^2 + \tau_{zy}^2 + \tau_{zz}^2$$

or

$$\left. \begin{array}{l} 9 + 4 + 1 = 14 \\ 4 + 4 + 1 = 9 \\ 1 + 1 + 0 = 2 \end{array} \right\} \quad \Rightarrow \quad 25$$

(d) $\vec{v}\vec{v} = \sum_i \sum_j \vec{\delta}_i \vec{\delta}_j v_i v_j = 25\vec{\delta}_1\vec{\delta}_1 + 15\vec{\delta}_1\vec{\delta}_2 + 35\vec{\delta}_1\vec{\delta}_3$
$$+ 15\vec{\delta}_2\vec{\delta}_1 + 9\vec{\delta}_2\vec{\delta}_2 + 21\vec{\delta}_2\vec{\delta}_3$$
$$+ 35\vec{\delta}_3\vec{\delta}_1 + 21\vec{\delta}_3\vec{\delta}_2 + 49\vec{\delta}_3\vec{\delta}_3$$

(e) $\{\vec{\vec{\tau}} \cdot \vec{\delta}\} = \sum_i \sum_l \vec{\delta}_i \vec{\delta}_l (\sum_j \tau_{ij} \delta_{jl}) = \vec{\vec{\tau}} = 3\vec{\delta}_1\vec{\delta}_1 + 2\vec{\delta}_1\vec{\delta}_2 - 1\vec{\delta}_1\vec{\delta}_3$
$$+ 2\vec{\delta}_2\vec{\delta}_1 + 2\vec{\delta}_2\vec{\delta}_2 + 1\vec{\delta}_2\vec{\delta}_3$$
$$- 1\vec{\delta}_3\vec{\delta}_1 + 1\vec{\delta}_3\vec{\delta}_2 + 0\vec{\delta}_3\vec{\delta}_3$$

Further relevant material and examples can be found in Aris (1962), Lin and Segal (1975), and Bird et al. (1987).

Basic Equations in Rectangular, Cylindrical, and Spherical Coordinates

Cylindrical System

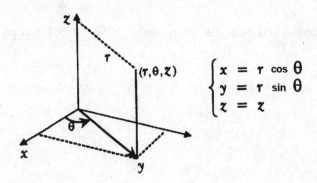

$$\begin{cases} x = r \cos \theta \\ y = r \sin \theta \\ z = z \end{cases}$$

Spherical System

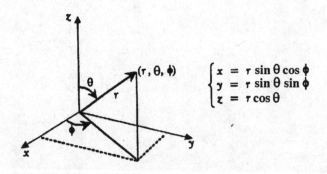

$$\begin{cases} x = r \sin \theta \cos \phi \\ y = r \sin \theta \sin \phi \\ z = r \cos \theta \end{cases}$$

B.1 The Continuity and Momentum Equations*

B.1.1 Continuity Equation

A *Rectangular Coordinates* (x, y, z)

$$\frac{\partial \rho}{\partial t} + \frac{\partial}{\partial x}(\rho v_x) + \frac{\partial}{\partial y}(\rho v_y) + \frac{\partial}{\partial z}(\rho v_z) = 0$$

---

* *Source:* Bird et al. (1987).

B   *Cylindrical Coordinates* (r, θ, z)

$$\frac{\partial \rho}{\partial t} + \frac{1}{r}\frac{\partial}{\partial r}(\rho r v_r) + \frac{1}{r}\frac{\partial}{\partial \theta}(\rho v_\theta) + \frac{\partial}{\partial z}(\rho v_z) = 0$$

C   *Spherical Coordinates* (r, θ, ϕ)

$$\frac{\partial \rho}{\partial t} + \frac{1}{r^2}\frac{\partial}{\partial r}(\rho r^2 v_r) + \frac{1}{r \sin\theta}\frac{\partial}{\partial \theta}(\rho v_\theta \sin\theta) + \frac{1}{r \sin\theta}\frac{\partial}{\partial \phi}(\rho v_\phi) = 0$$

B.1.2   *Equation of Motion in Terms of $\bar{\bar{\tau}}$*

A   *Rectangular Coordinates* (x, y, z)

$$\rho\left(\frac{\partial v_x}{\partial t} + v_x\frac{\partial}{\partial x}v_x + v_y\frac{\partial}{\partial y}v_x + v_z\frac{\partial}{\partial z}v_x\right) = \left[\frac{\partial}{\partial x}\tau_{xx} + \frac{\partial}{\partial y}\tau_{yx} + \frac{\partial}{\partial z}\tau_{zx}\right] - \frac{\partial p}{\partial x} + \rho g_x$$

$$\rho\left(\frac{\partial v_y}{\partial t} + v_x\frac{\partial}{\partial x}v_y + v_y\frac{\partial}{\partial y}v_y + v_z\frac{\partial}{\partial z}v_y\right) = \left[\frac{\partial}{\partial x}\tau_{xy} + \frac{\partial}{\partial y}\tau_{yy} + \frac{\partial}{\partial z}\tau_{zy}\right] - \frac{\partial p}{\partial y} + \rho g_y$$

$$\rho\left(\frac{\partial v_z}{\partial t} + v_x\frac{\partial}{\partial x}v_z + v_y\frac{\partial}{\partial y}v_z + v_z\frac{\partial}{\partial z}v_z\right) = \left[\frac{\partial}{\partial x}\tau_{xz} + \frac{\partial}{\partial y}\tau_{yz} + \frac{\partial}{\partial z}\tau_{zz}\right] - \frac{\partial p}{\partial x} + \rho g_z$$

B   *Cylindrical Coordinates* (r, θ, z)

$$\rho\left(\frac{\partial v_r}{\partial t} + v_r\frac{\partial v_r}{\partial r} + \frac{v_\theta}{r}\frac{\partial v_r}{\partial \theta} - \frac{v_\theta^2}{r} + v_z\frac{\partial v_r}{\partial z}\right)$$

$$= \left[\frac{1}{r}\frac{\partial}{\partial r}(r\tau_{rr}) + \frac{1}{r}\frac{\partial}{\partial \theta}\tau_{\theta r} + \frac{\partial}{\partial z}\tau_{zr} - \frac{\tau_{\theta\theta}}{r}\right] - \frac{\partial p}{\partial r} + \rho g_r$$

$$\rho\left(\frac{\partial v_\theta}{\partial t} + v_r\frac{\partial v_\theta}{\partial r} + \frac{v_\theta}{r}\frac{\partial v_\theta}{\partial \theta} + \frac{v_r v_\theta}{r} + v_z\frac{\partial v_\theta}{\partial z}\right)$$

$$= \left[\frac{1}{r^2}\frac{\partial}{\partial r}(r^2\tau_{r\theta}) + \frac{1}{r}\frac{\partial}{\partial \theta}\tau_{\theta\theta} + \frac{\partial}{\partial z}\tau_{z\theta} - \frac{\tau_{\theta r} - \tau_{r\theta}}{r}\right] - \frac{1}{r}\frac{\partial p}{\partial \theta} + \rho g_\theta$$

$$\rho\left(\frac{\partial v_z}{\partial t} + v_r\frac{\partial v_z}{\partial r} + \frac{v_\theta}{r}\frac{\partial v_z}{\partial \theta} + v_z\frac{\partial v_z}{\partial z}\right) = \left[\frac{1}{r}\frac{\partial}{\partial r}(r\tau_{rz}) + \frac{1}{r}\frac{\partial}{\partial \theta}\tau_{\theta z} + \frac{\partial}{\partial z}\tau_{zz}\right] - \frac{\partial p}{\partial z} + \rho g_z$$

C   *Spherical Coordinates* (r, θ, ϕ)

$$\rho\left(\frac{\partial v_r}{\partial t} + v_r\frac{\partial v_r}{\partial r} + \frac{v_\theta}{r}\frac{\partial v_r}{\partial \theta} + \frac{v_\phi}{r \sin\theta}\frac{\partial v_r}{\partial \phi} + \frac{v_\theta^2 + v_\phi^2}{r}\right)$$

$$= \left[\frac{1}{r^2}\frac{\partial}{\partial r}(r^2\tau_{rr}) + \frac{1}{r \sin\theta}\frac{\partial}{\partial \theta}(\tau_{\theta r}\sin\theta) + \frac{1}{r \sin\theta}\frac{\partial}{\partial \phi}\tau_{\phi r}\right.$$

$$\left. - \frac{\tau_{\theta\theta} + \tau_{\phi\phi}}{r}\right] - \frac{\partial p}{\partial r} + \rho g_r$$

$$\rho\left(\frac{\partial v_\theta}{\partial t} + v_r\frac{\partial v_\theta}{\partial r} + \frac{v_\theta}{r}\frac{\partial v_\theta}{\partial \theta} + \frac{v_\phi}{r\sin\theta}\frac{\partial v_\theta}{\partial \phi} + \frac{v_r v_\theta}{r} - \frac{v_\phi^2\cot\theta}{r}\right)$$

$$= \left[\frac{1}{r^3}\frac{\partial}{\partial r}(r^3\tau_{r\theta}) + \frac{1}{r\sin\theta}\frac{\partial}{\partial \theta}(\tau_{\theta\theta}\sin\theta) + \frac{1}{r\sin\theta}\frac{\partial}{\partial \phi}\tau_{\phi\theta}\right.$$

$$\left. - \frac{(\tau_{\theta r} - \tau_{r\theta}) - \tau_{\phi\phi}\cot\theta}{r}\right] - \frac{1}{r}\frac{\partial p}{\partial \theta} + \rho g_\theta$$

$$\rho\left(\frac{\partial v_\phi}{\partial t} + v_r\frac{\partial v_\phi}{\partial r} + \frac{v_\theta}{r}\frac{\partial v_\phi}{\partial \theta} + \frac{v_\phi}{r\sin\theta}\frac{\partial v_\phi}{\partial \phi} + \frac{v_\phi v_r}{r} + \frac{v_\theta v_\phi}{r}\cot\theta\right)$$

$$= \left[\frac{1}{r^3}\frac{\partial}{\partial r}(r^3\tau_{r\phi}) + \frac{1}{r\sin\theta}\frac{\partial}{\partial \theta}(\tau_{\theta\phi}\sin\theta) + \frac{1}{r\sin\theta}\frac{\partial}{\partial \phi}\tau_{\phi\phi}\right.$$

$$\left. + \frac{(\tau_{\phi r} - \tau_{r\phi}) - \tau_{\phi\theta}\cot\theta}{r}\right] - \frac{1}{r\sin\theta}\frac{\partial p}{\partial \phi} + \rho g_\phi$$

B.2 The Navier–Stokes Equations*

A Rectangular Coordinates (x, y, z)

$$\rho\left(\frac{\partial v_x}{\partial t} + v_x\frac{\partial v_x}{\partial x} + v_y\frac{\partial v_x}{\partial y} + v_z\frac{\partial v_x}{\partial z}\right) = \mu\left[\frac{\partial^2}{\partial x^2}v_x + \frac{\partial^2}{\partial y^2}v_x + \frac{\partial^2}{\partial z^2}v_x\right] - \frac{\partial p}{\partial x} + \rho g_x$$

$$\rho\left(\frac{\partial v_y}{\partial t} + v_x\frac{\partial v_y}{\partial x} + v_y\frac{\partial v_y}{\partial y} + v_z\frac{\partial v_y}{\partial z}\right) = \mu\left[\frac{\partial^2}{\partial x^2}v_y + \frac{\partial^2}{\partial y^2}v_y + \frac{\partial^2}{\partial z^2}v_y\right] - \frac{\partial p}{\partial y} + \rho g_y$$

$$\rho\left(\frac{\partial v_z}{\partial t} + v_x\frac{\partial v_z}{\partial x} + v_y\frac{\partial v_z}{\partial y} + v_z\frac{\partial v_z}{\partial z}\right) = \mu\left[\frac{\partial^2}{\partial x^2}v_z + \frac{\partial^2}{\partial y^2}v_z + \frac{\partial^2}{\partial z^2}v_z\right] - \frac{\partial p}{\partial z} + \rho g_z$$

B Cylindrical Coordinates (r, θ, z)

$$\rho\left(\frac{\partial v_r}{\partial t} + v_r\frac{\partial v_r}{\partial r} + \frac{v_\theta}{r}\frac{\partial v_r}{\partial \theta} - \frac{v_\theta^2}{r} + v_z\frac{\partial v_r}{\partial z}\right) = \mu\left[\frac{\partial}{\partial r}\left(\frac{1}{r}\frac{\partial}{\partial r}(rv_r)\right)\right.$$

$$\left. + \frac{1}{r^2}\frac{\partial^2 v_r}{\partial \theta^2} + \frac{\partial^2 v_r}{\partial z^2} - \frac{2}{r^2}\frac{\partial v_\theta}{\partial \theta}\right] - \frac{\partial p}{\partial r} + \rho g_r$$

$$\rho\left(\frac{\partial v_\theta}{\partial t} + v_r\frac{\partial v_\theta}{\partial r} + \frac{v_\theta}{r}\frac{\partial v_\theta}{\partial \theta} + \frac{v_r v_\theta}{r} + v_z\frac{\partial v_\theta}{\partial z}\right) = \mu\left[\frac{\partial}{\partial r}\left(\frac{1}{r}\frac{\partial}{\partial r}(rv_\theta)\right)\right.$$

$$\left. + \frac{1}{r^2}\frac{\partial^2 v_\theta}{\partial \theta^2} + \frac{\partial^2 v_\theta}{\partial z^2} + \frac{2}{r^2}\frac{\partial v_r}{\partial \theta}\right] - \frac{1}{r}\frac{\partial p}{\partial \theta} + \rho g_\theta$$

$$\rho\left(\frac{\partial v_z}{\partial t} + v_r\frac{\partial v_z}{\partial r} + \frac{v_\theta}{r}\frac{\partial v_z}{\partial \theta} + v_z\frac{\partial v_z}{\partial z}\right)$$

$$= \mu\left[\frac{1}{r}\frac{\partial}{\partial r}\left(r\frac{\partial v_z}{\partial r}\right) + \frac{1}{r^2}\frac{\partial^2 v_z}{\partial \theta^2} + \frac{\partial^2 v_z}{\partial z^2}\right] - \frac{\partial p}{\partial z} + \rho g_z$$

* *Source:* Bird et al. (1987).

C   *Spherical Coordinates* (r, θ, ϕ)

$$\rho\left(\frac{\partial v_r}{\partial t} + v_r\frac{\partial v_r}{\partial r} + \frac{v_\theta}{r}\frac{\partial v_r}{\partial \theta} + \frac{v_\phi}{r\sin\theta}\frac{\partial v_r}{\partial \phi} - \frac{v_\theta^2 + v_\phi^2}{r}\right)$$

$$= \mu\left[\frac{1}{r^2}\frac{\partial^2}{\partial r^2}(r^2 v_r) + \frac{1}{r^2\sin\theta}\frac{\partial}{\partial\theta}\left(\sin\theta\frac{\partial v_r}{\partial\theta}\right) + \frac{1}{r^2\sin^2\theta}\frac{\partial^2 v_r}{\partial\phi^2}\right] - \frac{\partial p}{\partial r} + \rho g_r$$

$$\rho\left(\frac{\partial v_\theta}{\partial t} + v_r\frac{\partial v_\theta}{\partial r} + \frac{v_\theta}{r}\frac{\partial v_\theta}{\partial\theta} + \frac{v_\phi}{r\sin\theta}\frac{\partial v_\theta}{\partial\phi} + \frac{v_r v_\theta}{r} - \frac{v_\phi^2\cot\theta}{r}\right)$$

$$= \mu\left[\frac{1}{r^2}\frac{\partial^2}{\partial r}\left(r^2\frac{\partial v_\theta}{\partial r}\right) + \frac{1}{r^2}\frac{\partial}{\partial\theta}\left(\frac{1}{\sin\theta}\frac{\partial}{\partial\theta}(v_\theta\sin\theta)\right) + \frac{1}{r^2\sin^2\theta}\frac{\partial^2 v_\theta}{\partial\phi^2}\right.$$

$$\left. + \frac{2}{r^2}\frac{\partial v_r}{\partial\theta} - \frac{2\cot\theta}{r^2\sin\theta}\frac{\partial v_\phi}{\partial\phi}\right] - \frac{1}{r}\frac{\partial p}{\partial\theta} + \rho g_\theta$$

$$\rho\left(\frac{\partial v_\phi}{\partial t} + v_r\frac{\partial v_\phi}{\partial r} + \frac{v\theta}{r}\frac{\partial v_\phi}{\partial\theta} + \frac{v_\phi}{r\sin\theta}\frac{\partial v_\phi}{\partial\phi} + \frac{v_\phi v_r}{r} + \frac{v_\theta v_\phi}{r}\cot\theta\right)$$

$$= \mu\left[\frac{1}{r^2}\frac{\partial}{\partial r}\left(r^2\frac{\partial v_\phi}{\partial r}\right) + \frac{1}{r^2}\frac{\partial}{\partial\theta}\left(\frac{1}{\sin\theta}\frac{\partial}{\partial\theta}(v_\phi\sin\theta)\right)\right.$$

$$\left. + \frac{1}{r^2\sin^2\theta}\frac{\partial^2 v_\phi}{\partial\phi^2} + \frac{2}{r^2\sin\theta}\frac{\partial v_r}{\partial\phi} + \frac{2\cot\theta}{r^2\sin\theta}\frac{\partial v_\theta}{\partial\phi}\right] - \frac{1}{r\sin\theta}\frac{\partial p}{\partial\phi} + \rho g_\phi$$

B.3   Component Equations*

B.3.1   Vorticity Vector

$$\vec{\zeta} = \nabla \times \vec{v}$$

Note: $\vec{\vec{\zeta}} = \nabla\vec{v} - (\nabla\vec{v})^{\text{tr}}.$

A   *Rectangular Coordinates*

$$\zeta_z = \frac{\partial v_y}{\partial x} - \frac{\partial v_x}{\partial y}$$

$$\zeta_y = \frac{\partial v_x}{\partial z} - \frac{\partial v_z}{\partial x}$$

$$\zeta_x = \frac{\partial v_z}{\partial y} - \frac{\partial v_y}{\partial z}$$

B   *Cylindrical Coordinates*

$$\zeta_z = \frac{1}{r}\frac{\partial}{\partial r}(r v_\theta) - \frac{1}{r}\frac{\partial v_r}{\partial\theta}$$

* *Source:* Bird et al. (1987).

$$\zeta_\theta = \frac{\partial v_r}{\partial z} - \frac{\partial v_z}{\partial r}$$

$$\zeta_r = \frac{1}{r}\frac{\partial v_z}{\partial \theta} - \frac{\partial v_\theta}{\partial z}$$

C Spherical Coordinates

$$\zeta_\phi = \frac{1}{r}\frac{\partial}{\partial r}(r v_\theta) - \frac{1}{r}\frac{\partial v_r}{\partial \theta}$$

$$\zeta_\theta = \frac{1}{r \sin \theta}\frac{\partial v_r}{\partial \phi} - \frac{1}{r}\frac{\partial}{\partial r}(r v_\phi)$$

$$\zeta_r = \frac{1}{r \sin \theta}\frac{\partial}{\partial \theta}(v_\phi \sin \theta) - \frac{1}{r \sin \theta}\frac{\partial v_\theta}{\partial \phi}$$

◆ $$\vec{\zeta} = 2\vec{\omega}$$

B.3.2 Rate of Deformation Tensor

$$\overset{\Rightarrow}{\dot{\gamma}} = \nabla \vec{v} + (\nabla \vec{v})^{\text{tr}}$$

Note: $\overset{\Rightarrow}{\dot{\gamma}} = 2\overset{\Rightarrow}{\dot{\varepsilon}}$.

A Rectangular Coordinates (x, y, z)

$$\dot{\gamma}_{xx} = 2\frac{\partial v_x}{\partial x}$$

$$\dot{\gamma}_{yy} = 2\frac{\partial v_y}{\partial y}$$

$$\dot{\gamma}_z = 2\frac{\partial v_z}{\partial z}$$

$$\dot{\gamma}_{xy} = \dot{\gamma}_{yx} = \frac{\partial v_y}{\partial x} + \frac{\partial v_x}{\partial y}$$

$$\dot{\gamma}_{yz} = \dot{\gamma}_{zy} = \frac{\partial v_z}{\partial y} + \frac{\partial v_y}{\partial z}$$

$$\dot{\gamma}_{zx} = \dot{\gamma}_{xz} = \frac{\partial v_x}{\partial z} + \frac{\partial v_z}{\partial x}$$

B Cylindrical Coordinates (r, θ, z)

$$\dot{\gamma}_{rr} = 2\frac{\partial v_r}{\partial r}$$

$$\dot{\gamma}_{\theta\theta} = 2\left(\frac{1}{r}\frac{\partial v_\theta}{\partial \theta} + \frac{v_r}{r}\right)$$

$$\dot{\gamma}_{zz} = 2\frac{\partial v_z}{\partial z}$$

$$\dot{\gamma}_{r\theta} = \dot{\gamma}_{\theta r} = r\frac{\partial}{\partial r}\left(\frac{v_\theta}{r}\right) + \frac{1}{r}\frac{\partial v_r}{\partial\theta}$$

$$\dot{\gamma}_{\theta z} = \dot{\gamma}_{z\theta} = \frac{1}{r}\frac{\partial v_z}{\partial\theta} + \frac{\partial v_\theta}{\partial z}$$

$$\dot{\gamma}_{zr} = \dot{\gamma}_{rz} = \frac{\partial v_r}{\partial z} + \frac{\partial v_z}{\partial r}$$

C Spherical Coordinates (r, θ, ϕ)

$$\dot{\gamma}_{rr} = 2\frac{\partial v_r}{\partial r}$$

$$\dot{\gamma}_{\theta\theta} = 2\left(\frac{1}{r}\frac{\partial v_\theta}{\partial\theta} + \frac{v_r}{r}\right)$$

$$\dot{\gamma}_{\phi\phi} = 2\left(\frac{1}{r\sin\theta}\frac{\partial v_\phi}{\partial\phi} + \frac{v_r}{r} + \frac{v_\theta\cot\theta}{r}\right)$$

$$\dot{\gamma}_{r\theta} = \dot{\gamma}_{\theta r} = r\frac{\partial}{\partial r}\left(\frac{v_\theta}{r}\right) + \frac{1}{r}\frac{\partial v_r}{\partial\theta}$$

$$\dot{\gamma}_{\theta\phi} = \dot{\gamma}_{\phi\theta} = \frac{\sin\theta}{r}\frac{\partial}{\partial\theta}\left(\frac{v_\phi}{\sin\theta}\right) + \frac{1}{r\sin\theta}\frac{\partial v_\theta}{\partial\phi}$$

$$\dot{\gamma}_{\phi r} = \dot{\gamma}_{r\phi} = \frac{1}{r\sin\theta}\frac{\partial v_r}{\partial\phi} + r\frac{\partial}{\partial r}\left(\frac{v_\phi}{r}\right)$$

B.3.3 Stress Tensor for Incompressible Newtonian Fluids

$$\vec{\vec{\tau}} = \mu\dot{\vec{\vec{\gamma}}} = \mu[\nabla\vec{v} + (\nabla\vec{v})^{tr}]$$

A Rectangular Coordinates (x, y, z)

$$\tau_{xx} = \mu 2\frac{\partial v_x}{\partial x}$$

$$\tau_{yy} = \mu 2\frac{\partial v_y}{\partial y}$$

$$\tau_z = \mu 2\frac{\partial v_z}{\partial z}$$

$$\tau_{xy} = \tau_{yx} = \mu\left[\frac{\partial v_x}{\partial y} + \frac{\partial v_y}{\partial x}\right]$$

$$\tau_{yz} = \tau_{zy} = \mu\left[\frac{\partial v_y}{\partial z} + \frac{\partial v_z}{\partial y}\right]$$

$$\tau_{zx} = \tau_{xz} = \mu\left[\frac{\partial v_z}{\partial x} + \frac{\partial v_z}{\partial z}\right]$$

B   Cylindrical Coordinates (r, θ, z)

$$\tau_{rr} = \mu 2\frac{\partial v_r}{\partial r}$$

$$\tau_{\theta\theta} = \mu 2\left(\frac{1}{r}\frac{\partial v_\theta}{\partial \theta} + \frac{v_r}{r}\right)$$

$$\tau_{zz} = \mu 2\frac{\partial v_z}{\partial z}$$

$$\tau_{r\theta} = \tau_{\theta r} = \mu\left[r\frac{\partial}{\partial r}\left(\frac{v_\theta}{r}\right) + \frac{1}{r}\frac{\partial v_r}{\partial \theta}\right]$$

$$\tau_{\theta z} = \tau_{z\theta} = \mu\left[\frac{\partial v_\theta}{\partial z} + \frac{1}{r}\frac{\partial v_z}{\partial \theta}\right]$$

$$\tau_{zr} = \tau_{rz} = \mu\left[\frac{\partial v_z}{\partial r} + \frac{\partial v_r}{\partial z}\right]$$

C   Spherical Coordinates (r, θ, ϕ)

$$\tau_{rr} = \mu 2\frac{\partial v_r}{\partial r}$$

$$\tau_{\theta\theta} = \mu 2\left(\frac{1}{r}\frac{\partial v_\theta}{\partial \theta} + \frac{v_r}{r}\right)$$

$$\tau_{\phi\phi} = \mu 2\left(\frac{1}{r\sin\theta}\frac{\partial v_\phi}{\partial \phi} + \frac{v_r}{r} + \frac{v_\theta\cot\theta}{r}\right)$$

$$\tau_{r\theta} = \tau_{\theta r} = \mu\left[r\frac{\partial}{\partial r}\left(\frac{v_\theta}{r}\right) + \frac{1}{r}\frac{\partial v_r}{\partial \theta}\right]$$

$$\tau_{\theta\phi} = \tau_{\phi\theta} = \mu\left[\frac{\sin\theta}{r}\frac{\partial}{\partial \theta}\left(\frac{v_\phi}{\sin\theta}\right) + \frac{1}{r\sin\theta}\frac{\partial v_\theta}{\partial \phi}\right]$$

$$\tau_{\phi r} = \tau_{r\phi} = \mu\left[\frac{1}{r\sin\theta}\frac{\partial v_r}{\partial \phi} + r\frac{\partial}{\partial r}\left(\frac{v_\phi}{r}\right)\right]$$

Note: Occasionally, $\vec{\vec{\tau}} = -\mu\vec{\vec{\gamma}}$, indicating that the stress acts opposite to the fluid motion as experienced by the wall.

B.4   The Two-Dimensional Navier–Stokes Equations in Curvilinear Coordinates

Consider an orthogonal system of coordinates that follow the shape of the body

$R(x) \triangleq$ radius of curvature at position x

$$R(x) \qquad > 0 \qquad \text{for convex (i.e., outward) walls}$$
$$R(x) \qquad < 0 \qquad \text{for concave walls}$$
$$R_0 \qquad \triangleq \qquad \text{radius of curvature at stagnation point}$$
$$k(x) = R_0/R(x) \triangleq \qquad \text{dimensionless curvature of surface}$$

Note: $\mathrm{Re} = u_\infty R_0/\nu$.

x-Momentum

$$\frac{\partial u}{\partial t} + \frac{R}{R+y} u \frac{\partial u}{\partial x} + v \frac{\partial u}{\partial y} + \frac{vu}{R+y}$$

$$= -\frac{R}{R+y} \frac{1}{l} \frac{\partial p}{\partial x} + \nu \left\{ \frac{R^2}{(R+y)^2} \frac{\partial^2 u}{\partial x^2} + \frac{\partial^2 u}{\partial y^2} + \frac{1}{R+y} \frac{\partial u}{\partial y} - \frac{u}{(R+y)^2} \right.$$

$$\left. + \frac{2R}{(R+y)^2} \frac{\partial v}{\partial x} - \frac{R}{(R+y)^3} \frac{dR}{dx} v + \frac{Ry}{(R+y)^3} \frac{dR}{dx} \frac{\partial u}{\partial x} \right\}$$

y-Momentum

$$\frac{\partial v}{\partial t} + \frac{R}{R+y} u \frac{\partial v}{\partial x} + v \frac{\partial v}{\partial y} - \frac{u^2}{R+y}$$

$$= -\frac{1}{l} \frac{\partial p}{\partial y} + \nu \left\{ \frac{\partial^2 v}{\partial y^2} - \frac{2R}{(R+y)^2} \frac{\partial u}{\partial x} + \frac{1}{R+y} \frac{\partial v}{\partial y} + \frac{R^2}{(R+y)^2} \frac{\partial^2 v}{\partial x^2} \right.$$

$$\left. - \frac{v}{(R+y)^2} + \frac{R}{(R+y)^3} \frac{dR}{dx} u + \frac{Ry}{(R+y)^3} \frac{dR}{dx} \frac{\partial v}{\partial x} \right\}$$

Continuity

$$\frac{R}{R+y} \frac{\partial u}{\partial x} + \frac{\partial v}{\partial y} + \frac{v}{R+y} = 0$$

The stress components are

$$\pi_{xx} = -p + 2\mu \left(\frac{R}{R+y} \frac{\partial u}{\partial x} + \frac{v}{R+y} \right)$$

$$\pi_{yy} = -p + 2\mu \frac{\partial v}{\partial y}$$

$$\pi_{xy} = -\mu \left(\frac{\partial u}{\partial y} - \frac{u}{R+y} + \frac{R}{R+y} \frac{\partial v}{\partial x} \right)$$

B.5   Thermal Energy Equations with Constant Material Properties

A   Rectangular Coordinates (x, y, z)

$$\rho c_p \left(\frac{\partial T}{\partial t} + u \frac{\partial T}{\partial x} + v \frac{\partial T}{\partial y} + w \frac{\partial T}{\partial z} \right) = k \left(\frac{\partial^2 T}{\partial x^2} + \frac{\partial^2 T}{\partial y^2} + \frac{\partial^2 T}{\partial z^2} \right) + \phi$$

where $\phi = 2\mu\left[\left(\dfrac{\partial u}{\partial x}\right)^2 + \left(\dfrac{\partial v}{\partial y}\right)^2 + \left(\dfrac{\partial w}{\partial z}\right)^2\right]$

$$+\mu\left[\left(\frac{\partial u}{\partial y} + \frac{\partial v}{\partial x}\right)^2 + \left(\frac{\partial u}{\partial z} + \partial w\partial x\right)^2 + \left(\frac{\partial v}{\partial z} + \frac{\partial w}{\partial y}\right)^2\right]$$

B Cylindrical Coordinates (r, θ, z)

$$\rho c_p\left(\frac{\partial T}{\partial t} + v_r\frac{\partial T}{\partial r} + \frac{v_\theta}{r}\frac{\partial T}{\partial \theta} + v_z\frac{\partial T}{\partial z}\right) = k\left(\frac{1}{r}\frac{\partial}{\partial r}\left(r\frac{\partial T}{\partial r}\right) + \frac{1}{r^2}\frac{\partial^2 T}{\partial \theta^2} + \frac{\partial^2 T}{\partial z^2}\right) + \phi$$

where $\phi = \mu\left\{\left(\dfrac{\partial v_\theta}{\partial z} + \dfrac{1}{r}\dfrac{\partial v_z}{\partial \theta}\right)^2 + \left(\dfrac{\partial v_z}{\partial r} + \dfrac{\partial v_r}{\partial z}\right)^2 + \left[\dfrac{1}{r}\dfrac{\partial v_r}{\partial \theta} + r\dfrac{\partial}{\partial r}\left(\dfrac{v_\theta}{r}\right)\right]^2\right\}$

$$+2\mu\left\{\left(\frac{\partial v_r}{\partial r}\right)^2 + \left[\frac{1}{r}\left(\frac{\partial v_\theta}{\partial \theta} + v_r\right)\right]^2 + \left(\frac{\partial v_z}{\partial z}\right)^2\right\}$$

C Spherical Coordinates (r, θ, ϕ)

$$\rho c_p\left(\frac{\partial T}{\partial t} + v_r\frac{\partial T}{\partial r} + \frac{v_\theta}{r}\frac{\partial T}{\partial \theta} + \frac{v_\phi}{r\sin\theta}\frac{\partial T}{\partial \phi}\right) = k\nabla^2 T + \mu\Phi$$

where $\nabla^2 T = \dfrac{1}{r^2}\dfrac{\partial}{\partial r}\left(r^2\dfrac{\partial T}{\partial r}\right) + \dfrac{1}{r^2\sin\theta}\dfrac{\partial}{\partial \theta}\left(\sin\theta\dfrac{\partial T}{\partial \theta}\right) + \dfrac{1}{r^2\sin^2\theta}\dfrac{\partial^2 T}{\partial \phi^2}$

and

$$\Phi = \left[r\frac{\partial}{\partial r}\left(\frac{v_\theta}{r}\right) + \frac{1}{r}\frac{\partial v_r}{\partial r}\partial\theta\right]^2 + \left[r\frac{1}{r\sin\theta}\frac{\partial v_r}{\partial \phi} + r\frac{\partial}{\partial r}\left(\frac{v_\phi}{\partial r}\right)\right]^2$$

$$+\left[\frac{\sin\theta}{r}\frac{\partial}{\partial \theta}\left(\frac{v_\phi}{\sin\theta}\right) + \frac{1}{r\sin\theta}\frac{\partial v_\theta}{\partial \psi}\right]^2$$

$$+2\left[\left(\frac{\partial v_r}{\partial r}\right)^2 + \left(\frac{1}{r}\frac{\partial v_\theta}{\partial \theta} + \frac{v_r}{r}\right)^2 + \left(\frac{1}{r\sin\theta}\frac{\partial v_\phi}{\phi} + \frac{v_r}{r} + \frac{v_\theta\cot\theta}{r}\right)^2\right]$$

Simplified Governing Equations

C.1 Generalized Transport Equation in Differential Form

$$\frac{\partial \psi}{\partial t} + \nabla \cdot (\vec{v}\psi) = \nabla \cdot \vec{\Omega} + \vec{\Sigma}$$

Note: Obtain basic transfer equations by specifying *arbitrary* system variables ψ as well as associated fluxes Ω and forces or net sources Σ.

C.2 Field Equations in Terms of Fluxes for the Transport of Any Fluid

$$\frac{\partial \rho}{\partial t} + \nabla \cdot (\rho \vec{v}) = 0 \qquad \text{(continuity equation)}$$

$$\frac{\partial}{\partial t}(\rho \vec{v}) + \nabla \cdot (\rho \vec{v}\vec{v}) = -\nabla p + \nabla \cdot \vec{\tau} + \rho \vec{\tau} \qquad \text{(equation of motion)}$$

$$\frac{\partial c}{\partial t} + \nabla \cdot (c\vec{v}) = -\nabla \cdot \vec{j_c} + S_c \qquad \text{(mass transfer equation)}$$

$$\frac{\partial}{\partial t}(\rho U) + \nabla \cdot (\rho U \vec{v}) = -\nabla \cdot \vec{q}_H - \left(\vec{\vec{\pi}} : \nabla \vec{v}\right) \qquad \text{(internal energy equation)}$$

$$\vec{\vec{\pi}} = -p\vec{\vec{\delta}} + \vec{\vec{\tau}} \qquad \text{(total stress tensor)}$$

Note: Obtain problem-specific equations by specifying $\tau = \tau[\eta(\dot{\gamma}), \dot{\gamma}]$ for non-Newtonian fluids; also determine $\rho = \rho[p, U(T)]$, $\vec{j_c} \propto \nabla c$, and $\vec{q}_H \propto \nabla T$.

Examples:

$$\vec{\vec{\tau}} = \eta \vec{\vec{\gamma}} = m\dot{\gamma}^{n-1}\vec{\vec{\gamma}} \qquad \text{(power-law fluid)}$$

$$\rho = p/RT \qquad \text{(ideal gas law)}$$

$$\vec{j_c} = D\nabla c \qquad \text{(Fick's law)}$$

$$\vec{q}_H = -k\nabla T \quad \text{(Fourier's law)}$$

A   *Conservation of Mass, Momentum, and Energy Written in Tensor Notation*

$$\frac{\partial \rho}{\partial t} + \frac{\partial}{\partial x_j}[\rho u_j] = 0 \qquad (i \text{ and } j = 1, 2, 3; \text{Einstein convention})$$

$$\frac{\partial(\rho u_j)}{\partial t} + \frac{\partial}{\partial x_j}[\rho u_i u_j + p\delta_{ij} - \tau_{ij}] = 0$$

$$\frac{\partial(\rho e)}{\partial t} + \frac{\partial}{\partial x_j}[\rho u_j(e + u_k u_i/2) + pu_j - q_j - u_i \tau_{ij}] = 0$$

where for Newtonian fluids

$$\tau_{ij} = \mu\left(\frac{\partial u_i}{\partial x_j} + \frac{\partial u_j}{\partial x_i}\right) + \left(\delta_{ij}\lambda\frac{\partial u_j}{\partial x_j}\right) \qquad (\lambda \triangleq \text{coefficient of bulk viscosity})$$

$$q_i = -k\frac{\partial T}{\partial x_j} := -\frac{\gamma}{Pr}\mu\frac{\partial e}{\partial x_j} \qquad (\gamma \triangleq \text{specific heat ratio})$$

C.3   Momentum, Heat, and Mass Transfer Equations with Constant Parameters μ, k, and D

C.3.1   *Primitive Variables*

$$\nabla \cdot \vec{v} = 0$$

$$\rho\frac{D\vec{v}}{Dt} = -\nabla p + \mu\nabla^2\vec{v} + \vec{g}$$

$$\rho c_p\frac{DT}{Dt} = k\nabla^2 T + S_h$$

$$\frac{Dc}{Dt} = D\nabla^2 c + S_c$$

Where, for example,

velocity vector $\vec{v} = \hat{i}u + \hat{j}v + \hat{k}w$

$$\frac{D\bullet}{Dt} \equiv \frac{\partial\bullet}{\partial t} + (\vec{v} \cdot \nabla)\bullet \qquad \text{(Stokes derivative)}$$

and

S_h, $S_c \triangleq$ net source terms

C.3.2 Governing Equations for Two-Dimensional Analyses in Rectangular Coordinates

A   Equation of Motion and Conservation of Fluid Mass

Basic Flow Equations (2-D Navier–Stokes)

$$\frac{\partial u}{\partial t} + u\frac{\partial u}{\partial x} + v\frac{\partial u}{\partial y} = -\frac{1}{\rho}\frac{\partial p}{\partial x} + \nu\left(\frac{\partial^2 u}{\partial x^2} + \frac{\partial^2 u}{\partial y^2}\right)$$

$$\frac{\partial v}{\partial t} + u\frac{\partial v}{\partial x} + v\frac{\partial v}{\partial y} = -\frac{1}{\rho}\frac{\partial p}{\partial y} + \nu\left(\frac{\partial^2 v}{\partial x^2} + \frac{\partial^2 v}{\partial y^2}\right)$$

$$\frac{\partial u}{\partial x} + \frac{\partial v}{\partial y} = 0$$

Nondimensionalization of the primitive equations is usually based on the advective time scale L/U, where L is a characteristic length and U is a characteristic velocity of the problem.

$$\tilde{u} = \frac{u}{U} \qquad \tilde{v} = \frac{v}{U}$$

$$\tilde{x} = \frac{x}{L} \qquad \tilde{y} = \frac{y}{L}$$

$$\tilde{t} = \frac{t}{(L/U)} \qquad \tilde{p} = \frac{p}{(\rho U^2)}$$

$$\mathrm{Re} = \frac{UL}{\nu}$$

Normalized Conservation Form of the Basic Flow Equations

$$\frac{\partial \tilde{u}}{\partial \tilde{t}} + \frac{\partial(\tilde{u}^2)}{\partial \tilde{x}} + \frac{\partial(\tilde{u}\tilde{v})}{\partial \tilde{y}} = -\frac{\partial \tilde{p}}{\partial \tilde{x}} + \frac{1}{\mathrm{Re}}\left(\frac{\partial^2 \tilde{u}}{\partial \tilde{x}^2} + \frac{\partial^2 \tilde{u}}{\partial \tilde{y}^2}\right)$$

$$\frac{\partial \tilde{v}}{\partial \tilde{t}} + \frac{\partial(\tilde{u}\tilde{v})}{\partial \tilde{x}} + \frac{\partial(\tilde{v}^2)}{\partial \tilde{y}} = -\frac{\partial \tilde{p}}{\partial \tilde{y}} + \frac{1}{\mathrm{Re}}\left(\frac{\partial^2 \tilde{v}}{\partial \tilde{x}^2} + \frac{\partial^2 \tilde{v}}{\partial \tilde{y}^2}\right)$$

Using the definition of the stream function in *rectangular* coordinates

$$u = \frac{\partial \psi}{\partial y} \quad \text{and} \quad v = -\frac{\partial \psi}{\partial x}$$

the pressure gradients can be eliminated by cross-differentiation and subtraction of the resulting equations.

Stream Function Equation

$$\frac{\partial}{\partial t}\nabla^2\psi + \frac{\partial \psi}{\partial y}\frac{\partial}{\partial x}\nabla^2\psi - \frac{\partial \psi}{\partial x}\frac{\partial}{\partial y}\nabla^2\psi = \nu\nabla^4\psi$$

Note: Rather than solving a nonlinear, fourth-order PDE, the combined stream function–vorticity transport approach is employed.

- In Rectangular Coordinates (x, y, z) (2-D flow, $w = 0$)

$$u = \frac{\partial \psi}{\partial y}, \qquad v = -\frac{\partial \psi}{\partial x} \quad \text{and} \quad \zeta_z = \frac{\partial^2 \psi}{\partial x^2} + \frac{\partial^2 \psi}{\partial y^2} = \nabla^2 \psi$$

so that

$$\frac{\partial \zeta_z}{\partial t} + \left(\frac{\partial \psi}{\partial y}\right) \frac{\partial \zeta_z}{\partial x} + \left(\frac{\partial \psi}{\partial x}\right) \frac{\partial \zeta_z}{\partial y} = \nu \nabla^2 \zeta_z$$

where $\qquad \nabla^2 \zeta_z = \psi_{xxxx} + 2\psi_{xxyy} + \psi_{yyyy}$

- In Polar Coordinates (r, θ, z) (2-D flow, $v_z = 0$)

$$v_r = \frac{1}{r}\frac{\partial \psi}{\partial \theta}, \qquad v_\theta = -\frac{\partial \psi}{\partial r} \quad \text{and} \quad \zeta_z = \frac{\partial^2 \psi}{\partial r^2} + \frac{1}{r}\frac{\partial \psi}{\partial r} + \frac{1}{r^2}\frac{\partial^2 \psi}{\partial \theta^2}$$

or

$$\zeta_z = \frac{1}{r}\frac{\partial}{\partial r}\left(r \frac{\partial \psi}{\partial r}\right) + \frac{1}{r^2}\frac{\partial^2 \psi}{\partial \theta^2}$$

so that

$$\frac{\partial \zeta_z}{\partial t} + \left(\frac{1}{r}\frac{\partial \psi}{\partial \theta}\right)\frac{\partial \zeta_z}{\partial r} + \left(-\frac{\partial \psi}{\partial r}\right)\frac{1}{r}\frac{\partial \zeta_z}{\partial \theta} = \nu\left[\frac{1}{r}\frac{\partial}{\partial r}\left(r\frac{\partial \zeta_z}{\partial r}\right) + \frac{1}{r^2}\frac{\partial^2 \zeta_z}{\partial \theta^2}\right]$$

- In Cylindrical Coordinates (r, θ, z) (axisymmetric flow, $v_\theta = 0$)

$$v_r = -\frac{1}{r}\frac{\partial \psi}{\partial z}, \qquad v_z = \frac{1}{r}\frac{\partial \psi}{\partial r} \quad \text{and} \quad \zeta_\theta = \frac{\partial}{\partial r}\left(\frac{1}{r}\frac{\partial \psi}{\partial r}\right) + \frac{\partial^2}{\partial z^2}\left(\frac{\psi}{r}\right)$$

so that

$$\frac{\partial \zeta_\theta}{\partial t} + \left(-\frac{1}{r}\frac{\partial \psi}{\partial z}\right)\frac{\partial \zeta_\theta}{\partial r} + \left(\frac{1}{r}\frac{\partial \psi}{\partial r}\right)\frac{\partial \zeta_\theta}{\partial z}$$

$$= \left(-\frac{1}{r}\frac{\partial \psi}{\partial z}\right)\frac{\zeta_\theta}{r} + \nu\left[\frac{\partial}{\partial r}\left(\frac{1}{r}\frac{\partial}{\partial r}(r\zeta_\theta)\right) + \frac{\partial^2 \zeta_\theta}{\partial z^2}\right]$$

Additional stream function formulations may be found in Appendix D.

B   Constituent Transport Equations

Here, constituent c is equivalent to the concentration of dissolved or suspended material following the fluid motion as well as the fluid's internal energy per unit volume (i.e., basically the fluid temperature).

$$\frac{\partial c}{\partial t} + u\frac{\partial c}{\partial x} + v\frac{\partial c}{\partial y} = D\left(\frac{\partial^2 c}{\partial x^2} + \frac{\partial^2 c}{\partial y^2}\right) + S_c \qquad (D \triangleq \text{diffusion coeff.})$$

Normalized Heat Equation

$$\frac{\partial \tilde{T}}{\partial \tilde{t}} + \tilde{u}\frac{\partial \tilde{T}}{\partial \tilde{x}} + \tilde{v}\frac{\partial \tilde{T}}{\partial \tilde{y}} = \frac{1}{Pe}\left(\frac{\partial^2 \tilde{T}}{\partial \tilde{x}^2} + \frac{\partial^2 \tilde{T}}{\partial \tilde{y}^2}\right) + \frac{Ec}{Re}\tilde{\Phi}$$

where $\tilde{\Phi}$ is the dimensionless dissipation function, $Pe = Re\,Pr$ is the Peclet number, and $Ec = U^2/(c_p\Delta T)$ and is the Eckert number $(Re = UL/\nu)$.

Stream Function Formulation in Planar and Axisymmetric Coordinates

Table D.1. *Stream Function Equations*

Type of motion	Coordinate system	Velocity components
Two-dimensional (planar)	Rectangular with $v_z = 0$, $f_z = 0$ and no z-dependence	$v_x = -\dfrac{\partial \psi}{\partial y}$ $v_y = +\dfrac{\partial \psi}{\partial x}$
	Cylindrical with $v_z = 0$, $f_z = 0$ and no z-dependence	$v_r = -\dfrac{1}{r}\dfrac{\partial \psi}{\partial \theta}$ $v_\theta = +\dfrac{\partial \psi}{\partial r}$
Axisymmetrical	Cylindrical with $v_\theta = 0$, $f_\theta = 0$ and no θ-dependence	$v_z = -\dfrac{1}{r}\dfrac{\partial \psi}{\partial r}$ $v_r = +\dfrac{1}{r}\dfrac{\partial \psi}{\partial z}$
	Spherical with $v_\phi = 0$, $f_\phi = 0$ and no ϕ-dependence	$v_r = -\dfrac{1}{r^2 \sin\theta}\dfrac{\partial \psi}{\partial \theta}$ $v_\theta = +\dfrac{1}{r \sin\theta}\dfrac{\partial \psi}{\partial r}$

Source: Bird et al. (1987).

*Here the Jacobians are designated by $\partial(g, h)/\partial(x, y) = \begin{vmatrix} \partial g/\partial x & \partial g/\partial y \\ \partial h/\partial x & \partial h/\partial y \end{vmatrix}$.

Differential equations $\rho(D/D_t)v = -\nabla p + \mu\nabla^2 v + f$	Expressions for operators
$\rho\left(\dfrac{\partial}{\partial t}(\nabla^2\psi) + \dfrac{\partial(\psi, \nabla\psi)^*}{\partial(x, y)}\right) = \mu\nabla^4\psi + [\nabla \times f]_z$	$\nabla^2 \equiv \dfrac{\partial^2}{\partial x^2} + \dfrac{\partial^2}{\partial y^2}$
	$\nabla^4\psi \equiv \nabla^2(\nabla^2\psi)$
	$\equiv \left(\dfrac{\partial^4}{\partial x^4} + 2\dfrac{\partial^4}{\partial x^2\partial y^2} + \dfrac{\partial^4}{\partial y^4}\right)\psi$
	$[\nabla \times f]_z = \dfrac{\partial}{\partial x}f_y - \dfrac{\partial}{\partial y}f_x$
$\rho\left(\dfrac{\partial}{\partial t}(\nabla^2\psi) + \dfrac{1}{r}\dfrac{\partial(\psi, \nabla^2\psi)^*}{\partial(r, \theta)}\right) = \mu\nabla^4\psi + [\nabla \times f]_z\nabla^2$	$\equiv \dfrac{\partial^2}{\partial r^2} + \dfrac{1}{r}\dfrac{\partial}{\partial r} + \dfrac{1}{r^2}\dfrac{\partial^2}{\partial\theta^2}$
	$[\nabla \times f]_z = \dfrac{1}{r}\dfrac{\partial}{\partial r}(rf_\theta) - \dfrac{1}{r}\dfrac{\partial}{\partial\theta}f_r$
$\rho\left(\dfrac{\partial}{\partial t}(E^2\psi) - \dfrac{1}{r}\dfrac{\partial(\psi, E^2\psi)^*}{\partial(r, z)} - \dfrac{2}{r^2}\dfrac{\partial\psi}{\partial z}E^2\psi\right)$ $= \mu E^4\psi + r[\nabla \times f]_\theta$	$E^2 \equiv \dfrac{\partial^2}{\partial r^2} - \dfrac{1}{r}\dfrac{\partial}{\partial r} + \dfrac{\partial^2}{\partial z^2}$ $E^4\psi \equiv E^2(E^2\psi)$ $[\nabla \times f]_\theta = \dfrac{\partial}{\partial z}f_r - \dfrac{\partial}{\partial r}f_z$
$\rho\left(\dfrac{\partial}{\partial t}(E^2\psi) + \dfrac{1}{r^2\sin\theta}\dfrac{\partial(\psi, E^2\psi)^*}{\partial(r, \theta)}\right)$ $- \dfrac{2E^2\psi}{r^2\sin^2\theta}\left(\dfrac{\partial\psi}{\partial r}\cos\theta - \dfrac{1}{r}\dfrac{\partial\psi}{\partial\theta}\sin\theta\right)$ $= \mu E^4\psi + r\sin\theta[\nabla \times f]_\phi$	$E^2 \equiv \dfrac{\partial^2}{\partial r^2} + \dfrac{\sin\theta}{r^2}\dfrac{\partial}{\partial\theta}\left(\dfrac{1}{\sin\theta}\dfrac{\partial}{\partial\theta}\right)$ $[\nabla \times f]_\phi = \dfrac{1}{r}\dfrac{\partial}{\partial r}(rf_\theta) - \dfrac{1}{r}\dfrac{\partial}{\partial\theta}f_r$

Physical Properties of Gases and Liquids

Table E.1. *Physical Properties of Gases at Atmospheric Pressure*

T, K	ρ, kg/m^3	c_p, kJ/kg·°C	μ, kg/m·s	ν, m^2/s ×10^6	k, W/m·K	α, m^2/s ×10^4	Pr
Air							
100	3.60.10	1.0266	0.6924×10^{-5}	1.923	0.009246	0.02501	0.770
150	2.3675	1.0099	1.0283	4.343	0.013735	0.05745	0.753
200	1.7684	1.0061	1.3289	7.490	0.01809	0.10165	0.739
250	1.4128	1.0053	1.488	10.53	0.02227	0.13161	0.722
300	1.1774	1.0057	1.983	16.84	0.02624	0.22160	0.708
350	0.9980	1.0090	2.075	20.76	0.03003	0.2983	0.697
400	0.8826	1.0140	2.286	25.90	0.03365	0.3760	0.689
450	0.7833	1.0207	2.484	31.71	0.03707	0.4222	0.683
500	0.7048	1.0295	2.671	37.90	0.04038	0.5564	0.680
550	0.6423	1.0392	2.848	44.34	0.04360	0.6532	0.680
600	0.5879	1.0551	3.018	51.34	0.04659	0.7512	0.680
650	0.5430	1.0635	3.177	58.51	0.04953	0.8578	0.682
700	0.5030	1.0752	3.332	66.25	0.05230	0.9672	0.684
750	0.4709	1.0856	3.481	73.91	0.05509	1.0774	0.686
800	0.4405	1.0978	3.625	82.29	0.05779	1.1951	0.689
850	0.4149	1.1095	3.765	90.75	0.06028	1.3097	0.692
900	0.3925	1.1212	3.899	99.3	0.06279	1.4271	0.696
950	0.3716	1.1321	4.023	108.2	0.06525	1.5510	0.699
1000	0.3524	1.1417	4.152	117.8	0.06752	1.6779	0.702
1100	0.3204	1.160	4.44	138.6	0.0732	1.969	0.704
1200	0.2947	1.179	4.69	159.1	0.0782	2.251	0.707
1300	0.2707	1.197	4.93	182.1	0.0837	2.583	0.705
1400	0.2515	1.214	5.17	205.5	0.0891	2.920	0.705
1500	0.2355	1.230	5.40	229.1	0.0946	3.262	0.705
1600	0.2211	1.248	5.63	254.5	0.100	3.609	0.705
1700	0.2082	1.267	5.85	280.5	0.105	3.977	0.705
Steam (H$_2$O vapor)							
380	0.5863	2.060	12.71×10^{-6}	21.6	0.0246	0.2036	1.060
400	0.5542	2.014	13.44	24.2	0.0261	0.2338	1.040
450	0.4902	1.980	15.25	31.1	0.0299	0.307	1.010
500	0.4405	1.985	17.04	38.6	0.0339	0.387	0.996
550	0.4005	1.997	18.81	47.0	0.0379	0.475	0.991
600	0.3652	2.026	20.67	56.6	0.0422	0.573	0.986
650	0.3380	2.056	22.47	64.4	0.0464	0.666	0.995
700	0.3140	2.085	24.26	77.2	0.0505	0.772	1.000
750	0.2931	2.119	26.04	88.8	0.0549	0.883	1.005
800	0.2739	2.152	27.86	102.0	0.0592	1.001	1.010
850	0.2579	2.186	29.69	115.2	0.0637	1.130	1.019

Table E.2. *Physical Properties of Saturated Liquids*

T, °K	ρ, kg/m³	c_p, kJ/kg · °C	ν, m²/s × 10⁶	k, W/m · K	α, m²/s × 10⁴	Pr
Water H₂O						
0	1,002.28	4.2178	1.788×10^{-6}	0.552	1.308	13.6
20	1,000.52	4.1818	1.006	0.597	1.430	7.02
40	994.59	4.1784	0.658	0.628	1.512	4.34
Engine Oil (unused)						
0	899.12	1.796	0.00428	0.147	0.911	47,100
20	888.23	1.880	0.00090	0.145	0.872	10,400
40	876.05	1.964	0.00024	0.144	0.834	2,870
60	864.04	2.047	0.839×10^{-4}	0.140	0.800	1,050
80	852.02	2.131	0.375	0.138	0.769	490
100	840.11	2.219	0.203	0.137	0.738	276
120	828.96	2.307	0.124	0.135	0.710	175
140	816.94	2.395	0.080	0.133	0.686	116
160	805.89	2.483	0.056	0.132	0.663	84

Numerical Tools and Program Listings

F.1 Nonlinear Algebraic Equations (Newton–Raphson Method)

The Newton–Raphson method provides a powerful algorithm for identifying the roots of nonlinear algebraic equations. In Newton's method the equation $y = f(x)$ must be differentiable. The first derivative $f'(x = x_0)$ provides the slope of the tangent at a given point. This information allows for an estimate of efficient convergence. Given a continuous function that crosses the x-axis only once over a given range, the equation of the tangent line is the following:

$$y = f'(x_0)(x - x_0) + f(x_0)$$

where $f'(x)$ is the slope at $x = x_0$, x is a neighboring value, and $f(x_0)$ is the magnitude of f at $x = x_0$. If a point on the x-axis $(x_1, 0)$ is to be computed, the following simplifications can be made:

$$x_1 = x_0 - \frac{f(x_0)}{f'(x_0)}$$

From this, the next guess for the root may be made. This process has to be continued until the required convergence is met. In general

$$x_{n+1} = x_n - \frac{f(x_n)}{f'(x_n)}$$

This information lends itself to developing a simple computer algorithm that is rather powerful. For a computer program, the following approximation is made:

$$f'(x_n) \cong \frac{f(x_n + \delta x_n) - f(x_n)}{\delta x_n}$$

where δx is a small interval. Of particular importance in this algorithm is the initial guess. A poor initial guess may make convergence more difficult or may lead to an incorrect answer. Therefore, an "educated" guess is a good starting point.

In the following, a program listing for a solver using the Newton–Raphson method is provided. In order to take advantage of this program, the user must enter the function in subroutine function at the designated point. Also, an initial guess and maximum number of iterations may be entered. The output of the program shown previously, providing the solution for the following function,

$$f = x^2 - 4x + 4$$

with an initial guess of 10, is shown later.

♦   *The solution is 2.0000.*

486

The solution to a system of nonlinear algebraic equations is obtained by first an outer iteration, that is, quasi-linearization (cf. Newton–Raphson method), and then an inner iteration (cf. Gauss–Seidel method).

Extending the previous root finding for a scalar equation to a vector equation, we solve the nonlinear system

$$\vec{f}(\vec{x}) = \vec{0}$$

by starting with an initial approximation $\vec{x}^0$ in the recurrence relation

$$\vec{x}^{(i+1)} = \vec{x}^{(i)} - [\mathbf{J}(x^{(i)})]^{-1}\vec{f}(x^{(i)})$$

Here, $\mathbf{J}(x^{(i)}$ is the Jacobian matrix to be inverted, representing the first derivatives evaluated at $\vec{x}^{(i)}$:

$$\mathbf{J}(x^{(i)})|_{m,n} = \frac{\partial f_m}{\partial x_n}\Big|_{\vec{x}^{(i)}}$$

Example:

Given a nonlinear function $f(y)$ in the ODE $y'' = f(y)$, $y(0) = y(1) = 0$. If we expand $f(y)$ about $y^{(i)}$ and keep only the linear term, we obtain

$$y'' \approx f(y^{(i)}) + (y - y^{(i)})f'(y^{(i)})$$

This boundary-value problem (BVP) is linear in the unknown $y(x)$ and could be solved by using the Runge–Kutta method (cf. App. F.3). The resulting approximate solution $y(x)$ is actually $y^{(i+1)}$, which is used in the recurrence formula

$$[y^{(i+1)}]'' = f(y^{(i)}) + (y^{(i+1)} - y^{(i)})f'(y^{(i)})$$

subject to

$$y^{(i)}(0) = y^{(i)}(1) = 0$$

This relation is repeatedly employed until convergence is achieved.

```
C*
C*  THIS PROGRAM USES THE NEWTON-RAPHSON METHOD TO SOLVE
C*  FOR THE ROOTS OF A POLYNOMIAL.
C*  THE USER MUST ENTER THE FOLLOWING INFORMATION:
C*
C*    F: THE FUNCTION FOR WHICH THE ROOTS ARE TO BE FOUND
C*    G: THE FIRST DERIVATIVE OF THIS FUNCTION
C*   MAX: THE MAXIMUM NUMBER OF ITERATIONS
C*    X: THE INITIAL GUESS
C*   DX: THE SMALL INTERVAL USED TO CALCULATE THE
C*      DERIVATIVE
C*   X2: THE POSITION, F(X+DX) AT WHICH THE FUNCTION IS
C*      CALCULATED TO COMPARE TO F(X).
C*
```

```
      IMPLICIT REAL*8(A-H,O-Z)
C*  ENTER THE MAXIMUM NUMBER OF ITERATIONS HERE
      MAX=100
      X=10.0D0
      DO 10 N=1,MAX
      DX=X*1.0D-3
      X2=DX+X
      CALL FUNCTION(FX,X)
      CALL FUNCTION(FDX,X2)
      G=(FDX-FX)/DX
      X=X-FX/G
   10 CONTINUE
      WRITE(*,20)X,FX
   20 FORMAT (2X,'THE SOLUTION IS ',F8.4,' AT WHICH, F(X)=',F8.4)
      END


C*
C*  SUBROUTINE FUNCTION
C*
      SUBROUTINE FUNCTION(F,X)
      IMPLICIT REAL*8(A-H,O-Z)

C*  ENTER THE FUNCTION HERE
      F=X**2-4.0D0*X+4.0D0


      RETURN
      END
```

F.2   System of Linear Algebraic Equations

F.2.1   A. Inversion of a Tridiagonal Matrix (The Thomas Algorithm)

The Thomas algorithm is used to find the solution vector of a set of simultaneous equations that form a tridiagonal matrix. A *tridiagonal matrix* is one in which only the center and the two off-center diagonals have nonzero values (cf. Burden et al. 1988). The following is the form of a tridiagonal matrix:

$$
\begin{bmatrix}
B_1^* & C_1^* & 0 & 0 & 0 & 0 & 0 \\
A_2 & B_2 & C_2 & 0 & 0 & 0 & 0 \\
0 & A_3 & B_3 & C_3 & 0 & 0 & 0 \\
0 & 0 & \backslash & \backslash & \backslash & 0 & 0 \\
0 & 0 & 0 & \backslash & \backslash & \backslash & C_{N-1} \\
0 & 0 & 0 & 0 & 0 & A_N^* & B_N^*
\end{bmatrix}
\begin{bmatrix}
u_1 \\
u_2 \\
u_3 \\
| \\
u_{N-1} \\
u_N
\end{bmatrix}
=
\begin{bmatrix}
D_1^* \\
D_2 \\
D_3 \\
| \\
D_{N-1} \\
D_N
\end{bmatrix}
$$

This matrix may be solved by other simultaneous equation solvers such as Gauss–Seidel. However, the Thomas algorithm will solve this system much more quickly, a great advantage in larger systems. Tridiangular matrices are often developed in finite-difference analyses. The algorithm transforms the tridiagonal matrix into an upper triangular matrix with coefficients of 1 along the diagonal and $-E_n$ on the line above the diagonal and zeroes everywhere else. The right-hand vector is transformed to F_n (these are for $n = 1, 2, 3, \ldots N$). The coefficients E_n and F_n are defined in the following manner for $n > 1$:

$$E_n = -\frac{C_n}{\text{DENO}} \qquad F_n = \frac{D_n - A_n F_{n-1}}{\text{DENO}}$$

where $\quad \text{DENO} = A_n E_{n-1} + B_n$

For $n = 1$,

$$E_1 = -\frac{C_1^*}{B_1} \qquad F_1 = \frac{D_1^*}{B_1}$$

Since this matrix is upper triangular, u_N can be determined directly

$$u_N = F_N$$

and u_n can be determined sequentially by using the back-substitution formulas

$$u_n = E_n u_{n+1} + F_n$$

This algorithm can easily be transcribed into a computer program. Below, a listing of the FORTRAN source code for the Thomas Algorithm is provided. This program will prompt the user for the size of the matrix and the components of the matrix. The program is written to accept linear systems of up to 100 equations. Consider the following example system of equations:

$$\begin{bmatrix} -5 & 1 & 0 & 0 \\ 2 & -5 & 1 & 0 \\ 0 & 2 & -5 & 1 \\ 0 & 0 & 2 & -5 \end{bmatrix} \begin{bmatrix} x_1 \\ x_2 \\ x_3 \\ x_4 \end{bmatrix} = \begin{bmatrix} -3 \\ -5 \\ -7 \\ -14 \end{bmatrix}$$

The input and output for this example are as follows:

INPUT

```
PLEASE ENTER THE NUMBER OF EQUATIONS N (<100)
4
PLEASE ENTER B(1)
-5.0
PLEASE ENTER C(1)
1.0
PLEASE ENTER D(1)
-3.0
PLEASE ENTER A(  2)
2.0
PLEASE ENTER B(  2)
-5.0
PLEASE ENTER C(  2)
1.0
PLEASE ENTER D(  2)
-5.0
```

```
PLEASE ENTER A( 3)
2.0
PLEASE ENTER B( 3)
-5.0
PLEASE ENTER C( 3)
1.0
PLEASE ENTER D( 3)
-7.0
PLEASE ENTER A( 4)
2.0
PLEASE ENTER B( 4)
-5.0
PLEASE ENTER D( 4)
-14.0
```

OUTPUT

```
X( 1) = .10000000E+01
X( 2) = .20000000E+01
X( 3) = .30000000E+01
X( 4) = .40000000E+01

C*
C* THIS PROGRAM IS THE DRIVER OF THE ''THOMAS'' SUBROUTINE THAT
C* SOLVES TRIDIAGONAL SYSTEMS OF EQUATIONS OF THE FOLLOWING FORMAT:
C
C*  [ b1 c1 0   0    ...            ] [ x1 ] [ d1 ]
C*  | a2 b2 c2  0    ...            | | x2 | | d2 |
C*  | 0  a3 b3  c3   ...            | | x3 | | d3 |
C*  |           ...                | |.... |=|.... |
C*  |           ... aN-1 bN-1 cN-1 | |xN-1| |dN-1|
C*  [           ... 0    aN   bN   ] [ xN ] [ dN ]
C*
C* THE MATRIX HAS NONZERO ELEMENTS ONLY ON THE DIAGONAL PLUS OR
C* MINUS ONE COLUMN. THIS SPECIAL CASE OF LINEAR EQUATION SYSTEMS
C* OCCURS FREQUENTLY, NOTABLY IN FINITE DIFFERENCE ANALYSIS.
C* b1 MUST NOT EQUAL ZERO!
C*
C* THE USER IS PROMPTED TO ENTER:
C*  + N   : THE NUMBER OF EQUATIONS. N MUST BE LESS THAN 100.
C*  + A(I): THE FACTOR ''ai''
C*  + B(I): THE FACTOR ''bi''
C*  + C(I): THE FACTOR ''ci''
C*  + D(I): THE FACTOR ''di''
C*
C* THIS PROGRAM IS WRITTEN FOR SYSTEMS OF LESS THAN OR EQUAL TO
C* 100 EQUATIONS.
C*

    DIMENSION A(100),B(100),C(100),D(100),X(100)
    WRITE(*,2)
```

```
2 FORMAT(1X,'PLEASE ENTER THE NUMBER OF EQUATIONS N (<100)')
  READ(*,*) N
  WRITE(*,3)
3 FORMAT(1X,'PLEASE ENTER B(1)')
  READ(*,*) B(1)
  WRITE(*,4)
4 FORMAT(1X,'PLEASE ENTER C(1)')
  READ(*,*) C(1)
  WRITE(*,5)
5 FORMAT(1X,'PLEASE ENTER D(1)')
  READ(*,*) D(1)
  DO 10 I=2,N-1
   WRITE(*,6) I
   READ(*,*) A(I)
   WRITE(*,7) I
   READ(*,*) B(I)
   WRITE(*,8) I
   READ(*,*) C(I)
   WRITE(*,9) I
   READ(*,*) D(I)
6 FORMAT(1X,'PLEASE ENTER A(',I3,')')
7 FORMAT(1X,'PLEASE ENTER B(',I3,')')
8 FORMAT(1X,'PLEASE ENTER C(',I3,')')
9 FORMAT(1X,'PLEASE ENTER D(',I3,')')
10 CONTINUE
  WRITE(*,11) N
11 FORMAT(1X,'PLEASE ENTER A(',I3,')')
  READ(*,*) A(N)
  WRITE(*,12) N
12 FORMAT(1X,'PLEASE ENTER B(',I3,')')
  READ(*,*) B(N)
  WRITE(*,13) N
13 FORMAT(1X,'PLEASE ENTER D(',I3,')')
  READ(*,*) D(N)

  CALL THOMAS(A,B,C,D,X,N)

  DO 15 I=1,N
   WRITE(*,14) I,X(I)
14 FORMAT(1X,'X(',I3,') = ',E15.8)
15 CONTINUE
  STOP
  END

C*
C* SUBROUTINE THOMAS
C*
  SUBROUTINE THOMAS(A,B,C,D,X,N)
  DIMENSION A(100),B(100),C(100),D(100),X(100),BETA(100),DELTA(100)
  BETA(1)=B(1)
  DELTA(1)=D(1)
```

```
      DO 10 I=2,N
       EPSILON=A(I)/BETA(I-1)
       BETA(I)=B(I)-EPSILON*C(I-1)
  10  DELTA(I)=D(I)-EPSILON*DELTA(I-1)
       X(N)=DELTA(N)/BETA(N)
       DO 20 I=N-1,1,-1
  20  X(I)=(DELTA(I)-C(I)*X(I+1))/BETA(I)
       RETURN
       END
```

F.2.2   Inversion of a Linear Matrix (The Gauss–Seidel Method)

Often there is a need to solve a large system of coupled linear algebraic equations. In this situation it may be best to use an iterative algorithm. Gauss–Seidel provides a powerful manner of solving such a system. Given a system of linear algebraic equations

$$\sum_{j=1}^{n} a_{ij}x_j = a_{i,n+1} \qquad \text{for } i = 1, 2, 3, \ldots, n$$

it can be represented in the following matrix form:

$$\begin{bmatrix} a_{11} & a_{12} & \cdots & a_{1_n} \\ a_{21} & \cdots & \cdots & a_{2_n} \\ | & \cdots & \cdots & | \\ a_{n1} & \cdots & \cdots & a_{nn} \end{bmatrix} \begin{bmatrix} x_1 \\ x_2 \\ | \\ x_n \end{bmatrix} = \begin{bmatrix} a_{1,n+1} \\ \cdots \\ \cdots \\ a_{n,n+1} \end{bmatrix}$$

Rewriting this form by solving for x we find that

$$x_i = \frac{1}{a_{ii}} \left(a_{i,n+1} - \sum_{j=1}^{n} a_{ij}x_j \right) \qquad \text{where } j \neq i$$

This form will allow for iterative solving, using the information gained each time to solve for the next guess, x_i. Therefore, for a single iteration, k, there must be n calculations. By taking advantage of the information gained in each of the calculations within the iteration, it seems that the number of iterations required could be less. This is the idea of the Gauss–Seidel algorithm. The following formula is used (cf. Burden et al. 1988):

$$x_i^{k+1} = \frac{1}{a_{ii}} \left(a_{i,n+1} - \sum_{j=1}^{i-1} a_{ij}x_j^{k+1} - \sum_{j=i+1}^{n} a_{ij}x_j^{k} \right) \qquad \text{where } i = 1, 2, 3, \ldots, n$$

It minimizes the number of iterations required to find a suitable value for each of the unknowns. This information can be transformed into a computer program such as the FORTRAN source code given later. In order to use this program, the user must create an input file, "INFILE," in which a matrix is entered in the following form:

```
 N,GUESS
 a,b,c,d,C1
 e,f,g,h,C2
```

```
i,j,k,l,C3
m,n,o,p,C4
```

Here, N is the total number of nodes, and GUESS is the initial guess for the unknown vector. The matrix should be entered in double precision. The answers, along with the entered matrix for checking purposes, will be printed into an output file, "OUTFILE." The following is a sample "OUTFILE" for a given "INFILE," assuming a 4×4 matrix problem with four unknown temperatures.

"INFILE"

```
4,50.0D0
12.0D0,1.0D0,15.0D0,0.0D0,4.0D0
1.0D0,3.0D0,1.0D0,16.0D0,0.0D0
0.0D0,2.0D0,-4.0D0,11.0D0,6.0D0
0.0D0,3.0D0,1.0D0,-42.0D0,0.0D0
```

"OUTFILE"

```
Number of nodes = 4          Initial guess =     50.000

Input matrix is :
     12.000    1.000    15.000       .000    4.000
      1.000    3.000     1.000     16.000     .000
       .000    2.000    -4.000     11.000    6.000
       .000    3.000     1.000    -42.000     .000
```

The solutions are

Node #	Temperature
1	2.370
2	−.030
3	−1.627
4	−.041

```
C*
C* THIS PROGRAM USES THE GAUSS-SEIDEL ALGORITHM TO SOLVE A SET
C* OF SIMULTANEOUS EQUATIONS ITERATIVELY.  THE USER IS ASKED TO
C* CREATE AN INPUT FILE, 'INFILE' WITH THE FOLLOWING FORMAT:
C*
C*   N,GUESS
C*   AB(1,1),AB(1,2),AB(1,3)
C*   AB(2,1),AB(2,2),AB(2,3)
C*   AB(3,1),AB(3,2),AB(3,3)
C*
C*   N:    NUMBER OF NODES
C*   GUESS:   AN INITIAL GUESS
C*   AB (I,J): THE MATRIX TO BE SOLVED
C*             NOTE:THE MATRIX SHOULD BE ENTERED WITH THE
C*                  FAR RIGHT COLUMN EQUAL TO THE B-VALUES.
C*
```

```
C* THE ANSWERS WILL BE PRINTED TO THE OUTPUT FILE 'OUTFILE'
C*
C*
      IMPLICIT REAL*8(A-H,O-Z)
      COMMON/MATR/AB(40,41),X(40)
      COMMON/NUM/N,NN
      COMMON/CONV/EPS,ITER

C***EPS DENOTES THE CONVERGENCE CRITERIA AND ITER DENOTES
C***THE MAXIMUM NUMBER OF ITERATIONS
      EPS=0.001D0
      ITER=100

C***INPUT FILE NAME=INFILE
C***OUTPUT FILE NAME=OUTFILE
      OPEN(UNIT=1,FILE='INFILE',STATUS='OLD')
      OPEN(UNIT=3,FILE='OUTFILE',STATUS='UNKNOWN')

C***INPUT FILE INCLUDES THE NUMBER OF NODES N, ONE GUESS
C***TEMPERATURE GUESS, AND THE MATRIX [A|B] DENOTED BY AB.
C***X IS THE UNKNOWN NODE TEMPERATURE
      READ(1,*)N,GUESS
      NN=N+1
      DO 20 I=1,N
      READ(1,*)(AB(I,J),J=1,NN)
      X(I)=GUESS
  20 CONTINUE

C***PRINT THE INPUT DATA FOR CHECKING PURPOSES.
      WRITE(3,10)N,GUESS
  10  FORMAT(1X,'Number of nodes = ',I2,5X,'Initial guess = ',F9.3)
      WRITE(3,15)
  15  FORMAT(//,1X,'Input matrix is :  ')
      DO 30 I=1,N
       WRITE(3,40)(AB(I,J),J=1,NN)
  40  FORMAT(//,10X,13F9.3)
  30  CONTINUE
C***CALL SUBROUTINE TO SOLVE THE SIMULTANEOUS EQUATION
      CALL GS

C***PRINT THE SOLUTION
      WRITE(3,50)
  50 FORMAT(//,1X,'The solutions are :  ')
      WRITE(3,60)
  60 FORMAT(//,31X,'Node #',5X,' Temperature',/)
      DO 70 I=1,N
       WRITE(3,80)I,X(I)
  70 CONTINUE
  80 FORMAT(29X,I5,7X,F9.3)
      STOP
      END
```

```
C
C* SUBROUTINE GS
C*
   SUBROUTINE GS
   IMPLICIT REAL*8(A-H,O-Z)
   COMMON/MATR/AB(40,41),X(40)
   COMMON/NUM/N,NN
   COMMON/CONV/EPS,ITER

C***THIS SUBROUTINE SOLVES THE SIMULTANEOUS EQUATIONS BY
C***USING THE GAUSS-SEIDEL METHOD.
   DO 1 K=1,ITER
    NEXTK=0
    DO 2 I=1,N
      XOLD=X(I)
      SUM=0.0D0
      DO 3 J=1,N
      IF(I.EQ.J) GOTO 3
      SUM=SUM+AB(I,J)*X(J)
 3    CONTINUE
      X(I)=(AB(I,NN)-SUM)/AB(I,I)
      IF(DABS(XOLD-X(I)).GT.EPS) NEXTK=1
 2    CONTINUE
      IF(NEXTK.EQ.0) GOTO 5
 1  CONTINUE
    WRITE(3,10) ITER
10  FORMAT(4X,'Sorry, cannot converge within ',I3,' iterations.')
 5  CONTINUE
    RETURN
    END
```

F.3   An Ordinary Differential Equation Solver for Initial Value Problems (Runge–Kutta Method)

It sometimes becomes necessary to solve linear or nonlinear m^{th} order ODE's. The Runge–Kutta method provides a powerful algorithm that can be used to solve such initial value problems (IVPs). In case of two-point boundary value problems (BVPs), a *second initial* condition is guessed, and *iteratively refined*, so that the solution matches the second boundary condition (cf. shooting method, e.g., Burden et al. 1988 or Ralston and Rabinowitz 1978). Consider the following initial conditions:

$$a \leq t \leq b \qquad u_i(a) = \alpha_i \qquad i = 1, 2, \ldots, m$$

and select a positive integer N to divide the interval into N subintervals

$$t_i = a + ih \qquad 0, 1, 2, \ldots, N \qquad h = \frac{b - a}{N}$$

$$\text{where} \qquad w_{ij} = u_i(t_j) \qquad j = 0, 1, 2, \ldots N \qquad i = 1, 2, \ldots, m$$

If $w_{1j}, \ldots, w_{mj}$ can be calculated, $w_{1,j+1}, \ldots, w_{m,j+1}$ can be calculated as follows:

$$\left. \begin{aligned} k_1 &= hf_i(t_j, w_{1i}, \ldots, w_{mj}) \\ k_2 &= hf_i(t_j + h/2, w_{1j} + k_{11}/2, \ldots, w_{mj} + k_{1m}/2) \\ k_3 &= hf_i(t_j + h/2, w_{1j} + k_{21}/2, \ldots, w_{mj} + k_{2m}/2) \\ k_4 &= hf_i(t_j, w_{1j} + k_{31}, \ldots, w_{mj} + k_{3m}) \end{aligned} \right\} \quad i = 1, 2, \ldots, m$$

where $\qquad w_{i,j+1} = w_{ij} + (k_{1i} + 2k_{2i} + 2k_3i + k_{4i}) \qquad i = 1, 2, \ldots, m$

Using this relationship, if initial conditions are known, w_{ij} can be calculated iteratively, therefore approximating $u_i(t_i)$ for $i = 1, 2, \ldots, m$ and $j = 1, 2, \ldots, N$. Therefore, in order to solve an mth order differential equation of the form:

$$y^{(m)}(t) = f(t, y, y', \ldots, y^{(m-1)}) \qquad a \le t \le b$$

with known initial conditions, we convert to a system of first order equations by defining the system as follows:

$$u_1(t) = y(t), \qquad u_2 = y'(t), \ldots, u_m(t) = y^{(m-1)}(t)$$

Now, the Runge–Kutta algorithm is used to solve the system. The method is a very powerful algorithm that may be converted into FORTRAN source code as given on p. 498. The program prompts the user for the order of the equation, the boundaries, and the initial conditions. The user must enter the equation into the code at the place noted before running the program. Each of the values will be printed into a file named "OUTPUT." The following are a sample run and output file for the Blasius problem $f''' + ff'' = 0$:

INPUT

```
PLEASE ENTER THE LOWER BOUNDARY XO
0.0
PLEASE ENTER THE UPPER BOUNDARY XLIM
1.0
PLEASE ENTER THE STEP SIZE H
0.1
PLEASE ENTER THE ORDER OF THE DIFFERENTIAL EQUATION
3
PLEASE ENTER THE INITIAL VALUE OF Y( 1)
0.469600
PLEASE ENTER THE INITIAL VALUE OF Y( 2)
0.0
PLEASE ENTER THE INITIAL VALUE OF Y( 3)
0.0
```

OUTPUT

X	Y(1)	Y(2)	>>> Y(N)
.10000000E+00	.46956320E+00	.46958130E-01	.23478710E-02
.20000000E+00	.46930610E+00	.93903390E-01	.93909990E-02
.30000000E+00	.46860880E+00	.14080280E+00	.21126630E-01

```
.40000000E+00    .46725430E+00    .18760130E+00    .37547610E-01
.50000010E+00    .46503060E+00    .23422270E+00    .58640220E-01
.60000010E+00    .46173490E+00    .28056980E+00    .84382060E-01
.70000010E+00    .45717820E+00    .32652580E+00    .11474000E+00
.80000010E+00    .45119110E+00    .37195590E+00    .14966830E+00
.90000020E+00    .44362960E+00    .41670970E+00    .18910710E+00
.10000000E+01    .43438120E+00    .46062370E+00    .23298060E+00
```

```
C*
C* THIS PROGRAM USES THE RUNGE-KUTTA METHOD ALGORITHM
C* TO SOLVE N-th ORDER LINEAR AND NON-LINEAR O.D.E.'S,
C* KNOWING THE INITIAL VALUES OF ALL THE LOWER ORDER
C* DERIVATIVES.
C* IF YOUR N-th ORDER DIFFERENTIAL EQUATION IS:
C*    y(N)=f(x,y,y',y'',.....,y(N-2),y(N-1))  XO \< x \<  XLIM
C* DEFINE: Y(1)=y(N-1), Y(2)=y(N-2),..,Y(I)=y(N-I),..,Y(N-1)=y',Y(N)=y
C*
C* THE USER IS REQUIRED TO ENTER:
C*   + X   :   THE LOWER BOUNDARY VALUE
C*   + XLIM:   THE UPPER BOUNDARY VALUE
C*   + H   :   THE STEP SIZE
C*   + N   :   THE ORDER OF THE DIFFERENTIAL EQUATION. IN THIS CASE,
C*       IT   IS LIMITED TO A MAXIMUM OF 10.  THIS LIMIT CAN HOWEVER
C*       BE   INCREASED BY INCREASING THE SIZE OF THE ARRAYS Y & F
C*   + Y(I):  FOR I=1,N.THE INITIAL VALUES OF THE VARIOUS DERIVATIVES
C*       AT x=XO.  THE PROMPT WILL ASK FOR: y(0)=y, y(1)=y',etc...
C*
C* THE USER MUST ENTER THE EXPRESSION OF THE DIFFERENTIAL EQ'N
C* AT''F(1)'' (LEVEL 10) USING THE DEFINITION OF ''Y''. F(I) REPRESENTS
C* THE DERIVATIVE OF Y(I). THEREFORE,
C*  F(1)=f(X,Y(N),Y(N-1),...,Y(2),Y(1))
C* FOR EXAMPLE, IF THE DIFFERENTIAL EQUATION IS:
C*    y'''=x**2 - y**2 + 3*y'*y''
C* THEN,       F(1)=X**2-Y(3)**2+3.*Y(2)*Y(1)
C*

  DIMENSION Y(10),F(10)
  OPEN(UNIT=10,FILE='OUTPUT',STATUS='UNKNOWN')
  M=0
4 FORMAT(9X,'X',12X,'Y(1)',12X,'Y(2)',6X,'> > > Y(N)')
5 FORMAT(1X,10E15.8)
  WRITE(*,21)
  READ(*,*) X
  WRITE(*,22)
  READ(*,*) XLIM
  WRITE(*,23)
  READ(*,*) H
  WRITE(*,24)
  READ(*,*) N
  WRITE(10,4)
  DO 2 I=1,N
```

```
      WRITE(*,25) I
      READ(*,*) Y(I)
2     CONTINUE
8     IF(X-XLIM) 6,6,7
6     CALL RUNGE(N,Y,F,X,H,M,K)
      GO TO (10,20),K

C*
C* YOU MUST TYPE IN THE APPROPRIATE EXPRESSION OF F(1), WHERE F(1) IS
C* EQUAL TO YOUR N-th ORDER y:y(N)=f(x,y,y',y'',y'',...,y(N-2),y(N-1))
C* PLEASE REMEMBER THAT: Y(1)=y(N-1), Y(2)=y(N-2),..,Y(N-1)=y',Y(N)=y
C* AS AN EXAMPLE, IN THIS CASE,N=3 & THE DIFFERENTIAL EQUATION IS:
C*      y''' = -y*y''
C*


10    F(1)=-Y(1)*Y(3)
      DO 15 I=2,N
       F(I)=Y(I-1)
15    CONTINUE
      GO TO 6
20    WRITE(10,5) X,(Y(I),I=1,N)
      GO TO 8
21    FORMAT(1X,'PLEASE ENTER THE LOWER BOUNDARY X0')
22    FORMAT(1X,'PLEASE ENTER THE UPPER BOUNDARY XLIM')
23    FORMAT(1X,'PLEASE ENTER THE STEP SIZE H')
24    FORMAT(1X,'PLEASE ENTER THE ORDER OF THE DIFFERENTIAL EQUATION')
25    FORMAT(1X,'PLEASE ENTER THE INITIAL VALUE OF Y(',I2,')')
7     CLOSE(10)
      STOP
      END

C*
C* SUBROUTINE RUNGE
C*
      SUBROUTINE RUNGE(N,Y,F,X,H,M,K)
      DIMENSION Y(10),F(10),Q(10)
      M=M+1
      GO TO (1,4,5,3,7),M
1     DO 2 I=1,N
2     Q(I)=0
      A=0.5
      GO TO 9
3     A=1.7071067811865475244
4     X=X+0.5*H
5     DO 6 I=1,N
       Y(I)=Y(I)+A*(F(I)*H-Q(I))
6     Q(I)=2.0*A*H*F(I)+(1.0-3.0*A)*Q(I)
      A=0.292832188134524756
      GO TO 9
7     DO 8 I=1,N
8      Y(I)=Y(I)+H*F(I)/6.0-Q(I)/3.0
      M=0
```

```
      K=2
      GO TO 10
9     K=1
10 RETURN
   END
```

F.4   Parabolic Partial Differential Equation Solver (Keller–Box Method)

In modern fluid mechanics and convective heat transfer, the need to solve partial differential equations (PDEs) which describe boundary-layer-type problems, often arises. Numerical methods provide one family of (approximate) solution techniques. With the ready availability of powerful computers, solutions can be found quickly using finite difference, finite element, or control volume methods. One such numerical technique is Keller's box method, developed by H. B. Keller (1970). There are four basic steps to Keller's box method:

1. Reduce the governing equations to first-order systems.
2. Write difference equations using central differencing.
3. Linearize the resulting equations (cf. Appendix F.1).
4. Write the equations in matrix-vector form and solve the linear system by the block tridiagonal elimination method.

For this discussion, the time-averaged (turbulent) mass, momentum, and energy conservation equations in rectangular coordinates, will be considered:

$$\frac{\partial}{\partial x}(\overline{\rho u}) + \frac{\partial}{\partial y}(\overline{\rho v}) = 0 \tag{1}$$

$$\overline{\rho u}\frac{\partial \bar{u}}{\partial x} + \overline{\rho v}\frac{\partial \bar{v}}{\partial y} = -\frac{\partial \bar{p}}{\partial x} + \frac{\partial}{\partial y}\left(\mu\frac{\partial \bar{u}}{\partial y} - \overline{\rho' v'}\right) \tag{2}$$

$$\overline{\rho u}\frac{\partial \bar{H}}{\partial x} + \overline{\rho v}\frac{\partial \bar{H}}{\partial y} = k\frac{\partial \bar{T}}{\partial y} - c_p\rho\overline{T'v'} + \bar{u}\left(\mu\frac{\partial \bar{u}}{\partial y} - \overline{\rho' v'}\right) \tag{3}$$

Equations (1)–(3) can be reduced to the following mixed differential equations (cf. Cebeci and Bradshaw 1988; Bradshaw et al., 1981):

$$(bf'')' + m_1 ff'' + m_2[c - (f')^2] = x\left(f'\frac{\partial f'}{\partial x} - f''\frac{\partial f}{\partial x}\right) \tag{4}$$

$$(eg' + df'f'')' + m_1 fg' = x\left(f'\frac{\partial g}{\partial x} - g'\frac{\partial f}{\partial x}\right) \tag{5}$$

The derivation of Equations (4) and (5) is outlined in Sect. 4.1.1, as well as in Cebeci and Bradshaw (1988).

In step 1, these equations are reduced to a first-order system by introducing three dependent variables, $u(x, \eta)$, $v(x, \eta)$, and $p(x, \eta)$, so that the transformed momentum and energy equations can be written with

$$f' = u$$
$$u' = v$$
$$g' = p$$

as

$$(bv)' + m_1 f_v + m_2[c - u^2] = x\left(u\frac{\partial u}{\partial x} - v\frac{\partial f}{\partial x}\right)$$

$$(ep + duv)' + m_1 fp = x\left(u\frac{\partial g}{\partial x} - p\frac{\partial f}{\partial x}\right)$$

where $g \equiv H/H_e$ is the dimensionless total enthalpy. The boundary conditions should be written in terms of the new dependent variables.

In step 2, the box method is employed to write the finite difference approximations. The points of the box are as follows:

$$x_0 = 0, \qquad x_n = x_{n-1} + k_n \qquad n = 1, 2, 3, \ldots N$$

$$\eta_0 = 0, \qquad \eta_j = \eta_{j-1} + h_j \qquad j = 1, 2, 3, \ldots J$$

where k_n and η_j are grid spacings in the x- and y-directions, respectively. The finite difference approximations for the first-order ordinary differential equations are written for the midpoint of segment $P_1 P_2(x_n, \eta_{j-1}/2)$. The finite difference approximations for the first-order partial differential equations are also written, but for the midpoint of the $P_1 P_2 P_3 P_4$ rectangle, which is at $(x_{n-1/2}, \eta_{j-1/2})$. The boundary conditions are also written in terms of the new dependent variables u, v, and p. This procedure produces $(5J + 5)$ nonlinear equations for the same number of unknowns.

In step 3, the $(5J + 5)$-nonlinear system of equations is linearized by using Newton's method. This procedure leads to $(5J + 5)$-linear equations that are quite cumbersome.

In step 4, these equations are written in the form of a matrix and solved. When placed in matrix vector form, the equations form tridiagonal boxes and can be solved by an elimination method. The method basically performs a forward and a backward sweep to calculate the unknowns. For more information on Keller's box method, consult Bradshaw et al. (1984).

To demonstrate this method, two programs were developed and the accuracy checked. The first program solves flow over a flat plate with various boundary conditions. The second program investigates flow in a pipe. The programs can be easily modified to obtain a variety of information.

F.4.1 Keller's Box Method for Flow over a Flat Plate

```
C*****************************************************************
C* PROGRAM PLATE
C*
C* THIS PROGRAM EMPLOYS A FINITE DIFFERENCE SCHEME TO COMPUTE THE
C* SOLUTIONS TO BOUNDARY LAYER FLOW OVER A FLAT PLATE.  THE BOUNDARY
C* CONDITIONS CAN BE VARIED TO INCLUDE DIFFERENT TEMPERATURES AT THE
```

```
C* WALL AND FREE STREAM AND EITHER SUCTION OR INJECTION OF FLUID AT
C* THE PLATE.  THIS PROGRAM TAKES ADVANTAGE OF THE KELLER-BOX METHOD
C* FOR SOLVING PARTIAL DIFFERENTIAL EQUATIONS.
C*
C*********************************************************************

C* DECLARE COMMON BLOCK VARIABLES
      COMMON /INPT2/ ETAE,VGP
      COMMON /BLC0/ NP,NPT,NX,NXT,IT
      COMMON /AK1/ RMUI,TI,RMI,UI,PR,HE
      COMMON /GRD/ X(60),ETA(61),DETA(61),A(61)
      COMMON /BLC3/ DELF(61),DELU(61),DELV(61),DELG(61),DELP(61)
      COMMON /BLC1/ F(61,2),U(61,2),V(61,2),B(61,2),G(61,2),P(61,2),
     &              C(61,2),D(61,2),E(61,2),RMU(61),BC(61)
      COMMON /EDGE/ UE(60),TE(60),RHOE(60),RMUE(60),PE(60), P1(60),P2(60)
      COMMON /INPUT1/ WW(60),ALFA0,ALFA1
      COMMON /VALUES2/ FWALL,TWALL,TFILM
      COMMON /VALUES/ RXA(61),CFA(61),ANUXA(61),HA(61),RTHETAA(61),
     &              STXA(61),THETAA(61),DELSA(61),NXA(61)


      NPT = 61
      ITMAX = 10
      NX = 1
      OPEN(UNIT=10,FILE='PLAT_INP.TXT',STATUS='OLD')
      OPEN(UNIT=12,FILE='PLAT_ANS.TXT',STATUS='UNKNOWN')
      CALL INPUT

C* GRID GENERATION
      IF((VGP-1.0) .LE. 0.001) GO TO 5
      NP = ALOG((ETAE/DETA(1))*(VGP-1.0)+1.0)/ALOG(VGP) + 1.0001
      GO TO 10
5     NP = ETAE/DETA(1) + 1.0001
10    IF(NP .LE. 61) GO TO 15
      WRITE(6,9000)
      STOP
15    ETA(1)= 0.0

      DO 20 J=2,NPT
       DETA (J) =VGP*DETA (J-1)
       A(J) =  0.5*DETA(J-1)
       ETA(J)=  ETA(J-1)+DETA(J-1)
20    CONTINUE
      CALL IVPL

30    CONTINUE
      IT =0
40    IT = IT +1

C* FLUID PROPERTIES

      DO 50 J=1,NP
```

```
          T=TFILM
          RMU(J) = 1.45E-6*(T**1.5)/(T+110.33)
          C(J,2) = T/TE(NX)
          BC(J) = RMU(J)/RMUE(NX)/C(J,2)
          D(J,2) = BC(J)*(UE(NX)**2)*(1.0 - 1.0/PR)/HE
          B(J,2) = BC(J)
          E(J,2) = BC(J)/PR
50     CONTINUE
          IF(IT .LE. ITMAX) GO TO 60
          WRITE(6,2500)
          STOP

60     CALL COEF
          CALL SOLV5

C* LAMINAR FLOW
          IF(ABS(DELV(1)) .GT. 1.0E-05) GO TO 40
          GO TO 100

C* CHECK FOR GROWTH
100    IF(V(1,2) .LT. 0.0) THEN
              WRITE(6,*)'AT ',X(NX)
              WRITE(6,*)'FLOW SEPARATION - PROGRAM HALTED'
              STOP
          END IF
          IF(NP .EQ. NPT) GO TO 120
          IF(ABS(V(NP,2)) .LE. 1.0E-03) GO TO 120
          NPO = NP
          NP1 = NP+1
          NP  = NP+1
          IF(NP   .GT. NPT) NP = NPT

C* DEFINITION  OF PROFILES FOR NEW NP
          DO 110  L=1,2
          DO 110  J=NP1,NP
          F(J,L)  = F(NPO,L) + ETA(J) -ETA(NPO)
          U(J,L)  = 1.0
          V(J,L)  = V(NPO,L)
          B(J,L)  = B(NPO,L)
          G(J,L)  = G(NPO,L)
          P(J,L)  = P(NPO,L)
          C(J,L)  = C(NPO,L)
          D(J,L)  = D(NPO,L)
          E(J,L)  = E(NPO,L)
110    CONTINUE
          IT = 0
          GO TO 40
120    CALL OUTPUT
          GO TO 30

2500   FORMAT(1H0,16X,25HITERATIONS EXCEEDED ITMAX)
```

```
9000  FORMAT(1H0,'  NP EXCEEDED NPT --- PROGRAM TERMINATED')
9100  FORMAT(1H0,4HNX =,I3,5X,3HX =,F10.3)
      END

C**********************************************************************
C* SUBROUTINE INPUT
C* THIS SUBROUTINE SPECIFIES:
C* THE WALL BOUNDARY CONDITIONS FOR THE ENERGY EQUATION
C* (NTX)--THE TOTAL NUMBER OF X SATATIONS
C* (m2) THE DIMENSIONLESS PRESSURE GRADIENT AT THE FIRST X STATION P2(1)
C* (h1) AND (K) WHICH ARE VARIABLE GRID PARAMETERS
C* FREESTREAM PARAMETERS: (RMI) MACH NUMBER
C*                        (TI) TEMPERATURE
C*                        (PI) PRESSURE
C* (PR) MOLECULAR PRANDTL #
C**********************************************************************

      SUBROUTINE INPUT
      COMMON  /INPT2/ ETAE,VGP
      COMMON  /AK1/ RMUI,TI,RMI,UI,PR,HE
      COMMON  /INPUT1/ WW(60),ALFA0,ALFA1
      COMMON  /BLCO/ NP,NPT,NX,NXT,IT
      COMMON  /GRD/ X(60),ETA(61),DETA(61),A(61)
      COMMON  /EDGE/ UE(60),TE(60),RHOE(60),RMUE(60),PE(60),P1(60),P2(60)
      COMMON  /VALUES2/ FWALL,TWALL,TFILM

      ETAE=8.0
C* PROMPT USER FOR PLATE LENGTH
      WRITE(6,*)'ENTER THE LENGTH OF THE PLATE IN METERS'
      READ(5,*)PLENGTH
      WRITE(6,*)' '
C* PROMPT USER FOR NUMBER OF X STATIONS
      WRITE(6,*)'ENTER THE NUMBER OF X-STATIONS'
      READ(5,*)NXT
      WRITE(6,*)' '

C* FROM GIVEN PLATE LENGTH AND NUMBER OF X-STATIONS DETERMINE WHERE
C* EACH X-STATION IS LOCATED
      DELTAX=PLENGTH/NXT
      P2(1)=DELTAX
      X(1)=DELTAX
      DO 10 I=2,NXT
       X (I)=X(I-1)+DELTAX
10    CONTINUE

C* GRID PARAMETERS DETA(1)=.2 VGP(1)=1.0 FOR LAMINAR FLOWS
      DETA(1)=.2
      VGP=1.0
11    CONTINUE

C* CONDITIONS FOR HEAT TRANSFER CONDITIONS AT WALL
```

```
C* ALFAO = 1.0 , ALFA1 = 0.0 ------> SPECIFIED WALL TEMPERATURE
C* ALFAO = 0.0 , ALFA1 = 1.0 ------> SPECIFIED HEAT FLUX

      WRITE(6,*)'IT IS ASSUMED THAT THERE IS CONSTANT WALL TEMPERATURE'
      WRITE(6,*)'BOUNDARY CONDITION'
        ALFAO=1.0
        ALFA1=0.0
      WRITE(6,*)'ENTER THE WALL TEMPERATURE IN K'
      READ(5,*)TWALL
      WRITE(6,*)' '

C* FRESTREAM PARAMETERS
C* FREE-STREAM TEMPERATURE IN K
      WRITE(6,*)'ENTER THE FREE-STREAM TEMPERATURE IN K'
      READ(5,*)TI
      TEMP=TI
      WRITE(6,*)' '

C* FREE-STREAM VELOCITY IN M/SEC
      WRITE(6,*)'ENTER THE FREESTREAM VELOCITY IN M/SEC'
      READ(5,*)UI
      WRITE(6,*)' '

C* FREE-STREAM MACH NUMBER
      RMI=.00576

C* ASSUME FREE-STREAM PRESSURE = 1 ATM ===> 84850 Pa
      PI=84850.
      TFILM=(TI+TWALL)/2
      TI=TFILM

C* VISCOSITY IN kg/m/s
      RMUI=1.45E-6*(TI**1.5)/(TI+110.33)

C* DENSITY IN kg/m**3
      RHOI = PI/TI/287

C* ENTHALPY m**2/s**2
      HE = 1004.3*TI + 0.5*UI**2

C* PRANDTL NUMBER
      PR=.708

C* EDGE VELOCITY AND DIMENSIONLESS WALL TEMPERATURE  OR THE
C* DIMENSIONLESS TEMPERATURE GRADIENT AT EACH X STATION
C* NOTE: UE(I) IN m/s
C* WW EITHER G(1,2) FOR WALL TEMPERATURE SPECIFIED
C* OR  P(1,2) FOR HEAT FLUX SPECIFIED

      DO 15 I=1,NXT
         UE(I)=.995*UI
```

```
          WW(I)=0.0
15     CONTINUE
       WRITE(6,*)'IN FLOWS WITH SUCTION OR BLOWING SPECIFY THE '
       WRITE(6,*)'DIMENSIONLESS WALL VELOCITY.'
       WRITE(6,*)'       NOTE: VWALL NEAGTIVE ---> SUCTION'
       WRITE(6,*)'             VWALL POSITIVE ---> INJECTION'
       WRITE(6,*)'             VWALL ZERO -----> IMPERMEABLE'
       WRITE(6,*)' '
       WRITE(6,*)'ENTER THE DIMENSIONLESS SUCTION OR INJECTION PARAMETER'
       READ(5,*)FWALL

C* EDGE CONDITIONS SET EQUAL TO THE FREE-STREAM VALUES
       DO 20 I=1,NXT
         TE(I)=TI
         RMUE(I)=RMUI
         PE(I)=PI
         RHOE(I)=RHOI
20     CONTINUE
         TI=TEMP

C* CALCULATION OF PRESSURE-GRADIENT PARAMETER P=P2
       P1(1) = 0.5*(P2(1)+1.0)
       DO 80 I=2,NXT
         IF(I .EQ. NXT) GO TO 60
         A1=(X(I)-X(I-1))*(X(I+1)-X(I-1))
         A2=(X(I)-X(I-1))*(X(I+1)-X(I))
         A3=(X(I+1)-X(I))*(X(I+1)-X(I-1))
         DUDS=-(X(I+1)-X(I))/A1*UE(I-1)+(X(I+1)-2.0*X(I)+X(I-1))/
     &         A2*UE(I)+(X(I)-X(I-1))/ A3*UE(I+1)
         GO TO 70
60       A1=(X(I-1)-X(I-2))*(X(I)-X(I-2))
         A2=(X(I-1)-X(I-2))*(X(I)-X(I-1))
         A3=(X(I)-X(I-1))*(X(I)-X(I-2))
         DUDS=(X(I)-X(I-1))/A1*UE(I-2) - (X(I)-X(I-2))/A2*UE(I-1) +
     &        (2.0*X(I)-X(I-2)-X(I-1))/A3*UE(1)
70       P2(I)=(X(I)/UE(I))*DUDS
         P1(I)=0.5*(1.0+P2(I)+X(I)*(RHOE(I)*RMUE(I) - RHOE(I-1)*
     &         RMUE(I-1))/(X(I)-X(I-1))/RHOE(I)/RMUE(I))
80     CONTINUE
       RETURN

8000   FORMAT(2I3,3F10.0)
8100   FORMAT(8F10.0)
8200   FORMAT(3F10.0)
9000   FORMAT(1H0,6HNXT =,I3,14X,6HNTR = ,6HDETA1=,E14.6,3X,
     &        6HVGP =,E14.6,3X,3HPR=,F7.3)
9010   FORMAT(1H0,30X,'INPUT DATA'///)
9100   FORMAT(1H0,'MACH NO =',E12.5,3X,'T INF. =',E12.5,3X,'PI = ',
     &        E12.5,/1H0,'UI =',E12.5,3X,'RHOI =',E12.5,3X,'HE = ',
     &        E12.5)
9200   FORMAT(1H0,2HNX,8X,2HUE,13X,2HTE,12X,4HRHOE,11X,4HRMUE,12X,2HP1,
```

```
      &         13X,2HP2/(1H ,I2,6E15.5))
      END


C************************************************************************
C* SUBROUTINE IVPL
C*
C* THIS SUBROUTINE IS USED TO GENERATE THE INITIAL
C* VELOCITY PROFILES FOR SIMILAR, COMPRESSIBLE LAMINAR FLOW
C************************************************************************
      SUBROUTINE IVPL
      COMMON /AK1/ RMUI,TI,RMI,UI,PR,HE
      COMMON /BLCO/ NP,NPT,NX,NXT,IT
      COMMON /GRD/ X(60),ETA(61),DETA(61),A(61)
      COMMON /BLC1/ F(61,2),U(61,2),V(61,2),B(61,2),G(61,2),P(61,2),
      &              C(61,2),D(61,2),E(61,2),RMU(61),BC(61)
      COMMON /BLC3/ DELF(61),DELU(61),DELV(61),DELG(61),DELP(61)
      COMMON /INPUT1/ WW(60),ALFA0,ALFA1
      COMMON /VALUES2/ FWALL,TWALL,TFILM


C* GENERATE INITIAL PROFILE BY SOLVING THE INCOMPRESSIBLE FLOW
      ETANPQ=0.25*ETA(NP)
      ETAU15=1.5/ETA(NP)
      DO 30 J=1,NP
       ETAB=ETA(J)/ETA(NP)
       ETAB2=ETAB**2
       F(J,2)=ETANPQ*ETAB2*(3.0-.5*ETAB2)
       U(J,2)=0.5*ETAB*(3.0-ETAB2)
       V(J,2)=ETAU15*(1.0-ETAB2)
       G(J,2)=ETAB
       P(J,2)=1.0/ETA(NP)
       B(J,2)=1.0
       C(J,2)=1.0
       E(J,2)=1.0/PR
       D(J,2)=0.0
30     CONTINUE
       IT=0
50     IT = IT + 1
       IF (IT .LE. 8) GO TO 70
       WRITE(6,9900)
       STOP
70     CONTINUE
       CALL COEF
       CALL SOLV5
       IF(ABS(DELV(1)) .GT. 1.0E-05) GO TO 50
       RETURN

9900   FORMAT(1H0,'INCOMPRESSIBLE, DID NOT CONVERGE')
       END


C************************************************************************
C* SUBROUTINE SOLV5
```

```
C*
C* SUBROUTINE TO SOLVE A LINEAR SYSTEM CONSISTING OF FIVE FIRST ORDER
C* EQUATIONS USING THE BLOCK ELIMINATION METHOD
C*********************************************************************
      SUBROUTINE SOLV5
      COMMON /INPUT1/ WW(60),ALFAO,ALFA1
      COMMON /BLCO/ NP,NPT,NX,NXT,IT
      COMMON /GRD/ X(60),ETA(61),DETA(61),A(61)
      COMMON /BLC1/ F(61,2),U(61,2),V(61,2),B(61,2),G(61,2),P(61,2),
     &              C(61,2),D(61,2),E(61,2),RMU(61),BC(61)
      COMMON /BLC3/ DELF(61),DELU(61),DELV(61),DELG(61),DELP(61)
      COMMON /BLC6/ S1(61),S2(61),S3(61),S4(61),S5(61),S6(61),S7(61),
     1              S8(61),B1(61),B2(61),B3(61),B4(61),B5(61),B6(61),
     2              B7(61),B8(61),B9(61),B10(61),R(5,61)
      COMMON /VALUES2/ FWALL,TPLATE,TFILM
      COMMON /AK1/ RMUI,TI,RMI,UI,PR,HE
      COMMON /EDGE/ UE(60),TE(60),RHOE(60),RMUE(60),PE(60),P1(60),P2(60)
      DIMENSION    A11(61),A12(61),A13(61),A14(61),A15(61),A21(61),
     1             A22(61),A23(61),A24(61),A25(61),A31(61),A32(61),
     2             A33(61),A34(61),A35(61),G11(61),G12(61),G13(61),
     3             G14(61),G15(61),G21(61),G22(61),G23(61),G24(61),
     4             G25(61),G31(61),G32(61),G33(61),G34(61),G35(61),
     5             W1(61),W2(61),W3(61),W4(61),W5(61)

C* ELEMENTS OF TRIANGLE MATRIX
      A11(1)=1.0
      A12(1)=0.0
      A13(1)=0.0
      A14(1)=0.0
      A15(1)=0.0
      A21(1)=0.0
      A22(1)=1.0
      A23(1)=0.0
      A24(1)=0.0
      A25(1)=0.0
      A31(1)=0.0
      A32(1)=0.0
      A33(1)=0.0
      A34(1)=ALFAO
      A35(1)=ALFA1

C* ELEMENTS O W-VECTOR
      W1(1)=R(1,1)
      W2(1)=R(2,1)
      W3(1)=R(3,1)
      W4(1)=R(4,1)
      W5(1)=R(5,1)

C* FORWARD SWEEP
C* DEFINITIONS
      DO 30 J=2,NP
```

```
      AA1=A(J)*A24(J-1)-A25(J-1)
      AA2=A(J)*A34(J-1)-A35(J-1)
      AA3=A(J)*A12(J-1)-A13(J-1)
      AA4=A(J)*A22(J-1)-A23(J-1)
      AA5=A(J)*A32(J-1)-A33(J-1)
      AA6=A(J)*A14(J-1)-A15(J-1)
      AA7=A(J)*S6(J)-S2(J)
      AA8=S8(J)*A(J)
      AA9=A(J)*B6(J)-B10(J)
      AA10=A(J)*B8(J)-B2(J)


C* ELEMENTS OF TRIANGLE MATRIX
      DET=A11(J-1)*(AA4*AA2-AA1*AA5)-A21(J-1)*(AA3*AA2-AA5*AA6)+
     &     A31(J-1)*(AA3*AA1-AA4*AA6)
      G11(J)=(-(AA4*AA2-AA5*AA1)+A(J)**2*(A21(J-1)*AA2-A31(J-1)*AA1))
     &         /DET
      G12(J)=((AA3*AA2-AA5*AA6)-A(J)**2*(A11(J-1)*AA2-A31(J-1)*AA6))/
     &        DET
      G13(J)=(-(AA3*AA1-AA4*AA6)+A(J)**2*(A11(J-1)*AA1-A21(J-1)*AA6))
     &         /DET
      G14(J)=G11(J)*A12(J-1)+G12(J)*A22(J-1)+G13(J)*A32(J-1)+A(J)
      G15(J)=G11(J)*A14(J-1)+G12(J)*A24(J-1)+G13(J)*A34(J-1)
      G21(J)=(S4(J)*(AA2*AA4-AA1*AA5)+A31(J-1)*(AA1*AA7-AA4*AA8)+
     &         A21(J-1)*(AA5*AA8-AA7*AA2))/DET
      G22(J)=(A11(J-1)*(AA2*AA7-AA5*AA8)+A31(J-1)*(AA3*AA8-AA6*AA7)+
     &         S4(J)*(AA5*AA6-AA2*AA3))/DET
      G23(J)=(A11(J-1)*(AA4*AA8-AA1*AA7)+S4(J)*(AA3*AA1-AA4*AA6)+
     &         A21(J-1)*(AA7*AA6-AA3*AA8))/DET
      G24(J)=G21(J)*A12(J-1)+G22(J)*A22(J-1)+G23(J)*A32(J-1)-S6(J)
      G25(J)=G21(J)*A14(J-1)+G22(J)*A24(J-1)+G23(J)*A34(J-1)-S8(J)
      G31(J)=(B4(J)*(AA4*AA2-AA5*AA1)-AA9*(A21(J-1)*AA2-A31(J-1)*AA1
     &         )+AA10*(A21(J-1)*AA5-A31(J-1)*AA4))/DET
      G32(J)=(-B4(J)*(AA3*AA2-AA5*AA6)+AA9*(A11(J-1)*AA2-A31(J-1)*AA6
     &         )-AA10*(A11(J-1)*AA5-A31(J-1)*AA3))/DET
      G33(J)=(B4(J)*(AA3*AA1-AA4*AA6)-AA9*(A11(J-1)*AA1-A21(J-1)*AA6
     &         )+AA10*(A11(J-1)*AA4-A21(J-1)*AA3))/DET
      G34(J)=G31(J)*A12(J-1)+G32(J)*A22(J-1)+G33(J)*A32(J-1)-B6(J)
      G35(J)=G31(J)*A14(J-1)+G32(J)*A24(J-1)+G33(J)*A34(J-1)-B8(J)


C* ELEMENTS OF TRIANGLE MATRIX
      A11(J)=1.0
      A12(J)=-A(J)-G14(J)
      A13(J)=A(J)*G14(J)
      A14(J)=-G15(J)
      A15(J)=A(J)*G15(J)
      A21(J)=S3(J)
      A22(J)=S5(J)-G24(J)
      A23(J)=S1(J)+A(J)*G24(J)
      A24(J)=-G25(J)*S7(J)
      A25(J)=A(J)*G25(J)
      A31(J)=B3(J)
```

```
        A32(J)=B5(J)-G34(J)
        A33(J)=B9(J)+A(J)*G34(J)
        A34(J)=B7(J)-G35(J)
        A35(J)=B1(J)+A(J)*G35(J)

C* ELEMENTS OF W-VECTOR
        W1(J)=R(1,J)-G11(J)*W1(J-1)-G12(J)*W2(J-1)-G13(J)*W3(J-1)-
     &          G14(J)*W4(J-1)-G15(J)*W5(J-1)
        W2(J)=R(2,J)-G21(J)*W1(J-1)-G22(J)*W2(J-1)-G23(J)*W3(J-1)-
     &          G24(J)*W4(J-1)-G25(J)*W5(J-1)
        W3(J)=R(3,J)-G31(J)*W1(J-1)-G32(J)*W2(J-1)-G33(J)*W3(J-1)-
     &          G34(J)*W4(J-1)-G35(J)*W5(J-1)
        W4(J) =R(4,J)
        W5(J) =R(5,J)
30      CONTINUE

C* BACKWARD SWEEP
        J=NP
C* DEFINITIONS OF VALUES AT J=NP (BOUNDARY LAYER EDGE)
        DP=-(A31(J)*(A13(J)*W2(J)-W1(J)*A23(J))-A32(J)*(A11(J)*
     &    W2(J)-W1(J)*A21(J)) + W3(J)*(A11(J)*A23(J)-A13(J)*A21(J)))
        DV=-(A31(J)*(W1(J)*A25(J)-W2(J)*A15(J))-W3(J)*(A11(J)*A25(J)
     &    -A15(J)*A21(J))+A35(J)*(A11(J)*W2(J)-W1(J)*A21(J)))
        DF=-(W3(J)*(A13(J)*A25(J)-A23(J)*A15(J))-A33(J)*(W1(J)*A25(J)
     &    -A15(J)*W2(J))+A35(J)*(W1(J)*A23(J)-A13(J)*W2(J)))
        D1=-(A31(J)*(A13(J)*A25(J)-A23(J)*A15(J))-A33(J)*(A11(J)*
     &    A25(J)-A21(J)*A15(J))+A35(J)*(A11(J)*A23(J)-A21(J)*A13(J)))
C* ELEMENTS OF DELTA-VECTOR FOR J=NP
        DELP(J)=DP/D1
        DELV(J)=DV/D1
        DELF(J)=DF/D1
        DELG(J)=0.0
        DELU(J)=0.0

40      J = J-1
C* DEFINITIONS OF VALUES FROM J=NP-1 TO J=1
        BB1=DELU(J+1)-A(J+1)*DELV(J+1)-W4(J)
        BB2=DELG(J+1)-A(J+1)*DELP(J+1)-W5(J)
        CC1=W1(J)-A12(J)*BB1-A14(J)*BB2
        CC2=W2(J)-A22(J)*BB1-A24(J)*BB2
        CC3=W3(J)-A32(J)*BB1-A34(J)*BB2
        DD1=A13(J)-A12(J)*A(J+1)
        DD2=A23(J)-A22(J)*A(J+1)
        DD3=A33(J)-A32(J)*A(J+1)
        EE1=A15(J)-A14(J)*A(J+1)
        EE2=A25(J)-A24(J)*A(J+1)
        EE3=A35(J)-A34(J)*A(J+1)
        DETT=A11(J)*DD2*EE3+A21(J)*DD3*EE1+A31(J)*DD1*EE2
     &      -A31(J)*DD2*EE1-A21(J)*DD1*EE3-A11(J)*DD3*EE2
C* ELEMENTS OF DELTA-VECTOR
        DELF(J)=(CC1*DD2*EE3+CC2*DD3*EE1+CC3*DD1*EE2-CC3*DD2*EE1
```

```
      &                -CC2*DD1*EE3-CC1*DD3*EE2)/DETT
         DELV(J)=(A11(J)*CC2*EE3+A21(J)*CC3*EE1+A31(J)*CC1*EE2-
      &           A31(J)*CC2*EE1-A21(J)*CC1*EE3-A11(J)*CC3*EE2)/DETT
         DELP(J)=(A11(J)*CC3*DD2+A21(J)*CC1*DD3+A31(J)*CC2*DD1-
      &           A31(J)*CC1*DD2-A21(J)*CC3*DD1-A11(J)*CC2*DD3)/DETT
         DELU(J)=BB1-A(J+1)*DELV(J)
         DELG(J)=BB2-A(J+1)*DELP(J)
         IF(J .GT. 1)goto 40

C*  NEW VALUES OF F,U,G,P
         DO 50 J=1,NP
         F(J,2)=F(J,2)+DELF(J)
         U(J,2)=U(J,2)+DELU(J)
         V(J,2)=V(J,2)+DELV(J)
         G(J,2)=G(J,2)+DELG(J)
         P(J,2)=P(J,2)+DELP(J)
50       CONTINUE

C* SET BOUNDARY CONDITIONS
         U(1,2)=0.0
         F(1,2)=FWALL
         RETURN
         END

C****************************************************************
C* SUBROUTINE COEF
C*
C* THIS SUBROUTINE CONTAINS THE COEFFICIENTS OF THE LINEARIZED MOMENTUM
C* AND ENERGY EQUATIONS
C****************************************************************
         SUBROUTINE COEF
         COMMON /INPUT1/ WW(60),ALFA0,ALFA1
         COMMON /BLC0/ NP,NPT,NX,NXT,IT
         COMMON /BLC1/ F(61,2),U(61,2),V(61,2),B(61,2),G(61,2),P(61,2),
      &                C(61,2),D(61,2),E(61,2),RMU(61),BC(61)
         COMMON /GRD/ X(60),ETA(61),DETA(61),A(61)
         COMMON /EDGE/ UE(60),TE(60),RHOE(60),RMUE(60),PE(60),P1(60),P2(60)
         COMMON /BLC6/ S1(61),S2(61),S3(61),S4(61),S5(61),S6(61),S7(61),
      &                S8(61),B1(61),B2(61),B3(61),B4(61),B5(61),B6(61),
      &                B7(61),B8(61),B9(61),B10(61),R(5,61)

         IF(IT .GT. 1) GO TO 5
         CEL=0.0
         IF(NX .GT. 1) CEL=0.5*(X(NX)+X(NX-1))/(X(NX)-X(NX-1))
         P1P=P1(NX) + CEL
         P2P=P2(NX) + CEL
         IF(ALFA0 .GT. 0.1) G(1,2)=WW(NX)
         IF(ALFA1 .GT. 0.1) P(1,2)=WW(NX)
5        CONTINUE

         DO 100 J=2,NP
```

```
C* PRESENT STATION
      USB=0.5*(U(J,2)**2+U(J-1,2)**2)
      FVB=0.5*(F(J,2)*V(J,2)+F(J-1,2)*V(J-1,2))
      FPB=0.5*(F(J,2)*P(J,2)+F(J-1,2)*P(J-1,2))
      UGB=0.5*(U(J,2)*G(J,2)+U(J-1,2)*G(J-1,2))
      UB=0.5*(U(J,2)+U(J-1,2))
      VB=0.5*(V(J,2)+V(J-1,2))
      FB=0.5*(F(J,2)+F(J-1,2))
      GB=0.5*(G(J,2)+G(J-1,2))
      PB=0.5*(P(J,2)+P(J-1,2))
      CB=0.5*(C(J,2)+C(J-1,2))
      DERBV = (B(J,2)*V(J,2)-B(J-1,2)*V(J-1,2))/DETA(J-1)
      DEREP = (E(J,2)*P(J,2)-E(J-1,2)*P(J-1,2))/DETA(J-1)
      DRDUV = (D(J,2)*U(J,2)*V(J,2) - D(J-1,2)*U(J-1,2)*V(J-1,2))/
     &          DETA(J-1)
      IF(NX .GT. 1) GO TO 10

C* PREVIOUS STATION
      CFB= 0.0
      CVB= 0.0
      CPB= 0.0
      CUB= 0.0
      CGB= 0.0
      CUGB= 0.0
      CFPB= 0.0
      CFVB= 0.0
      CUSB= 0.0
      CDERBV= 0.0
      CDEREP= 0.0
      CRB= -P2(NX)*CB
      CTB= 0.0
      GO TO 20
10    CFB=0.5*(F(J,1)+F(J-1,1))
      CVB=0.5*(V(J,1)+V(J-1,1))
      CPB=0.5*(P(J,1)+P(J-1,1))
      CUB=0.5*(U(J,1)+U(J-1,1))
      CGB=0.5*(G(J,1)+G(J-1,1))
      CFVB=0.5*(F(J,1)*V(J,1)+F(J-1,1)*V(J-1,1))
      CFPB=0.5*(F(J,1)*P(J,1)+F(J-1,1)*P(J-1,1))
      CUGB=0.5*(U(J,1)*G(J,1)+U(J-1,1)*G(J-1,1))
      CUSB=0.5*(U(J,1)**2+U(J-1,1)**2)
      CCB=0.5*(C(J,1) + C(J-1,1))
      CDERBV=(B(J,1)*V(J,1)-B(J-1,1)*V(J-1,1))/DETA(J-1)
      CDEREP=(E(J,1)*P(J,1)-E(J-1,1)*P(J-1,1))/DETA(J-1)
      CDRDUV=(D(J,1)*U(J,1)*V(J,1)-D(J-1,1)*U(J-1,1)*V(J-1,1))/
     &          DETA(J-1)
      CLB=CDERBV + P1(NX-1)*CFVB + P2(NX-1)*(CCB-CUSB)
      CRB=-CLB - P2(NX)*CB - CEL*CUSB + CEL*CFVB
      CMB=CDEREP + CDRDUV + P1(NX-1)*CFPB
      CTB=-CMB + CEL*(CFPB-CUGB)
```

```
C* COEFFICIENTS OF THE DIFFERENCED-MOMENTUM EQ.
20     CONTINUE
       S1(J)=B(J,2)/DETA(J-1) +0.5*P1P*F(J,2) -0.5*CEL*CFB
       S2(J)=-B(J-1,2)/DETA(J-1) + 0.5*P1P*F(J-1,2) - 0.5*CEL*CFB
       S3(J)=0.5*(P1P*V(J,2) + CEL*CVB)
       S4(J)=0.5*(P1P*V(J-1,2) + CEL*CVB)
       S5(J)=-P2P*U(J,2)
       S6(J)=-P2P*U(J-1,2)
       S7(J)=0.0
       S8(J)=0.0
       R(2,J)=CRB - (DERBV+P1P*FVB - P2P*USB + CEL*(FB*CVB-VB*CFB))

C* COEFFICIENTS OF DIFFERENCED ENERGY EQ.
       B1(J)=E(J,2)/DETA(J-1) + 0.5*P1P*F(J,2) - 0.5*CEL*CFB
       B2(J)=-E(J-1,2)/DETA(J-1) + 0.5*P1P*F(J-1,2) - 0.5*CEL*CFB
       B3(J)=0.5*(P1P*P(J,2) + CEL*CPB)
       B4(J)=0.5*(P1P*P(J-1,2) + CEL*CPB)
       B5(J)=D(J,2)*V(J,2)/DETA(J-1) - 0.5*CEL*(G(J,2)-CGB)
       B6(J)=-D(J-1,2)*V(J-1,2)/DETA(J-1) - 0.5*CEL*(G(J-1,2)-CGB)
       B7(J)=-0.5*CEL*(U(J,2)+CUB)
       B8(J)=-0.5*CEL*(U(J-1,2)+CUB)
       B9(J)=D(J,2)*U(J,2)/DETA(J-1)
       B10(J)=-D(J-1,2)*U(J-1,2)/DETA(J-1)
       R(3,J)=CTB - (DEREP + DRDUV + P1P*FPB - CEL*(UGB-CGB*UB+CUB*GB)+
      &         CEL*(CPB*FB-CFB*PB))

C* DEFINITIONS OF RJ
       R(1,J)=F(J-1,2)-F(J,2)+DETA(J-1)*UB
       R(4,J-1)=U(J-1,2)-U(J,2)+DETA(J-1)*VB
       R(5,J-1)=G(J-1,2)-G(J,2)+DETA(J-1)*PB
100    CONTINUE
       R(1,1)=0.0
       R(2,1)=0.0
       R(3,1)=0.0
       R(4,NP)=0.0
       R(5,NP)=0.0
       RETURN
       END


C*******************************************************************
C* SUBROUTINE OUTPUT
C*
C* PRINTS RESULTS INTO THE OUTPUT FILE
C*
C*******************************************************************

       SUBROUTINE OUTPUT
       COMMON/OTPT1/ RX,CNUE
       COMMON/AK1/ RMUI,TI,RMI,UI,PR,HE
       COMMON/BLCO/ NP,NPT,NX,NXT,IT
       COMMON/GRD/ X(60),ETA(61),DETA(61),A(61)
```

```
      COMMON/BLC1/ F(61,2),U(61,2),V(61,2),B(61,2),G(61,2),P(61,2),
     &            C(61,2),D(61,2),E(61,2),RMU(61),BC(61)
      COMMON/EDGE/ UE(60),TE(60),RHOE(60),RMUE(60),PE(60),P1(60),P2(60)
      COMMON/INPUT1/ WW(60),ALFA0,ALFA1
      COMMON/VALUES/ RXA(61),CFA(61),ANUXA(61),HA(61),RTHETAA(61),
     &               STXA(61),THETAA(61),DELSA(61),NXA(61)
      COMMON/VALUES2/ FWALL,TWALL,TFILM
      DIMENSION ACNU(61)
      IF(NX.EQ.NXT) THEN
      WRITE(12,5300)NX,X(NX)
      WRITE (12,5400)
      DO 15 J=1,NP
       IF ( V(J,2) .GT. 0.0005 .AND. U(J,2) .LT. 1.001)THEN
         WRITE (12,5500) ETA(J),F(J,2),U(J,2),V(J,2),G(J,2),P(J,2)
       END IF
15    CONTINUE
      END IF

      IF(NX .EQ. 1) GO TO 210
      TERMP = 0.0
      SUM = 0.0
      SUMC = 0.0

      DO 20 J=2,NP
       SUMC = SUMC + A(J)*(C(J,2) + C(J-1,2))
       TERM = U(J,2)*(1.0 - U(J,2))
       SUM = SUM + A(J)*(TERM + TERMP)
       TERMP = TERM
20    CONTINUE

      DELS = X(NX)/SQRT(RX)*(SUMC - F(NP,2))
      RDELS = SQRT(RX)*(SUMC - F(NP,2))
      THETA = X(NX)/SQRT(RX)*SUM
      RTHETA = SQRT(RX)*SUM

      IF (PASS .EQ. 1) GOTO 30
      IF(RX .GE. 6000000) THEN
        NTR=NX
        XNTR=X(NX)
        PASS=1
      END IF
30    CONTINUE

      H = DELS/THETA
      CF2 = BC(1)*V(1,2)/SQRT(RX)
      CF=2*CF2
      IF(ABS(G(1,2)-1.0) .LT. 0.00001) GO TO 200
      ANUX = P(1,2)*BC(1)*SQRT(RX)/(1.0 - G(1,2))
      STX = ANUX/RX/PR

C* STORE VALUES FOR EACH X STATION IN ARRAYS
```

```
        STXA(NX)=STX
        ANUXA(NX)=ANUX
200     CONTINUE
        RXA(NX)=RX
C       RXA(NXT)=UE(NXT)*X(NXT)/CNUE
        CFA(NX)=CF
        NXA(NX)=NX
        HA(NX)=H
        RTHETAA(NX)=RTHETA
        THETAA(NX)=THETA
        DELSA(NX)=DELS

C* WRITE VALUES TO SCREEN AND FILE
210     NX = NX+1
        IF(NX .GT. NXT) THEN
c           WRITE(12,5650)PR,UI,NTR
c           WRITE(12,5655)TI,TWALL
c           WRITE(12,5660)FWALL
c           WRITE(12,*)'VALUES FOR X-STATIONS'
c           WRITE(12,5750)
c           DO I=2,NXT
c               ACNU(I)=0.332*(PR**0.3333)*(UI*X(I)/16.84E-6)**0.5
c               WRITE(12,5825)X(I),RXA(I),ANUXA(I),CFA(I),STXA(I),ACNU(I)
c           END DO
        CLOSE(12)
        WRITE(6,*)'*************************************'
        WRITE(6,*)'*************=====PROGRAM COMPLETED=====*************'
        WRITE(6,*)'*************************************'
        STOP
        END IF


        CNUE = RMUE(NX)/RHOE(NX)
        RX = UE(NX)*X(NX)/CNUE


C* SHIFT PROFILES
        DO 250 J=1,NP
        F(J,1)= F(J,2)
        U(J,1)= U(J,2)
        V(J,1)= V(J,2)
        G(J,1)= G(J,2)
        P(J,1)= P(J,2)
        E(J,1)= E(J,2)
        C(J,1)= C(J,2)
        D(J,1)= D(J,2)
        B(J,1)= B(J,2)
250     CONTINUE

        RETURN

4300    FORMAT(1X,'X-STATION # = ',I2,3X,'LOCATION = ',F6.2)
4400    FORMAT(T3,'eta',T15,'f(eta)',T25,'f''(eta)',T35,'f''''(eta)',T50,
```

```
      &        'H/He',T60,'g(eta)''')
C4500  FORMAT(T3,F6.4,T15,F7.5,T25,F7.5,T35,F7.5,t50,f7.5,t60,f7.5)
5300   FORMAT(I2,',',F6.2)
5400   FORMAT('eta,f(eta),f''(eta),f''''(eta),H/He,g(eta)')
5500   FORMAT(F6.4,',',F7.5,',',F7.5,',',F7.5,',',F9.5,',',F9.5)
5600   FORMAT(1X,'Rx=',',',E9.3,',',',NUx=',',',E10.4,',',',Cf=',',',
      &        E10.4,',',',Stx=',',',E10.4)
5650   FORMAT(1X,'Pr # =',',',F10.4,',',',UI = ',',',F5.2,',',',NTR = ',
      &        ',', I3)
5655   FORMAT(1X,'TI = ',',',F6.2,',',',TWALL = ',',',F6.2)
5660   FORMAT(1X,'FWALL = ',',',F8.4)
5700   FORMAT(T2,'LOC',',',T5,'X',',',T12,'Rx',',',T24,'NUx',',',T38,
      &        'Cf',',',T50,'STx')
5750   FORMAT('   X        Rx        NUx        Cf        STx        NUexp')
5800   FORMAT(T2,I3,',',T5,F5.2,',',T12,E10.4,',',T25,E10.4,',',T38,E10.4
      &        ,',',T50,E10.4)
5825   FORMAT(F5.3,',',E10.4,',',E10.4,',',E10.4,',',E10.4,',',E10.4)
5850   FORMAT(1X,'NTR = ',I3,',',',UI = ',
      &        F5.2)

9000   FORMAT(T2,'LOCAL',T15,'LOCAL',T37,'LOCAL')
9010   FORMAT(T2,'REYNOLDS',T15,'NUSSELT',T25,'SKIN',T37,'STANTON')
9020   FORMAT(T2,'NUMBER',T15,'NUMBER',T25,'FRICTION',T38,'NUMBER')
9100   FORMAT(T1,E9.3,T12,E10.4,T23,E10.4,T36,E10.4)
       END
```

Sample Calculations

The following is a sample input for the preceding program:

```
ENTER THE LENGTH OF THE PLATE IN METERS
1.5

ENTER THE NUMBER OF X-STATIONS
60

IT IS ASSUMED THAT THERE IS CONSTANT WALL TEMPERATURE
BOUNDARY CONDITION
ENTER THE WALL TEMPERATURE IN K
250

ENTER THE FREE-STREAM TEMPERATURE IN K
350

ENTER THE FREE-STREAM VELOCITY IN M/SEC
2

IN FLOWS WITH SUCTION OR BLOWING SPECIFY THE
DIMENSIONLESS WALL VELOCITY.
        NOTE: VWALL NEGATIVE ---> SUCTION
              VWALL POSITIVE ---> INJECTION
```

```
          VWALL ZERO -----> IMPERMEABLE

ENTER THE DIMENSIONLESS SUCTION OR INJECTION PARAMETER
0
```

For this sample input, the following relationship was developed for the local Nusselt number along a flat horizontal plate (cf. graphs).

Local Nusselt Number vs. x / L for Flow over Flat Plates

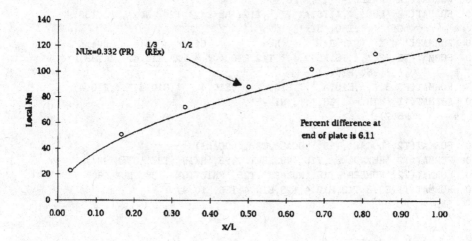

Local Nusselt Number vs. x / L for Flow over Flat Plates

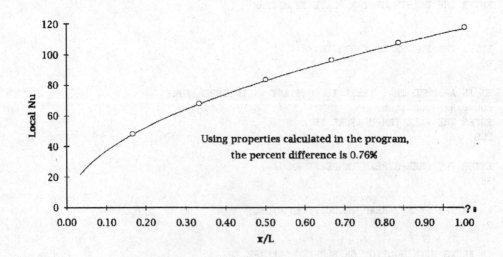

The graphs show the expected results. For further confirmation of the usefulness of this program, the velocity profiles for suction and injection were plotted (cf. graph).

Velocity Profiles over a Flat Plate

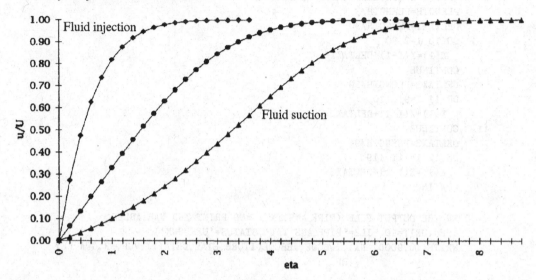

F.4.2   Keller's Box Method for Pipe Flow

```
C************************************************************************
C* PROGRAM PIPE
C*
C* THIS PROGRAM COMPUTES THE FINITE-DIFFERENCE SOLUTION OF THE
C* ENERGY EQUATION FOR LAMINAR AND TURBULENT FLOW IN A PIPE.
C* THE VELOCITY PROFILE IS ASSUMED TO BE FULLY DEVELOPED AND ONLY
C* AXISYMMETRIC FLOWS WITH UNIFORM WALL TEMPERATURE ARE CONSIDERED.
C*
C************************************************************************

        COMMON/BLC0/NXT,IWBCOE,IEBCOE,ITURB,ICOORD,INDEX,N,NP,PR,VGP,
       +            GWA,REY,RF,CEL,ETA(51),UP(51),DETA(51),A(51),YP(51),
       +            X(120),GW(120),PW(120),GE(120),G(51,2),P(51,2)
        COMMON/BLC1/   S1(51),S2(51),S3(51),R1(51),R2(51),A1(51,2),A2(51,2)

C* VARIABLES TO BE ENTERED BY THE USER
        NXT=119.
        VGP=1.15
        ETAE=10.0
        GWA=0.2E-7
        DETA(1)=0.025
        ITURB=0.
        PR=.02
        REY=0.5E5
        RF=0.0
        IWBCOE=1.
        IEBCOE=1.
        PLENGTH=.5
```

```
C* IDENTIFY X-STATIONS (MUST BE MODIFIED FOR CHANGES IN NXT)
      PLENGTH=PLENGTH/3.
      DELTAX1=PLENGTH/90
      DO 10 I=2,90
         X(I)=X(I-1)+DELTAX1
10    CONTINUE
      DELTAX2=PLENGTH/20
      DO 11 I=91,110
         X(I)=X(I-1)+DELTAX2
11    CONTINUE
      DELTAX3=PLENGTH/9
      DO 12 I=111,119
         X(I)=X(I-1)+DELTAX3
12    CONTINUE

C* OPEN THE OUTPUT FILE (PIPE_ANS.TXT) AND PRINT THE VARIABLES
      OPEN(UNIT=10,FILE='PIPE_ANS.TXT',STATUS='UNKNOWN')
      WRITE(10,9000) NXT,IWBCOE,IEBCOE,ITURB,ETAE,DETA(1),VGP,PR,GWA,REY
     +                  ,RF

      WRITE(10,*) 'X','Nu'
      IF((VGP-1.0) .GT. 0.0001) GO TO 20
      NP= ETAE/DETA(1)+1.0001
      GO TO 30
20    NP=ALOG((ETAE/DETA(1))*(VGP-1.0)+1.0)/ALOG(VGP)+1.0001
30    IF(NP .LE. 51) GO TO 40
      STOP
40    ETA(1)=0.0
      YP(1)=0.0

C* INITIAL TEMPERATURE PROFILE AT X=XO
      G(1,2)= 1.0
      P(1,2)= 0.0
      DO  50 J= 2,51
      YP(J)= 0.0
      G(J,2)= 1.0
      P(J,2)= 0.0
      A1(J,2)= 1.0
      A2(J,2)= 0.0

C* GENERATION OF GRID SYSTEM
      DETA(J)=DETA(J-1)*VGP
      ETA(J)=ETA(J-1)+DETA(J-1)
50    A(J) = 0.5*DETA(J-1)
      N=1
      ICOORD = 1
      INDEX =0
      GW(1) = 1.0
60    CONTINUE
      IF(N .EQ. 1) GO TO 80
      IF(ICOORD.EQ. 2) CEL = 1.0/(X(N)-X(N-1))
```

```
      IF(ICOORD.EQ. 1) CEL = 0.5*(X(N)+X(N-1))/(X(N)-X(N-1))
      IGROW= 0
70    CALL COEF
      CALL SOLV2

C* CHECK FOR BOUNDARY LAYER GROWTH
      IF(ICOORD.EQ.2 .OR. NP.EQ.51) GO TO 80
      IF(IGROW.GE.3.OR. ABS(P(NP,2)).LT.1.E-04) GO TO 80
      XSWITCH= 1.0/ETA(NP+1)**2
      IF(X(N).GE.XSWITCH) GO TO 80
      NP= NP+1
      IGROW=IGROW+1
      G(NP,1)=G(NP-1,1)
      P(NP,1)=0.0
      GO TO 70
80    CALL OUTPUT
      GO TO 60

9000 FORMAT(1H0,7HNXT =,I3,14X,7HIWBCOE=,I3,14X,7HIEBCOE=,I3,14X,
     1       7HITURB =,I3/1H ,7HETAE=,E14.6,3X,7HDETA1 =,E14.6,3X,
     2       7HVGP   =,E14.6,3X,7HPR  =,E14.6/1H ,7HGWA =,E14.6,3X
     3       7HREY   =,E14.6,3X,7HRF  =,E14.6/)
9100 FORMAT(1H0,35HNP EXCEEDED 51-- PROGRAM TERMINATED)
9200 FORMAT(/1H0,2HN=,I3,5X,3HX =,E14.6)
      END

C*********************************************************************
C* SUBROUTINE COEF
C*
C* THIS SUBROUTINE DEFINES THE COEFFICIENTS OF THE
C* FINITE-DIFFERENCE ENERGY EQUATION
C*********************************************************************

      SUBROUTINE COEF

      REAL K,KH
      COMMON/BLC0/ NXT,IWBCOE,IEBCOE,ITURB,ICOORD,INDEX,N,NP,PR,VGP,
     1             GWA,REY,RF,CEL,ETA(51),UP(51),DETA(51),A(51),YP(51),
     2             X(120),GW(120),PW(120),GE(120),G(51,2),P(51,2)
      COMMON/BLC1/ S1(51),S2(51),S3(51),R1(51),R2(51),A1(51,2),A2(51,2)
      DIMENSION C(5),EDV(51),DUDY(51)
      DATA K,KH/0.4,0.44/
      DATA IGWALL,APLUS,C/0,26.0,34.96,28.79,33.95,6.3,-1.186/

      SWITCH=0.0
      XP1=1.0
      JJ=NP
      IF(ICOORD.EQ.2) GO TO 15
      SWITCH=1.0
      XP1=SQRT(X(N))
      GE(N)=1.0
```

```
      DO 5 J=1,51
      YP(J)=XP1*ETA(J)
      IF(YP(J).GT.1.0) GO TO 10
    5 CONTINUE
      JJ= 51
      GO TO 15
   10 JJ=J-1
   15 IF(ITURB.EQ.1) GO TO 25


C* VELOCITY PROFILE & COEFFS. OF ENERGY EQ. FOR LAMINAR FLOW
      DO 20 J=1,JJ
      YP(J)=XP1*ETA(J)
      RP=1.0-YP(J)
      UP(J)=2.0*(1.0-RP**2)
      A1(J,2)=RP
   20 A2(J,2)=UP(J)*RP
      GO TO 45


C*    VELOCITY PROFILE, EDDY VISCOSITY & TURBULENT PRANDTL NUMBER
C*    FOR TURBULENT FLOW
   25 CONTINUE
      IF(RF .EQ. 0.)  GO TO 26
      F=1./(1.8*ALOG1 0(1./(.0675*RF+3.25/REY)))**2
      CKSPS=REY*SQRT(F/8.)*RF
      IF(CKSPS .LT. 5.)   GO TO 26
      IF(CKSPS .LT. 70.)  DYP=.9*(SQRT(CKSPS)-CKSPS*EXP(-CKSPS/6.))
     +                     /REY/SQRT(F/8.)
     , IF(CKSPS .GE. 70.)  DYP=.7*CKSPS**.58/REY/SQRT(F/8.)
      GO TO 27
   26 F=0.3164/(2.*REY)** 0.25
      DYP=0.
   27 ALOGPR=ALOG10(PR)
      SUM=C(1)
      DO 30 I=2,5
   30 SUM=SUM+C(I)*ALOGPR**(I-1)
      BPLUS=SUM/SQRT(PR)
      PRT=K/KH*BPLUS/APLUS
      UP(1)=0.0
      CYOA=0.5*REY/APLUS*SQRT(0.5*F)
      CYOB=CYOA*APLUS/BPLUS
      DO 40 J=1,JJ
      YP(J)=XP1*ETA(J)
      RP=1.0-YP(J)
      RP1=RP-DYP
      IF(RP1 .LT. 0.)  RP1=0.
      YOA=YP(J)*CYOA
      YOB=YP(J)*CYOB
      EXPYOA=0.0
      EXPYOB=0.0
      IF(YOA.LT.50.0) EXPYOA=EXP(-YOA)
```

```
      IF(YOB.LT.50.0) EXPYOB=EXP(-YOB)
      CMIX=(0.14-RP1**2*(0.08+0.06*RP1**2))*(1.-EXPYOA
     +      *EXP(-DYP*CYOA))
      DUDY(J)=0.25*REY*F*RP/(1.+SQRT(1.+0.5*F*RP*(REY*CMIX)**2))
      CMIX=(0.14-RP**2*(0.08+0.06*RP**2))*(1.-EXPYOA)
      EDV(J)=CMIX**2*REY*DUDY(J)
      IF(J.EQ.1)GO TO 35
      UP(J)=UP(J-1)+0.5*(DUDY(J)+DUDY(J-1))*(YP(J)-YP(J-1))
      PRT=K/KH*(1.-EXPYOA)/(1.0-EXPYOB)
C*   COEFFS. OF THE ENERGY EQ. FOR TURBULENT FLOW
   35 A1(J,2)=RP*(1.0+PR/PRT*EDV(J))
      A2(J,2)=RP*UP(J)
   40 CONTINUE
   45 GW(N)=0.0
      IF(IGWALL.EQ.1) GO TO 50
      GW(N)=0.5*(1.0+COS(3.14159*(X(N)-X(1))/GWA))
      IF(X(N).LT.(X(1)+GWA)) GO TO 50
      GW(N)=0.0
      IGWALL=1
C*   COEFFS OF THE FINITE DIFFERENCE EQUATIONS
   50 DO 55 J=2,NP
      ETAB=0.5*(ETA(J)+ETA(J-1))
      CGB=0.5*(G(J,1)+G(J-1,1))
      CPB=0.5*(P(J,1)+P(J-1,1))
      A2B=0.5*(A2(J,2)+A2(J-1,2))
      CA2B=0.5*(A2(J,1)+A2(J-1,1))
      DERA1P=(A1(J,1)*P(J,1)-A1(J-1,1)*P(J-1,1))/DETA(J-1)
      S1(J)=A1(J,2)/DETA(J-1)+0.25*ETAB*A2B*SWITCH
      S2(J)=-A1(J-1,2)/DETA(J-1)+0.25*ETAB*A2B*SWITCH
      S3(J)=-0.5*(CA2B+A2B)*CEL
      R1(J)=2.0*S3(J)*CGB-DERA1P-0.5*CA2B*ETAB*CPB*SWITCH
   55 R2(J-1)=0.0
      RETURN
   60 FORMAT(1H0,51HTHIS PROGRAM IS NOT APPLICABLE ON TRANSITION REGION)
      END


C**********************************************************************
C*   SUBROUTINE OUTPUT
C*
C*   THIS SUBROUTINE PRINTS OUT THE PROFILES FOR THE SHEAR LAYER
C*   AND CALCULATES THE MIXED-MEAN TEMPERATURE AND THE NUSSELT
C*   NUMBER.
C**********************************************************************

      SUBROUTINE OUTPUT
      COMMON/BLCO/ NXT,IWBCOE,IEBCOE,ITURB,ICOORD,INDEX,N,NP,PR,VGP,
     +            GWA,REY,RF,CEL,ETA(51),UP(51),DETA(51),A(51),YP(51),
     +            X(120),GW(120),PW(120),GE(120),G(51,2),P(51,2)
      COMMON/BLC1/ S1(51),S2(51),S3(51),R1(51),R2(51),A1(51,2),A2(51,2)
```

```
C*  PRINT OUT DIMENSIONLESS TEMPERATURE PROFILE
C       WRITE(10,9100) (J,ETA(J),G(J,2),P(J,2),UP(J),YP(J),J=1,NP)
        IF(N.EQ.1) GO TO 15

C CALCULATE & PRINT OUT DIMENSIONLESS MIX TEMPERATURE & NUSSELT NUMBER
        CNUXO=2.0*P(1,2)
        C=0.0
        IF(ICOORD.EQ.2) GO TO 5
        C=1.0
        CNUXO=CNUXO/SQRT(X(N))
     5  GMIX=0.0
        F2=0.0
        DO 10 J=2,NP
        F1=F2
        F2=2.*(1.-YP(J))*UP(J)*(G(J,2)-C)
        GMIX=GMIX+0.5*(F1+F2)*(YP(J)-YP(J-1))
    10  CONTINUE
        GMIX=GMIX+C
        CNUXM=CNUXO/(GMIX-G(1,2))
        WRITE(10,*)X(N),CNUXM
    15  IF(N.EQ.NXT) STOP
        IF(INDEX.EQ.1) GO TO 25
        N=N+1
        IF(ICOORD.EQ.2) GO TO 35
        XSWITCH = 1./ETA(NP)**2
        IF(X(N).LT.XSWITCH) GO TO 35
        DO 20 II=N,NXT
        I=NXT-II+N
        X(I+1)=X(I)
    20  CONTINUE
        X(N)=XSWITCH
        NXT=NXT+1
        INDEX=1
        GO TO 35

C*  SWITCH TO PRIMITIVE VARIABLES
    25  SQX=SQRT(X(N))
        DO 30 J=1,51
        ETA(J)=ETA(J)*SQX
        DETA(J)=DETA(J)*SQX
        P(J,2)=P(J,2)/SQX
        IF(J.EQ.1) GO TO 30
        A(J)=0.5*DETA(J-1)
    30  CONTINUE
        ICOORD=2
        IEBCOE=0
C       WRITE(10,9200)
C       WRITE(10,9100) (J,ETA(J),G(J,2),P(J,2),UP(J),YP(J),J=1,NP)
        CALL COEF
        N=N+1
        INDEX=0
```

```
C*  SHIFT PROFILES FOR NEXT STATION CALCULATION
    35 DO 40 J=1,51
       G(J,1)=G(J,2)
       P(J,1)=P(J,2)
       A1(J,1)=A1(J,2)
       A2(J,1)=A2(J,2)
    40 CONTINUE
       RETURN

9100 FORMAT(1H0,2X,1HJ,3X,3HETA,10X,1HG,13X,1HP,12X,2HUP,12X,2HYP/
    1         (1H ,I3,F10.5,4E14.6))
9200 FORMAT(1H0,31H***** PRIMITIVE VARIABLES *****)
     END


C***********************************************************************
C*  SUBROUTINE SOLV2
C*  THIS SUBROUTINE SOLVES A LINEAR SYSTEM USING THE BLOCK-ELIMINATION
C*  SCHEME
C***********************************************************************
     SUBROUTINE SOLV2
     COMMON/BLCO/ NXT,IWBCOE,IEBCOE,ITURB,ICOORD,INDEX,N,NP,PR,VGP,
    1             GWA,REY,RF,CEL,ETA(51),UP(51),DETA(51),A(51),YP(51),
    2             X(120),GW(120),PW(120),GE(120),G(51,2),P(51,2)
     COMMON/BLC1/ S1(51),S2(51),S3(51),R1(51),R2(51),A1(51,2),A2(51,2)
     DIMENSION G11(51),G12(51),A11(51),A12(51),A21(51),A22(51),W1(51),
    1             W2(51),DEN(51)

     IF(IWBCOE.EQ.0) GO TO 10

C*  CONDITIONS FOR HEAT TRANSFER CONDITIONS AT WALL
C*  ALFA0 = 1.0 , ALFA1 = 0.0 ------> SPECIFIED  WALL TEMPERATURE
C*  ALFA0 = 0.0 , ALFA1 = 1.0 ------> SPECIFIED HEAT FLUX

C*  SPECIFIED WALL TEMPERATURE
     ALFA0=1.0
     ALFA1=0.0
     G(1,2)= GW(N)
     P(1,2)= 0.0
     GO TO 20

C*  SPECIFIED WALL HEAT FLUX
    10 ALFA0=0.0
     ALFA1=1.0
     G(1,2)=0.0
     P(1,2)=PW(N)
    20 GAMMA0=ALFA0*G(1,2)+ALFA1*P(1,2)
     R1(1)=GAMMA0
     IF(IEBCOE.EQ.0) GO TO 30
```

```
C*    SPECIFIED EDGE TEMPERATURE
      BETA0=1.0
      BETA1=0.0
      G(NP,2)=GE(N)
      P(NP,2)=0.0
      GO TO 40

C*    SPECIFIED EDGE TEMPERATURE GRADIENT
    30BETA0=0.0
      BETA1=1.0
      G(NP,2)=0.0
      P(NP,2)=0.0

    40GAMMA1=BETA0*G(NP,2)+BETA1*P(NP,2)
      R2(NP)=GAMMA1

C* W-ELEMENTS FOR J=1
      W1(1)=R1(1)
      W2(1)=R2(1)

C* ALFA ELEMENTS FOR J=1
      A11(1)=ALFA0
      A12(1)=ALFA1
      A21(1)=-1.0
      A22(1)=-0.5*DETA(1)

C* GAMMA ELEMENTS FOR J=2
      DET=ALFA1-0.5*DETA(1)*ALFA0
      G11(2)=(S2(2)-0.5*DETA(1)*S3(2))/DET
      G12(2)=(ALFA0*S2(2)-ALFA1*S3(2))/DET

C* FORWARD SWEEP
      DO 60 J=2,NP
      DEN(J)=A11(J-1)*A22(J-1)-A21(J-1)*A12(J-1)
      IF(J.EQ.2) GO TO 50
      G11(J)=(S3(J)*A22(J-1)-S2(J)*A21(J-1))/DEN(J)
      G12(J)=(S2(J)*A11(J-1)-S3(J)*A12(J-1))/DEN(J)
    50 A11(J)=S3(J)-G12(J)
      A12(J)=S1(J)+A(J)*G12(J)
      A21(J)=-1.0
      A22(J)=-A(J+1)
      W1(J)=R1(J)-G11(J)*W1(J-1)-G12(J)*W2(J-1)
      W2(J)=R2(J)
    60 CONTINUE
C* BACKWARD SWEEP
      DENO=A11(NP)*BETA1-A12(NP)*BETA0
      G(NP,2)=(W1(NP)*BETA1-W2(NP)*A12(NP))/DENO
      P(NP,2)=(W2(NP)*A11(NP)-BETA0*W1(NP))/DENO
      J=NP
    70 J=J-1
      E1=W2(J)-G(J+1,2)+A(J+1)*P(J+1,2)
```

```
G(J,2)=(W1(J)*A22(J)-E1*A12(J))/DEN(J+1)
P(J,2)=(E1*A11(J)-W1(J)*A21(J))/DEN(J+1)
IF(J.GT.1) GO TO 70
RETURN
END
```

Sample Calculations

For the following input values

NXT=119.	(total number of X-stations)
VGP=1.15	(step increment multiplication factor; use values near 1.0)
ETAE=10.0	(transformed thermal boundary-layer thickness, η_e; a recommended value is 10)
GWA=0.2E-7	(wall heating length)
DETA(1)=0.025	(initial η spacing, for ETAE = 10.0, choose greater than 0.02 for DETA(1).
ITURB=0.	(0 for laminar, 1 for turbulent)
PR=.02	(Prandtl number, doesn't matter for laminar flow calculations)
REY=0.5E5	(Reynolds's number, doesn't matter for laminar flow calculations)
RF=0.0	(pipe roughness)
IWBCOE=1.	(use 1 for constant wall temperature boundary condition, 0 for constant wall heat flux condition)
IEBCOE=1.	(edge boundary condition, use 1 for specified edge temperature)
PLENGTH=.5	(length of pipe)

The locations of the X-stations must be chosen carefully to obtain sufficient coverage of the pipe. Results were obtained for the local Nusselt number and plotted (cf. graph).

Local Nusselt Number vs. x/L for Flow Inside Pipe

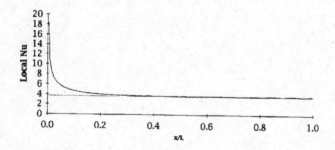

As expected, the Nusselt number quickly approaches 3.657 as the thermal flow develops in the pipe.

References and Further Reading Material

Aris, R. 1962. *Vectors, Tensors and the Basic Equations of Fluid Mechanics*, Prentice-Hall. Englewood Cliffs, NJ.

Bird, R. B., R. C. Armstrong, and O. Hassager. 1987. *Dynamics of Polymeric Liquids*. Volume I. *Fluid Mechanics*. Wiley-Interscience, New York.

Bradshaw, P., T. Cebeci, and J. H. Whitelaw. 1981. *Engineering Calculation Methods for Turbulent Flow*. McGraw-Hill, New York.

Burden, R. L., J. D. Faires, and A. C. Reynolds. 1988. *Numerical Analysis*. Prindle, Weber and Schmidt, Boston, MA.

Cebeci, T., and P. Bradshaw. 1988. *Physical and Computational Aspects of Convective Heat Transfer*. Springer Verlag, New York.

Currie, I. G. 1993. *Fundamental Mechanics of Fluids*. McGraw-Hill, New York.

Keller, H. B. 1970. "A New Difference Scheme for Parabolic Problems," In: *Numerical Solution of Partial-Differential Equations*, Vol. II, J. Bramble (ed.). Academic Press, New York.

Lin, C. C., and L. A. Segal. 1975. *Mathematics Applied to Deterministic Problems in the Natural Sciences*. Macmillan, New York.

Ralston, R., and P. Rabinowitz. 1978. *A First Course in Numerical Analysis*. McGraw-Hill, New York.

Index